Emerging pests and vector-borne diseases in Europe

Emerging pests and vector-borne diseases in Europe

Ecology and control of vector-borne diseases

Volume 1

edited by:

Willem Takken
and
Bart G.J. Knols

Wageningen Academic
Publishers

ISBN 978-90-8686-053-1
ISSN 1875-0699

First published, 2007

Wageningen Academic Publishers
The Netherlands, 2007

Ecology and control of vector-borne diseases

In the past century, many advances were made in the control of vector-borne diseases. Malaria disappeared from the northern hemisphere, diseases such as typhus, Bartonella and yellow fever were seriously reduced in prevalence and in many countries effective methods of disease control contributed to a greatly reduced incidence of such diseases. Most of these advances were beneficial to the industrialized world, whereas underdeveloped countries continued to suffer much as before. Indeed, several diseases such as malaria, Rift Valley fever and African sleeping sickness are still highly prevalent in specific countries. 'New' vector-borne diseases, dengue and West Nile fever, have emerged and are invading previously disease-free regions. The discovery of new drugs and vaccines has made great advances and allows for the effective treatment and control of many diseases. In contrast, vector control has lagged behind in development, even though it is realized that effective vector control would allow for an immediate interruption of the transmission of disease, and aid in disease control and eradication. In the last 20 years new initiatives on vector control have been undertaken, leading to a rapid development of effective and lasting methods of vector control. The development of molecular genetics has provided new insight in vector biology and behaviour, which is being used for developing new strategies of vector control. Advances in geographic information systems allow for precision targeting of interventions. The collective information on new developments in Vector Ecology and Control of Vector-borne Diseases is scattered over numerous periodicals and electronic databases. This book series intends to bring together this information in annual volumes arranged around selected themes that are currently of interest. Forthcoming themes will include 'Emerging pests and vector-borne diseases in Europe', 'Surveillance methods of vectors of infectious diseases', 'Mating behaviour of disease vectors', 'Recent advances in biological control of mosquitoes'. Other topics will be added as perceived relevant.

Willem Takken and Bart Knols are well-known experts in the field of Medical and Veterinary Entomology. Both have extensive experience from field work in the tropics and ecological studies in the laboratory and field. They also have a wide experience in publishing and other methods of research dissemination. Willem Takken is Professor in Medical and Veterinary Entomology at Wageningen University. Bart Knols is senior lecturer at the same institution.

Table of contents

Preface

This book is the result of many years of collaboration between the editors. Our collective work, inspired by many evenings of discussion and friendly, sometimes hotly debated, arguments, has caused us to wonder why so little of the vast amount of information on pests and vector-borne diseases has been put together in a series where scientists and students can find information arranged around one specific theme. Scientific output has increased dramatically in the last decade, accompanied by a rapid increase in journal publications. The internet and literature databases such as Web-of-Science, PubMed and Scopus give access to a sheer endless, and therefore insurmountable, resource of information that is still difficult to grasp for those in search of an update on a specific topic.

The current book is the first volume of a new book series titled 'Ecology and Control of Vector-borne Diseases', which we hope will become a platform for presenting rapid overviews on important topics in this field. Ecology is our focus, as the rapid changes in environment and climate are likely to affect arthropod vectors of disease in ways that were not foreseen not very long ago. The 19th Century was an age of discovery, when many vector-borne diseases were identified and a start was made with the introduction of public health. This was followed by rapid developments of effective drugs and pesticides throughout the 20th Century. It was believed that vector-borne diseases would be eradicated, or at least, become things of the past. At the start of the 21st Century the situation has changed and infectious vector-borne diseases, as well as pests, are resurging once again or advance their distribution. This necessitates re-addressing the factors that determine the force of infection and transmission in natural ecosystems, where most of these arthropods originate from.

In the coming years we intend to address a series of topics in vector ecology that are considered 'hot' and/or sufficiently important that they merit a special issue, such as the book in front of you. In order to make the book accessible to a wide readership, we have arranged, with kind agreement of the publisher, that it will appear as a free e-publication one year after publication.

In this volume we focus on emerging pests and vector-borne diseases in Europe. More than seventy scientists have contributed to this highly relevant topic, and we thank them for their contribution and having stayed within the time frame required for the production of this book. All chapters have been reviewed by independent reviewers, and the comments made have been used for improvement of the manuscripts. This anonymous reviewing work takes time, and we are very grateful to those colleagues who have accepted our request for review. We thank Mike Jacobs and Marijn van der Gaag of Wageningen Academic Publishers for having worked efficiently and in a friendly manner to make publication of this book possible. Hubert Krekels of Wageningen University and Research Centre has made arrangements for the e-publication, and we very much appreciate his assistance. We thank Margriet Huisman for secretarial support and Hans Smid for the cover design, which inspired us to start thinking of Volume 2. Special thanks to Françoise Kaminker, who has language-edited several chapters, and provided useful advice.

Wageningen, 2 October 2007

Willem Takken and Bart G.J. Knols

1. Alarm bells ringing: more of the same, and new and novel diseases and pests

Bart G.J. Knols and Willem Takken

Abstract

Global change has increased the interest of politicians and the public alike to seek ways and means to preserve the planet and mitigate a looming human health catastrophe. Europe has experienced, during recent years, the introduction of vector-borne diseases from tropical regions, notably Africa. Although it has been argued that climate change is the key responsible factor for more northerly distributions of vectors, their competence to transmit pathogens, and observed invasions, other drivers, notably travel and trade, and insecticide resistance have also facilitated these processes. Examples are given of recent vector-borne disease outbreaks (Chikungunya in Italy and Bluetongue virus in The Netherlands). It is concluded that increased occurrence of vector-borne diseases in endemic regions and more frequent 'contact' with Europe will lead to an increasing number of invasions and disease outbreaks in the foreseeable future.

Keywords: Europe, pest, vector-borne disease, climate change, transport, travel, insecticide resistance

Introduction

Never before in the history of mankind has the impact of globalisation and industrialisation been more felt and debated than at the start of the third millennium. Not only have the consequences of our changing planet featured more prominently on the agenda of scientists and politicians during recent years, also the broader public has been exposed and involved more intensively, not in the least due to Al Gore's 'An inconvenient truth' (Gore 2006). The world has witnessed the sudden outbreak of severe acute respiratory syndrome (SARS) in 2003, as well as its rapid decline, with alarm and disbelief (Elston 2005, Yu and Sung 2004). Suddenly, a new disease, potentially fatal to humans, had emerged seemingly from nowhere, and for reasons that were not well understood. Around the same time, outbreaks of Ebola virus in Africa were reported with increasing frequency, without a clear explanation for the rise in incidence of this highly fatal disease (Legrand *et al.* 2007, Walsh *et al.* 2005). Since 2005 the world is in the grip of avian influenza, caused by the influenza viral strain H5N1, which not only affects birds, but can jump to humans where it causes extraordinarily high fatalities in patients (Elston 2005, Peiris *et al.* 2007).

The realisation and acceptance that global change, and more particularly so climate change (Figure 1), are very likely the result of human activities have led to rising concern over how this will affect human health (WHO 1996, McMichael 2001, IPCC 2007). The already observed and likely to increase phenomena related to climate change are multifold and entail, besides global warming, more extreme weather conditions. Warmer winters, increased precipitation, and warm spells/heat waves are likely not only to affect humans (Kovats *et al.* 2005, Campbell-Lendrum and Woodruff 2006) but entire ecosystems and organisms that occupy these. Rising temperatures will have the highest impact on the biology of poikilothermic species (i.e. species for which body temperature varies with the temperature of its surroundings), which includes all arthropods that feature in this volume. Arctic regions and deserts are hostile regions for most arthropods, whereas these

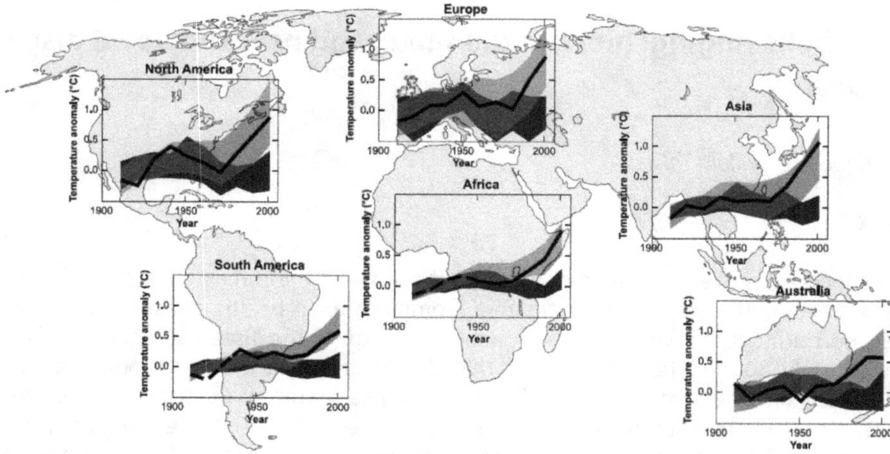

Figure 1. Changes in the observed (black line) and modeled temperature anomalies (5 and 95% confidence limits for the period 1906-2005. Light grey: confidence limits incorporating 14 climate models that include anthropogenic change. Dark grey: confidence limits of 5 models excluding human-induced change (IPCC 2007).

organisms thrive in regions where temperatures are tropical but rainfall tempers the impact of heat. The worlds biodiversity hotspots are located in tropical regions, where insect life in particular is abundant and characterised by high diversity (Orme *et al.* 2005). In general, arthropods respond to increased temperatures by expanding their distribution range to occupy new suitable habitats, by displaying altered physiological processes that include higher reproductive rates, by having an increased longevity and higher food consumption, and through benefiting the transmission of pathogens because of reduced extrinsic incubation periods. All of these factors are therefore likely to increase the abundance and spatio-temporal distribution of arthropods that cause nuisance to humans and livestock as well as those that transmit disease (Githeko *et al.* 2000, Khasnis and Nettleman 2005). Consequently, we may expect a geographical redistribution of vector-borne diseases and pests depending on the response of these organisms to the changing environmental conditions (Patz and Kovats 2002, Patz and Olson 2006, Rogers and Randolph 2006).

It has been argued that the focus on global warming as the sole cause for increased problems with arthropod-borne diseases is (at least partially) misleading as other anthropogenic and environmental changes also contribute to the observed phenomena (Molyneux 2003, Sutherst 2004, Zell 2004, Takken *et al.* 2005a, Sumilo *et al.* 2007). In the case of malaria, for example, well-documented land use changes such as hydrological, urbanisation, mining, agricultural, and forest-related impacts (exploitation activities, road construction, deforestation and population movements) are having significant impact on the global burden of this disease. In the case of many settings, for instance the Amazon region of Brazil, all these factors interplay (Takken *et al.* 2005b) and continue to aggravate the situation (Arruda *et al.* 2007, WHO 2005).

The above example relates to a tropical disease, but similar problems are being faced within the confines of the European continent. Unlike most people inhabiting developing countries, who have adapted both from a socio-economic and health perspective to the ever-present burden of arthropod-borne infectious disease, Europeans have become used to the absence of such threats, progressively so since the 1960s. The gradual decline of diseases like malaria, largely attributed to improved health care and application of the residual insecticide dichloro-diphenyl-trichloroethane (DDT) for vector control (e.g. Snowden 2006, Bruce-Chwatt and De Zulueta 1980), resulted in 'Anophelism without malaria', when potential vectors persisted in the environment, but the parasite reservoir was eradicated (Jetten and Takken 1994). Most Europeans now consider malaria a disease of the tropics and merely a risk associated with travel to endemic regions. Yet, in south-eastern Europe (Turkey) a contrasting situation is unfolding, whereby agricultural expansion through irrigation schemes, coupled with a migrant labour force, is causing grave concern and indeed more frequent malaria epidemics (Ejov 2001). With increasing frequency, these developments spark discussions about the possible return of malaria to much of southern and even northern Europe. Although the risk of focal autochthonous transmission is not zero, historic endemicity levels will not recur given the likely and adequate responses of public health authorities to such possible events (see Chapters 2 and 3).

From models to outbreaks

Much different is the situation regarding arboviral diseases such as the West Nile, Usutu, dengue and Chikungunya viruses. Mosquito-borne, these diseases originate from the African continent and have become established in the northern hemisphere. Of key importance here is the fact that these viruses have non-human hosts, in which the virus can amplify, and through which they can conquer a lot of territory over a short period of time (e.g. through bird migration). West Nile virus reached New York in 1999 (Briese *et al.* 1999) and moved across the USA to reach California, following hundreds of fatal and thousands of seriously ill patients, a mere five years later. The massive impact of this disease introduction in a naïve host population (Komar 2003) has recently been attributed to genetic changes affecting its virulence (Brault *et al.* 2007). Likewise, the Usutu virus reached Central Europe, possibly through a migratory bird returning from Africa, and has since sparked an outbreak in bird populations in Austria (Weissenböck *et al.* 2002). It has spread to neighbouring countries since, causing local outbreaks in Budapest, Zürich and northern Italy (see Chapter 9).

The current availability of powerful computing tools and remote sensing technology has led to numerous attempts to predict the distribution of vectors and disease in response to change, from country, to regional, to global levels (Martens and McMichael 2003). Biological or empirical models that incorporate measured relationships between vector and/or pathogen development, or are based on correlations between spatio-temporal distributions in relation to environmental variables (statistical models) have sought to define the likelihood for establishment of vectors and the stability of the host-pathogen-vector system once in place (Campbell-Lendrum and Woodruff 2006, Kiszewski *et al.* 2004). The basic reproductive rate, R_0, integrates several important variables, and is indicative for the absence (if <1) or presence (if >1) of potential spread of disease (Anderson and May 1992).

Regarding Chikungunya virus, recent calculations of this latter parameter yielded values of close to 1 for southern Europe in the case of transmission by the Asian tiger mosquito *Aedes albopictus* (Skuse). The establishment of this competent vector in southern Europe (mainly in Italy), following its global spread facilitated by the used-tyre trade (Reiter 1998) coupled with the return of virus-infected human hosts from various Indian Ocean islands in 2005/6, where a massive epidemic of this disease raged at that time (Pialoux *et al.* 2007), caused concern for an outbreak in Europe (Fusco *et al.* 2006). As recent as August 2007 this indeed happened when the first outbreak of this tropical virus was recorded in two small towns near Ravenna, in the Emilia Romagna province, on the Adriatic coast. Believed to be introduced by an infected traveller returning from India, the local tiger mosquito population vectored the disease effectively to some 150 inhabitants of the Ravenna region (Enserink 2007, see also Chapters 10 and 14 for more details). This event clearly demonstrates that the likelihood of vector-borne disease establishment can be predicted accurately and may indeed happen.

Invasion through travel and trade

The aforementioned case of Chikungunya introduction in north-eastern Italy demonstrates the importance of another driver for change in vector-borne disease occurrence, namely travel and trade. Since the 1950s, air transportation has seen a growth of nearly 9% per year, and shipping traffic has grown nearly 30% in that time (Upham *et al.* 2003, Zachcial and Heideloff 2003). Africa alone has seen an increase in tourist travelers from 6.7 to nearly 17 million within a decade (1990-2000) (Franco-Parades and Santos-Preciado 2006). Not only has the number of travelers returning to Europe with infectious pathogens increased (Jelinek 2000), also vectors have been shown capable of hitchhiking on long-haul flights (Tatem *et al.* 2006a) or in shipping containers (Tatem *et al.* 2006b). A striking example entails the introduction and possible establishment of *Ae. albopictus* in The Netherlands in 2005 through container shipments of Lucky Bamboo (*Dracaena sanderiana*) arriving from southern China (Scholte *et al.* 2007, see also Chapter 14).

The devastating outcome of the introduction of bluetongue virus in The Netherlands in August 2006 provides yet another example of how trade and travel can introduce a vector-borne disease in areas far from where it is endemic. In September 2007, a year after its introduction, Dutch farmers were suffering the loss of some 3000 sheep per week (see Chapters 6 and 7).

More of the same...

Two interesting examples of human and livestock pests feature in this book because of another major driving force: insecticide resistance. Head lice have long played a negligible role until resistance to commonly used products surfaced and led to increased prevalence and sometimes epidemics in the late 1990s (Hill 2006, see also Chapter 17). A similar problem has been reported regarding the resurgence of the sheep scab mite in the UK, which again appears to be closely linked to the loss of insecticidal efficacy of widely used veterinary pesticides (see Chapter 13). The story is not different for bed bugs that can now be found in five star hotels in major European cities (see Chapter 15).

And new and novel...

Throughout this volume it surfaces that Europe can no longer be considered without risk of introduction of vector-borne disease. Since the start of this millennium many introductions have taken place, and most recently this included an African virus, transmitted by an Asian mosquito species, in Europe. Of the various drivers of change discussed, climate change, travel and trade, and insecticide resistance are predominantly reported to have caused more of the same, or new and novel. The presence of Lyme borreliosis and Tick-borne encephalitis in Europe is not new, but certainly considered to be increasing. In the Netherlands alone, cases have tripled since 1994 (estimated at nearly seventeen thousand in 2006) (Hofhuis *et al.* 2006). Although the changing distribution and overwintering of ticks in larger numbers and different life stages has been attributed to climate change (Lindgren and Jaenson 2006), this again appears to be only part of the cause for increased prevalence of tick-borne disease (Sumilo *et al.* 2007). Changes in human behaviour, and in particular increased outdoor activity, are considered additional drivers (see Chapters 11 and 12).

Worldwide a large number of diseases are transmitted by arthropods. Increased prevalence in regions where these are endemic, coupled with increased interaction with Europe are making their introduction and possible establishment ever more likely. This volume details, for a number of arthropods and vector-borne diseases, the factors that underpin invasion success and elaborates on how these can be mitigated. Alarm bells, though, are ringing.

References

Anderson, RM and May M (1992) Infectious Diseases of Humans - Dynamics and Control. Oxford University Press, Oxford.

Arruda ME, Zimmerman RH, Souza RM and Oliveira-Ferreira J (2007) Prevalence and level of antibodies to the circumsporozoite protein of human malaria parasites in five states of the Amazon region of Brazil. Mem Inst Oswaldo Cruz 102: 367-371.

Brault AC, Huang CY, Langevin SA, Kinney RM, Bowen RA, Ramey WN, Panella NA, Holmes EC, Powers AM and Miller BR (2007) A single positively selected West Nile viral mutation confers increased virogenesis in American crows. Nat Genet 39:1162-1166.

Briese T, Jia X, Huang C, Grady L and Lipkin WL (1999) Identification of a Kunjin/West Nile-like flavivirus in brains of patients with New York encephalitis. Lancet 354: 1261-1262.

Bruce-Chwatt LJ and De Zulueta J (1980) The rise and fall of malaria in Europe. Oxford University Press, Oxford, UK.

Campbell-Lendrum D and Woodruff R (2006) Comparative risk assessment of the burden of disease from climate change. Environ Health Perspect 114: 1935-1941.

Ejov M (2001) Malaria situation in European Region. In: Proceedings First Balkan Conference, Malaria and Mosquito Control, Serres, Greece, pp. 20-26.

Elston DM (2005) New and emerging infectious diseases. J Am Acad Dermatol 52: 1062-1068.

Enserink M (2007) Tropical disease follows mosquitoes to Europe. Science 317: 1485.

Franco-Parades C and Santos-Preciado JI (2006) Problem pathogens: prevention of malaria in travellers. Lancet Infect Dis 6: 139-149.

Fusco MF, Puro V, Di Caro A, Nicastri E, Carannante N, Faella FS, Barzon L, Di Cesare S, Palu G, Capobianchi MR and Ippolito G (2006) Casi di febbre Chikungunya in Italia in viaggiatori di ritorno dall'Oceano indiano e rischio di introduzione nel territorio Italiano. Infez Med 4: 238-245.

Githeko AK, Lindsay SW, Confalonieri UE and Patz JA (2000) Climate change and vector-borne diseases: a regional analysis. Bull World Health Organ 78: 1136-1147.

Gore A (2006) An Inconvenient truth: The planetary emergency of global warming and what we can do about it. Rodale Books, USA. 328 pp.

Hill N (2006) Control of head lice: past, present and future. Expert review of anti-infective therapy 4: 887-894.

Hofhuis A, van der Giessen JW, Borgsteede FH, Wielinga PR, Notermans DW and van Pelt W (2006) Lyme borreliosis in the Netherlands: strong increase in GP consultations and hospital admissions in past 10 years. Euro surveill 11: E060622 060622.

IPCC (2007) Summary for Policymakers. In: Climate Change 2007: The Physical Science Basis. Contribution of Working Group I to the Fourth Assessment Report of the Intergovernmental Panel on Climate Change (Solomon S, Qin D, Manning M, Chen Z, Marquis M, Averyt KB, Tignor M and Miller HL (eds.). Cambridge University Press, Cambridge, UK and New York, NY, USA. 18 pp.

Jelinek T (2000) Dengue Fever in International Travelers. Clin Infect Dis 31: 144-147.

Jetten TH and Takken W (1994) Anophelism without Malaria in Europe; A review of the ecology and distribution of the genus *Anopheles* in Europe. Wageningen Agricultural University Papers 94-5, Wageningen, The Netherlands.

Khasnis AA and Nettleman MD (2005) Global warming and infectious disease. Arch Med Res 36: 689-696.

Kiszewski A, Mellinger A, Spielman A, Malaney P, Sachs SE and Sachs J (2004) A global index representing the stability of malaria transmission. Am J Trop Med Hyg 70: 486-498.

Komar N, Langevin S, Hinten S, Nemeth N, Edwards E, Hettler D, Davis B, Bowen R and Bunning M (2003) Experimental infection of North American birds with the New York 1999 strain of West Nile virus. Emerg Infect Dis 9: 311-322.

Kovats RS, Campbell-Lendrum D and Matthies F (2005) Climate change and human health: estimating avoidable deaths and disease. Risk Anal 25: 1409-1418.

Legrand J, Grais RF, Boelle PY, Valleron AJ and Flahault A (2007) Understanding the dynamics of Ebola epidemics. Epidemiol Infect 135: 610-621.

Lindgren EJ and Jaenson TG (2006) Lyme borreliosis in Europe: Influences of climate and climate change, epidemiology, ecology and adaptation measures. World Health Organization, Copenhagen.

Martens P and McMichael AJ (eds.) (2003). Environmental change, climate and health: Issues and research methods. Cambridge University Press,UK. 352 pp.

McMichael T (2001) Human frontiers, environments and disease: Past patterns, uncertain futures. Cambirdge University Press, Cambridge, UK. 413 pp.

Molyneux DH (2003) Common themes in changing vector-borne disease scenarios. Trans R Soc Trop Med Hyg 97: 129-132.

Orme CD, Davies RG, Burgess M, Eigenbrod F, Pickup N, Olson VA, Webster AJ, Ding TS, Rasmussen PC, Ridgely RS, Stattersfield AJ, Bennett PM, Blackburn TM, Gaston KJ and Owens IP (2005) Global hotspots of species richness are not congruent with endemism or threat. Nature 436: 1016-1019.

Patz JA and Kovats RS (2002) Hotspots in climate change and human health. BMJ 325: 1094-1098.

Patz JA and Olson SH (2006) Climate change and health: global to local influences on disease risk. Ann Trop Med Parasitol 100: 535-549.

Peiris JSM, de Jong MD and Guan Y (2007) Avian influenza virus (H5N1): a threat to human health. Clin Microbiol Rev 20: 243-267.

Pialoux G, Gauzere BA, Jaureguiberry S and Strobel M (2007) Chikungunya, an epidemic arbovirosis. Lancet Infect Dis 7: 319-27.

Reiter P (1998) *Aedes albopictus* and the world trade in used tires, 1988-1995: the shape of things to come. J Am Mosq Control Assoc 14: 83-94.

Rogers DJ and Randolph SE (2006) Climate change and vector-borne diseases. Adv Parasitol 62: 345-381.

Scholte EJ, Jacobs F, Linton YM, Dijkstra E, Fransen J and Takken W (2007a) First record of *Aedes* (Stegomyia) *albopictus* in the Netherlands. Eur Mosq Bull 22: 5-9.

Snowden FM (2006) The conquest of malaria - Italy, 1900-1962. Yale University Press, New Haven.

Sumilo D, Asokliene L, Bormane A, Vasilenko V, Golovljova I and Randolph SE (2007) Climate change cannot explain the upsurge of tick-borne encephalitis in the Baltics. PLoS ONE 2: e500.

Sutherst RW (2004) Global change and human vulnerability to vector-borne diseases. Clin Microbiol Rev 17: 136-173.

Takken W, Martens P and Bogers RJ (eds.) (2005a) Environmental change and malaria risk. Wageningen Universirty Frontis Series, Vol. 9. Springer, Dordrecht, The Netherlands. 139 pp.

Takken W, de Tarso Vilarinhos P, Schneider P and dos Santos F (2005b) Effects of environmental change on malaria in the Amazon region of Brazil. In: Takken W, Martens P and Bogers RJ (eds.) (2005a) Environmental change and malaria risk. Wageningen Universirty Frontis Series, Vol. 9. Springer, Dordrecht, The Netherlands, pp 113-123.

Tatem AJ, Hay SI and Rogers DJ (2006b) Global traffic and disease vector dispersal. Proc Nat Acad Sci USA 103: 6242-6247.

Tatem AJ, Rogers DJ and Hay SI (2006a). Estimating the malaria risk of African mosquito movement by air travel. Malar J 5: 57.

Upham P, Thomas C, Gillingwater D and Raper D (2003) J Air Transport Manag 9: 145-151.

Walsh PD, Biek R and Real LA (2005) Wave-like spread of Ebola Zaire. PLoS Biol 3: e371.

Weissenböck H, Kolodziejek J, Url A, Lussy H, Rebel-Bauder B and Nowotny N (2002) Emergence of Usutu virus, an African mosquito-borne Flavivirus of the Japanese encephalitis virus group, central Europe. Emerg Infect Dis 8: 652-656.

World Health Organisation (1996) Climate change and human health (McMichaeil AJ, Haines A, Slooff R and Kovats S (eds.). WHO/EHG/96.7, Geneva, Switzerland. 297 pp.

World Health Organisation (2005) World malaria report 2005. World Health Organization, Geneva, Switzerland (available from http://www.rbm.who.int/wmr2005/index.html (Accessed 17 September 2007).

Yu ITS and Sung JJY (2004) The epidemiology of the outbreak of severe acute respiratory syndrome (SARS) in Hong Kong - what we do know and what we don't. Epidemiol Infect 132: 781-786.

Zachcial M and Heideloff C (2003) ISL Shipping Statistics Yearbook 2003 (Inst. of Shipping Economics and Logistics, Bremen, Germany).

Zell R (2004) Global climate change and the emergence/re-emergence of infectious diseases. Int J Med Microbiol 293: 16-26.

Malaria and *Leishmania* parasites

2. Will malaria return to North-West Europe?

Willem Takken, Piet A. Kager and Jan Peter Verhave

Abstract

Historically, malaria was widespread in Europe, reaching as far north as the United Kingdom, the Netherlands and even Sweden and Finland. In southern Europe, malaria was caused by *Plasmodium falciparum*, *P. vivax* and *P. malariae*, but further north only the latter two parasite species were found. The main vector species was *Anopheles atroparvus* Van Thiel, although *An. messeae* Falleroni was suspected to be a vector in Central Europe and Scandinavia. The disease was eradicated from Europe in the second half of the 20th century with DDT and modern drugs. In spite of a near-constant and large number of cases of imported malaria, as well as the proven presence of anopheline vectors, the return of local malaria transmission has not been observed, presumably because of a lack of a sufficiently dense cluster of parasite carriers and an efficient health care system that is able to diagnose and treat malaria patients successfully. It is considered whether climate change might alter this situation, allowing for higher mosquito densities and more favourable environmental conditions for *Plasmodium* development in the mosquito vectors. We argue that under climate change, causing higher temperatures and intensified precipitation, conditions for local malaria transmission will improve, but that the public health measures will preclude the building up of an infectious parasite reservoir. So people might be exposed to more mosquito bites, but the chance of such mosquitoes becoming infected with malaria parasites will not alter significantly compared to the current situation. Therefore, climate change is unlikely to affect the malaria-free state of North-West Europe.

Keywords: malaria, North-West Europe, history, climate change, transmission, *Anopheles*, mosquito

Introduction

Current attention to malaria is focused on the impact of the disease in tropical countries, as the chief burden of this disease is experienced in this region and in spite of huge efforts to control the disease, millions of people are still affected annually (Greenwood *et al.* 2005). Until far into the 20th Century, many parts of Europe suffered from malaria as well, with focal areas in southern Europe and the coastal areas of North-West Europe (Bruce-Chwatt and De Zulueta 1980, Snowden 2006), although malaria was also recorded from Central Europe and the western areas of modern Russia. In southern Europe malaria was caused by *Plasmodium falciparum*, *P. vivax* and *P. malariae*, whereas in North-West Europe only *P. vivax* and *P. malariae* occurred. Following the discovery of the *Plasmodium* life cycle, and the role of anopheline mosquitoes in malaria transmission, rapid progress was made in developing effective intervention strategies (Bruce-Chwatt and De Zulueta 1980) so that by the end of the 1940s malaria was eradicated from much of Europe. Indeed, by 1978 Europe was formally declared 'free' of malaria by the World Health Organisation.

Each year, however, numerous malaria cases are still recorded in Europe in travellers who have returned from a visit to a malaria-endemic country (Janssens 1983, Behrens and Curtis 1993, Jelinek *et al.* 2002, www.euro.who.int/cisid). In addition, thousands of migrants and refugees enter Europe from disease-endemic countries, and many of them carry malaria parasites with them (Janssens 1983, Martens and Hall 2000, Leder *et al.* 2004, Askling *et al.* 2005, Health Protection

Agency 2007, Anonymous 2007). Whereas *P. falciparum* cases will most probably be detected shortly after arrival in Europe, *P. vivax* and *P. malariae* infections may go undetected for many months or years because of the resting stages of the parasites in the liver (Cook 1996) or survival in the blood at submicroscopical density.

The discussion about climate change and large population movements has provoked interest in the question whether malaria could become re-established in Europe under changing environmental and demographic circumstances. This suggestion is fuelled by the recent spread of malaria in Central Asian Republics following the collapse of the USSR and by persistent malarial infections in South Korea along the boundary with North Korea (Feighner *et al.* 1998, Sabatinelli 2000, Severini *et al.* 2004, Yeom *et al.* 2005). Here, conditions of local transmission are not unlike those experienced in many parts of Europe, although the anopheline species responsible for the transmission are different. In this chapter we review the current situation of malaria in north-western Europe as well as recent knowledge on the ecology of anopheline vectors in this region. We then assess the probability that malaria could become re-established in north-western Europe under future climatic conditions.

History of malaria in North-West Europe

Historically, malaria was widely present in Europe, dating back to Roman times (Bruce-Chwatt and De Zulueta 1980). Until the late middle ages little is known about the incidence of malaria in Europe (Reiter 2000). There is no doubt about the presence of malaria in north-western Europe in the period from the middle ages to the 19th century (Scott 1942, Bruce-Chwatt and De Zulueta 1980, Dobson 1989, Reiter 2000, Knottnerus 2002, Kuhn *et al.* 2003, Hutchinson and Lindsay, 2006), but how many of 'the agues', the fevers, and the deaths described since the middle ages can actually be attributed to malaria remains difficult to assess. The diagnosis of malaria purely on clinical grounds is unreliable nowadays and more so in the historical records where the terminology was different, imprecise and variable. In Great Britain, malaria was often called 'ague', and it is said that several Kings and noblemen died of this affliction (Dobson 1989, Kuhn *et al.* 2003, Hutchinson and Lindsay, 2006). Ague originally meant any acute febrile illness (the French word aigue means acute, raging), especially if accompanied by shaking or shivering, but we must realise that the definition of fever (a body temperature above 38 °C, measured with a thermometer), is not applicable to the past, when it meant a general disease condition with many different forms. (Hutchinson and Lindsay 2006). 'Ague' was used more restrictively to describe 'intermittent fever' only from the 18th century onwards. Swellengrebel and De Buck remark that the term 'intermittent fevers' has one meaning now but had a variety of meanings in the past 'without the need of being intermittent or even being a fever (the double tertians of Walcheren 1809 (see below) were rather of a continued kind)'. In their opinion 'the pernicious intermittent fevers must be regarded with the gravest doubts' and they feel that 'it is advisable to reject mortality figures as a source of evidence' (Swellengrebel and De Buck 1938), an opinion shared by recent investigators (Hutchinson and Lindsay 2006). Two malaria epidemics in the Netherlands with high death rates are regularly cited in the literature: on the island of Walcheren in 1809 and in the city of Groningen in 1826 (Bruce-Chwatt and De Zulueta 1980). About the Walcheren epidemic, Swellengrebel and De Buck note that it was an epidemic of the British troops on their overcrowded ships, not of the Dutch army and of the local population and that 42 autopsies were performed that showed enlarged spleens, testifying of malaria, but other findings were abscesses and tubercles in spleen, liver and lungs and ulcerative processes in the intestines, not consistent with malaria. Of 134 autopsies performed in those who died during the Groningen epidemic an enlarged spleen was found in 26 and the small intestine was inflamed and sometimes ulcerated in 25, the colon in 48, also pointing to causes of death other than malaria (Swellengrebel and De Buck 1938). Without excluding the occurrence of tertian malaria, the Dutch fever epidemics must have been of typhoid nature (enteric fever), which does explain the autopsy findings and the impressive death rates. Malaria was present in north-western Europe as were many other diseases, especially in the marshy areas and among the poor. It remains difficult to establish how much malaria contributed to severe disease conditions but the idea that malaria was an important cause of death must be rejected (Swellengrebel and De Buck 1938, Hutchinson and Lindsay 2006). Response to specific treatment may be used as an argument for malaria as the cause of 'ague' and 'intermittent fevers'. The bark of *Chinchona officinalis* as an effective anti-malarial drug was already known in the 17th century (Bruce-Chwatt 1988, Honingsbaum 2001), and Carl Linnaeus conducted his PhD thesis on the treatment of malaria with botanical extracts (Linnaeus 1735). The bark first appeared in the London Pharmacopeia of 1677 as *Cortex peruanus* (Scott 1942). Response to treatment may have happened in some or many instances but this might also have been the natural course of the disease. In many instances the bark did not help, probably because it was given for a wrong diagnosis, at the wrong time or incorrectly dosed, while the concentration of the active ingredient was variable in the different sources of the bark and there was no way of assessing this. Not all physicians favoured the drug; for example Gideon (not William) Harvey wrote in 1683 'I could wish these fathers (Jesuits) had kept their Indian bark to themselves, and sure I am hundreds would be on this side of the Grave, whose Bones are now turned into their first element…' (Scott 1942). Only after the discovery of the malaria parasite by Laveran in 1880 and the unravelling of the malaria life cycle (Manson 1900), confirmed data about the extent of malaria in Europe became available and data on adequate treatment could be collected.

Current malaria cases in Europe

As stated in the introduction, each year many people return to Europe from the tropics carrying *Plasmodium* parasites as a souvenir (Bottieau *et al.* 2006, Jelinek *et al.* 2002, Leder *et al.* 2004, Ranque *et al.* 2005). On the website of WHO (www.euro.who.int (Accessed 3 September 2007) click data and publications, data, cisid, malaria) data can be found for all European countries from 1994 to (inclusive) 2004 (Table 1).

Whereas public health authorities and travel clinics provide detailed information on methods for prevention of malaria (Franco-Paredes and Santos-Preciado 2006, Shanks and Edstein 2005), numerous travellers do not follow this advice or consider it an unnecessary burden. Some even refuse preventive measures because of perceived side effects of chemoprophylaxis (Chen *et al.* 2007). Thus, more than 30 years after the eradication of malaria in Europe (De Zulueta 1973), the disease is still often encountered by physicians, requiring hospitalisation of many patients. The disappearance of malaria was largely accomplished with the aid of modern insecticides, but the anopheline mosquitoes were not eradicated. During the eradication campaign, vector populations were suppressed long enough to allow public health workers to identify 'healthy' parasite carriers who were subsequently treated with a drug to clear the parasites from their body (De Zulueta 1990, 1998, Verhave 2000). Today, anopheline populations are widely present in formerly malarious areas of Europe, creating a situation termed as 'anophelism without malaria' (Lindsay and Thomas 2001, Poncon *et al.* 2007, Romi *et al.* 1997, Romi 1999, Takken *et al.* 2002). Occasionally, malaria parasite carriers in Europe are the source of local transmission, infecting people through infections of European malaria vectors (Baldari *et al.* 1998, Cuadros *et al.* 2002, Kampen *et al.* 2002, Kruger *et al.* 2001, Doudier *et al.* 2007). In no case, however, did the disease become established, even

Table 1. Reported malaria cases in some Western European countries (WHO, CISID[1]).

Year	1996	2000	2002	2004
Norway	101	79	45	49
Sweden	189	132	132	102
Finland	53	38	31	26
Denmark	191	202	135	106
United Kingdom	2,500	2,069	1,945	1,660
The Netherlands	308	691[a]	395	307
Belgium	nk	337	299	212
Germany	1,021	732	861	708
Austria	87	62	65	54
Switzerland	292	317	239	229

[1]www.euro.who.int (data and publications; data; cisid; malaria).
[a]After change of reporting system.
nk = not known.

though climatic circumstances were favourable for *Plasmodium* development in the mosquito vector and the obvious presence of infectious mosquitoes. As most of these infections occurred in August/September, it is assumed that these small epidemics died out with the advent of winter. In addition to these autochthonous infections, malaria is also distributed by *Plasmodium*-infected mosquitoes that arrive with aircraft from malaria-endemic areas (Karch *et al.* 2001, Tatem *et al.* 2006, Thang *et al.* 2002). Such cases of 'airport' malaria are fairly common, notably in France and Italy, but have not caused an epidemic situation because the imported vectors did not become established around the airports and the diseased persons were not bitten by a sufficient number of local mosquitoes before their malaria was confirmed and treated. It can therefore be assumed that many malaria cases arrive in Europe annually, but hitherto these have not been incriminated in the establishment of a focus of malaria transmission in any of the European countries.

Ecology of malaria vectors in northern Europe

Several anopheline species associated with malaria transmission have a natural distribution in Europe (Table 2).

Members of the *Anopheles maculipennis* sibling complex are common throughout the continent, but with distinctly different distributions (Jetten and Takken 1994a). Of these, *Anopheles atroparvus* Van Thiel has the widest distribution, from Scandinavia to southern Italy, and from Ireland to the Caucasus. *An. messeae* Falleroni has a similarly wide distribution, except for the Iberian peninsula, but its status as a vector is uncertain (Jaenson *et al.* 1986, Sokolova and Snow 2002). This species was recently discovered to consist of two separate species: *An. messeae* and *An. daciae* (Nicolescu *et al.* 2004), and it may be that the latter species is a vector, whereas the former is not. In southern Europe, *An. labranchiae* Falleroni and *An. sacharovi* Favr are important vectors, with further distributions into the Mediterranean (Jetten and Takken 1994a). Other members of the *An. maculipennis* complex in Europe have an uncertain status as *Plasmodium* vectors. *Anopheles claviger* Meigen has possibly the widest distribution in Europe. Its status as a malaria vector has been confirmed from the Mediterranean, but not from other temperate climate zones (Schaffner

Table 2. Anopheles species recorded from Europe, their status as malaria vector and geographic distribution.

Species	Status as malaria vector	Geographic distribution
Anopheles algeriensis Theobald	+	Mediterranean countries
Anopheles claviger (Meigen)	+	all of Europe
Anopheles hyrcanus (Pallas)	??	along Mediterranean coast
Anopheles maculipennis complex:		
An. maculipennis s.s. Meigen	??	temperate regions of Europe
An. atroparvus van Thiel	++	temperate regions of Europe, including Iberian peninsula
An. beklemishevi Stegnii & Kabanova	??	Sweden, Finland, Russia (Siberia)
An. daciae Linton, Nicolescu & Harbach	??	Romania, UK, other countries??
An. labranchiae Falleroni	+++	Southern Italy, Sardinia, Sicily, Corsica, Dalmatian coast
An. melanoon Hackett		Mediterranean coast
An. messeae Falleroni	??	all of Europe except Iberian peninsula, Southern Italy and Balkans/Greece
An. sacharovi Favre	+++	Southern Italy, Sardinia, Sicily, Corsica, Balkan, Greece, Turkey
An. superpictus Grassi	??	Corsica, Italy, Balkans, Greece
Anopheles petragnani del Vecchio	??	along Mediterranean coast
Anopheles plumbeus Stephens	??	temperate regions
Anopheles sergentii (Theobald)	+++	Sicily

+,++,+++Proven malaria vecors, in ascending importance; ??Malaria vector status uncertain.

et al. 2003). *Anopheles plumbeus* Stevens is the other species that has been incriminated with malaria transmission (Blacklock and Carter 1920, Kruger *et al.* 2001, Shute and Williams 1954).

With the exception of *An. plumbeus*, which is a tree-hole breeder, all European anopheline species use ditches and natural pools as larval habitat. Such water bodies often carry submerged aquatic vegetation, particularly in the summer. In coastal areas, *An. atroparvus*, *An. labranchiae* and *An. sacharovi* breed in brackish water, but all three species are also found in freshwater areas. Eggs are laid in water exceeding 12 °C, and typically larval development rate is temperature dependent (Mosna 1937). In the Netherlands the first larvae appear in March. Further south this can be earlier in the season. Members of the *An. maculipennis* complex overwinter as adults, when there is gonotrophic dissociation. This quiescence is triggered by day length, and broken by temperature in the spring. *An. atroparvus* continues to feed on blood during the winter, but sympatric *An. messeae* undergo a complete diapause as adult, not feeding from September to March.

The blood sources of this large and heterogenic group of anophelines vary, but all feed mostly on mammals, not expressing a particular preference for any species. It was reported that *An. atroparvus* would often bite people, hence its role as a malaria vector, but in choice situations, equal numbers of mosquitoes blood fed on a pig as on a human host (Van Thiel 1939). In Italy, *An. labranchiae* was reported to have strong anthropophilic behaviour (Van Thiel 1939), but also this species would readily feed on domestic animals. Many European anopheline species are associated with livestock, on which they feed readily. Stables and other animal shelters are very often used as resting sites, particularly in the winter. Because of a strong tendency for gonotrophic dissociation in the winter months, malaria parasites are rarely acquired at this time. Many *Plasmodium* infections in mosquitoes are acquired in the autumn, when the mosquitoes seek shelter, and may accidentally bite humans, but the parasites gradually degenerate at lowering temperature (Swellengrebel and De Buck 1938). There are few reports on the density of adult anopheline mosquitoes and the success of malaria transmission in Europe. Swellengrebel conducted a series of studies on the ecology of *An. atroparvus* and *An. messeae* in the Netherlands from 1920 onwards (Swellengrebel and De Buck 1938). In these studies, estimates of mosquito population density were made, based on both larval and adult densities. Swellengrebel then estimated the number of *Plasmodium*-infected people and the number of mosquitoes biting each day. From these data he could make an estimate of the transmission intensity in North Holland province of the Netherlands (Swellengrebel 1924, Swellengrebel *et al.* 1938). The researchers found that only one anopheline species (*An. atroparvus*) was responsible for the transmission (Box 1) and that malaria transmission was accomplished because of the large number of healthy carriers with cryptic forms of *P. vivax* and, incidentally, *P. malariae*.

Box 1. Vectors and non-vectors within the *Anopheles maculipennis* group in the Netherlands.

In the Netherlands three members of the *Anopheles maculipennis* sibling group of mosquitoes are present: *An. atroparvus*, *An. messeae* and *An. maculipennis s.s.* The latter species is relatively rare, with a limited distribution. *An. atroparvus* and *An. messeae* had an overlapping distribution in much of the country, with *An. messeae* more abundant than *An. atroparvus* (Van de Torren 1935). *An. atroparvus* was adapted to brackish water, although it would often be found in fresh water as well. Only *An. atroparvus* has been incriminated as a malaria vector. It was discovered that only this mosquito species was responsible for the fall/winter transmission of *Plasmodium vivax* and *P. malariae*, because of its habit to continue blood feeding in the winter months (Swellengrebel *et al.* 1938). *An. messeae* entered complete hibernation, as adult, resting in outdoor shelters or stables. Moreover, *An. messeae* rarely fed on human blood, whereas *An. atroparvus* would blood feed on humans by entering homes when the nights cooled down in the fall. Elsewhere in Europe *An. messeae* has been incriminated as a malaria vector, but in the Netherlands this has not been observed. Today, *An. messeae* has become the dominant anopheline species in the Netherlands, and *An. atroparvus* is present in few locations only, and in very low densities (Takken *et al.* 2002). It is suggested that the lack of suitable winter resting sites for *An. atroparvus*, caused by a change in house design and modernisation of livestock farms, has led to a sharp fall in abundance and spread of *An. atroparvus*.

These served as unintended infectious reservoir, contributing to a higher rate of transmission than could be expected based on the number of patients seeking medical treatment (Swellengrebel and De Buck 1938). These accurate recordings of malaria infection data and mosquito biting studies are rare in Europe, and one of the few other countries where such information is available from is Italy (Snowden 2006). Here, malaria was much more prevalent, and with many more casualties than in the Netherlands, even forcing people to flee malaria-infested zones. Unlike in the Netherlands, transmission was less seasonal, and accomplished by a different group of anopheline vectors.

Moreover, *P. falciparum* was an important parasite in addition to *P. vivax* and *P. malariae*. As we have seen, this parasite was not present in the Netherlands or elsewhere in North-West Europe.

Does climate change pose a risk for reintroduction of malaria in North-West Europe?

Malaria is now found in 88 countries covering 27% of the land surface of the world with 48% of the world population, almost 3 x 10^9 persons. In 1900 these data were as follows: 53% of the land surface of the world, 140 countries with 77% of the world population, then almost 9 x 10^8 persons. About 85% of the morbidity and mortality of malaria are now in Africa. (Hay *et al.* 2004) The current distribution limits of *P. falciparum* malaria, the potentially fatal infection, lie entirely within the tropics (Snow *et al.* 2005) while *P. vivax* and *P. malariae* are also found in subtropical areas (Hay *et al.* 2004) The reasons for the restriction of *P. falciparum* to the tropics and the heavy burden of malaria as a disease on tropical countries as compared to other areas are the less favourable climatic conditions for parasite development in the mosquito vectors under temperate conditions and the low level of socio-economic development in many tropical countries. These factors conspire for the maintenance of a continuous reservoir of malaria parasites and mosquito vectors and an environment that allows for effective transmission. Parasite transmission is accomplished at times of the year when sufficient numbers of anopheline mosquitoes are present, often favoured by rain and the availability of dark shelter(s). The latter are used as resting sites by blood fed female mosquitoes while they are developing eggs. Thus, environmental factors contribute to the epidemic status of malaria in the tropics. The widespread poverty in many tropical countries is an additional factor favouring malaria, because of inadequate health care and housing that allows easy access for malaria mosquitoes.

Malaria can only survive naturally when climatic conditions are suitable for growth and development of anopheline mosquitoes and for the establishment and development of the parasite in the mosquitoes. Tropical *Plasmodium* parasites require minimum temperatures of 16 and 18 °C in the case of *P. vivax* and *P. falciparum*, respectively (Gilles and Warrell 1993). For this reason, transmission of these parasite species is not expected in temperate climates with severe winters. As an exception to this rule, *P. vivax* had adapted to survival in Europe by a unique development process: the summer temperatures were sufficiently high for *P. vivax* development in the mosquito. Infection of future patients occurred in September/October, a time when numerous mosquitoes entered homes in search of shelter for the cooler nights. Once inside a house, the mosquitoes would remain there benefiting from these artificial shelters as protection from the harsh outdoor climate (Shute *et al.* 1977, Swellengrebel 1924, Swellengrebel and De Buck 1938). The parasites had adapted to the phenology of the mosquito: parasites that had entered the human host in the fall would remain dormant in the liver until the following spring at a time that coincided with the appearance of a new generation of adult anopheline mosquitoes. These *P. vivax* parasites generally had an incubation time of 6 to 9 months in the human host. In the tropics, by contrast, the incubation time is 10 days. Today these winter-adapted *P. vivax* parasites remain active in the Koreas, Armenia and possibly Eastern Turkey.

In southern Europe, *Plasmodium falciparum* was a common disease in Greece, Italy, Spain and Portugal until its eradication in the 20th century. It is assumed that this strain of *P. falciparum* was adapted to European anopheline vectors, and may have been genetically distinct from tropical *P. falciparum* strains (see Chapter 3). Studies in which European anophelines were exposed to tropical *P. falciparum* strains repeatedly failed to infect them, suggesting complete refractoriness to the tropical strains (Marchant *et al.* 1998, Ramsdale and Coluzzi 1975, Shute 1940). However, recent research with *Anopheles atroparvus* from Portugal demonstrates susceptibility of these mosquitoes to *P. falciparum* from Africa (A. Luty, personal communication), and the results of the study suggest that specific environmental conditions need to be met for *An. atroparvus* to be competent as vector of tropical *P. falciparum*.

Apart from *P. vivax* and *P. falciparum*, *P. malariae* was also observed in patients in Europe. With the disappearance of the former two *Plasmodium* spp from Europe, *P. malariae* was also eradicated.

Although malaria has disappeared from continental Europe, the vectors are still widely present (see previous section) and Turkey still reports thousands of malaria cases each year (Ardic *et al.* 2005). Because of the high frequency of human movement between Turkey and the European continent, it may be expected that malaria cases are regularly arriving from this region. Under the current environmental conditions it seems unlikely, though, that these imported cases of malaria will create a large malaria epidemic or even a permanent focus of malaria transmission.

The question arises, however, whether climate change will cause a shift in the status quo concerning malaria in North-West Europe, or indeed, the entire continent. As we have seen, anophelines are present across a wide belt of Europe, with several species known as competent for development and transmission of *P. vivax* and, possibly, *P. malariae*. Higher temperatures and/or increased rainfall intensity may make larger areas favourable for malaria transmission and model predictions suggest that the countries along the Mediterranean already are suitable for parasite transmission (Jetten and Takken 1994b, Kuhn *et al.* 2002). Therefore, it is suggested that under climate change the environmental conditions will become more favourable for malaria transmission than at present. Whereas this may apply to southern Europe, the situation for the north-western part of the continent is different: here only one vector species (*Anopheles atroparvus*) is the likely vector. Recent studies have shown that this species has suffered much from environmental change to its habitat and resting sites (Takken *et al.* 2002) and is unlikely to pose a threat in the reestablishment of malaria. *An. messeae*, which is still abundantly present in many areas of north-western Europe (Hulden *et al.* 2005, Sokolova and Snow 2002) rarely bites man, and is therefore considered a poor vector. In addition, for malaria to become re-established a sizeable parasite reservoir must be present, and it is unlikely that clusters of infected persons will arise from whom mosquitoes can become infected in numbers required for continued parasite transmission (Jetten *et al.* 1996, Kuhn *et al.* 2003). Even with continued increase in international travel and continued immigration of people from endemic countries, it is unlikely that a sufficient reservoir of parasite carriers will be formed. This has not happened in the past with returning soldiers and evacuees from colonies after World War II to Britain and the Netherlands, with immigrants from Central America into the USA and it has not happened in recent years due to immigration. For reestablishment of endemic malaria, a fairly large number of suitable vectors and a good number of gametocyte carriers are necessary and they still have to meet under favourable conditions for parasite development. A subsequently infected person will become symptomatic. In the present state of health care in Europe, this person will most likely be diagnosed and treated and this will end the cycle. As Janssens already noted in 1983, claims of re-establishment of malaria in Europe are unfounded and based on insufficient knowledge of the man - parasite - vector cycle and its intricacies (Janssens 1983). Moreover, there is a tendency towards less import malaria in all Western European countries in recent years (Table 1) and we still hope that all efforts to reduce malaria in endemic countries will lead to some success and thus reduced risk for travelers. Thus, even if climate change would result in higher anopheline densities or abundance, the absence of an infectious reservoir of *Plasmodium* parasites would preclude the return to malaria endemicity. The high standards of health care are expected to identify newly arriving patients at an early stage of infection before they have had an opportunity to infect anopheline mosquitoes. Europe has witnessed a few cases of local transmission in the last decade, but each time these have not caused secondary cases or an epidemic. The probability for malaria to become once more established in North West Europe seems therefore (highly?) unlikely.

Discussion

The rising interest of the scientific community and general public in emerging infectious diseases has created heightened interest in malaria in Europe. This interest is fuelled by reports of the historic endemicity of the disease in numerous European countries, including the north-western countries. The anopheline mosquitoes are still widely present and many malaria cases are entering Europe each year. The combination of these data with those of climate predictions suggests that Europe is heading for a renewed establishment of malaria foci. In this paper we provide information that should refute this suggestion and we argue that the poor role of *An. atroparvus* as malaria vector and the high standard of health care make such scenarios highly unlikely. By contrast, however, the tree hole breeding mosquito *An. plumbeus* might be a much more likely vector under climate change. This species has been incriminated with indigenous malaria transmission in Germany, and laboratory studies have demonstrated its vector competence for *Plasmodium falciparum* (Kruger *et al.* 2001 and C.F. Curtis, personal communication). We can only speculate about its role as vector under rising temperatures and intensified rainfall. Recent reports that this species has adapted to sewage ditches as larval habitat are disturbing, and studies should be undertaken to understand the ecology of this species under variable environmental conditions. However, even if *An. plumbeus* can act as an effective vector, the health care levels would remain high, identifying malaria patients rapidly, thus preventing the building up of an infectious human reservoir of *Plasmodium* parasites. We conclude that the return of malaria to North-West Europe as a result of climate change is unlikely, but that this scenario may be possible in southern Europe, where more mosquito species are present and the conditions for parasite development are better than in north-western Europe. Still, also in southern Europe the public health infrastructure is expected to prevent the reestablishment of endemic malaria.

References

Anonymous (2007) De onbeschermde reiziger (the non protected traveller). Vaste Prik no 2: 4-7.

Ardic N, Tanyuksel M, Ozyurt M and Araz E (2005) Is the incidence of malaria decreasing in endemic area of Turkey? New Microbiol 28: 277-280.

Askling HH, Nilsson J, Tegnell A, Janzon R and Ekdahl K (2005) Malaria risk in travellers. Emerging Infect Dis 11: 436-441.

Baldari M, Tamburro A, Sabatinelli G, Romi R, Severini C, Cuccagna G, Fiorilli G, Allegri MP, Buriani C and Toti M (1998) Malaria in Maremma, Italy. Lancet 351: 1246-1247.

Behrens RH and Curtis CF (1993) Malaria in travellers: Epidemiology and prevention. Br Med Bull 49: 363-381.

Blacklock B and Carter HF (1920) Further experiments with *Anopheles plumbeus*, Stephens; Its infection with *P. falciparum* in England; also notes on the apparatus and technique employed. Ann Trop Med Parasit 14: 275-283.

Bottieau E, Clerinx J, Van Den Enden E, Van Esbroeck M, Colebunders R, Van Gompel A and Van Den Ende J (2006) Imported non-*Plasmodium falciparum* malaria: A five-year prospective study in a European referral center. Am J Trop Med Hyg 75: 133-138.

Bruce-Chwatt LJ and De Zulueta J (1980) The rise and fall of malaria in Europe, A historico-epidemiological study. Oxford University Press, Oxford.

Bruce-Chwatt LJ (1988) Three hundred and fifty years of the Peruvian fever bark. BMJ (Clin Res Ed) 296: 1486-1487.

Chen LH, Wilson ME and Schlagenhauf P (2007) Controversies and misconceptions in malaria chemoprophylaxis for travelers. JAMA 297: 2251-2263.

Cook GC (1996) Manson's Tropical Diseases. WB Saunders Company Ltd, London.

Cuadros J, Calvente MJ, Benito A, Arevalo J, Calero MA, Segura J and Rubio JM (2002) *Plasmodium ovale* malaria acquired in central Spain. Emerg Infect Dis 8: 1506-1508.

De Zulueta J (1973) Malaria eradication in Europe. The achievements and the difficulties ahead. In: Ninth International Congress on Tropical Medicine and Malaria. Athens, pp 279-282.

De Zulueta J (1990) Forty years of malaria eradication in Sardinia: a new appraisal of a great enterprise. Parassitologia 32: 231-236.

De Zulueta J (1998) The end of malaria in Europe: an eradication of the disease by control measures. Parassitologia 40: 245-246.

Dobson MJ (1989) History of Malaria in England. J R Soc Med 82: 3-7.

Doudier B, Bougreau H, DeVries A, Ponçon N, Stauffer III WM, Fontenille D, Rogier C and Parola P (2007) Possible autochthonous malaria from Marseille to Minneapolis. Emerg Infect Dis 13:1236-1238.

Feighner BH, Pak SI, Novakoski WL, Kelsey LL and Strickman D (1998) Reemergence of *Plasmodium vivax* malaria in the republic of Korea. Emerg Infect Dis 4:295-297.

Franco-Paredes C and Santos-Preciado JI (2006) Problem pathogens: prevention of malaria in travellers. Lancet Infect Dis 6: 139-149.

Gilles HM and Warrell DA (1993) Bruce-Chwatt's essential malariology. Edward Arnold, London.

Greenwood BM, Bojang K, Whitty CJ and Targett GA (2005) Malaria. Lancet 365: 1487-1498.

Hay SI, Guerra CA, Tatem AJ, Noor AM and Snow RW (2004) The globel distribution and population at risk of malaria: past, present, and future. Lancet Infect Dis 4: 327-336.

Health Protection Agency (2007) Malaria reported in the United Kingdom in 2006: Implications for those advising travellers. Health Protection report. Vol 1, no 18. 1 June. www.hpa.org.uk/hpr/news/default.htm#malaria (Accessed 30 August 2007).

Honingsbaum M (2001) The fever trail: in search of the cure for malaria. Macmillan, London.

Hulden L, Hulden L and Heliovaara K (2005) Endemic malaria: an 'indoor' disease in northern Europe. Historical data analysed. Malar J 4: 19.

Hutchinson RA and Lindsay SW (2006) Malaria and deaths in the English marshes. Lancet 367: 1947-1951.

Jaenson TGT, Lokki J and Saura A (1986) *Anopheles* (Diptera: Culicidae) and malaria in northern Europe, with special reference to Sweden. J Med Entomol 23: 68-75.

Janssens PG (1983) Malaria, wereldwijd gezondheidsprobleem. Mogelijke terugslag op de eigen Belgische gezondheidsproblematiek. Belgisch Arch Soc Geneesk 41: 257-320.

Jelinek T, Mühlberger N, Harms G, Corachán M, Grobusch MP, Knobloch J, Bronner U, Laferl H, Kapaun A, Bisoffi Z, Clerinx J, Puente S, Fry G, Schulze M, Hellgren U, Gjørup I, Chalupa P, Hatz C, Matteelli A, Schmid M, Nielsen LN, da Cunha S, Atouguia J, Myrvang B, Fleischer K and European Network on Surveillance of Imported Infectious Diseases (2002) Epidemiology and Clinical Features of Imported Dengue Fever in Europe: Sentinel Surveillance Data from TropNetEurop. Clin Infect Dis 35: 1047-1052.

Jetten TH and Takken W (1994a) Anophelism without Malaria in Europe. A review of the ecology and distribution of the genus *Anopheles* in Europe. Wageningen Agricultural University Papers 5.

Jetten TH and Takken W (1994b) Impact of climate change on malaria vectors. Change 18: 10-12.

Jetten TH, Martens WJM and Takken W (1996) Model simulations to estimate malaria risk under climate change. J Med Entomol 33: 361-371.

Kampen H, Maltezos E, Pagonaki M, Hunfeld KP, Maier WA and Seitz HM (2002) Individual cases of autochthonous malaria in Evros Province, northern Greece: serological aspects. Parasitol Res 88: 261-266.

Karch S, Dellile MF, Guillet P and Mouchet J (2001) African malaria vectors in European aircraft. Lancet 357: 235.

Knottnerus OS (2002) Malaria around the North Sea: a survey. In Wefer F, Berger WH, Behre KE, Jansen E, eds Climatic Development and history of the North Atlantic realm: Hanse conference report. Berlin, Heidelberg, Springer Verlag, pp 339-353.

Kruger A, Rech A, Su XZ and Tannich E (2001) Short communication: Two cases of autochthonous *Plasmodium falciparum* malaria in Germany with evidence for local transmission by indigenous Anopheles plumbeus. Trop Med Int Health 6: 983-985.

Kuhn KG, Campbell-Lendrum DH and Davies CR (2002) A continental risk map for malaria mosquito (Diptera: Culicidae) vectors in Europe. J Med Entomol 39: 621-630.

Kuhn KG, Campbell-Lendrum DH, Armstrong B and Davies CR (2003) Malaria in Britain: past, present, and future. Proc Natl Acad Sci USA 100: 9997-10001.

Leder K, Black J, O'Brien D, Greenwood Z, Kain KC, Schwartz E, Brown G and Torresi J (2004) Malaria in travelers: A review of the GeoSentinel surveillance network. Clin Infect Dis 39: 1104-1112.

Lindsay SW and Thomas CJ (2001) Global warming and risk of vivax malaria in Great Britain. Glob Change Hum Health 2: 80-84.

Linneaus C (1735) Hypothesis nova de febrium intermittentium causa. In: University of Harderwijk, Harderwijk, the Netherlands, pp 19.

Manson P (1900) Experimental proof of the mosquito-malaria theory. Lancet ii: 923-925.

Marchant P, Eling W, Van Gemert GJ, Leake CJ and Curtis CF (1998) Could British mosquitoes transmit falciparum malaria. Parasitol Today 14: 344-345.

Martens P and Hall L (2000) Malaria on the move: human population movement and malaria transmission. Emerg Infect Dis 6: 103-109.

Martin JA and Mak DB (2006) Changing faces: a review of infectious disease screening of refugees by the Migrant Health Unit, Western Australia in 2003 and 2004. Med J Aust 185: 607-610.

Mosna E (1937) On the thermal characteristics of the Breeding Places of *A. maculipennis*. Riv Parassit 1: 139-155.

Nicolescu G, Linton YM, Vladimirescu A, Howard TM and Harbach RE (2004) Mosquitoes of the *Anopheles maculipennis* group (Diptera: Culicidae) in Romania, with the discovery and formal recognition of a new species based on molecular and morphological evidence. Bull Entomol Res 94: 525-535.

Poncon N, Toty C, L'Ambert G, Le Goff G, Brengues C, Schaffner F and Fontenille D (2007) Biology and dynamics of potential malaria vectors in Southern France. Malar J 6: 18.

Ramsdale CD and Coluzzi M (1975) Studies on the infectivity of tropical African strains of *Plasmodium falciparum* to some southern European vectors of malaria. Parassitologia XVII 39-48.

Ranque S, Marchou B, Malvy D, Adehossi E, Laganier R, Tissot Dupont H, Lotte A, Dydymsky S, Durant J, Stahl JP, Bosseray A, Gaillat J, Sotto A, Cazorla C, Ragneau JM, Brouqui P and Delmont J (2005) Treatment of imported malaria in adults: a multicentre study in France. QJM 98: 737-743.

Reiter P (2000) from Shakespeare to Defoe: Malaria in England in the Little Ice Age. Emerg Infect Dis 6 (1): 1-11.

Romi R, Pierdominici G, Severini C, Tamburro A, Cocchi M, Menichetti D, Pili E and Marchi A (1997) Status of malaria vectors in Italy. J Med Entomol 34: 263-271.

Romi R (1999) *Anopheles labranchiae*, an important malaria vector in Italy, and other potential malaria vectors in Southern Europe. Eur Mosq Bull 4: 8-10.

Sabatinelli G (2000) Contextual determinants of malaria in the WHO European Region. in: Contextual Determinants of Malaria. Lausanne May 15-19 2000. Centre for Integrated Study of the Human Dimensions of Global Change, Carnegic Mellon University, USA.

Schaffner F, Marquine M, Pasteur N and Raymond M (2003) Genetic differentiation of *Anopheles claviger s.s.* in Europe. J Med Entomol 40: 865-875.

Scott HH (1942) A history of tropical medicine, 2nd ed. London, Edward Arnold & Co.

Severini C, Menegon M, Di Luca M, Abdullaev I, Majori G, Razakov SA and Gradoni L (2004) Risk of Plasmodium vivax malaria reintroduction in Uzbekistan: genetic characterization of parasites and status of potential malaria vectors in the Surkhandarya region. Trans R Soc Trop Med Hyg 98: 585-592.

Shanks GD and Edstein MD (2005) Modern malaria chemoprophylaxis. Drugs 65: 2091-2110.

Shute PG (1940) Failure to infect English specimens of *Anopheles maculipennis* var. *atroparvus* with certain strains of *Plasmodium falciparum* of tropical origin. J Trop Med Hyg 43: 175-178.

Shute PG and Williams REO (1954) Indigenous *P. vivax* malaria in London believed to have been transmitted by *Anopheles plumbeus*. Office of the Ministry of health and the public Health Laboratory Service Vd 13: 48-51.

Shute PG, Lupascu G, Branzei P, Maryon M, Constantinescu P, Bruce-Chwatt LJ, Draper CC, Killick-Kendrick R and Garnham PC (1977) A strain of *Plasmodium vivax* characterized by prolonged incubation: the effect of numbers of sporozoites on the length of the prepatent period. Trans R Soc Trop Med Hyg 70: 474-481.

Snow RW, Guerra CA, Noor AM, Myint HY and Hay SI (2005) The global distribution of clinical episodes of *Plasmodium falciparum* malaria. Nature 434: 214-217.

Snowden FM (2006) The conquest of malaria - Italy, 1900-1962. Yale University Press, New Haven.

Sokolova MI and Snow KR (2002) Malaria vectors in European Russia. Eur Mosq Bull 12: 1-6.

Swellengrebel NH (1924) Nadere bijzonderheden over de herfst- en winterbesmetting met malariaparasieten van *A. maculipennis* om Amsterdam. Ned. Tijdschr Geneesk 68: 750-763.

Swellengrebel NH and De Buck A (1938) Malaria in the Netherlands. Scheltema & Holkema Ltd., Amsterdam.

Swellengrebel NH, De Buck A and Kraan H (1938) Mechanism of malaria transmission in the province of north Holland. J Hyg 38: 62-74.

Takken W, Geene R, Adam W, Jetten TH and Van der Velden JA (2002) Distribution and dynamics of larval populations of *Anopheles messeae* and *A-atroparvus* in the delta of the rivers Rhine and Meuse, The Netherlands. Ambio 31: 212-218.

Tatem AJ, Rogers DJ and Hay SI (2006) Estimating the malaria risk of African mosquito movement by air travel. Malar J 5: 57.

Thang HD, Elsas RM and Veenstra J (2002) Airport malaria: report of a case and a brief review of the literature. Neth J Med 60: 441-443.

Van de Torren, G (1935) De zoölogische verspreiding van Anopheles maculipennis atroparvus en *Anopheles maculipennis messeae* in Westelijk Nederland met het oog op 'Species-assaineering'. In: Faculteit der Geneeskunde. Universiteit van Amsterdam, Amsterdam, pp 77.

Van Thiel PH (1939) On zoophilism and anthropophilism of *Anopheles* biotypes and species. Riv Malariol 18: 95-124.

Verhave JP (2000) The disappearance of Dutch malaria and the Rockefeller Foundation. Parassitologia 42: 111-115.

Yeom JS, Ryu SH, Oh S, Lee WJ, Kim TS, Kim KH, Kim YA, Ahn SY, Cha JE and Park JW (2005) Status of *Plasmodium vivax* malaria in the Republic of Korea during 2001-2003. Am J Trop Med Hyg 73: 604-608.

3. Malaria in Southern Europe: resurgence from the past?

Bülent Alten, Helge Kampen and Didier Fontenille

Abstract

Malaria is an ancient diseasFe with historic records of high endemicity in southern Europe. With the exception of Turkey, the disease was eliminated from this region in the 20th Century by a mixture of targeted control, habitat modification and improvements in general living standards. Here we examine whether the disease might resurge in southern Europe under the environmental and climatic changes anticipated for the future. Populations of malaria vectors remain high in many countries of the continent, and their presence poses a risk of renewed transmission, should infected human hosts be available. The great plains of many southern European countries provide ideal habitats for malaria vectors. In the past 50 years, particularly in Italy, Greece and Turkey, the expansion of rice cultivation has resulted in great increases of the population densities of potential malaria vectors and current environmental conditions are suitable for malaria transmission. In addition, malaria remains endemic in the coastal and internal plains of the Asian parts of Turkey. This region also boarders on malaria-endemic countries such as Syria, Iran, Azerbaijan and Armenia. The increased agricultural activities and associated human migration in search of labour may serve to spread malaria across Europe and into currently malaria-free zones. This concerns mostly the upsurge of *Plasmodium vivax*, currently the only malaria parasite causing autochthonous cases in southern Europe. Increased international travel has caused an upsurge in cases of imported malaria in all European regions. However, the high level of health care and tendency of people to settle in urban areas, where malaria vector densities are low, are likely to prevent renewed outbreaks. Records of autochthonous transmission of (imported) malaria are rare and unlikely to cause renewed epidemics. It is concluded that under environmental and climate change the southern European region remains vulnerable to malaria transmission, and that locally outbreaks of malaria may occur, favoured by conditions that promote transmission of the disease. It remains to be seen whether the vector-parasite system is capable to adapt to the predicted changes and can adjust with evolutionary changes.

Keywords: malaria, *Anopheles*, historical and current situation, country profile, Southern Europe, climatic changes

An historical overview of malaria in Southern Europe

The role of the geography of Southern Europe in malaria occurrence

The first part of this chapter is concerned with geography: geography of a particular kind, with special emphasis on human factors. But it is also an attempt to convey a particular kind of history.

Geography in this context is no longer an end in itself but a means to an end. It helps us to rediscover the slow unfolding of structural realities, to see things in the perspective of the very long term (Braudel 1958). Geography, here, helps us to discover the almost imperceptible movement of history, if only we are prepared to follow its lesson and accept its categories and divisions.

Southern Europe, particularly the Mediterranean, has at least two faces. In the first place it is composed of a series of compact, mountainous peninsulas, interrupted by vital plains affecting malaria during history: Italy, the Balkan peninsula and Asia Minor, which experienced indigenous malaria in the past and present, besides the Iberian peninsula. Second, between these miniature continents lie vast, complex, and fragmented stretches of sea, for the Mediterranean is not so much a single entity as a 'complex of seas'. On one side, to the South, the Mediterranean is a near neighbour of the great desert that runs uninterrupted from the Atlantic Sahara to the Gobi Desert and up to the Gates of Beijing. From southern Tunisia to southern Syria, the desert directly borders the sea. The relationship is not casual; it is intimate, sometimes difficult, and always demanding. So desert is not a mere peculiarity of southern Europe; on the contrary, it composes an important face of it. On the other side, to the North, lies Europe, which is often affected by Mediterranean influences and has had an equally great and sometimes decisive influence on southern Europe.

The four peninsulas of the inland sea are very similar. If one thinks of their topography these are regularly divided between mountains - the largest part - a few plains, occasional hills, and wide plateaux. With respect to the malaria situation in southern Europe, the plains are more important than other geographical features. So rather than to consider each peninsula as an autonomous entity, we focus on the analogies between the plains that make them up.

It is even easier to be mistaken about the role of the plain in southern Europe. If we say mountains, it suggests austerity, harness, backwardness, and a scattered population. If we say plain, it suggests abundance, ease, wealth, and good living. In the period we are studying, and with reference to countries of southern Europe, the suggestion is likely to be misleading (Braudel 1978).

There are many large and small plains in southern Europe, especially in the Mediterranean, which are situated between the Pyrenean and Alpine mountain ranges, often resulting from a collapse followed by silting up: the age-long work of lakes, rivers or seas. It is hardly necessary to stress that, whether they are great or small, or near or far from the sea, the plains have the same light, the same colours, the same calendar, or even the same problems such as malaria. They must have been easy to bring under control when they were of limited proportions. Man immediately took possession of the rising ground, strategic hillocks, the fluvial terraces (Marçais 1936) where he established his large, compact villages, sometimes even towns. But at the lowest point of a geological basin, always threatened by the waters, a dispersed habitat was often the rule. When they were larger, the Mediterranean plains were far more difficult to conquer. For a long time they were only very imperfectly utilised by humans. It was not until after 1922 that Greek colonisation eventually triumphed over the marshes in the plain of Saloniki (Ancel 1930). And it was on the eve of World War II that work was finished on draining of the Ebro Delta and the Pontine Marshes (Dobby 1936). In the past, then, the large plains were not all wealthy. By an apparent paradox, they frequently presented a spectacle of misery, desolation and many human health problems such as malaria and other vector-borne diseases.

Between the 16th and 19th Century, first, there was the problem of flooding in southern Europe. Mountain regions provide a source of water that normally collects in the lower regions during the rainy season, when the plains become inundated. When the waters reach level terrain, they do not always run smoothly to the sea (Suyev 1953). In Italy, waters from the Alban and Volscian hills formed a stagnant stretch of about 30 kilometres wide, the Pontine Marshes. The reasons for this were the flatness of the plain, the slow flow of water, and the high line of sand dunes barring the way to the sea. There was not a plain in the Mediterranean, from Portugal to Southeast Turkey that was not threatened by the danger of floodwaters. In fact there was almost always stagnant water at these low levels. And the outcome was always the same: *'Aqua, ora vita, ora morte'* (Braudel 1978). Where it stands it creates vast stretches of reeds and rushes. At the very least, in summer it maintains the dangerous humidity of the marshes and riverbeds, from which came the terrible swamp fever - malaria -, the scourge of the plains in the hot season.

Before the use of quinine, malaria was often a fatal disease in Europe. It was a disease that directly resulted from the geographical environment. Plague, carried from India and China by long-distance travellers, although to be feared because of its extreme virulence, was only a passing visitor in southern Europe. But malaria was permanently established there. It constituted the background to Mediterranean pathology itself in that time interval.

In order to conquer the plains the unhealthy water had to be dealt with and malaria reduced. One of the problems arising from choosing Ankara as the Turkish Capital, in the 1920s, was the prevalence of malaria in the neighbouring plains (Roger 1930). The next task was to obtain fresh water for irrigation.

It does seem, however, that there have been, in the history of malaria, periods of greater and lesser virulence (Jones 1907). There may well have been an increase in marsh fevers at the end of the Roman Empire. There was another increase towards the end of the 15th Century. Fresh pathogenic elements made their appearance at this time. Along with the *Treponema pallidum,* the recently discovered Americas contributed to the old Mediterranean world new malaria parasites. Pope Alexander VI himself was possibly one of its first victims in 1503 (Hiltebrandt 1940).

It is very difficult to pronounce with certainty that the Antiquity and the Middle Ages must have witnessed a disease like malaria, but fevers were probably less dangerous, since Horace crossed the Pontine Marshes without harm in spite of mosquito bites; and more particularly, since in September 1494, the army of Charles VIII encamped safely around Ostia, a particularly dangerous site (Hiltebrandt 1940). We need more precise documentation of the history of malaria than we possess at the moment. We do know that Alexandretta (Iskenderun in the modern Republic of Turkey), which served as a port for Aleppo from 1593, had to be abandoned later because of malaria. We know that Baiae, on the gulf of Naples, which in Roman times was a resort of leisure for the high society, and which Petrarch described as a charming place in a letter to Cardinal Giovanni Colonna in 1343, was deserted by its population, fleeing from fever, in the 16th Century. On the other hand we should note that about twenty years before Columbus, in 1473, the Venetian fleet, which was operating along the Albanian coast during the first siege of Scutari, was decimated by fever and had to move into Cattaro to recover (Braudel 1978).

Nevertheless, one cannot escape the impression that there were fresh outbreaks of malaria during these periods. Perhaps it was because at this time man was running ahead of the enemy. During the whole of the 16th Century, as indeed during the previous century, humans were in search of new land. Where was there a more promising prospect than in shifting marshes of the plains? And precisely the greatest danger lied in the first disturbance of malaria-infested regions. To colonise a plain often meant to die there: we know how many times villages established in plains in southern Europe had to be resettled after being founded, before the plain was won over from fever in the painful struggle of the 19th Century. Moreover the internal colonisation, which occurred throughout the Mediterranean during at least three centuries, also took its toll.

As mentioned above malaria was widespread in the whole of Europe until the 20th century. However, already in the late 19th century the drainage of swamps, moats, ditches and other stagnant waters had been initiated. This large-scale elimination of mosquito breeding sites resulted in a slow but general decrease in malaria incidence all over Europe. Further environmental modification, but particularly the availability of efficient synthetic anti-malarial drugs and improved mosquito control activities including application of dichloro-diphenyl-trichloroethane (DDT) spraying after World War II, eventually led to the disappearance of malaria from Europe (Bruce-Chwatt and De Zulueta 1980). Mosquito populations, and especially so *Anopheles* populations, were significantly reduced by control methods, although, in most cases, not eradicated. Most European countries became free of malaria between the middle of the 20th century and the late 1960s, and, according to the World Health Organisation (WHO), the last autochthonous European malaria case was documented in 1975 in Greek Macedonia (Bruce-Chwatt and De Zulueta 1980)(see also Chapter 2). In a part of southern Europe, malaria has never been completely eradicated, although, due to the mosquito control campaigns, numbers of new infections considerably declined over time. In Southeast Anatolia, Turkey, for example, *P. vivax* malaria was decimated until the 1970s but reached high prevalence again after cessation of the control activities (Caglar and Alten 2000). For this reason we describe the malaria situation of Turkey, particularly that of Southeast Anatolia, in more detail below. This situation is still a threat to the rest of Europe.

Malaria vectors in Southern Europe

The compilation of data on the occurrence and distribution of former and present malaria vectors in Southern Europe is a difficult task. First, there are several *Anopheles* species belonging to species complexes such as the sibling species of the *An. maculipennis* and the *An. claviger* complexes which are not or hardly distinguishable by morphological features. It is sometimes questionable whether old records refer to the correct sibling species, provided the differentiation is made at all. Particularly with the *An. claviger* complex, even literature produced at a time when the sibling species composition of that complex was known does not necessarily give clues to what is meant by '*An. claviger*', the complex or the sibling. Second, some taxa still await final clarification on their taxonomic status. For example, *An. subalpinus* Hackett & Lewis is sometimes listed as a true sibling species of the *An. maculipennis* complex but sometimes as a geographic variant of *An. melanoon* Hackett only. While its egg texture and the results of isoenzyme analyses performed on Italian specimens (Cianchi *et al.* 1987) argue for separate species status, cytogenetic studies (White 1978) and, more recently, DNA analyses are in favour of conspecificity (e.g. Linton *et al.* 2002, Boccolini *et al.* 2003, Kampen, unpublished data). Third, it is not clear whether the presence of formerly vector-competent European *Anopheles* species can be correlated with a certain risk of present malaria transmission after infection of the mosquitoes, even if the climatic conditions are beneficial to parasite development in mosquitoes. While several indigenous *Anopheles* species readily transmitted autochthonous *Plasmodium* strains until the middle of the 20th century as a result of co-evolution and adaptation, they do not appear to be generally susceptible to tropical strains of *P. falciparum* (Shute 1940, De Zulueta *et al.* 1975, Ramsdale and Coluzzi 1975, Daškova and Rasnicyn 1982). On the other hand, one of the latest experiments was performed by Sousa *et al.* (2007) on this subject. The standard membrane feeding assay developed by researchers at Radboud University, Nijmegen, the Netherlands, has proven its utility for experimental assessments of the vector competence of different European anophelines such as *An. superpictus* Grassi from Turkey, *An. atroparvus* van Thiel from Portugal, *An. melanoon* and *An. hyrcanus* (Pallas) from France with respect to a strain of *P. falciparum* originating in tropical Africa. They proved *An. superpictus* females collected from southeast Anatolia could transmit *Plasmodium falciparum* under laboratory conditions. The results obtained to date with this assay suggest that there is a theoretical, although nevertheless negligibly small, possibility of malaria transmission in some areas by certain anopheline species currently found with local and, for the most part, seasonally abundant distributions. The theoretical risk of renewed transmission of plasmodia in areas of Europe where the parasites were formerly endemic, as a result of predicted rising global temperatures and greater population mobility, obviously presupposes, among other pivotal factors, the existence of sufficient vectorial competence.

Despite these problems, efforts to analyse the sometimes-confusing scientific literature produced very helpful compilations of the European mosquitoes including the genus *Anopheles* (e.g. Dahl and White 1978, Minář 1990, Snow and Ramsdale 1999, Ramsdale and Snow 2000, Ramsdale *et al.* 2001, Schaffner *et al.* 2001, Becker *et al.* 2003). These usually contain records from both times of malaria endemicity and times following malaria eradication. This is important to note because malaria eradication programmes in southern Europe mainly carried out after World War II usually included intense mosquito control not only by insecticide application but also by environmental modification, which sometimes resulted in changes in the *Anopheles* fauna. In contrast, Bruce-Chwatt and De Zulueta (1980) in their book 'The rise and fall of malaria in Europe' describe the situation of malaria and its vectors at endemic times (Table 1).

According to the most recent checklists (Ramsdale and Snow 2000, Ramsdale *et al.* 2001, Schaffner *et al.* 2001, Becker *et al.* 2003), there are presently 18 *Anopheles* species occurring in Europe (Table 2), with *An. melanoon* and *An. subalpinus* regarded as two distinct species. Fourteen of these are known to be malaria vectors, with the most important former vector species belonging to the

Table 1. Anopheles *species in Southern Europe according to Bruce-Chwatt and de Zulueta (1980) during malaria-endemic times.*

	Portugal	Spain	France	Italy	Romania	Bulgaria	F.Yugoslavia	Albania	Greece	Turkey*
An. sacharovi			++	++	++	++	++	++	++	++
An. atroparvus	+	+		++	++	+				
An. labranchiae	+	+	++	++						
An. maculipennis s.s.	+	+	+		++	+	++	++	++	
An. messeae			+	+	++	+	++	+		
An. melanoon/subalpinus				+				+		+
An. claviger s.s	+	+	+		+	+	+	+	+	
An. superpictus				++		++	++	++	++	++
An. hyrcanus			+			+	+	+	+	
An. algeriensis			+	+				+	+	
An. plumbeus/nigripes			+	+			+	+	+	
An. marteri				+				+	+	
An. cinereus/hispaniola				+						

*European part; + present; ++ present and important malaria vector.

Table 2. Anopheles *species in Southern Europe according to Ramsdale and Snow (2000), Ramsdale* et al. *(2001), Schaffner* et al. *(2001) and Becker* et al. *(2003).*

	Portugal	Spain	France	Italy	Romania	Bulgaria	Slovenia	Croatia	Boznia-Her.	Montenegro	Serbia	Macedonia	Albania	Greece	Turkey*
An. sacharovi			+			+		+		+			+	+	+
An. atroparvus		+	+	+	+	+	+	+	+		+				
An. labranchiae			+	+				+							
An. maculipennis s.s.	+	+	+	+	+	+	+	+	+			+	+	+	+
An. messeae			+	+	+	+	+	+	+			+	+	+	
An. melanoon/subalpinus	+	+	+	+				+		+		+	+	+	+
An. claviger s.s	+	+	+	+	+	+	+	+	+			+	+	+	+
An. petragnani	+	+	+	+											
An. superpictus			+	+		+		+				+	+	+	+
An. multicolor		+													
An. hyrcanus		+	+	+	+	+		+				+	+	+	+
An. algeriensis	+	+	+	+				+					+	+	+
An. plumbeus/nigripes	+	+	+	+	+	+	+	+	+			+	+	+	+
An. marteri	+	+	+	+		+							+	+	+
An. cinereus/hispaniola	+	+													
An. sergentii				+											

* European part; + present; ++ present and important malaria vector.

Anopheles maculipennis complex: *An. sacharovi* Favr, *An. labranchiae* Falleroni and *An. atroparvus*. Less efficient in malaria transmission during endemic times were *An. messeae* Falleroni and *An. maculipennis s.s.* Meigen of the same complex, *An. claviger s.s.* (Meigen) of the *An. claviger* complex and *An. superpictus*. Although vectors in some cases, all further *Anopheles* species did not play significant roles in the malaria epidemiology in Europe.

Efficient malaria vectors used to occur in southern Europe, and these same species still occur, although in some countries or regions species compositions may have changed due to environmental changes/modifications promoting the development of some species and inhibiting that of others.

Country profiles in Southern Europe

The situation and vector species of southern European countries for which important surveillance data are available will be discussed below.

Portugal and Spain: By 1950, Romeo Viamonte described 13 *Anopheles* species for the Iberian Peninsula. Of these, *An. atroparvus* and, to a much lesser extent and only in the more arid areas, *An. labranchiae* were considered to be the major malaria vectors although some other species may locally also have contributed to disease transmission (Bruce-Chwatt and De Zulueta 1977).

In Portugal, rice cultivation was clearly associated with malaria until the beginning of the 20th century as the rice fields provided unlimited breeding sites for *An. atroparvus* (Cambournac 1939, Cambournac 1942). DDT spraying in combination with the introduction of new anti-malaria drugs resulted in the eradication of malaria at the end of the 1950s. In 1988, Ribeiro *et al.* (1988) presented a checklist of the mosquitoes of continental Portugal describing ten *Anopheles* species including *An. atroparvus* and *An. maculipennis s.s.*, but not *An. labranchiae*.

In Spain, widespread distribution of drugs was much more responsible for the decline of malaria after the Spanish Civil War and World War II than mosquito control, which was only applied in specific areas (Bruce-Chwatt and De Zulueta 1980). According to a newer listing prepared by Eritja *et al.* (2000) on the basis of personal records and published notes, 13 *Anopheles* species have been registered for continental Spain. The occurrence of two of these, however, appears to be doubtful. While the former description of *An. superpictus* most likely accounted to misidentifications of *An. cinereus* Theobald, *An. labranchiae* obviously disappeared from Spain due to the abandonment of the rice fields.

France: According to the literature analysed by Bruce-Chwatt and De Zulueta (1980) until 1980, the anopheline population of France was supposed to be composed of *An. sacharovi, An. labranchiae, An. atroparvus, An. messeae, An. maculipennis* and *An. melanoon* of the *An. maculipennis* complex, as well as *An. claviger, An. hyrcanus, An. plumbeus* Stephens, *An. marteri* Senevet and Prunelle and *An. algeriensis* Theobald. Malaria transmission particularly occurred in southern continental France and Corsica, where *An. atroparvus* and *An. messeae* (continental France) and *An. labranchiae* and *An. sacharovi* (Corsica), respectively, were the main vectors.

It appears that various developments contributed to the eradication of malaria from France: (1) replacement of *An. atroparvus* by more zoophilic species of the *An. maculipennis* complex, (2) vector deviation from humans by keeping cattle in stables during the winter (Juillard 1953), (3) elimination of *Anopheles* breeding sites by drainage of marshes and general water sanitation, (4) loss of the ability of regional plasmodial strains to produce infective gametocytes (Sautet 1947, Sautet 1953), (5) general improvements in socio-economic conditions, rural housing and agricultural techniques, migration to urban areas and widespread use of quinine (Brumpt, 1944/1945). Mosquito control by DDT spraying was implemented after World War II and particularly successful in Corsica (Jaujou, 1954).

In a revised checklist of the mosquitoes of France by Schaffner (1998), 13 anophelines are listed, including those mentioned in a former list of the French Culicidae prepared by Callot and Rioux (1965) only a few years after malaria disappeared from France. In his new checklist, Schaffner added *An. petragnani* Del Vechhio, a sibling to *An. claviger s.s.*, and *An. subalpinus*, whose species status is not clarified. Both taxa are irrelevant to malaria transmission.

In a recent publication based on mosquito collections and blood meal identification, Ponçon *et al.* (2007) concluded that in the Camargue *An. hyrcanus* presently seems to be the only culicid mosquito to potentially play a role in malaria transmission as it is abundant and anthropophilic.

Italy: Until the beginning of DDT application, the main malaria vectors in Italy were *An. superpictus* as well as two species of the *An. maculipennis* complex, *An. labranchiae* and *An. sacharovi* (Hackett and Missiroli 1935). In fact, it was Falleroni (1924) who found in Italy that biological races of *An. maculipennis* could be distinguished by their egg surface structure and morphology which eventually led to the solution of the 'malaria puzzle' and the explanation of the phenomenon of 'anophelism without malaria' (Fantini 1994). By 1969, out of 13 *Anopheles* species that had been described for Italy only five were still present on the mainland excluding *An. superpictus, An. labranchiae* and *An. sacharovi* (Sacca *et al.* 1969). In Sardinia, however, *An. superpictus* and *An. labranchiae* continued to exist.

In a survey by published by Romi *et al.* (1997), only *An. labranchiae* and *An. superpictus* could be found. While *An. labranchiae* formerly used to breed in various types of waters, rice fields as newly established in the 1970s nowadays represent its most important larval habitats in central Italy (Bettini *et al.* 1978, Romi *et al.* 1992). In Grosseto province, *An. labranchiae* replaced *An. melanoon,* which had accounted for 100% of the anopheline fauna in 1970 (Majori *et al.* 1970).

Balkans: Generally, in the Balkan states (Bulgaria, Romania, former Yugoslavia, and Albania) *An. sacharovi* used to be the main malaria vector in coastal areas and *An. superpictus* and *An. maculipennis s.s.* in inland areas due to the specific adaptations of their pre-imaginal stages (Hackett 1937, Hadjinicolaou and Betzios 1973, Bruce-Chwatt and De Zulueta 1980). *An. sacharovi* larvae and pupae are tolerant against brackish water but not against salt water. *An. superpictus* breeds in slowly flowing waters in hilly areas while *An. maculipennis s.s.* breeds in stagnant inland waters (Jetten and Takken 1994). Due to the often reduced socio-economic conditions and subtropical climates in many Balkan regions, malaria persisted longer in these than in other southern European countries.

As elsewhere in southern Europe, DDT spraying was the main reason for malaria decline and eradication (e.g. Belios 1958). Additionally, in some coastal areas of Albania marsh lagoons were flooded by seawater to increase the salt concentration to an unsuitable level for *An. sacharovi* breeding (Hackett 1937).

Romania: In Romania, *An. sacharovi, An. maculipennis s.s., An. messeae* and *An. atroparvus* contributed to endemic malaria. Although locally important elsewhere on the Balkans, *An. superpictus* did not occur in this country. *An. messeae* played a prominent role as a malaria vector in the Danube valley and delta. *An. maculipennis s.s.* was mainly responsible for malaria transmission in the Rumanian plains, and *An. sacharovi* and *An. atroparvus* were vectors on the Black Sea coast (Zotta 1938, Ciuca 1966).

Greece: Malaria was a big problem during the First World War, but also in the years thereafter, when a huge population shift took place. One million Turkish Greeks had to be accommodated in temporary refugee camps and malaria was rampant among them, with 7% deaths (report on Greek Macedonia, by the Malaria Committee of the League of Nations, 1924).

The main malaria vectors in Greece were *An. sacharovi, An. superpictus* and *An. maculipennis s.s.*, in order of importance. DDT larviciding, house spraying and air spraying together with epidemiological surveillance and malaria treatment eventually resulted in malaria eradication from Greece (Belios 1955, Belios 1978). In 1993, Samanidou-Voyadjoglou & Darsie (1993) described 12 *Anopheles* species for Greece, including *An. sacharovi, An. superpictus, An. maculipennis s.s.* as well as *An. atroparvus* and *An. messeae.*

Turkey: Turkey is an impressive example for how control programmes and the respective infrastructure may influence malaria epidemiology in a country. Situated on the edge of South-

East Europe and Western Asia, Turkey is the eastern-most country in the temperate climate zone of the European continent. It encompasses areas differing widely both in their original endemicity and in their receptivity to re-introduction of the diseases.

History of malaria in Turkey

The first list of the mosquitoes of Turkey is to be found in Martini's (1929-31) *Die Fliegen der Palaearktischen Region*. A succession of workers, notably Akalin (1936), Irdem (1940, 1941, 1943), Suyev (1953), Parrish (1959), Gökberk (1961, 1970), Erel (1967), Postiglione *et al.* (1970, 1972, 1973), Coluzzi *et al.* (1974), Lane (1982, 1985, 1992), Alten *et al.* (2000) and Ramsdale *et al.* (2001), have since added to or amended this list as new data became available. According to Ramsdale *et al.* (2001), thirteen anopheline species were identified in Turkey: *An. algeriensis, An. claviger, An. hyrcanus, An. maculipennis, An. marteri, An. plumbeus, An. pulcherrimus* Theobald, *An. sacharovi, An. subalpinus, An. superpictus* and as doubtful and unconfirmed records *An. melanoon, An. multicolor* Cambouliou, *An. sergentii* (Theobald). *An. sacharovi* and *An. superpictus* (Sousa *et al.* 2007) are the most important vectors of malaria in Turkey. Other species of the *An. maculipennis* complex, *An. claviger* and *An. hyrcanus* may play a role in malaria transmission in certain situations but none approach *An. sacharovi* in importance as vectors.

Epidemic malaria has played an important role in the collapse of many antique civilisations on the Aegean and Mediterranean coast in the history of Anatolia. There are many examples from western- and southwestern Anatolian civilisations, situated in coastal plains such as Ephesus, Caunos and Pergesus, in whose decline malaria played an important role. Towards the end of the 19th Century malaria began to be taken seriously. Unfortunately due to fanaticism of the Sultans regime, malaria continued to cause great damage in areas near the shores, around rivervalleys and lakes. The mortality rate increased year after year. According to unofficial data, almost half a million people were treated with quinine in the Ottoman Empire during the First World War (Suyev 1953). The malaria situation is well illustrated by the malaria incidence in the German army, which was allied to the Ottoman Army in 1917-1918. According to this, the malaria incidence among soldiers in the Balkans and Turkey was estimated at 132.4 and 183.7 cases per 1000 soldiers, respectively (Jetten and Takken, 1994).

Before the National Malaria Eradication Programme, one of the most important fluctuations in malaria cases appeared between 1940 and 1951. While the number of malaria cases was 146,077 in 1942, it gradually decreased to 4,211 by 1950 caused by activities of the Anti-Malaria Campaign. Of 20,132 positive blood samples found in the Malaria Campaign Areas in 1951, 14,560 were parasitologically identified as *Plasmodium vivax*, 3,449 were *P. falciparum* and 584 were *P. malariae* (Suyev 1953).

After many years of malaria control in rural as well as urban situations, the National Malaria Eradication Programme was introduced in 1957, and became operational in 1960. From being a common and widely distributed disease, the incidence of malaria decreased annually, malarious areas became progressively more focal, and the transmission of *P. falciparum* ceased. Unfortunately, an agricultural development programme initiated in the Cukurova Plain in the 1970s caused a substantial migration of workers from eastern areas where malaria at that time was more prevalent. This population movement together with the industrial expansion that took place resulted in a serious epidemic of *P. vivax* malaria in 1977 in the provinces of Cukurova, where 115,512 cases were reported. Through concentrated efforts and at considerable cost, the incidence of the disease in the plains began to recede in 1978, and 29,324 cases of malaria were reported in 1979. During 1980, however, the malaria situation deteriorated again and the number of cases reached over 56,000, and this tendency remained unchanged, with 66,681 cases in 1983. Since 1990, when only 8,680 cases were reported, there has been a marked deterioration in the malaria situation in the country. The number of cases reached its peak with 84,345 in 1994. During 1995-1999, there was a steady decline in the malaria incidence, and the number of cases dropped from 82,096 to about 20,000 (Ejov 2001). In the following years, malaria cases were reduced to target level from 19,000 in 2000 to less than 1,000 in 2006 (Table 3).

Since 1985, a serious malaria epidemic has been building up in the south-eastern provinces. The gravitational centre of the disease has now moved from the Cukurova to the Southeastern Agricultural Project Areas (GAP) in Southeast Anatolia and beyond. The indicator of this movement is that 89% of total cases (36,842) in 1998 was confined to the GAP region.

At present, malaria is one of the most important health problems in Turkey, as it was in the past. More than 15 million people or 23% of the total population of Turkey still live in areas where malaria is endemic. A rather large proportion of the total population (nearly 44%) lives in unstable non-epidemic areas, where the risk of the explosive resumption of focal transmission of malaria, producing an outbreak, remains high.

Table 3. Annual distribution of malaria in Turkey between 1925 and 2006 (Turkish Ministry of Health, Malaria Institute).

Year	Malaria incidence	Year	Malaria incidence	Year	Malaria incidence	Year	Malaria incidence	Year	Malaria incidence
1925	1434	1945	16,739	1965	4,587	1985	47,311	2005	2,084
1926	14791	1946	10,373	1966	3,793	1986	37,899	2006	944
1927	10190	1947	5,979	1967	3,975	1987	20,134		
1928	9928	1948	7,298	1968	3,318	1988	16,245		
1929	36,186	1949	4,973	1969	2,173	1989	12,112		
1930	45,653	1950	4,211	1970	1,263	1990	8,680		
1931	61,241	1951	20,132	1971	2,046	1991	12,218		
1932	72,500	1952	8,400	1972	2,892	1992	18,676		
1933	50,609	1953	5,227	1973	2,438	1993	47,210		
1934	48,744	1954	2,489	1974	2,877	1994	84,345		
1935	40,482	1955	1,494	1975	9,828	1995	82,096		
1936	62,466	1956	1,573	1976	37,320	1996	60,884		
1937	69,850	1957	5,536	1977	115,512	1997	35,456		
1938	81,702	1958	11,213	1978	87,867	1998	41,395		
1939	120,060	1959	7,305	1979	29,324	1999	20,963		
1940	115,683	1960	3,092	1980	56,000	2000	19,150		
1941	94,534	1961	3,498	1981	54,415	2001	10,812		
1942	146,077	1962	3,594	1982	62,038	2002	10,224		
1943	115,546	1963	4,365	1983	66,681	2003	9,222		
1944	80,387	1964	5,081	1984	55,020	2004	5,302		

Epidemiological properties of malaria in Turkey

Three parasite species have been observed as indigenous cases in the history of Turkey (Akdur 1977). Today, although occasional imported cases of *P. malariae* and *P. falciparum* are observed, all indigenous cases of malaria are *P. vivax*. But there remains always a possibility for the other species to become re-established in Turkey. At present, resistance to anti-malarial drugs has not been reported in local parasite species, although *P. vivax* in the Far East has begun to show resistance, necessitaing a 28 d treatment regimen of Primaquine. *Plasmodium vivax* is expected to develop resistance to chloroquine. Occasional imported cases of *P. falciparum* and resistant *P. falciparum* cases are being observed in Turkey. Even if it is unlikely, there is some concern about this species causing indigenous cases of malaria. There is also a possibility of some changes in behaviour and parasite competence of mosquitoes in Turkey. The problem of insecticide resistance is also important. Considering these factors, malaria will maintain its position as one of the most important infectious diseases in Turkey.

Factors affecting malaria epidemics in Turkey

This sections relates to factors like transhumance and climate in Anatolia that have affected the historic malaria endemicity with impact on continental Europe. Information presented here relies on the work by Ramsdale and Haas (1978) presented in 'Some aspects of epidemiology of resurgent malaria in Turkey'. Their findings are compared with present data and updated, and some of the original maps have been re-drawn.

Factors related to the parasite

Formerly *P. falciparum* occurred in all the countries of the Mediterranean basin, but its requirement of relatively high temperatures for completion of the extrinsic cycle limited its northerly distribution. By contrast, the distribution of *P. vivax* extended to northern Europe (see Chapter 3) and, in favourable years, transmission of this parasite species has occurred even within the Arctic Circle (Russell 1952). Temperature limits under which development of the parasite in the mosquito is halted lie between 14.5 °C and 16 °C for *P. vivax*, and 16 °C and 19 °C for *P. falciparum*. However, MacDonald (1957) considered that the risk of *P. vivax* transmission remains slight until mean temperatures reach 21° C. Similarly because of the higher temperature requirements of *P. falciparum*, the risk of transmission of this parasite in Turkey remains small until temperatures would increase by 3 to 4 °C.

According to data supplied by the Turkish Meteorological Institute, in all parts of Turkey the mean temperature decreases during September and October. However, malaria transmission could continue after the temperature has fallen below thresholds for development of the parasites because of the existence of already infective mosquitoes. Nevertheless, it is obvious that areas experiencing longer and hotter summers will be more receptive to re-introduced transmission with either parasite, and will also support transmission at a higher level (Ramsdale and Haas, 1978).

With exception of the Aras Valley near the Russian and Iranian borders, areas where temperatures are suitable for the transmission of *P. falciparum* are restricted to the Marmara, Aegean and Mediterranean coastal region, and to the relatively low-lying regions of Southeast Anatolia, which lie at the northern extremity of the great plain of Syria and Iraq. Possible areas of penetration and establishment of this parasite into the interior of the country are strictly limited, being confined to the large river valleys, principally that of the Euphrates. The critical temperature for parasite development is never reached in the greater part of the country, and the relative ease with which *P. falciparum* was eradicated from Turkey indicates that even the hotter parts of the country were areas of marginal distribution. Although occasional imported cases of this parasite are still encountered, all autochthonous malaria at the present time is caused by *P. vivax*.

Transmission of *P. vivax* is possible over a much greater part of Turkey than that of *P. falciparum*. However, conditions favourable for transmission for more than three months occur in a more restricted area. In addition to the region also suitable for the transmission of *P. falciparum*, this includes some Black Sea coastal plains and the valleys of some of the larger rivers penetrating the central plateau Summer temperatures in large tracts of northern and eastern Turkey are never high enough to support transmission of *P. vivax*, and the summer is so short in areas where the critical temperature of 20 °C is not reached until August, that the parasite would have great difficulty in establishing and maintaining itself. The rest of the central plateau, the eastern highlands and the Black Sea coastal areas, where the critical temperature is reached in July, can be regarded as marginal for *P. vivax*, with outbreaks of malaria amenable to remedial measures. This is supported by events, the many foci of transmission, which have been found on the plateau having been eliminated without difficulty, despite an abundance of vector mosquitoes and a high rate of importation of the parasite in at least one area.

Factors related to human ecology

Population movements in Turkey take many forms (Figure 1). A steady drift from the countryside to larger urban centres, principally to Istanbul and Ankara, is as old as the Turkish Republic. It is well documented by the Turkish Institute of Statistics but is not of great importance in the spread of malaria. On the other hand, traditional movements by nomads following seasonal grazing, or by semi-nomads taking their flocks to upland pastures for the summer (transhumance), have been the means by which malaria is continually carried from the low-lying areas of the coast of south-eastern Anatolia to the central plateau and eastern highlands.

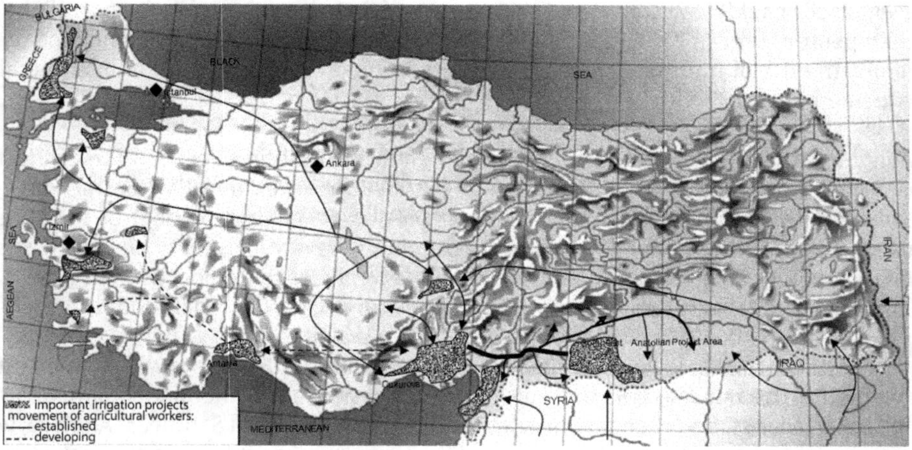

Figure 1. Schematic representation of seasonal population movements in Turkey (taken from Ramsdale and Haas (1978); modified and updated by Alten in 2006).

By far the most important category of population movements from an epidemiological point of view is the seasonal migration of agricultural workers to the large irrigation projects. The origins of the present situation in the plains of Cukurova and Hatay can be traced to the malarious sources of migrant labourers in south-eastern Anatolia. The movement of these labourers is in two directions and they are now re-exporting parasites from the plains of Cukurova and Hatay to their places of origin (Ramsdale and Haas 1978). After the Agricultural Development Project mentioned above was established, malaria once more became a big problem in parts of southeastern Anatolia. Along with this development exists the threat of a potential increase in the incidence of malaria, as occurred during the late-mid 1970s in the Cukurova Valley. The resurgence of malaria in this region presents a new and much greater problem than in the past. Unlike the previous situation in which the problem was located in a relatively small area, where malaria was practically eradicated at the time when developmental activities started, the present epidemic threat covers a much larger area. This area is presently covered by vast areas of wetland and numerous isolated villages providing ample habitat for mosquito breeding and for malaria transmission. There is, therefore, a real danger that with the completion of the extensive irrigation system, a much larger epidemic may take place, adversely affecting the development of this fertile region. The use of irrigation in agriculture will be more widespread with the start of the Southeast Anatolian Project, and the climate, which for many years was arid and dry, will slowly become more temperate and humid. Therefore the potential new breeding sites for the already present mosquito populations will increase significantly. In addition global climate changes, the different effects of which for different continents are already being felt (see Chapter 1), will speed up this process and will be effective in creating new temperate bridge zones between countries.

Recently there has been a tendency amongst large groups of emigrants in southeastern Anatolia to leave from the Agricultural Project Area of Sanliurfa Province and settle on the outskirts of the towns and cities, in substandard buildings constructed on undrained sites, where excavations create ideal vector breeding places. Nevertheless, there is still a constant interchange of population between southeastern Anatolia and the Cukurova Plain, but also between other agricultural project areas in Mediterranean, Aegean and European Turkey, with the result that malaria has now become a serious problem in these areas.

The irrigation projects of the plain of Cukurova and the Southeast Anatolian Project (GAP) are examples of many similar undertakings in the Mediterranean, Aegean and Marmara coastal plains that enhance the risk of malaria. As water becomes increasingly available new land is constantly being brought under cultivation, and the demand for seasonal labour continually increases. Already seasonal workers from southeastern Anatolia are being attracted to the coastal plains of the Mediterranean Region. This tendency will inevitably increase yearly, and will soon include other areas southern and western Turkey, thus facilitating dispersal of the parasite to new receptive areas, as shown in Figure 1. Increased migration of the population from endemic to non-endemic regions in search of labour may also contribute to enhanced risks of malaria establishment and transmission.

A further class of migrant is the international traveller. Uncontrolled international travel of all kinds is continuously increasing, and political frontiers do not present serious obstacles to dispersal of the parasite. Malaria transmission in the country is now a matter of concern to neighbouring countries, and also to other more distant countries having close contact with Turkey and between which there is an important amount of travel passing through European countries. Most of the resurgent malaria cases are due to *P. vivax,* although there have been an increasing number of cases of *P. falciparum*. In view of the mobility of the human population, the increased number of malaria cases constitutes a threat to the plains of Mediterranean Europe (Ejov 2001).

In conclusion, given the presence of formerly vector-competent *Anopheles* species in southern Europe and the limited activities of mosquito control in the region, an increase of locally-acquired malaria would be a logical consequence of the increasing number of imported malaria cases, the enhanced mobility of the population and rising temperatures due to global warming. However, while the formerly indigenous *Plasmodium* strains were efficiently transmitted by indigenous *Anopheles* populations, those *Plasmodium* strains became extinct in Europe due to the massive malaria eradication campaigns after World War II. These parasites were not eradicated from various eastern areas close to Europe, such as some states of the former Soviet Union (e.g. Armenia, Azerbaijan, and Tajikistan) and south-eastern Turkey where malaria is still endemic. Although the European *Anopheles* mosquitoes do not appear to be adapted to tropical *P. falciparum* strains, numbers of cases of autochthonous malaria will most likely increase in southern Europe with changing ecological factors beneficial to *Anopheles* and *Plasmodium* development. Hygienic and socio-economic factors, however, are probably much more important determinants to prevent malaria from becoming endemic again in southern Europe. Trained medical entomologists, regular mosquito monitoring and availability of action plans for mosquito control are a prerequisite to avoid this scenario.

Current situation of malaria in Southern Europe

Recent autochthonous malaria cases in Southern Europe

As mentioned above, malaria is an ancient disease of Europe and northern countries where the disease was eradicated by a mixture of targeted control, habitat modification and improvements in general living standards. However, anopheline populations remain present in almost all ecosystems, and malaria regularly occurs from sporadic autochthonous cases to localised outbreaks. On the other hand the incidence of malaria in sub-Saharan Africa has increased during the last decades. The world is dramatically changing as demonstrated by environmental and climate change. All these changes are likely to cause modifications in mosquito- and disease distribution. The risk for malaria to resurge in Europe seems negligible, although sporadic autochthonous cases regularly occur, showing that a malariogenic potential remains present. This malariogenic potential of a given area depends on its receptivity (i.e. vectorial capacity of the anopheline population: density, feeding behaviour, longevity), infectivity (competition of the sporogonic cycle of *Plasmodium* species within a vector species or population), exposure and vulnerability (imported malaria cases from endemic regions during the period in which malaria transmission is possible). Apart from Turkey, as seen above, only very few endemic cases have occurred in southern Europe in the time since authochtonous malaria became extinct in Europe.

Greece: During the recent years several autochthonous cases have been observed in the Evros province in northern Greece. At least four cases (one *Plasmodium falciparum*, one *P. malariae* and two *P. vivax*) were diagnosed from blood smears in 1994 and 1995. Serological investigation among 1,102 samples concluded that probably 11 patients got malaria locally in the preceding years. Two more *P. vivax* cases were detected in the summer of 1998 in the same region (Kampen *et al.* 2002). More recently, in July 2000, two German tourists were infected with *P. vivax* in Chalkidiki (ProMED 2000). The more plausible explanation for these cases in these regions close to Turkish and Bulgarian borders was the presence of numerous refugees from Armenia, Georgia, Azerbaijan and also, via Turkey, from Pakistan, Iraq and Iran, who may have introduced parasites in this receptive area where *An. maculipennis s.s., An. melanoon, An. sacharovi* and *An. superpictus* have been recorded. It is interesting to note that this region was highly infected by malaria at the beginning of the 20th century.

Italy: Italy was a malaria-endemic country in the past and very efficient vectors remain present (Romi 1999). The question of malaria reappearance is regularly asked since the official eradication in 1970 (Romi *et al.* 2001). The last autochthonous malaria case in Italy occurred in August 1997 in Maremma in the Grosseto region, when a 62 year old woman with no evidence of travel outside the country, was admitted to hospital. This was clearly a secondary malaria case. Investigations in an area three kilometres around the patients house showed that the parasite (*P. vivax*) had been introduced from India by a young Indian girl who arrived in May 1997. Parasitological investigations conducted in August 2006 showed that the girl had *P. vivax* blood forms. The distance between the houses of the primary and secondary case was only few hundred meters. There was a marshy river, with *An. labranchiae* larval development sites, between the two houses. Moreover *An. labranchiae* females were collected biting human volunteers or in animal shelters (Baldari *et al.* 1998). The Marremma marshy region was a highly endemic malaria area until malaria eradication in the 1950s, and this imported case demonstrated that malariogenic potential remains in some Italian areas.

France: The last outbreaks of *P. vivax* malaria in continental France occurred in 1943 in the Camargue, in the south of France. A small outbreak took place in Corsica between 1970 and 1972. In spite of an average of 5,000 annual imported cases, (Legros *et al.* 2006), as well as several airport malaria cases in the last 30 years (Guillet *et al.* 1998), no local case has been observed until 2006. In August 2006 one local case of *P. vivax* was detected in Corsica (Armengaud *et al.* 2007) and two other autochthonous cases (*P. falciparum* and *P. vivax*) were suspected near Marseille (Doudier *et al.* 2007).

The patient from the Corsica case stayed in the town of Porto on the west coast of Corsica from early to mid summer 2006. He had never travelled in malaria-endemic areas and had not been inside any airport for at least 10 years. Investigation around this patient revealed another patient who had been hospitalised for *P. vivax* malaria, after a recent visit to Madagascar. It is likely that this latter first patient was bitten by a local anopheline mosquito, which survived to the sporogonic cycle of *P. vivax* (estimated to 11-12 days during this period), and bit the second patient later on. Potential malaria vectors, such as *An. labranchiae*, *An. atroparvus*, *An. sacharovi* and *An. claviger*, are present in Corsica. Entomological investigations only recorded *An. claviger* larvae in the area where the transmission occurred.

During early spring and early summer 2006, two cases of *P. falciparum* malaria were detected in Southern France in a French woman and an American man of Togolese origin who reported no recent travel to malaria-endemic countries. Both infections occurred after a stay near Marseille, which raises the possibility of autochthonous transmission. In southern France the last malaria cases in patients without any recent travel in malaria-endemic areas were reported in 1994 and 2000 (Delmont *et al.* 1994, Baixench *et al.* 2000).

The first case, a 41-year-old woman, was admitted to Marseille hospital at the end of June. She had never travelled outside Europe and had no history of blood transfusion or drug use by injections. Her residence was >20 km from the nearest airport and 10 km from the seaport. Her neighbourhood included families of Comorian descent, in which members recently travelled to the Comoros where malaria is endemic. Entomological surveillance carried out four weeks after she became ill failed to find any *Anopheles* mosquitoes. The presence of urban anophelines, like *An. plumbeus*, which may transmit malaria, can however not be excluded (Krüger *et al.* 2001, Eling *et al.* 2003).

The second case is more problematic. It is a 34-year-old man who had lived in Togo, until he emigrated to Minnesota in April 2000. He visited Togo in December 2004 for the last time. He travelled in France in Paris and in the Marseille region in April 2006. He reported a day trip to the Camargue, a marshy area having many potential malaria vectors (Ponçon *et al.* 2007), where he sustained multiple mosquito bites. This patient had no illness to suggest an untreated infection since his last trip to Togo, and onset of acute symptoms and parasitaemia level of 3% suggests a recent contamination. Moreover, if he had been persistently infected with *P. falciparum* since its last travel in Africa, this case would still represent an extremely long incubation period from at least 16 months.

Spain: Malaria was officially eradicated from Spain in 1964. The main vectors were *An. atroparvus*, which remains currently abundant in several areas, and *An. labranchiae*. In March 2001, a 75-year-old woman was admitted to Madrid hospital. Blood parasitological and molecular biology investigations concluded to the presence of *P. ovale*. The patient never travelled outside Spain. No entomological study was carried out in the vicinity of the patient. While a locally-acquired infection was undoubted, it remained unclear if it was autochthonous or airport malaria. Actually the patient was living close to two international airports (4 and 18 kilometres), which cannot exclude this possibility (Cuadros *et al.* 2002).

The malariogenic potential in Southern Europe

According to the World Health Organisation, the malariogenic potential of a given area can be considered as a function of its receptivity and of vulnerability and infectivity of mosquitoes (WHO, Technological Report Series. 1966: 324). The receptivity takes into account all parameters of the vectorial capacity of anopheline population (density, trophic behaviour, longevity, duration of sporogonic cycle of *Plasmodium*). The infectivity (i.e. the vectorial competence) is the ability of individual mosquitoes to replicate and transmit a given *Plasmodium* species or strain. Vulnerability corresponds to imported malaria cases from endemic regions during the period in which malaria transmission is possible. Since malaria eradication in the 50s malariogenic potential of most of European regions remains very low.

Receptivity

Research projects on mosquito biology conducted between 2004 and 2007 showed that potential malaria vectors are still present and abundant in different European areas (European Union 'Emerging diseases in a changing European environment (EDEN)' project [http://www.eden-fp6project.net/] (Accessed 5 September 2007). Density, trophic behaviour and longevity depend on location and species. Theoretical duration of the sporogonic cycles of *P. vivax* and *P. falciparum* depends on temperature. The relationship between this duration and average temperature has been well studied. *Plasmodium vivax* completes its cycle in 10 and 13 days at 28 °C and 22 °C, respectively, while it lasts 16 and 20 days for *P. falciparum* at the same temperatures. These temperatures are common in southern Europe, outdoors in the summer and indoors during spring and autumn. Differences between *P. vivax* and *P. falciparum* explain why *P. falciparum* was more easily eradicated than *P. vivax* in Europe.

It is not possible to go into details in all the studies conducted in Europe, but the few following examples taken in south-western Europe show that receptivity might be locally high.

In France, in the Camargue historical malaria region, *An. hyrcanus* is the most abundant *Anopheles* species. *Anopheles melanoon* is less abundant, and *An. atroparvus* and *An. algeriensis* are rare. Only 245 out of 127,000 anopheline specimens collected during an in depth study conducted in 2005 were *An. atroparvus* (Ponçon *et al.* 2007). *Anopheles hyrcanus* and *An. melanoon* are present in the summer, whereas *An. atroparvus* is present in mainly in autumn and winter. A large number of *An. hyrcanus* females was collected on humans (density reaching sometimes more than 100 bites per hour per human) whereas almost exclusively animals attracted *An. melanoon*. Based on an enzyme-linked immunosorbent assay, almost 90% of *An. melanoon* blood meals analysed had been taken on horse or bovine. The parity rates of *An. hyrcanus* and *An. melanoon* varied highly during the season for each species. They were regularly low, most often between 0.32 and 0.67, suggesting a relatively short life expectancy. Since the near disappearance of *An. atroparvus,* the historical malaria vector, *An. hyrcanus* seems to be the only culicid likely to play a role in malaria transmission in the Camargue, as it is abundant and anthropophilic. Density variation of this species during the last 60 years depended on rice field surfaces in the Camargue and on insecticide use against rice pests (Ponçon *et al.*, unpublished results). In Corsica, 13 anopheline species have been recorded. The main historical malaria vectors are *An. labranchiae, An. atroparvus, An. superpictus, An. plumbeus* and *An. sacharovi*. Despite mosquito control *An. labranchiae* remains abundant and anthropophilic in several localities, particularly in the plains, while the four other potential vectors are rare or have almost disappeared. *Anopheles claviger*, another species breeding in small artificial pools in the village has a large distribution in the island, and should also be considered as a potential vector (Barré, personal communication).

In Italy, malaria was eradicated just after the second World War. The three main vectors were *An. labranchiae, An. sacharovi* and *An. superpictus.* Romi *et al.* (2001) studied the distribution and the density of main potential vectors, and evaluated their vectorial capacity in five regions of central and southern Italy. Mosquitoes were collected in animal shelters and on human volunteers. In the Maremma region in Central Italy, six *Anopheles* species are present: *An. claviger, An. plumbeus, An. labranchiae, An. maculipennis s.s., An. melanoon* and *An. atroparvus*. This region is a historical malaria area where the last autochthonous case occurred. The main potential vector is *An. labranchiae,* which became dominant among species of the maculipennis complex since the introduction of rice cultivation in the 1970s. The vectorial capacity for *P. vivax* varied from 0.82 to 8.9 with an average of 16 bites per human per night while it was up to 32.5 in 1994, in the rice field area, with more than 200 infected bites per night (Romi *et al.* 1997, Romi 1999). The vectorial capacity was lower in the other regions studied due to a lower mosquito density.

In Spain, malaria is historically related to rice cultivation mainly (Cambournac 1939). Recent studies in the Ebro Delta confirmed that *An. atroparvus* is abundant and remains the only potential vector. Its larval development sites are the rice fields, which are abundant in the area (Bargues, personal communication). In Portugal, vectorial capacity of *An. atroparvus* has been evaluated in the Comporta Region, 60 km south of Lisbon. *An. atroparvus* may bite humans as well as other vertebrates. The Man Biting Rate evaluated in the summer was around 11 bites per human per night. Parous rates were most often below 0.6. All the field parameters suggest a low vectorial capacity of the anopheline population in this area (Sousa *et al.*, unpublished results). *An. labranchiae*, which was previously also implicated in malaria transmission, seems to have disappeared from the Iberian Peninsula.

Infectivity

Many papers show that if European anopheline species were able to replicate and transmit *P. vivax*, most of the species and populations tested experimentally were refractory to African or Asian *P. falciparum* (Daskova and Rasnicyn 1982, Ramsdale and Coluzzi 1975). However, *P. falciparum* malaria was transmitted in Europe in the past and autochthonous cases show that this transmission may occur suggesting that refractoriness of European anopheline species to *P. falciparum* is not absolute (Eling *et al.* 2003). As mentioned above, this hypothesis is currently being tested in Nijmegen, the Netherlands, within the EDEN European project.

Vulnerability

Quality of data on imported malaria varies depending on countries, several countries under-reporting this disease. According to WHO (Malaria in the WHO European Region, www.euro.who.int/malaria/ctryinfo/ctryinfotop (Accessed on 2 September 2007), between 10,000 and 12,000 cases of imported malaria are notified in the European Union each year (Sabatinelli *et al.* 2001). About 75% of the cases are diagnosed in continental France, the United Kingdom, Germany and Italy. Since the early 1970s there has been a ten-fold increase in the number of imported cases, from about 1,500 in 1972 to more than 15,000 in 2000. A decrease of the number of cases has been observed in all countries since 2000, probably due to the use of more efficient drugs against resistant malaria strains and to better information to travelers (see Chapter 2). Almost all the cases are acquired while visiting endemic areas. The proportion of *P. falciparum* malaria was less than 30% before the early 1980s and increased to almost 80% after the year 2000. Between 1990 and 2003, almost 900 people died from imported *P. falciparum* malaria in the WHO European Region.

In metropolitan France there is an average of more than 5,000 imported cases every year. It was 5,940 in 1998 to 7,127 in 1999 and 8,056 in 2000, and 7,223 in 2001 (Danis *et al.* 2002). Sixty-three percent of cases involved people of African origin and 37% involved 'Westerners'. Ninety-five percent of cases were contracted in Africa. In total 83% of cases involved *P. falciparum*, 6% *P. vivax*, 6.5% *P. ovale* and 1.3% *P. malariae*. Because most of these cases occurred in cities where anopheline mosquitoes are rare and the risk of initiating secondary cases was negligible. In total, 96 deaths have been recorded among the 21,888 *P. falciparum* malaria cases reported between 1996 and 2003 (Legros *et al.* 2007).

In Italy, like in France, the number of imported malaria cases is decreasing since 2000. It was 977, 889, 737, 688, and 671 respectively from 2000 to 2004. Pattern of *Plasmodium* species and origin of infection is the same as in France: 88% of cases was acquired in Africa, *P. falciparum* accounting for 77.5% of the imported cases. Short-term travellers, long-term visitors and immigrants from endemic areas, represented 42%, 26.4% and 27% respectively of cases (Spinazzola *et al.* 2007).

Fewer imported cases are recorded in the Iberian Peninsula. An average of 69 and 309 malaria cases have been registered per year during the last 10 years in Portugal and Spain respectively (http://data.euro.who.int/cisid (Accessed 6 September 2007)

Malariogenic potential

The malariogenic potential of an area corresponds to the risk of local transmission of malaria and is the result of the simultaneous presence of the *Plasmodium* parasite, competent anopheline mosquitoes, susceptible human hosts, and contact between these three components. In Europe, potential anopheline vectors are still present, sometimes in high numbers and with a preference to bite humans. Historical *P. vivax* transmission demonstrates that they are competent for transmitting this parasite. Ability to transmit *P. falciparum*, as well as other *Plasmodium* species is more controversial. *Plasmodium falciparum* was locally transmitted in the past, and rare autochthonous cases occurred recently, suggesting that this species could be transmitted.

In Europe, where endemic malaria does not occur anymore, only imported malaria is recorded, with very exceptional autochthonous cases. The malariogenic potential is low due to a low receptivity and infectivity, as well as low vulnerability in almost all European countries. Even in France where the majority of imported cases are diagnosed, and where potential vectors are present, local cases are registered only exceptionally. Moreover, socio-economic development and health services efficiency in the diagnosis and treatment of imported case generate a very low malariogenic potential in almost all southern European regions. However, as pointed out by WHO, when importation of malaria parasite coincides with socio-economic degradation and the collapse of health services, large-scale malaria epidemics might take place. Such epidemics occurred at the end of the 20th century in the eastern part of Turkey, and are now under control. Expected future climate change may probably modify the distribution of potential vector species (i.e. *An. sacharovi*, *An. atroparvus*, *An. labranchiae*, *An. superpictus*), and may change their vectorial capacity, however this should not generate a malariogenic potential over the epidemic threshold.

Acknowledgements

The authors wish to thank the Eden project (EU grant GOCE-2003-010284) for sharing results. This publication is catalogued by the EDEN Steering Committee (see http://www.eden-fp6project.net/ (Accessed 5 September 2007). The contents of this publication are the responsibility of the authors and don't necessarily reflect the views of the European Commission.

References

Akalin MS (1936) Anadolu makulipennis'leri. Unpublished internal document of the Adana Malaria Institute Ministry of Health, Ankara, Turkey.

Akdur R (1997) Sitma Egitim Notlari. T.C. Saglik Bakanligi Publishing, Ankara, Turkey.

Alten B, Caglar SS and Ozer N (2000) Malaria and its vectors in Turkey. Eur Mosq Bull 7: 27-33.

Ancel J (1930) La plaine de Salonique. In: Braudel F (1978) The Mediterranean and the Mediterranean World in the Age of Philip II. 3rd imp. William and Colins Sons and Co. Ltd.Press, London, UK, p 61.

Armengaud A, Legros F, Quatresous I, Barre H, Valayer P, Fanton Y, D'Ortenzio E and Schaffner F (2006) A case of autochthonous *Plasmodium vivax* malaria, Corsica, August 2006. Euro Surveill 11: E061116-061113

Baixench MT, Suzzoni-Blatger J, Magnaval JF, Lareng MB and Larrouy G (1998) Two cases of inexplicable autochthonous malaria in Toulouse, France. Med Trop 58: 62–64.

Baldari MA, Tamburro G, Sabatinelli R, Romi C, Severini G, Cuccagna G, Fiorilli MP, Allegri CB and Toti M (1998) Malaria in Maremma, Italy. Lancet 351: 1246-1247.

Becker N, Petrić D, Zgomba M, Boase C, Dahl C, Lane J and Kaiser A (2003) Mosquitoes and their Control. Kluwer Academics/Plenum Publishers, New York, Boston, Dordrecht, London, Moscow.

Belios GD (1955) Recent course and current pattern of malaria in relation to its control in Greece. Riv Malariol 34: 1-24.

Belios GD (1958) Malaria eradication in Greece. In:Proceedings 6th International Congress for Tropical Medicine and Malaria, Lisbon, 7: 665.

Belios GD (1978) From malaria control to eradication: problems and solutions. Arheion Hyg 27: 54-59.

Bettini S, Gradoni L, Cocchi M and Tamburro A (1978) Rice culture and *Anopheles labranchiae* in central Italy. WHO/VBC series, 78.686, WHO/MAL/78.897, WHO, Geneva.

Boccolini D, di Luca M, Marinucci M and Romi R (2003) Further molecular and morphological support for the formal synonymy of *Anopheles subalpinus* Hackett & Lewis with *An. melanoon* Hackett. Eur Mosq Bull 16: 1-5.

Braudel F (1958) Histoire et sciences, la longue dureé. In: Annales E.S.C., pp 725-753.

Braudel F (1978) The Mediterranean and the Mediterranean World in the Age of Philip II. 3rd imp. William and Colins Sons and Co. Ltd.Press, London, UK.

Bruce-Chwatt LJ and De Zulueta J (1977) Malaria eradication in Portugal. Trans R Soc Trop Med Hyg 71: 232-240.

Bruce-Chwatt LJ and De Zulueta J (1980) The Rise and Fall of Malaria in Europe. Oxford University Press, Oxford, UK.

Brumpt E (1944/1945) Anophelisme sans paludisme et régression spontanée du paludisme. Ann Parasitol Hum Comp 20: 67-91.

Caglar SS and Alten B (2000) Malaria situation and its vectors in Turkey. In: Proceedings of the 13th European SOVE Meeting (SS Caglar, B Alten and N Özer, eds.) Ankara, pp 234-245.

Callot J and Rioux JA (1965) Liste sommaire des culicides de France. Ann Parasitol Hum Comp 40: 242-245.

Cambournac FJC (1939) A method for determining the larval *Anopheles* population and its distribution in rice fields. Riv Malariol 1: 17-22.

Cambournac FJC (1942) Sobre a epidemiologia do sezonismo en Portugal. Sociedade Industrial e Tipografia, Lisboa.

Cianchi R, Sabatini A, Boccolini B, Bullini L and Coluzzi M (1987) Electrophoretic evidence of reproductive isolation between sympatric populations of *Anopheles melanoon* and *An. subalpinus*. Third International Congress on Malaria and Babesiosis, France, p 1560.

Ciuca M (1966) L'eradication de Paludisme en Roumanie. Editions Medicales, Bucarest.

Coluzzi M, Sabatini A, Bullini L and Ramsdale C (1974) Nuovi dati sulla distribuzzione delle specie del complesso *mariae* del genere *Aedes*. Riv Parasitol 35: 321-330.

Cuadros J, Calvente MJ, Benito A, Arevalo J, Calero MA, Segura J, and Rubio JM (2002) Plasmodium ovale malaria acquired in Central Spain. Emerg Infect Dis 8: 1506-1508.

Dahl C and White GB (1978) Culicidae. Limnofauna Europaea (J Illies, ed.) Gustav Fischer, Stuttgart, pp. 389-395.

Danis M, Legros F, Thellier M and Caumes E (2002) Données actuelles sur le paludisme en France métropolitaine. Med Trop 62: 214-218.

Daškova NG and Rasnicyn SP (1982) Review of data on susceptibility of mosquitoes in the USSR to imported strains of malaria parasites. Bull World Health Organ 60: 893-897.

De Zulueta J, Ramsdale CD and Coluzzi M (1975) Receptivity to malaria in Europe. Bull World Health Organ 52: 109-111.

Delmont J, Brouqui P, Poullin P and Bourgeade A (1994) Harbour-acquired *Plasmodium falciparum* malaria. Lancet 344: 330-331.

Dobby EHG (1936) The Ebro Delta. In: Braudel F (1978) The Mediterranean and the Mediterranean World in the Age of Philip II. 3rd imp. William and Colins Sons and Co. Ltd.Press, London, UK, p 62.

Doudier B, Bogreau H, DeVries A, Ponçon N, Stauffer III WM, Fontenille D, Rogier C and Parola P (2007) Autochthonous malaria from Marseille to Minneapolis. Emerg Infect Dis, in press

Ejov M (2001) Malaria situation in European Region. In: Proceedings First Balkan Conference, Malaria and Mosquito Control, Serres, Greece, pp. 20-26.

Eling W, Gemert GJV, Akinpelu O, Curtis J, Curtis CF and van Gemert GJ (2003) Production of Plasmodium falciparum sporozoites by Anopheles plumbeus. Eur Mosq Bull 7: 12-13.

Erel D (1967) Sivrisineklerin Morfolojisi ve Biyolojisi. T.C. Saglik ve Sosyal Yardim Bakanligi Hifzissihha Okulu Publishing No 28, Gursoy Pres Co., Ankara, Turkey.

Eritja R, Aranda C, Padrós J, Goula M, Lucientes J, Escosa R, Marquès E and Cáceres F (2000) An annotated checklist and bibliography of the mosquitoes of Spain (Diptera, Culicidae). Eur Mosq Bull 8: 10-18.

Falleroni D (1924) Studio sugli *A. maculipennis* delle Paludi Pontine. Tipografia Economica, Roma.

Fantini B (1994) Anophelism without malaria: an ecological and epidemiological puzzle. Parassitologia 36: 83-106.

Gökberk C (1961) *Anopheles sacharovi* (Favre, 1903) in Turkey. Mosq News 21: 101-102.

Gökberk C (1970) The first record of *Anopheles pulcherrimus* Theobald, 1902 in Turkey. Its morphology, ecology and distribution. Turk Hifzissihha Tecrube Bioloji Dergisi 30: 20-27.

Guillet P, Germain MC, Giacomini T, Chandre F, Akogbeto M, Faye O, Kone A, Manga L and Mouchet J (1998) Origin and prevention of airport malaria in France. Trop Med Int Health 3: 700-705.

Hackett LW (1937) Malaria in Europe: An Ecological Study. Oxford University Press, London.

Hackett LW and Missiroli A (1935) The varieties of *Anopheles maculipennis* and their relation to the distribution of malaria in Europe. Riv Malariol 14: 45-109.

Hadjinicolaou J and Betzios B (1973) Resurgence of *Anopheles sacharovi* following malaria eradication. Bull World Health Organ 48:699-703.

Hiltebrandt P (1940) Der kampf ums Mittelmeer. In: Braudel F. (1978), The Mediterranean and the Mediterranean World in the Age of Philip II. 3rd imp. William and Colins Sons and Co. Ltd. Press, London, UK, p 65.

Irdem E (1940) *Anopheles multicolor*, a mosquito new to Turkey. Sihhiye Mecmuasi 16: 86-94.

Irdem E (1941) *Anopheles sergentii* a mosquito new to Turkey. Sihhiye Mecmuasi 17: 296-297.

Irdem E (1943) *Anopheles marteri*, a first record from Turkey. Sihhiye Mecmuasi 17: 440-442.

Jaujou C (1954) La lutte antipaludique en Corse. Bull World Health Organ 11: 635-677.

Jetten TH and Takken W (1994) Anophelism Without Malaria in Europe – A review of the ecology and distribution of the genus *Anopheles* in Europe. Wageningen Agricultural University Papers, Wageningen, The Netherlands.

Jones WHS (1907) Malaria, a neglected factor in the history of Greece and Rome. In: Braudel F (1978), The Mediterranean and the Mediterranean World in the Age of Philip II. 3rd imp. William and Colins Sons and Co. Ltd. Press, London, UK, p 65.

Juillard E (1953) Du pâturage à la stabulation dans la plaine d'Alsace. Rev Pathol Gén Comp 53: 1316.

Kampen HJ, Proft S, Etti E, Maltezos M, Pagonaki WA Maier and Seitz HM (2003) Individual cases of autochthonous malaria in Evros Province, northern Greece: entomological aspects. Parasitol Res 89: 252-258.

Krüger A, Rech A, Su XZ and Tannich E (2001) Two cases of autochthonous *Plasmodium falciparum* malaria in Germany with evidence for local transmission by indigenous *Anopheles plumbeus*. Trop Med Int Health 6: 983-985.

Lane CJ (1982) *Aedes cretinus* Edwards 1921. Mosq Syst 14: 81-85.

Lane CJ (1985) *Aedes cretinus* Edwards 1921. Trans R Soc Trop Med Hyg 79: 278

Lane CJ (1992) *Aedes (Ochlerotatus) nigrocanus* Martini, its taxonomic status. Mosq Syst 24: 12-15.

Legros F, Arnaud A, El Mimouni B and Danis M. Paludismo d'importation en France métropolitaine données épidémiologiques 2001-2004. http:// www.invs.sante.fr/beh/2006/32/beh_32_2006.pdf (Accessed 5 September 2007)

Legros F, Bouchaud O, Ancelle Y, Arnaud A, Cojean S, Le Bras J, Danis M, Fontanet A and Durand R (2007) Risk factors for imported fatal Plasmodium falciparum malaria, France, 1996-2003. Emerg Infect Dis 13: 883-888.

Linton YM, Smith L and Harbach R (2002) Observations on the taxonomic status of *Anopheles subalpinus* Hackett & Lewis and *An. melanoon* Hackett. Eur Mosq Bull 13: 1-7.

Majori G, Maroli M, Bettini S and Pierdominici G (1970) Osservazioni sull'anofelismo residuo nel Grossetano. Riv Parassitol 31: 147-154.

Marçais G (1936) Tlemcen, ville d'art et d'historie. In: Braudel F (1978), The Mediterranean and the Mediterranean World in the Age of Philip II. 3rd imp. William and Colins Sons and Co. Ltd.Press, London, UK, p 61.

Martini E (1929-31) 11 und 12. Culicidae. Die Fliegen der Palaearktischen Region (ed. by Lindner E) 3, 1-144 (1929), 145-320 (1930), 321-398 (1931), Stuttgart, Germany.

MacDonald G (1957) The epidomiology and Control of Malaria. Oxford University Press, London, UK.

Minář J (1990) Family Culicidae. Catalogue of Palaearctic Diptera, Vol. 2 (Soós Á and Papp L, eds.) Akadémiai Kiadó, Budapest, pp. 74-113.

Parrish DW (1959) The mosquitoes of Turkey. Mosq News 19: 264-266.

Ponçon N, Toty C, L'Ambert G, Le Goff G, Brengues C, Schaffner F and Fontenille D (2007) Biology and dynamics of potential malaria vectors in Southern France. Malar J 6: 18.

Postiglione M, Bruno Smiraglia C, Lavagnino A, Gokberk C and Ramsdale CD (1970) A preliminary note on the occurrence in Turkey of the *subalpinus* form of the *A. maculipennis* Complex. Riv Parassitol 31: 155-158.

Postiglione M, Tabanli S and Ramsdale CD (1972) *Anopheles claviger* in Turkey. Riv Parassitol 33: 219-230.

Postiglione M, Tabanli S and Ramsdale CD (1973) The *Anopheles* of Turkey. Riv Parassitol 34: 127-159.

ProMED Digest (2000) Malaria, vivax – Germany ex Greece. ProMED-mail 20000713.0164

Ramsdale CD, Alten B, Caglar SS and Özer N (2001) A revised, annotated checklist of the mosquitoes (Diptera, Culicidae) of Turkey. Eur Mosq Bull 9: 18-28.

Ramsdale CD and Coluzzi M (1975) Studies on the infectivity of tropical African strains of *Plasmodium falciparum* to some southern European vectors of malaria. Parassitologia 17: 39-48.

Ramsdale CD and Haas E (1978) Some aspects of the epidemiology of resurgent malaria in Turkey. Trans R Soc Trop Med Hyg 72: 570-580.

Ramsdale C and Snow K (2000) Distribution of the genus *Anopheles* in Europe. Eur Mosq Bull 7: 1-26.

Ribeiro H, da Cunha Ramos H, Pires CA and Antunes Capela R (1988) An annotated checklist of the mosquitoes of continental Portugal (Diptera, Culicidae). In: Actas III Congreso Ibérico de Entomologia, Lisbon, pp. 233-254.

Roger N (1930) En Asie mineure. In: Braudel F (1978), The Mediterranean and the Mediterranean World in the Age of Philip II. 3rd imp. William and Colins Sons and Co. Ltd.Press, London, UK, p 64.

Romeo Viamonte JM (1950) Los anofelinos de España y de la zona española del Protectorado Marrueccos. Su relacion con la difusion del paludismo. Rev San E Hig Publ 24: 213-295.

Romi R (1999) *Anopheles labranchiae*, an important malaria vector in Italy and other potential malaria vectors in Southern Europe. Eur Mosq Bull: 8-10.

Romi R, Pierdominici G, Severini C, Tamburro A, Cocchi M, Menichetti D, Pili E and Marchi A (1997) Status of malaria vectors in Italy. J Med Entomol 34: 263-271.

Romi R, Sabatinelli G and Majori G (2001) Could malaria reappear in Italy? Emerg Infect Dis 7: 915-919.

Romi R, Severini C, Cocchi M, Tamburro A, Menichetti D, Pierdominici G and Majori G (1992) Anofelismo residuo in Italia: distribuzione nelle aree risicole delle provincie di Grosseto e Siena. Ann Ist Super Sanita 28: 527-531.

Russell P (1952) Malaria: Basic principles briefly stated. Blackwells, Oxford, UK.

Sabatinelli G, Ejov M and Joergensen P (2001) Malaria in the WHO European Region (1971-1999). Euro Surveill 6: 61-65.

Saccà G, Scirocchi A and Stella E (1969) Anofelismo residuo in provincia di Latina a 15 anni dall'eradicazione della malaria. Parassitologia 11: 189-204.

Samanidou-Voyadjoglou A and Darsie RF (1993) An annotated checklist and bibliography of the mosquitoes of Greece (Diptera: Culicidae). Mosq Syst 25: 177-185.

Sautet J (1947) A propos de regression spontanée du paludisme. Bull Soc Pathol Exot Fil 41: 31.

Sautet J (1953) Les fluctuation du paludisme dans le sud de la France et dans les départements insulaires. Rev Pathol Gén Comp 53: 1199-1208.

Schaffner F (1998) A revised checklist of the French Culicidae. Eur Mosq Bull 2: 1-9.

Schaffner F, Angel G, Geoffroy B, Hervy JP, Rhaiem A and Brunhes J (2001) The Mosquitoes of Europe/Les Moustiques d'Europe. An Identification and Training Programme (CD-Rom). IRD Éditions & EID Méditerranée, Montpellier, France.

Shute PG (1940) Failure to infect English specimens of *Anopheles maculipennis* var. *atroparvus* with certain strains of *Plasmodium falciparum* of tropical origin. J Trop Med Hyg 43: 175-178.

Snow K and Ramsdale C (1999) Distribution chart for European mosquitoes. Eur Mosq Bull 3: 14-31.

Sousa C, Rosario V, Ozer N, Poncon N, Alten B, Caglar SS, Kaynas S, Simsek F, Vegte-Bolmer M, van Gemert GJ, Fontenille D and Luty AJF (2007) Transmission of African *P. falciparum* by European anophelines. European Community FP6 EDEN Project Annual Meeting, Antalya, Turkey.

Spinazzola F, Nicastri E, Vlassi C, Ghirga P, De Marco M, Pittalis S, Paglia MG, Ferrari C and Narciso P (2007) Imported malaria at Italy's National Institute for Infectious Diseases Lazzaro Spallanzani, 1984-2003. European Journal of Clin Microbiol Infect Dis 26: 175-179

Suyev M (1953) Sitma Savasi Calismalari Albumu. Turkish Ministry of Health, Husunutabiat Publishing, Istanbul, Turkey.

White GB (1978) Systematic reappraisal of the *Anopheles maculipennis* complex. Mosq Syst 10: 13-44.

Zotta G (1938) Contribution à l'etude de la distribution des races d'*A. maculipennis* en Roumanie. Arch Roum Pathol Exp Microbiol 11: 209-246.

4. Avian malaria in Europe: an emerging infectious disease?

Silvie Huijben, Willem Schaftenaar, Adriaan Wijsman, Krijn Paaijmans and Willem Takken

Abstract

Emerging infectious diseases have had, and likely will have, a great impact on human and animal populations, and on biodiversity. Avian malaria is one such disease, which reached epidemic levels soon after its introduction in Hawaii and contributed to a massive decrease of avian biodiversity on the island group. Avian malaria is also a common infection amongst endemic and migratory birds in Europe, but appears to cause little harm among the native birds. However, frequent deaths of non-indigenous birds due to avian malaria are being reported. In this chapter, we consider whether avian malaria can reach epidemic proportions in Europe as an emerging infectious disease and as a result of climate change. In an ecological field study, we investigated the occurrence of avian malaria in Rotterdam zoo in the Netherlands during the summer of 2003. In a longitudinal approach mosquito dynamics were studied and a cross-sectional study was undertaken to estimate parasite prevalence in exotic birds kept in the zoo. Adult mosquitoes were caught throughout the study period and were examined for parasite infections. Of the 575 adult mosquitoes caught in traps, 61 percent was *Culex pipiens pipiens*. Six mosquitoes of this species were infected with malaria. Nine birds, belonging to eight different species, were infected with malaria, out of 81 birds that were examined. Twelve other birds died later that summer and were diagnosed with avian malaria, most of them being black-footed penguins. The peak in mortality occurred in the third week of August, two weeks after the peak in larval mosquito densities. It appeared that there is annual seasonal malaria transmission within Rotterdam zoo. We suggest that these infections result from a reservoir of parasites living within less susceptible exotic birds that are kept in the zoo and in endemic birds living within or near the zoo. These field data support our view that an outbreak of avian malaria among the endemic bird population of Europe is unlikely to happen under the current circumstances. Avian malaria is a common parasitic infection in the endemic and migratory bird populations, and after a long history of co-evolution, the parasites are causing what appears to be little harm. However, a change of climate or parasite mutational events could alter the equilibrium between parasite and host, potentially resulting in an epidemic. Frequent screening of the endemic and migratory bird populations should therefore be employed to prevent an outbreak such as happened in Hawaii.

Keywords: avian malaria, *Plasmodium*, bird, mosquito, Europe, emerging infectious disease

Introduction

Among vertebrates, infectious diseases can have a devastating effect on populations. Through co-evolution, most animals have developed an effective immune response to infections with pathogens and parasites[1] and they can therefore survive periodic outbreaks of diseases. Some infections express a greater virulence than others, depending on the frequency and intensity of parasite-host interactions as well as the genetic make-up of parasite and host (Daszak *et al.* 2000). Geographic isolation may sometimes prevent the exposure of animal populations to parasites, rendering such populations highly susceptible to introduced pathogens and parasites. Historic examples of such events are the accidental importation of plague and smallpox into Europe

[1] In this paper pathogens include viruses, fungi, protons and bacteria, and parasites include protozoa and nematodes.

(Harrison 2004) and measles into South America (Black *et al.* 1974), resulting in high mortality in the affected population.

Here we discuss avian malaria, a parasitic disease affecting birds. As with human malaria parasites, avian malaria parasites are transmitted from one vertebrate to another through mosquito bites (Valkiunas 2005). Although in human malaria only mosquitoes of the genus *Anopheles* transmit the parasites, avian malaria parasites are usually transmitted by other mosquito genera, of which *Culex* spp. are the most recognised vectors. We discuss the possibility of avian malaria becoming an infectious disease in Europe, whereby we use a recent outbreak of avian malaria in Hawaii as an example. Moreover, we show novel data from a recent study of avian malaria in the Rotterdam zoo. Could the occurrence of an avian malaria parasite reservoir and abundant vector populations result in epidemic levels of avian malaria in Europe?

An emerging disease is defined as 'an infection that newly appears in a population, or has existed but is (rapidly) increasing in incidence or geographic range' (Lederberg *et al.* 1992). There are many examples of infectious diseases, such as the plague that swept through Europe and Asia, killing millions of people during the Middle Ages (Harrison 2004), or 'Spanish influenza', killing an approximated 20 million people across the globe between 1918-1919 (Taubenberger *et al.* 1997). More recently, the vector-borne Chikungunya virus has rapidly spread across the islands of the Indian Ocean, with a peak impact on the island of Réunion, where almost one third of the human population was affected by the disease (Schuffenecker *et al.* 2006). Another example is the arbo-virus bluetongue, which showed up Italy in 1998, affecting millions of sheep (Baylis *et al.* 2001). Bluetongue continues to plague Europe with a new outbreak occurring in north-western Europe in 2006 (see Chapter 7), affecting sheep and cattle and causing huge economic losses (Mehlhorn *et al.* 2007). A final and well-known example is the West Nile virus, which was accidentally introduced in the United States in 1999 and spread rapidly across the North-American continent with numerous casualties and many hospitalisations (Tyler 2004). All these emerging diseases follow a common path where a highly virulent pathogen enters and rapidly spreads in a susceptible population whereby it kills a relatively high percentage of its hosts.

Avian malaria: the parasites

The term malaria has confusingly been used for either all *Haemosporidians* (class *Sporozoa*, order *Haemosporida*), a group of protozoans that use blood-sucking dipteran insects as vectors to complete their life-cycle, or strictly parasites of the genus *Plasmodium*. Although haemosporidian parasites are genetically closely related, the life-cycle, vector species, and epidemiology of parasites from different families are very different (Valkiunas *et al.* 2005). Therefore, in this chapter, when we refer to avian malaria we mean parasites of the genus *Plasmodium*.

Thirty-eight morphologically different avian *Plasmodium* species have been described. The parasite species that has received the most attention is *P. relictum*. This parasite is not only well known from avian malaria in Hawaii (Van Riper III *et al.* 1986) and avian malaria outbreaks in zoos all around the world (Cranfield *et al.* 1994, Fix *et al.* 1988, Graczyk *et al.* 1995), but it has also been used as a model species to study human malaria during the end of the 19th and beginning of the 20th century (Garnham 1966).

Unfortunately, there is little information available on the vector species of avian malaria. However, since much experimental work has been carried out with *P. relictum*, it is known that this parasite can complete its life cycle in 26 different species of the Culicidae family, including the genera *Aedes*, *Anopheles*, *Culex* and *Culiseta* (LaPointe *et al.* 2005, Valkiunas 2005). The life cycles of the different species of avian malaria parasites are generally similar. Figure 1 shows a schematic overview of the life cycle of avian malaria, using *P. relictum* as an example.

The sexual reproduction phase of the parasite takes place in the mosquito midgut, where male and female gametes fuse to form a zygote, which in turn, differentiates into the mobile ookinete. The ookinete penetrates the midgut to form an oocyst on the outer midgut wall of the mosquito. Within the oocyst, hundreds of sporozoites are being produced, until the oocyst eventually ruptures, resulting in a release of the sporozoites. These migrate to the salivary glands of the mosquito, after which the mosquito is able to transmit the sporozoites while blood-feeding on a bird. The time for the parasite to develop within the mosquito takes approximately seven days.

When blood-feeding on a bird, the mosquito injects its saliva, containing various enzymes to enhance blood uptake and prevent clotting, together with the sporozoites, which end up in

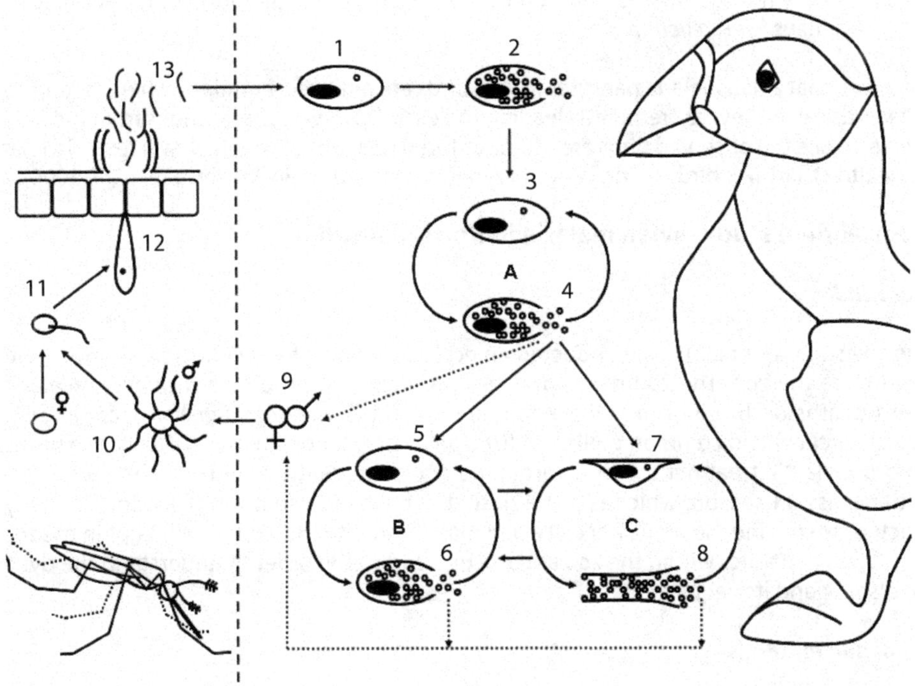

Figure 1. Life cycle of the avian malaria parasite Plasmodium relictum. On the right the asexual life cycle in the bird is depicted: 1) infected reticuloendothelial cell by sporozoites that are injected in the blood stream by the mosquito; 2) emerging cryptozoites; 3) infected macrophage; 4) emerging metacryptozoites; 5) infected erythrocyte; 6) emerging erythrocytic merozoites; 7) infected endothelial cell of the capillaries; 8) emerging phanerozoites; 9) male and female gametocytes. A) Primary exoerythrocytic cycle; B) Erythrocytic cycle; C) Secondary exoerythrocytic cycle. On the left the sexual life cycle in the mosquito is illustrated: 10) male and female gametes; 11) fertilised zygote; 12) ookinete moving through midgut wall, and 13) emerging sporozoites from oocyst.

the bird's blood stream. The sporozoites invade the reticular cells of various organs, such as the spleen, and tissues, such as skin tissue, where they develop into cryptozoites. The merozoites that are developing within the cryptozoites are, unlike the merozoites in human malaria, unable to infect red blood cells immediately, but undergo a second exo-erythrocytic cycle, during which they invade the macrophages in many organs. Part of the resulting metacryptozoites stay in the primary exo-erythrocytic cycle and infect new macrophages again. The remaining parasites either (a) enter the erythrocytic cycle, where they invade the red blood cells and multiply to a schizont; (b) invade a red blood cell to immediately develop into a gametocyte; (c) enter the secondary exo-erythrocytic cycle, where they invade the endothelial cells of many organs (including the brain and liver) in which the phanerozoites develop. A proportion of the merozoites from the erythrocytic cycle stays in the erythrocytic cycle and reinfects new erythrocytes. The remaining of the parasites can either infect a new red blood cell to develop into a gametocyte or enter the secondary exo-erythrocytic cycle. The phanerozoites of the secondary exo-erythrocytic cycle can also develop into gametocytes or enter the erythrocytic cycle or reinfect the endothelial cells, where they can stay for the remainder of the host's life. Periodic relapses, resulting from these dormant phanerozoites, often occur and are for instance related to a weakened immune system and environmental stress and often synchronised to the breeding season (Valkiunas 2005). The time to maturation of the first generation of metacryptozoites, the prepatent period, is usually less than five days for *P. relictum*.

It is believed that avian malaria parasites in wild birds are relatively harmless, whereas morbidity in captive birds can be severe, frequently leading to death. However, observations from the field may be skewed since the observed birds are often captured with mist nets which will result in relatively more healthy than sick birds being caught (Van Riper III *et al.* 1986, Westerdahl *et al.* 2005).

Ecological field study: avian malaria in Rotterdam zoo

Background

During the summer of 2003 we conducted an ecological field study in Rotterdam zoo, hereafter referred to as the zoo. The zoo has experienced a history of problems with avian malaria and frequent deaths occurred in their black-footed penguin (*Spheniscus demersus*), common puffin (*Fratercula arctica*) and common guillemot (*Uria aalge aalge*) populations. Usually, the birds were given prophylactic treatment with chloroquine and primaquine to control the disease during peak transmission season, which is not a desirable long-term solution. Therefore, a study was conducted to examine the prevalence of avian malaria among the exotic birds kept in the zoo and the mosquito dynamics within the zoo and its surroundings in order to understand the dynamics of the disease and its vectors.

Material and methods

The mosquito population was monitored from May to September 2003 within the zoo and its close surroundings. Twenty-four potential larval breeding sites were selected at the beginning of the study and were visited twice a week throughout the study period. Larvae were sampled randomly using a standard white dipper, as described in Service (1993). The number of dips that were taken depended on larval densities: 25 dips were taken in sites with a high larval abundance (>2.5 larvae per dip), 50 dips in sites with intermediate larval abundance (between 1 and 2.5 larvae per dip) and 100 dips in sites with low larval abundance (less than one larva per dip). Larvae were collected and morphologically identified.

The adult mosquito population was monitored with three mosquito traps (Mosquito Magnet®, American Biophysics Corporation, North Kingstown, RI, USA). One trap was placed in the residence of the puffins and guillemots, another trap was placed next to a large water source in the surroundings of the penguin enclosure, and the last was placed at a central location in the zoo (Figure 2). The traps were emptied twice a week and mosquitoes were stored on silica gel. In the laboratory the collected mosquitoes were morphologically identified (Van Haren and Verdonschot 1995) and counted. Later, all mosquitoes were checked for infections with avian malaria using a Polymerase Chain Reaction (PCR) as described by Waldenström *et al.* (2004), which identifies both *Plasmodium* spp. and *Haematoproteus* spp. As *Haematoproteus* spp. are not transmitted by mosquitoes, positive samples indicated *Plasmodium* spp. infections. DNA isolation was performed with a standard phenol extraction and ethanol precipitation (Snounou *et al.* 1993a).

A cross-sectional parasitological study was performed during 4-5 July 2003. Approximately ten percent of each bird population, representing 44 species, was randomly selected and from each bird 200 µl of blood was drawn, which was immediately mixed with EDTA. On 5 August, blood was taken from all black-footed penguins in the same way. DNA was extracted with the QIAamp DNA mini kit, with the adjustment that the blood was incubated with 200 µl lysis buffer and 20 µl proteinase K at 55 °C overnight, due to clotting of the blood. A general PCR to identify *Plasmodium* spp. was used on these blood samples, which detects all *Plasmodium* species, including human malaria (Snounou *et al.* 1993b).

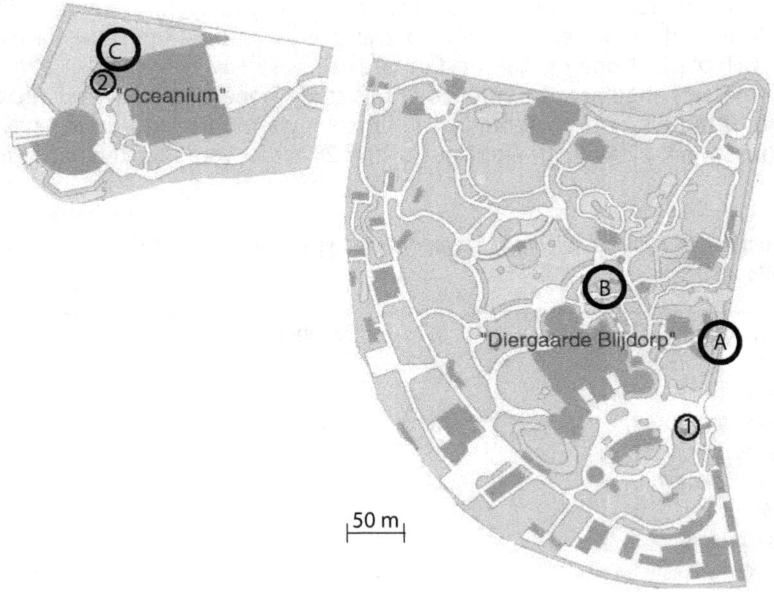

Figure 2. Map of Rotterdam zoo. Number 1 indicates the residence of the black-footed penguins, number 2 the residence of the puffins, guillemots and blacklegged kittiwakes. A, B and C indicate the location of the mosquito traps. The zoo is divided in two parts by railway tracks.

Results

Regarding the larval sampling, the most abundant mosquito species in the zoo was *Culex pipiens pipiens* (hereafter referred to as *Cx. pipiens*). Other frequently caught mosquito larvae in the zoo were *Culiseta annulata* and mosquitoes of the *Anopheles maculipennis* complex. Considering the adult catches, a total number of 575 mosquitoes were caught in the traps (Table 1). The most abundant species in the traps were again *Cx. pipiens* (60.5%), *Culiseta annulata* (20.5%) and *Anopheles maculipennis* spp. (8.5%). The mosquito traps experienced temporary malfunctioning, with all traps not operating during week 34, which caused a dip in the mosquito density measured during that week. Moreover, mosquito storage problems resulted in a decay of some mosquitoes beyond recognition and a loss of some of the collections. Therefore, it was not possible to follow adult mosquito dynamics accurately through time. The mosquitoes of the *An. maculipennis* complex were not identified to species by PCR, but it is assumed these mosquitoes were of the *An. messeae* species, which is the most abundant species of this complex in The Netherlands (Takken *et al.* 2002).

As for actual mortality of birds due to malaria, the zoo experienced 12 cases during the study period, between 20 July and 17 August: six black-footed penguins, three common guillemots, one black-legged kittiwake (*Rissa tridacttyla pollicaris*), one European eider (*Somateria mollissima*) and one demoiselle crane (*Anthropoides virgo*). These were all diagnosed with avian malaria by the veterinary surgeon using microscopy. The peak in mortality due to avian malaria occurred in week 33 (11 August). The peak in larval abundance of *Cx. pipiens* occurred two weeks earlier (in week 31, 28 July; Figure 3). The adult *Culex pipiens* population also reached its highest numbers between week 31 and week 33.

In the cross-sectional study, blood was taken from 81 individual birds, representing 42 different species. Nine birds, representing eight species, were found PCR-positive for *Plasmodium* spp. infections (Table 2). None of the infected birds showed any symptoms of a malaria infection. Six black-footed penguins were enrolled in this study, none of them were infected. At the next screening of the entire black-footed penguin colony a month later during peak transmission season (first week of August), eight out of the 28 birds were infected with *Plasmodium* spp.

Table 1. Cumulative number adult mosquito species that were caught in Rotterdam zoo in three mosquito traps between May and September 2003.

Mosquito species	Number
Culex pipiens	348
Culiseta annulata	118
Anopheles maculipennis	49
Coquillettidia richiardii	6
Culex modestus	3
Anopheles algeriensis	3
Culiseta suborchea	1
Culex terrestans	1
Unknown	46
Total	575

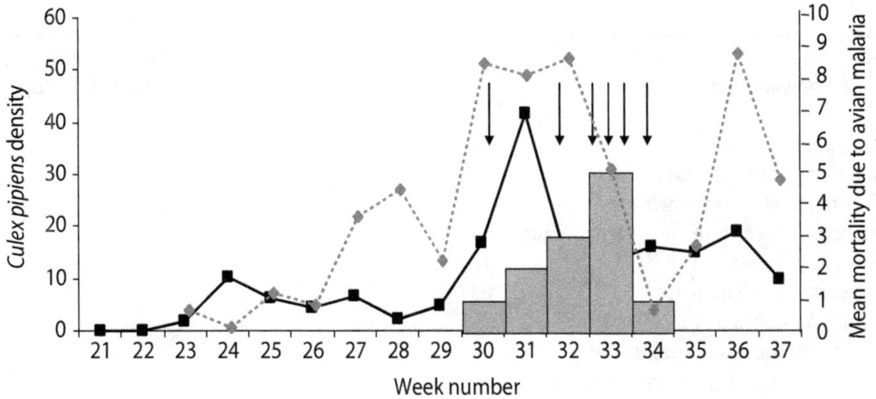

Figure 3. Larval (black line; number of larvae per dip) and adult (gray dotted line; cumulative number of adults) densities of Culex pipiens. *Larvae were sampled weekly from 24 selected larval habitats and adults were collected from three mosquito traps, both between May and September 2003 in Rotterdam zoo. The traps were malfunctioning in week 34. The grey bars indicate the numbers of birds in captivity that died per week due to avian malaria. The arrows indicate when infected mosquitoes were collected.*

Table 2. Bird species of which blood was examined during a cross sectional study on 4 and 5 July 2003 in Rotterdam zoo. Numbers of individuals that were infected with avian malaria are given.

Bird species English name *Scientific name*	Number of birds examined	Infected with *Plasmodium* spp.
Andean guan *Penelope montagnii montagnii*	1	
Bahama pintail *Anas bahamensis bahamensis*	1	
Black stork *Ciconia nigra*	1	
Black-footed penguin *Spheniscus demersus*	6	
Black-legged kittiwake *Rissa tridacttyla pollicaris*	5	
Blue-eyed cockatoo *Cacatua ophthalmica*	1	
Blue-winged kookaburra *Dacelo leachii*	1	
Caribbean flamingo *Phoenicopterus ruber rubber*	3	
Common goldeneye *Bucephala clangula*	1	
Common guillemot *Uria aalge aalge*	2	1
Common puffin *Fratercula arctica*	2	
Common shelduck *Tadorna tadorna*	3	1
Common shoveler *Anas clypeata*	1	
Dalmatian pelican *Pelecanus crispus*	2	
Demoiselle crane *Anthropoides virgo*	1	
Emu *Dromaius novaehollandiae*	2	
Eurasian eagle owl *Bubo bubo bubo*	1	
European eider *Somateria mollissima mollissima*	1	
European goosander *Mergus merganser merganser*	3	
Falcated duck *Anas falcata*	3	
Fulvous whistling duck *Dendrocygna bicolor*	4	1

Table 2. Continued.

Bird species English name *Scientific name*	Number of birds examined	Infected with *Plasmodium* spp.
Glossy ibis *Plegadis falcinellus*	1	1
Greater flamingo *Phoenicopterus roseus*	7	2
Helmeted guineafowl *Numida meleagris*	1	1
Long-billed corella *Cacatua tenuirostris tenuirostris*	1	
Mallard *Anas platyrhnchos*	1	
Manchurian crane *Grus japonensis*	2	1
Mandarin Duck *Aix galericulata*	3	
Marabou stork *Leptoptilos crumeniferus*	1	
Mexican military macaw *Ara miliaris mexicana*	1	
Mitchell's cockatoo *Cacatua leadbeateri*	1	
Ostrich *Struthio camelus*	1	
Palawan pacock pheasant *Polyplectron emphanum*	1	
Red junglefowl *Gallus gallus*	2	
Red-billed curassow *Crax blumenbachii*	1	
Red-crested pochard *Netta rufina*	2	
Scheepmaker's crowned pigeon *Goura scheepmakeri sclaterii*	1	
Smew *Mergus albellus*	4	
Snowy owl *Nyctea scandiaca*	1	1
South African crowned crane *Balearica regulorum regulorum*	1	
White-naped crane *Grus vipio*	2	
White-winged wood duck *Cairina scutulata*	1	
Total	81	9

parasites in their blood. Six days after sampling, one positive juvenile bird died with symptoms of avian malaria. The other seven birds, all adults, survived the summer. However, all black-footed penguins, puffins and guillemots were given prophylactic treatment the third week of August and were taken into quarantine, which probably prevented more fatalities.

Of all adult *Cx. pipiens*, *Culiseta annulata* and *Anopheles maculipennis* s.l. collected in the traps, only *Cx. pipiens* mosquitoes were infected with avian malaria parasites. Six specimens were found positive with *Plasmodium* (1.7%). These were all captured between 22 July and 19 August, which coincided with the peak in bird mortality (Figure 3). The infected mosquitoes were spread randomly throughout the zoo, as infected mosquitoes were found in each of the three traps.

Discussion

This field study shows the various components of an entire avian malaria cycle. Six infected *Cx. pipiens* were found (1.7 percent of all *Cx. pipiens* adults caught), whereas none of the other mosquito species were found to be infected. However, *Culiseta annulata* has been identified as a competent vector of *P. relictum* (Valkiunas 2005) and an infection within this species could have been missed due to the low overall infection rate that was observed. Overall, our findings suggest that *Cx. pipiens* was the main vector of avian malaria in the Rotterdam zoo. This species was also found to be an avian malaria vector in the Baltimore zoo, USA, together with *Culex restuans* (Beier and Stoskopf 1980). *Culex pipiens* was the dominant species in the zoo, probably due to many water bodies, such as water ponds, puddles and buckets within the zoo and water storage containers in the public vegetable gardens just outside the zoo. These are ideal breeding habitats for *Culex* spp., which frequently breed in artificial collections of water, whereas *Anopheles* spp. and *Culiseta* spp. more often breed in natural water bodies.

Unfortunately, due to some unforeseen storage problems and malfunctioning of the mosquito traps, there are some gaps in the data of the adult mosquito collections. Therefore, the larval density was used to compare vector and disease dynamics. The immature stages of *Cx. pipiens* reached their peak density at the end of July, just two weeks before the peak of avian mortality. The development of the parasite within the mosquito is dependent on weather conditions, but takes at least seven days during the summer. The incubation time for *P. relictum* from sporozoites to blood stages with clinical symptoms takes up to five days (Valkiunas 2005). Therefore, it is very likely that the sudden increase in numbers of adults of *Cx. pipiens* was a direct cause for the peak mortality observed two weeks later.

Eleven percent of the exotic birds in Rotterdam zoo were infected with *Plasmodium* spp. infections, which demonstrates that such infections may be quite common. These were probably chronic infections since all blood smears of the birds that appeared to be infected using PCR remained below detectable level using microscopy, except for the snowy owl (*Nyctea scandiaca*), who had a very low parasitaemia. Also, none of the infected birds showed symptoms of infection and this survey was done one month before peak transmission season. Apart from the common guillemot, there are no records of avian malaria within these infected species. Wild lesser flamingos (*Phoeniconaias minor*) have been reported with *P. relictum* (Peirce 2005). Unfortunately, we were not able to identify the avian malaria parasites from both infected birds and mosquitoes to species level.

Chronically infected birds living within the zoo perimeters can have important implications for the spread of avian malaria, since they can act as a reservoir. However, another possibility is that the source of infection is coming from infected endemic local birds. Unfortunately, there are no data available on the prevalence of the *Plasmodium* spp. parasites in endemic wild birds, resident in the zoo or in the Netherlands. In Baltimore zoo, the source of avian malaria infection among their black-footed penguin colony resulted from wild passerine birds living in the zoo (Cranfield *et al.* 1990). In Europe, it is well known that many bird species, especially passerines, harbour avian malaria parasites in their blood (Scheuerlein and Ricklefs 2004). *Plasmodium relictum*, most probably the causative agent for avian malaria in the zoo, has been reported in many different wild bird species, for instance in Spain, Bulgaria and Czech Republic (Blanco *et al.* 1997, Figuerola *et al.* 1999, Shurulinkov and Golemansky 2003, Votypka *et al.* 2003). It is therefore highly likely that the wild birds living in the zoo are also infected with avian malaria parasites. Therefore, the source of avian malaria infections could result from both the exotic birds kept in the zoo and the endemic wild birds living in its surroundings. It is possible that the parasite is relatively harmless in their reservoir hosts, and that only certain bird species are highly susceptible.

Migratory birds passing through The Netherlands could be another source of avian malaria. It is known that migratory birds carry avian malaria parasites (Perez-Tris and Bensch 2005, Shurulinkov and Golemansky 2003, Westerdahl *et al.* 2005). However, migratory birds pass by in the spring and autumn, whereas the peak in vector abundance and mortality caused by avian malaria occurs in the summer (see above). Moreover, studies on two migratory birds, the ruff (*Philomachus pugnax*) and little stint (*Calidris minuta*), which migrate from Africa to the high northern latitudes, have shown that these passerines can be heavily infected with malaria parasites in their African habitat. However, neither of these species were found infected when travelling through The Netherlands or breeding in the Arctics (Mendes *et al.* 2005). The authors proposed three possible explanations for this phenomenon. First, only the healthy birds start their journey. This is supported by a recent study which simulated migration flight in a wind tunnel with red knots (*Calidris canutus*); birds with a lower immunity level refused to fly (Hasselquist *et al.* 2007). Second, infected birds that do start the flight have a lower chance of survival due to the infection. However, the wind tunnel experiments revealed that a long distance flight did not have an effect on the immunity levels of the red knots (Hasselquist *et al.* 2007). Third, the immune system may have suppressed the infection before the birds reach the northern latitudes. This seems plausible, since the immune system is not as suppressed by the long flight, as has always been assumed (Hasselquist *et al.* 2007). With recent molecular techniques to study parasite lineages, it appears that 'hitchhiking' of parasites is not as common as previously thought (Hellgren *et al.* 2007). This study distinguished between parasite lineages on sub-morphological level of migratory and endemic birds in Africa and Europe, and showed that only one of the 65 encountered *Plasmodium* lineages were present in both Africa and Europe, meaning that the parasites from the migratory birds are hardly ever transmitted to European birds at stop-over or breeding sites. Apparently, parasites that hitchhike from Africa to Europe appear not to leave their ride.

In summary, it seems unlikely that the migratory birds are the cause for the annual outbreaks of avian malaria in zoos. Our results suggest that there is an indigenous reservoir of avian malaria parasites within the exotic and endemic birds that reside near or within the zoo perimeters.

Avian malaria: an emerging disease in Hawaii

In the beginning of the 20th century, avian malaria spread rapidly over the islands of Hawaii, contributing to the extinction of many endemic bird species. Avian malaria parasites were likely to have been present in migratory birds that visit Hawaii annually. However, without a suitable mosquito vector present on the islands, the parasites never came in contact with the endemic bird population. This all changed when a ship named 'Wellington' brought water from the west coast of Mexico to the island of Maui in 1826 and accidentally introduced the tropical house mosquito *Culex quinquefasciatus* as a new species in the local ecosystem (Warner 1968). Avian malaria did not reach epidemic levels in Hawaii before the turn of the 20th century. It is believed that by then the mosquito population was fully established and that the outbreak of avian malaria was initiated by the introduction of over 200 exotic birds in the early 1900s, any number of which may have served as a reservoir of the malaria parasite *P. relictum* (Van Riper III *et al.* 1986, Warner, 1968). During the last century, 23 of the 71 endemic bird species believed to inhabit the islands at the end of the 18th century, became extinct and another 30 are currently listed endangered (Atkinson *et al.* 1995). The emergence of avian malaria on Hawaii is believed to play a key role in these extinctions.

Key factors in the success of the avian malaria parasites in Hawaii are time and isolation. Hawaiian's avian population has evolved in almost near-isolation over a period of approximately 70 million years (Mac *et al.* 1998). The first Polynesian settlers arrived about a thousand years ago and the first Europeans arrived as recently as 1778. The Hawaiian bird population, being isolated for that long, did not have a history of co-evolution with avian malaria parasites and therefore the relatively benign parasites in the introduced exotic birds were highly virulent when infecting the naïve Hawaiian birds.

Avian malaria in Europe

Various studies on avian malaria parasites across Europe show that avian malaria parasites are common in endemic bird populations. *Plasmodium* spp. infections have been reported in at least 36 passerine European bird species, spread around Europe (reviewed in Scheuerlein and Ricklefs 2004), as well as in migratory birds passing through Europe (Perez-Tris and Bensch 2005, Shurulinkov and Golemansky 2003, Westerdahl *et al.* 2005). However, unlike Hawaii, there are no reports of extinction or high mortality of bird species as a result of an infection with avian malaria in Europe.

The vector, parasites and hosts of avian malaria have co-existed since ancient times in Europe and have thus co-evolved together. Therefore, an arms race between both the host and the parasite has been running since the first contact between parasite and host. Parasites will try to increase their transmission potential by being more infectious within their host, whereas the host will respond by trying to reduce the parasite exploitation (Ewald 1995). When such co-evolved parasites subsequently invade a naïve individual, they can be extremely virulent due to the lack of an evolved immune response of the new host to this parasite (Mackinnon and Read 2004). This is exactly what happened when avian malaria parasites were introduced in the Hawaiian bird population; the otherwise relatively benign parasites to the introduced exotic birds became extremely virulent in the naïve Hawaiian birds. Natural selection, however, quickly favours those hosts with some degree of immunity. This is already happening in Hawaii, where some of the endemic Hawaiian bird species are gaining resistance against the malaria parasites (Woodworth *et al.* 2005) and behavioural adaptations such as daily migration to the mosquito-free highlands to spend the night and the covering of vulnerable areas by plumage to reduce mosquito bites are examples of adaptations of the endemic bird population to the avian malaria parasite (Valkiunas 2005, Van Riper III *et al.* 1986).

The impact of avian malaria on endemic avian wildlife in Europe will thus be less. However, some exotic birds that are kept in European zoos are facing greater problems with the disease. These birds originate from areas where there is either no transmission or transmission of other malaria species and may come into contact with the local avian malaria parasites. There are many reports of avian malaria in zoos in Europe and the USA (Cranfield *et al.* 1994, 1990, Fix *et al.* 1988, Graczyk *et al.* 1995). The causative agents that are reported are *P. relictum* and *P. elongatum* (Cranfield *et al.* 1994). Mainly individuals of the black-footed penguins are infected with malaria, which often results in seasonal deaths. There are also lethal cases known among puffins (Loupal and Kutzer 1996), common guillemots, European eiders and kittiwakes (W. Schaftenaar, unpublished data). Many zoos have decided to give their penguin population *Plasmodium* prophylaxis during peak transmission season. These medications can have adverse effects on the birds, resulting in neurological dysfunction at high and repetitive dosages (Wunschmann *et al.* 2006).

We conducted a questionnaire survey on the prevalence of avian malaria between 1995 and 2003 in eight zoos in The Netherlands and Belgium. All of these zoos were asked for cases of malaria among their avian population during this time period. The survey revealed that all participating zoos experienced mortality due to avian malaria in their birds, predominantly in their penguin populations. Most cases occurred in August and September (Figure 4), though deaths occurred all year-round.

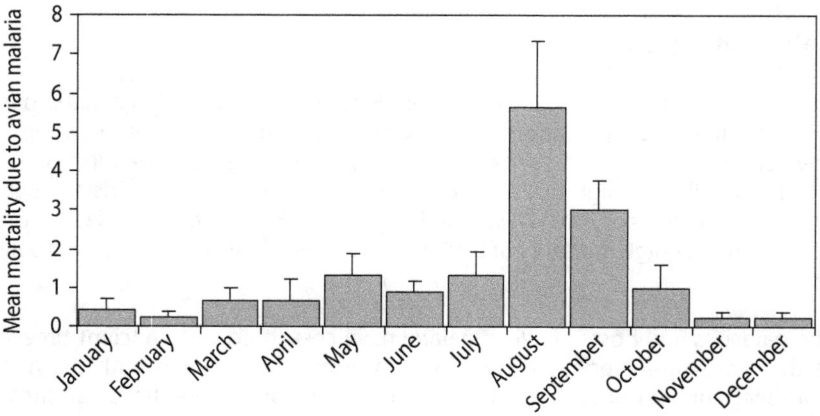

Figure 4. Monthly reported mortality caused by avian malaria (mean ± standard error) in bird populations in eight zoos in the Netherlands or Belgium from 1995 to 2003.

The juvenile birds were at highest risk: 41 percent of all observed deaths were among birds less than one year old. This is probably due to a lower immunity of these young birds (Lozano and Lank 2003). These findings are, however, in contrast with the infection percentages found in other studies, where the majority of infections were among adult birds (Mendes *et al.* 2005, Merila and Andersson 1999, Sol *et al.* 2000, Valkiunas 2005). However, an increased intensity of infection was reported among the juveniles compared to the adults, even though the prevalence was lower (Sol *et al.* 2000), which can explain an increased mortality observed in this age group. Remarkably, the annual cumulative number of cases of avian malaria reported in zoos increased over time (Linear regression, $p< 0.001$, Figure 5). Unfortunately, we do not have data on the growth of the bird population of each zoo during this time period. Therefore, the observed increase of reported avian malaria cases could be a reflection of a growing avian population. Also, an increased awareness towards an avian malaria diagnosis may have resulted in an increased number of reported avian

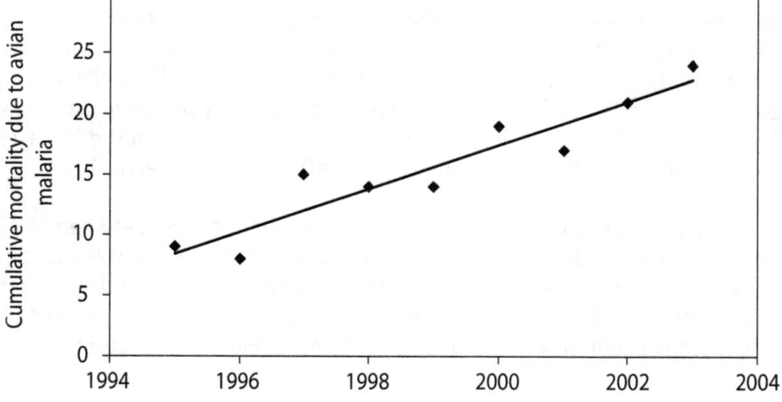

Figure 5. Annual cumulative mortality due to avian malaria in bird populations in eight zoos in the Netherlands or Belgium from 1995 to 2003.

malaria cases. However, it is also highly plausible that there was an increase in transmission rate of avian malaria between 1995 and 2003, which suggests that avian malaria is an emerging infectious disease amongst the avifauna in Dutch and Belgian zoos. We suggest a close monitoring of all avian populations within the zoos to establish whether the observed trend in this survey is still continuing and whether this is due to an actual increase in transmission.

Avian malaria: an emerging infectious disease in Europe?

In conclusion, could avian malaria potentially reach epidemic levels of avian malaria in Europe? Fact is that Hawaii is not a unique event; New-Zealand is currently struggling with avian malaria outbreaks after the introduction of *Cx. quinquefasciatus* three decades ago (Tompkins and Gleeson 2006). The same mosquito species also invaded the Galápagos Islands two decades ago, where the malaria parasite is, for the time being at least, still absent (Whiteman *et al.* 2005). However, with the increase of international travel, there is a high probability that parasites will be introduced to the Galápagos Islands at some point, which may potentially result in the same devastating effects as seen on the Hawaiian Islands. That would have devastating consequences for the biodiversity of these unique islands (Daszak *et al.* 2000).

An emerging disease only occurs when the hosts are highly susceptible. There are three scenarios in which a parasite is likely to have such an impact that a disease is gaining epidemic proportions. The first scenario is that a new and naïve host is introduced in a disease endemic area, as is the case with the black-footed penguins in our field study of Rotterdam zoo. Second, when a new parasite is introduced in a naïve bird population, such as on the Hawaiian Islands. Third, when due to adaptation or mutation, a parasite is suddenly capable of infecting a new host that it was unable to infect before or has gained an increased virulence. There are many examples of the latter in other diseases, such as the HIV/AIDS virus (Jaffar *et al.* 2004) and the Chikungunya virus in Réunion (Schuffenecker *et al.* 2006).

It is difficult to predict whether future introductions of avian malaria in Europe should be considered as a threat of an emerging infectious disease, since there are few data available about the prevalence and especially the mortality rate of avian malaria in Europe. Therefore, the impact of the disease on the avian fauna on this continent is difficult to estimate. At the moment, a future outbreak of avian malaria is unlikely to happen; all endemic birds appear to be adapted to the parasites with whom they come into contact. However, a mutation in the existing parasite population can render it highly virulent to avian hosts. Also, a change of abiotic factors can affect the equilibrium between parasite and host. A change in climate for instance, may result in a longer transmission season or an increased transmission pressure, to which the birds might not be able to adapt to quickly enough. Also, new favourable climatic conditions might result in the spread of the parasites to a larger geographic area, where they will come into contact with new and possibly susceptible naïve hosts. Like in Hawaii, an introduction of a new, more competent vector can largely increase transmission of avian malaria. In other words, in a stable environment with regular host-parasite interactions it is unlikely that an avian malaria epidemic will occur. However, a change of the equilibrium between parasite and host or the emergence of new and virulent parasite strains or competent vectors could result in an epidemic among European endemic birds. It is therefore important to screen the avian population regularly, both migratory and endemic, so that intervention methods can be put into action before an outbreak reaches epidemic proportions.

The ecology and evolution of infectious diseases is an emerging research field. Emerging infectious diseases have drawn much attention lately, since they affect our health or that of our crops or livestock (Weiss and McMichael 2004). It is important to understand the principles behind the spread of emerging infectious diseases and to study host-parasite co-evolution. The study of avian malaria parasites will contribute much to our knowledge, and can be used as a model for other emerging infectious diseases. The introduction and spread of the avian malaria parasites in Hawaii and the subsequent evolution between parasite and host is a schoolbook example of what can happen if a virulent parasite enters a susceptible population. The recent past and the near future of the endemic Hawaiian bird population will teach us how the birds evolve to gain resistance and adapt to their new pathogens. Nevertheless, it resulted in an irreversible loss in biodiversity.

Acknowledgements

We appreciate the advice and assistance provided by Dr. J.J. Verweij of Leiden University. We thank the staff of Rotterdam zoo, in particular Harald Schmidt, for their cooperation and provision of facilities. Dennis LaPointe is thanked for his useful comments to improve the manuscript. We thank all zoos that participated in the survey.

References

Atkinson CT, Woods KL, Dusek RJ, Sileo LS and Iko WM (1995) Wildlife disease and conservation in Hawaii: pathogenicity of avian malaria (*Plasmodium relictum*) in experimentally infected iiwi (*Vestiaria coccinea*). Parasitology 111: S59-69.

Baylis M, Mellor PS, Wittmann EJ and Rogers DJ (2001) Prediction of areas around the Mediterranean at risk of bluetongue by modelling the distribution of its vector using satellite imaging. Vet Rec 149: 639-643.

Beier JC and Stoskopf MK (1980) The epidemiology of avian malaria in black-footed penguins (*Spheniscus demersus*). J Zoo Anim Med 11: 99-105.

Black FL, Hierholz WJ, Pinheiro FD, Evans AS, Woodall JP, Opton EM, Emmons JE, West BS, Edsall G, Downs WG and Wallace GD (1974) Evidence for persistence of infectious agents in isolated human-populations. Am J Epidemiol 100: 230-250.

Blanco G, Merino S, Tella JL, Fargallo JA and Gajon A (1997) Hematozoa in two populations of the threatened red-billed chough in Spain. J Wildl Dis 33: 642-645.

Cranfield MR, Graczyk TK, Beall FB, Ialeggio DM, Shaw ML and Skjoldager ML (1994) Subclinical avian malaria infections in African black-footed penguins (*Spheniscus demersus*) and induction of parasite recrudescence. J Wildl Dis 30: 372-376.

Cranfield MR, Shaw ML, Beall FB, Skjoldager ML and Ialeggio DM (1990) A review and update of avian malaria in the African penguin (*Spheniscus demersus*). Proc Am Assoc Zoo Vet, 243-248.

Daszak P, Cunningham AA and Hyatt AD (2000) Wildlife ecology - Emerging infectious diseases of wildlife - Threats to biodiversity and human health. Science 287: 443-449.

Ewald PW (1995) The evolution of virulence - a unifying link between parasitology and ecology. J Parasitol 81: 659-669.

Figuerola J, Munoz E, Gutierrez R and Ferrer D (1999) Blood parasites, leucocytes and plumage brightness in the Cirl Bunting, Emberiza cirlus. Funct Ecol 13: 594-601.

Fix AS, Waterhouse C, Greiner EC and Stoskopf MK (1988) *Plasmodium relictum* as a cause of avian malaria in wild-caught magellanic penguins (*Spheniscus magellanicus*). J Wildl Dis 24: 610-619.

Garnham PCC (1966) Malaria parasites and other Haemosporidia. Blackwell. Oxford.

Graczyk TK, Cranfield MR, Brossy JJ, Cockrem JF, Jouventin P and Seddon PJ (1995) Detection of avian malaria infections in wild and captive penguins. J Helminthol Soc Wash 62: 135-141.

Harrison M (2004) Disease and the Modern World - 1500 to the present day. Polity Press. Cambridge, UK.

Hasselquist D, Lindstrom A, Jenni-Eiermann S, Koolhaas A and Piersma T (2007) Long flights do not influence immune responses of a long-distance migrant bird: a wind-tunnel experiment. J Exp Biol 210: 1123-1131.

Hellgren O, Waldenstrom J, Perez-Tris J, Szollosi E, Hasselquist D, Krizanauskiene A, Ottosson U and Bensch S (2007) Detecting shifts of transmission areas in avian blood parasites - a phylogenetic approach. Mol Ecol 16: 1281-1290.

Jaffar S, Grant AD, Whitworth J, Smith PG and Whittle H (2004) The natural history of HIV-1 and HIV-2 infections in adults in Africa: a literature review. Bull World Health Organ 82: 462-469.

LaPointe DA, Goff ML and Atkinson CT (2005) Comparative susceptibility of introduced forest dwelling mosquitoes in Hawaii to avian malaria, *Plasmodium relictum*. J Parasitol 91: 843-849.

Lederberg J, Shope RE and Oaks SC (1992) Emerging infections: microbial threats to health in the United States. National Academy Press. Washington.

Loupal G and Kutzer E (1996) Infections with Plasmodium spec in puffins (*Fratercula arctica*). Kleintier Prax 41, 901-906.

Lozano GA and Lank DB (2003) Seasonal trade-offs in cell-mediated immunosenescence in ruffs (*Philomachus pugnax*). Proc R Soc Lond B 270: 1203-1208.

Mac MJ, Opler PA, Pucket Haeker CE and Doran PD (1998) Hawaii and the pacific islands: Status and trends of the nation's biological resources, U.S. Department of the Interior, U.S. Geological Survey, Reston.

Mackinnon MJ and Read AF (2004) Virulence in malaria: an evolutionary viewpoint. Phil Trans R Soc Lond B 359: 965-986.

Mehlhorn H, Walldorf V, Klimpel S, Jahn B, Jaeger F, Eschweiler J, Hoffmann B and Beer M (2007) First occurrence of *Culicoides obsoletus*-transmitted bluetongue virus epidemic in central Europe. Parasitol Res 101: 219-228.

Mendes L, Piersma T, Lecoq M, Spaans B and Ricklefs RE (2005) Disease-limited distributions? Contrasts in the prevalence of avian malaria in shorebird species using marine and freshwater habitats. Oikos 109: 396-404.

Merila J and Andersson M (1999) Reproductive effort and success are related to haematozoan infections in blue tits. Ecoscience 6: 421-428.

Peirce MA (2005) Pathogenic subspecies of *Plasmodium relictum* found in African birds. Vet Rec 156: 328-328.

Perez-Tris J and Bensch S (2005) Dispersal increases local transmission of avian malarial parasites. Ecol Lett 8: 838-845.

Scheuerlein A and Ricklefs RE (2004) Prevalence of blood parasites in European passeriform birds. Proc R Soc Lond B 271: 1363-1370.

Schuffenecker I, Iteman I, Michault A, Murri S, Frangeul L, Vaney MC, Lavenir R, Pardigon N, Reynes JM, Pettinelli F, Biscornet L, Diancourt L, Michel S, Duquerroy S, Guigon G, Frenkiel MP, Brehin AC, Cubito N, Despres P, Kunst F, Rey FA, Zeller H and Brisse S (2006) Genome microevolution of Chikungunya Viruses causing the Indian Ocean outbreak. Plos Med 3: 1058-1070.

Service MW (1993) Mosquito Ecology - Field Sampling Methods. Elsevier Applied Science. London.

Shurulinkov P and Golemansky V (2003) Plasmodium and Leucocytozoon (Sporozoa: Haemosporida) of wild birds in Bulgaria. Acta Protozool 42: 205-214.

Snounou G, Viriyakosol S, Jarra W, Thaithong S and Brown KN (1993a) Identification of the 4 human malaria parasite species in field samples by the Polymerase Chain Reaction and detection of a high prevalence of mixed infections. Mol Biochem Parasitol 58: 283-292.

Snounou G, Viriyakosol S, Zhu XP, Jarra W, Pinheiro L, Dorosario VE, Thaithong S and Brown KN (1993b) High sensitivity of detection of human malaria parasites by the use of nested Polymerase Chain Reaction. Mol Biochem Parasitol 61: 315-320.

Sol D, Jovani R and Torres J (2000) Geographical variation in blood parasites in feral pigeons: the role of vectors. Ecography 23: 307-314.

Takken W, Geene R, Adam W, Jetten TH and van der Velden JA (2002) Distribution and dynamics of larval populations of *Anopheles messeae* and *An. atroparvus* in the delta of the rivers Rhine and Meuse, The Netherlands. Ambio 31: 212-218.

Taubenberger JK, Reid AH, Krafft AE, Bijwaard KE and Fanning TG (1997) Initial genetic characterization of the 1918 "Spanish" influenza virus. Science 275: 1793-1796.

Tompkins DM and Gleeson DM (2006) Relationship between avian malaria distribution and an exotic invasive mosquito in New Zealand. J R Soc N Zeal 36: 51-62.

Tyler KL (2004) West Nile virus infection in the United States. Arch Neurol 61: 1190-1195.

Valkiunas G (2005) Avian malaria parasites and other haemosporidia. CRC Press. Boca Raton, Florida.

Valkiunas G, Anwar AM, Atkinson CT, Greiner EC, Paperna I and Peirce MA (2005) What distinguishes malaria parasites from other pigmented haemosporidians? Trends Parasitol 21: 357-358.

Van Haren JMC and Verdonschot PFM (1995) Proeftabel Nederlandse Culicidae, pp. 1-106, Instituut voor Bos- en Natuuronderzoek.

Van Riper III C, Van Riper SG, Goff ML and Laird M (1986) The epizootiology and ecological significance of malaria in Hawaiian land birds. Ecol Monogr 56: 327-344.

Votypka J, Simek J and Tryjanowski P (2003) Blood parasites, reproduction and sexual selection in the red-backed shrike (Lanius collurio). Ann Zool Fenn 40: 431-439.

Waldenstrom J, Bensch S, Hasselquist D and Ostman O (2004) A new nested polymerase chain reaction method very efficient in detecting *Plasmodium* and *Haemoproteus* infections from avian blood. J Parasitol 90: 191-194.

Warner RE (1968) The role of introduced diseases in the extinction of the endemic Hawaiian avifauna. Condor 70: 101-120.

Weiss RA and McMichael AJ (2004) Social and environmental risk factors in the emergence of infectious diseases. Nature Med 10: S70-S76.

Westerdahl H, Waldenstrom J, Hansson B, Hasselquist D, von Schantz T and Bensch S (2005) Associations between malaria and MHC genes in a migratory songbird. Proc R Soc Lond B 272: 1511-1518.

Whiteman NK, Goodman SJ, Sinclair BJ, Walsh T, Cunningham AA, Kramer LD and Parker PG (2005) Establishment of the avian disease vector Culex quinquefasciatus Say, 1823 (Diptera: Culicidae) on the Galapagos Islands, Ecuador. Ibis 147: 844-847.

Woodworth BL, Atkinson CT, LaPointe DA, Hart PJ, Spiegel CS, Tweed EJ, Henneman C, LeBrun J, Denette T, DeMots R, Kozar KL, Triglia D, Lease D, Gregor A, Smith T and Duffy D (2005) Host population persistence in the face of introduced vector-borne diseases: Hawaii amakihi and avian malaria. Proc Nat Acad Sci US A 102: 1531-1536.

Wunschmann A, Armien A, Wallace R, Wictor M and Oglesbee M (2006) Neuronal storage disease in a group of captive Humboldt penguins (*Spheniscus humboldti*). Vet Pathol 43: 1029-1033.

5. The leishmaniases of Southern Europe

Marina Gramiccia and Luigi Gradoni

Abstract

Leishmaniases range over the intertropical zones of America and Africa, and extend into temperate regions of Latin America, Southern Europe and Asia. About 20 named *Leishmania* species and subspecies are pathogenic for humans and are annually responsible for 1-1.5 million cases of cutaneous (CL) and 0.5 million cases of visceral (VL) forms of the disease. Thirty sandfly species are proven vectors. The epidemiology is largely diverse, being the agents and related diseases grouped into several zoonotic or anthroponotic entities. Southern European countries are mainly affected by zoonotic VL and sporadic CL, both caused by different strains of *L. infantum*. About 4,700 leishmaniasis cases were reported in the region over the past 10 years, the highest number being recorded in Italy (1,677). Acute VL is diagnosed in children in a proportion varying from 30% (Southern France) to over 60% (Albania) of all recorded cases. The region was particularly affected by an increasing incidence of HIV/VL co-infections until the late 1990s, but after the introduction of the HAART therapy the incidence rates were considerably reduced. VL is also frequently recorded in adults with other immunosuppressive conditions, e.g. treatments for organ transplant or immunological disorders and malignancies. *L. infantum* is widespread in the canine reservoir, several thousands infected dogs being annually diagnosed in the endemic zones. The proven VL vectors in Southern Europe are five species of the subgenus *Phlebotomus* (*Larroussius*): *P. perniciosus* Newstead, *P. ariasi* Tonnoir, *P. perfiliewi* Parrot, *P. neglectus* Tonnoir and *P. tobbi* Adler and Theodor, differently distributed in the region. Some of them have increased in density and expanded their geographic range over the past 15 years, suggesting a northward spreading of *L. infantum* transmission. The burden of leishmaniases in endemic countries bordering Europe at the South (Afro-Mediterranean region) and South-East (Turkey, Near and Middle East) is much higher than in Southern Europe. Zoonotic CL caused by *L. major* and anthroponotic CL caused by *L. tropica* affect thousands individuals in rural and urban areas. These agents are frequently identified in imported cases and, because of the presence of susceptible vectors, the risk for the introduction of new parasites into leishmaniasis-endemic zones of Southern Europe should not be underestimated.

Keywords: zoonotic visceral leishmaniasis, sporadic cutaneous leishmaniasis, *Leishmania infantum, Leishmania tropica*; *Phlebotomus* (*Larroussius*) spp., *Phlebotomus* (*Paraphlebotomus*) *sergenti,* canine leishmaniasis

Overview

Leishmaniases are diseases caused by members of the genus *Leishmania*, protozoan parasites infecting numerous mammal species including humans, and transmitted by the bite of phlebotomine sandflies. Human leishmaniases have diverse clinical manifestations. Visceral leishmaniasis (VL), caused by *Leishmania donovani* in the Old World and *Leishmania infantum* (synonym: *L. chagasi)* in both the Old and New Worlds, is the most severe form which, if left untreated, invariably leads to death. A number of different species of *Leishmania* cause cutaneous (CL) or mucocutaneous (MCL) leishmaniases which, if not fatal, are responsible for considerable morbidity among a vast number of people from endemic foci. According to available figures, 66 Old World and 22 New World countries are endemic for human leishmaniases, with an estimated yearly incidence of 1-1.5 million cases of CL forms and 500,000 cases of VL forms (Desjeux 1996). Overall prevalence is 12 million people with Disability Adjusted Life Years (DALYs) burden of 860,000 for men and 1.2 million for women (http://www.who.int/tdr/diseases/leish/). The disease affects the poorest people in the poorest countries (Alvar *et al.* 2006): 72 are developing countries and 13 are among the least developed. The incidence is not uniformly distributed in endemic areas: about 90% of CL cases occur in seven countries only (Afghanistan, Algeria, Brazil, Iran, Peru, Saudi Arabia and Syria), whereas some 90% of VL cases occur in rural and suburban areas of five countries (Bangladesh, India, Nepal, Sudan and Brazil). These figures, however, must be regarded as approximate and, most probably, underestimated. Currently, it appears that the global incidence of human leishmaniases is higher than before: environmental and human behavioural factors contribute to the changing landscape of these diseases, which show a wider geographical distribution than previously known. Autochthonous *Leishmania* transmission is being recorded in traditionally non-endemic areas, such as several U.S. states and Canadian provinces (Duprey *et al.* 2006), Australia's Northern Territory (Rose *et al.* 2004), and in some parts of continental Europe (Koehler *et al.* 2002, Maroli *et al.* in press).

Leishmania life cycle and classification

Leishmania (Kinetoplastida: Trypanosomatidae) are dimorphic protozoa characterised by the intracellular presence of the kinetoplast, a network of maxi- and minicircles of mitochondrial DNA found close to the base of the flagellum. *Leishmania* shows two principal morphological stages: the amastigote, a round or oval body of 2.5-6.8 μ in diameter, is found in the phagolysosome of mononuclear phagocyte cells of the mammalian host (Figure 1A). The promastigote, an extracellular dividing form of 15-30 μ in length with a long anterior flagellum, is found in the gut of the phlebotomine vectors (Diptera: Psychodidae) (Figure 1B). Small, fast swimming and non-dividing promastigotes found in the foregut of the infected sandflies are metacyclic forms infective to the mammal (Sacks and Perkins 1984).

In nature, *Leishmania* are alternatively hosted by the sandfly and by mammalian hosts. When a female sandfly takes a blood meal from a *Leishmania*-infected mammal, amastigotes are ingested and, following at least one cycle of binary division, they transform into motile promastigotes which escape through the peritrophic membrane enveloping the blood meal. The promastigotes multiply intensively inside the intestinal tract of the sandfly. This development occurs in the

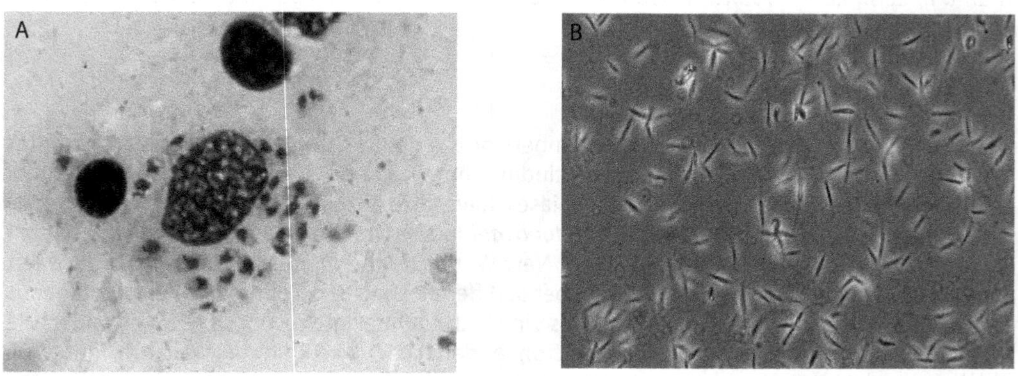

Figure 1. Leishmania *morphological stages: (A) amastigotes, (B) promastigotes.*

midgut (*Leishmania* subgenus) or in the hindgut and the midgut (*Viannia* subgenus). Whatever the multiplication site, the parasites subsequently migrate to the foregut (anterior cardia area and/or pharynx and proboscis) where they change into metacyclic forms. The time requested to complete the parasite cycle in the sandfly is variable, depending on both *Leishmania* and phlebotomine species, but is about five days on average.

Once the metacyclic promastigotes have been deposited in the mammal's dermis by the bite of the sandfly, they are rapidly phagocytosed by cells of the mononuclear phagocyte system. The ingested parasites change into the non-motile amastigote stage. The surviving amastigotes divide by mitosis within the macrophage's phagolysosome and the infection can spread in the mammalian host when heavily parasitised macrophages burst and amastigotes are ingested by other macrophages. The outcome of exposure to infection may not necessarily be overt disease and, in any case, the complex of parasite-cell biochemical interactions affects the course of the disease. The life cycle of *Leishmania* is completed when a female sandfly takes a blood meal containing *Leishmania*-infected cells. The inoculation of metacyclic promastigotes through the sandfly bite is the usual method of leishmaniasis transmission, other routes (e.g. transfusion or congenital transmission) being exceptional.

Since the naming of the genus *Leishmania* by Ross in 1903, the classification and nomenclature of the species and subspecies has for a long time and, sometime still is, a contentious matter. They show a homogeneous protozoan group mostly morphologically undistinguishable, for which the initial taxonomic criteria, like the human clinical picture and geographical distribution, were not enough for a correct taxonomical identification. Since the 1980s considerable efforts have been made to base the taxonomy of the genus *Leishmania* on scientific footings. The technique of multilocus enzyme electrophoresis (MLEE) has been applied for more than 25 years on several thousands parasite strains and still represents the gold standard for *Leishmania* identification and taxonomy. Strains are characterised by their enzymatic profiles and grouped into homogeneous taxonomic units, the zymodemes. Phylogenetic classification of zymodeme complexes reveals a parental relationship between different *Leishmania* taxa. A simplified isoenzyme-based *Leishmania* classification is shown in Table 1 (Rioux *et al.* 1990).

However MLEE shows some limitations mainly due to the time-consuming procedures and mass parasite culturing required, therefore DNA genotyping methods have been investigated as alternative techniques, e.g. multi-locus microsatellite typing (MLMT) and PCR-restriction fragment length polymorphism (RFLP) (Schönian *et al.* in press). Phylogenies based on nucleotide polymorphisms in different genomic targets have largely confirmed the taxonomy of the genus *Leishmania* by MLEE with some exceptions, e.g. the intra-zymodeme genetic polymorphism of *L. infantum* MON-1, or the taxonomic status of the *L. donovani* complex in East Africa (Kuhls *et al.* 2007, Lukeš *et al.* 2007).

Table 1. Simplified classification of Leishmania *genus including species pathogenic to humans.*

Subgenus	World distribution	Species
Leishmania	Old World	***L. donovani* complex**
Ross, 1903		*L. donovani* Laveran & Mesnil, 1903
		L. archibaldi Castellani & Chalmers, 1919
		L. infantum* complex
		L. infantum Nicolle, 1908 (syn. *L. chagasi* Cunha & Chagas, 1937)
		***L. tropica* complex**
		L. tropica Wright, 1903
		***L. killicki* complex**
		L. killicki Rioux, Lanotte & Pratlong, 1986
		***L. aethiopica* complex**
		L. aethiopica Bray, Ashford & Bray, 1973
		***L. major* complex**
		L. major Yakimoff & Shokhor, 1914
	New World	***L. mexicana* complex**
		L. mexicana Biagi, 1953 (syn. *L. pifanoi* Medina & Romero, 1959)
		***L. amazonensis* complex**
		L. amazonensis Lainson & Shaw, 1972 (syn. *L. garnhami* Scorza *et al.*, 1979)
		L. aristidesi Lainson & Shaw, 1979
Viannia	New World	***L. braziliensis* complex**
Lainson & Shaw, 1987		*L. braziliensis* Viannia, 1911
		L. peruviana Velez, 1913
		***L. guyanensis* complex**
		L. guyanensis Floch, 1954
		L. panamensis Lainson & Shaw, 1972
		L. shawi Lainson *et al.*, 1989
		***L. naiffi* complex**
		L. naiffi Lainson & Shaw, 1989
		***L. lainsoni* complex**
		L. lainsoni Silveira *et al.*, 1987

*This complex is also diffuse in the New World, with the local name of *L. chagasi*. It is believed that Mediterranean *L. infantum* was imported in South America by the Spanish conquerors (Lukeš *et al.* 2007).

Clinical aspects

The establishment of metacyclic parasites in the dermis of the mammalian skin is facilitated by the sandfly saliva, which enhances *Leishmania* infectivity. After phagocytosis by macrophages, amastigotes have the capacity to resist intracellular digestion as a result of several parasite and host cell factors.

When the intracellular development of the amastigotes remains localised at the inoculation site, various cytokines are released and cell reactions are generated, resulting in the development of a localised lesion of CL. In other instances, the parasites spread to organs rich in mononuclear phagocytes, giving rise to VL. Amastigotes may also spread to other cutaneous sites, as in diffuse cutaneous leishmaniasis (DCL), or to facial mucosae in the case of mucocutaneous leishmaniasis (MCL). The localisation of parasites in various tissues and organs is dependent on both intrinsic parasite tropism of a given *Leishmania* species and the immunological status of the host (Gradoni and Gramiccia 1994), resulting in the clinical expression of the disease.

Visceral leishmaniasis (VL)

Two species are usually responsible for VL: *L. donovani* in the Indian sub-continent, East Africa and Arabian peninsula, and *L. infantum* in the Mediterranean, Middle East, Central Asia and the Americas (Figure 2).

The incubation period is generally 2-6 months, but can range from 10 days (exceptional) to many years (more common). A classical VL syndrome includes fever, asthenia, weight loss, anaemia, splenomegaly, hepatomegaly and sometimes adenopathia. An intermittent and irregular fever is the major symptom. Splenomegaly appears early and is almost invariably present; anaemia is responsible for an extreme paleness of skin and mucosa. In India, patient skin has a greyish pigmentation which gives rise to the local name of the disease (kala-azar). If left untreated, VL is fatal in more than 90% of patients. Current treatment options based on pentavalent antimony salts, conventional or lipid-associated amphotericin B and miltefosine show a high cure rate in different endemic settings (den Boer and Davidson 2006). VL caused by *L. donovani* shows frequently a dermal manifestation known as post kala-azar dermal leishmaniasis (PKDL) occurring after an apparent cure or recovering of the visceral disease. Beginning as depigmented maculae, the PKDL lesions can extend to the whole body, playing an important role in the sandfly transmission.

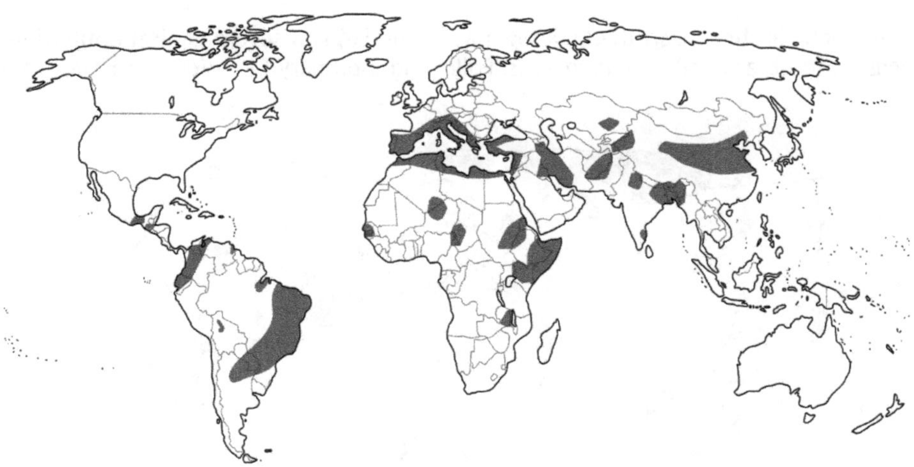

Figure 2. Global distribution of visceral leishmaniases.

Cutaneous leishmaniasis (CL)

The world distribution of all tegumentary leishmaniases is shown in Figure 3.

CL is a dermatological disease known since earliest antiquity (Oumeish 1999). It consists of one or more localised skin lesions (depending on the number of infecting bites) without mucosal involvement nor evidence of dissemination. Lesions occur on exposed parts of the body accessible to sandfly bites. All anthropophilic *Leishmania* species, including the viscerotropic ones, can be responsible for a localised CL, which presents as a mild self-healing infection. The incubation period varies between a week and several months. The mature lesion is well defined, generally round or oval with variable dimensions ranging 0.5-10 cm in diameter. The most common clinical feature is the ulcerative lesion with sloping sides and central ulcer. A 'wet' type is typical of zoonotic CL lesions caused by *L. major*, *L. mexicana*, *L. peruviana* and *L. braziliensis*. A 'dry' type, presenting as papulo-nodular lesions covered by scales, is the usual form of the anthroponotic CL caused by *L. tropica*. The clinical evolution of CL is chronic and leads to spontaneous cure in a time varying from few months (*L. major*, *L. mexicana*, *L. peruviana*) to few years (*L. aethiopica*, *L. infantum*, *L. tropica*, *L. guyanensis*, *L. panamensis*). The spontaneous cure always results in a disfiguring scar, while early treatment with pentavalent antimony salts can prevent such a condition.

Other tegumentary forms

Diffuse cutaneous leishmaniasis (DCL) is a severe form caused by a few *Leishmania* species, *L. aethiopica* in the Old World and *L. amazonensis* (rarely *L. mexicana*) in the New World, in patients who defect in cell-mediated immunity. The primary lesion is a non-ulcerated nodule rich in parasites. The nodules become numerous, disseminate to the whole body mimicking the presentation of lepromatous leprosy. The severity of DCL is shown by its resistance to anti-leishmanial drugs and it never cures spontaneously.

Mucocutaneous leishmaniasis (MCL), known as 'espundia', is a severe clinical entity due to *L. braziliensis* and occasionally *L. panamensis*. Following a primary CL lesion, secondary mucosal

Figure 3. Global distribution of tegumentary leishmaniases.

involvement occurs in a period between several weeks to many years. The mucosal involvement usually starts from the cartilaginous part of the nasal septum, which is rapidly destroyed. Mouth and lips mucosa is affected at a later stage of the disease which, in the advanced stage, leads to severe tissue necrosis and disfigurement. Death can occur following pulmonary super-infections. When treated and cured, MCL patients show disfiguring, sometimes retractile scars.

Epidemiology and distribution

Leishmaniases range over the intertropical zones of America and Africa, and extend into temperate regions of Latin America, Southern Europe and Asia. Their extension limits are latitude 45° North and 32° South. About 20 named *Leishmania* species and subspecies are pathogenic for humans and 30 sandfly species are proven vectors. Each parasite species circulate in natural foci of infection where susceptible phlebotomine species and mammals coexist. There are two main epidemiological entities:

- Zoonotic leishmaniases: where domestic or wild animal reservoirs are involved in the tramsmission cycle and humans play a role of an accidental host.
- Anthroponotic leishmaniases: where man is the sole reservoir and source of vector's infection.

Zoonotic leishmaniases

Thirteen well-recognised *Leishmania* species known to cause disease in humans have a zoonotic nature (Gramiccia and Gradoni 2005).

Zoonotic VL is the most widespread entity caused by a single parasite species, *L. infantum*. The disease occurs in several countries of the Mediterranean basin, Central Asia and Latin America. Sporadic CL infections are also found in the same endemic areas. Several vector species are involved in the Old World, mainly belonging to the subgenus *Phlebotomus (Larroussious)*, whereas in the New World the vector is *Lutzomyia longipalpis*. Dogs are the main domestic reservoirs, while foxes, jackals and wolves are the sylvatic ones. Most of the recent issues discussed in the second part of this article deal with this leishmaniasis entity.

Old World zoonotic CL is caused mainly by *L. major*. This parasite is widely distributed in arid and savannah areas, several rodent species acting as reservoir hosts. Proven vectors belong to the subgenus *Phlebotomus* (*Phlebotomus*). *P. papatasi* Scopoli is the principal vector over a geographical range extending from Northern Africa to India.

New World zoonotic CL are caused by several members of the *Viannia* and *Leishmania* subgenera and transmitted by different species of *Lutzomyia* vectors. Some species are not uncommonly diagnosed from imported CL cases in Europe, namely *L. braziliensis, L. amazonensis, L. guyanensis* and *L. panamensis*.

Anthroponotic leishmaniases

The only two species considered as having an exclusive or predominant anthroponotic transmission pattern are *L. donovani* (including *L. archibaldi*) and *L. tropica* (including *L. killicki*). However the presence of mammal reservoirs has recently been indicated in several endemic settings, such as

Eastern Sudan for *L. donovani* (Dereure *et al.* 2003), and Morocco (Dereure *et al.* 1991), Northern Israel (Jacobson *et al.* 2003) and Iran (Mohebali *et al.* 2005) for *L. tropica*.

On the Asian continent, anthroponotic VL caused by *L. donovani* is restricted to North-East India, Bangladesh and Nepal, where *P. (Euphlebotomus) argentipes* is the sole vector (Sudhakar *et al.* 2006). In East Africa (Kenya, Ethiopia, Somalia and Sudan) *L. donovani* shows a distribution associated to the sandflies *P. (Larroussius) orientalis* and *P. (Synphlebotomus) martini* (Killick-Kendrick 1999).

Anthroponotic CL caused by *L. tropica* is highly prevalent in semi-arid subtropical regions from south-east of Turkey to north-west of India. In well-established foci, CL is transmitted man-to-man through *P. (Paraphlebotomus) sergenti* in urban settings in Turkey, Syria, Jordan, Lebanon, Yemen, Iran, Iraq, Afghanistan and Pakistan. Large-scale urban migrations influence the *L. tropica* transmission patterns, so that infections may occur in outbreaks which last for some years. Small and discontinuous foci are also described in North Africa (Morocco and Tunisia), Greece, Saudi Arabia and in Central Asia. The occurrence of *L tropica* in sub-saharian Africa was reported in Kenya (Sang *et al.* 1994) and Ethiopia (Gebre-Michael *et al.* 2004, Hailu *et al.* 2006).

Leishmaniasis entities endemic in Southern Europe

Two main entities of leishmaniasis are endemic in Southern Europe. Zoonotic VL and sporadic CL, both caused by different strains of *L. infantum*, are widespread in all the endemic zones of the region. A third leishmaniasis entity, CL due to *L. tropica*, is rare having been identified only in Greece (Frank *et al.* 1993). The epidemiology of this disease form is still unclear, as it is not known whether it can be considered an anthroponosis, like for the classical urban type, or a Mediterranean zoonosis as recently hypothesised for Israel (Jacobson *et al.* 2003).

In general, when compared with the annual incidence encountered in the Indian subcontinent, Sudan and Brazil for VL (about 450,000 cases), and in Afghanistan, Algeria, Iran, Syria and Brazil for CL (about 1,300,000 cases) (Desjeux 1996) the disease burden of leishmaniases in endemic zones of Southern Europe appears to be much lower. Recent figures on human leishmaniasis notifications are now available at the Centralised Information System for Infectious Diseases (CISID) of the WHO Regional office for Europe (http://data.euro.who.int/CISID (Accessed 4 September 2007). This information, however, does not distinguish between disease form or *Leishmania* species involved. Table 2 shows the leishmaniasis situation in 12 reporting countries of Southern Europe where the disease is supposed to be currently endemic.

About 4,700 cases were reported in the region over the 1996-2005 period, the highest number of cases being recorded in Italy (1,677), Albania (1,148) and Spain (945). It must be considered, however, that in some countries (e.g. France) leishmaniases were made notifiable only recently. Furthermore, the disease is often underreported also in countries where notification is compulsory. Underreporting is especially common for the cutaneous forms, which usually do not require hospitalisation and are treated at private dermatology clinics (Velo *et al.* 2003).

There are countries where the disease was reported in the past, but the current endemic status is not ascertained, e.g. in former Yugoslavian countries of Serbia and Slovenia. Finally, numerous imported leishmaniasis cases are also diagnosed in Central and Northern European countries that are non-endemic. Most of these patients acquired a *Leishmania* infection during summer holidays spent in the European countries listed in Table 2. Over the above period, 80 cases were

Table 2. All leishmaniasis cases notified at CISID, WHO Regional Office for Europe.

Leishmaniasis endemic country	Leishmaniasis cases reported from 1996 through 2005
Albania	1,148
Bosnia and Herzegovina	15
Bulgaria	61
Croatia	11
Cyprus	2[a]
France	198[b]
Greece	390
Italy	1,677
Malta	121
Portugal	117
Spain	945
The former Yugoslav Republic of Macedonia	20[b]

[a]Reporting from 2004.
[b]Reporting from 2003.

notified in the United Kingdom, 16 in Czech Republic, 12 in Belgium and 3 in Finland. These figures are probably under-estimated: features of leishmaniasis importation to Germany were recently analysed for 58 cases diagnosed during a 2-year period (Harms *et al.* 2003).

Zoonotic visceral leishmaniasis

Human disease: There is strong evidence that the infections caused by *L. infantum* in humans are much more common in the region than previously thought, and that they mainly occur in asymptomatic condition (Le Fichoux *et al.* 1999). An estimate of current acute VL cases by country can be inferred by literature analysis (Grech *et al.* 2000, Maltezou *et al.* 2000, Gradoni *et al.* 2003, Velo *et al.* 2003, Campino *et al.* 2006, Pratlong *et al.* 2004, Bosnic *et al.* 2006, Cruz *et al.* 2006). About 200 cases are diagnosed annually in Italy and Albania; 120-150 cases in Portugal and Spain; 20-30 cases in Southern France and Greece; less than 10 cases in Croatia, Cyprus and Malta.

Cases are reported in children below five years in a proportion varying from about 30% (Southern France) to over 60% (Albania) of all VL cases. In general, the highest is the socio-economic development of an endemic area, the lower is the rate of paediatric cases, thus suggesting a strong correlation between child's nutritional status and the development of an acute VL. In immunocompetent patients the agent is usually identified as *L. infantum* zymodeme MON-1. A few zymodeme variants were described in geographically distinct areas, such as MON-72 in the Campania region of Italy (Gramiccia *et al.* 1992), MON-77 in Catalonia, Spain (Gallego *et al.* 2001), or MON-98 in Greece (Antoniou *et al.* 2004).

Immunosuppressed adults are also at high risk of acute VL. The region was particularly affected by the high incidence of HIV/VL co-infections during the AIDS epidemics (Desjeux and Alvar 2003). Over 1,900 cases were reported in adults (mean age 38.6 years, 95% CI 38.2-39.0; males: 84.8%) by early 2001, mostly from 4 countries: Spain (1,099 cases), Italy (335), France (318) and Portugal (159). Intravenous drug users were the main risk group (72%), and epidemiological and bio-analytical data suggested that artificial *Leishmania* transmission by syringe was common in these individuals. Parasites causing VL in HIV-infected patients were found highly polymorphic and genetically associated to the geographical origin of infection (Chicharro *et al.* 2003, Gramiccia 2003, Pratlong *et al.* 2003). About 30 *L. infantum* zymodemes were identified, of which 1/3 consisted of new zymodemes found exclusively in these patients. Overall, about 60% of patients were found infected with the common viscerotropic MON-1 zymodeme, 20% with dermotropic *L. infantum* zymodemes (see below) and 20% with the new zymodemes. After the introduction of Highly Active Anti-Retroviral Therapy (HAART) for the treatment of HIV infection in late 1990s, the incidence rates were considerably reduced. As shown in the graph of Figure 4, the mean incidence of cumulative cases recorded in France, Italy and Spain had a peak in 1997, followed by a reduction over 3 years and reaching a steady low incidence in subsequent years.

Immunosuppression conditions other than co-infection with HIV were found to influence the clinical outcome of VL. Immunosuppressive therapies following organ transplant (Basset *et al.* 2005), treatments for immunological disorders (Bassetti *et al.* 2006, Ossandon *et al.* 2006), a chronic status of hepatic cirrhosis (Pagliano *et al.* 2007) and probably pregnancy (Pagliano *et al.* 2005), were all hypothesised to favour the multiplication and dissemination of *L. infantum* in the patients.

Canine reservoir: Compared with humans, dogs appear more susceptible to *L. infantum* infection and develop severe disease (canine leishmaniasis, CanL) more frequently. This is characterised by elevated titres of anti-*Leishmania* serum antibodies, progressive weight loss, lymph node enlargement, skin alterations (exfoliative dermatitis, ulcers and alopecia) and renal failure (Figure 5).

Although CanL and human VL distribution overlap in the region, the incidence of the former is much higher, several thousand cases being diagnosed annually in dogs from all endemic zones. Reported seroprevalences range from 1.7% in Southern Cyprus to over 40% along the Tyrrhenian

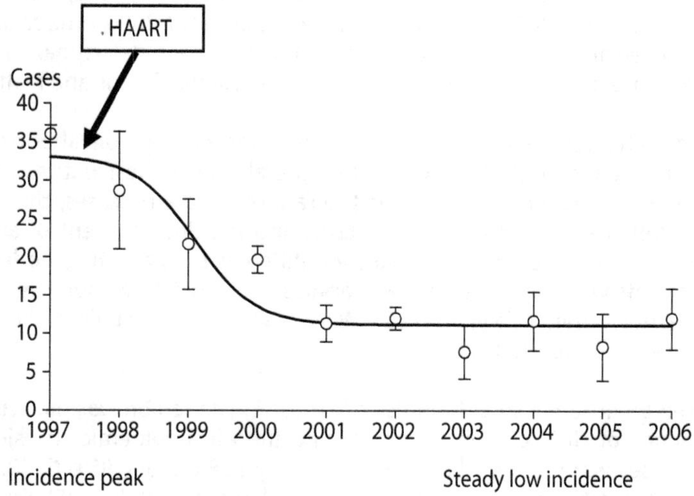

Figure 4. Cumulative incidence trend of HIV/VL co-infections recorded in France, Italy and Spain. In these countries, HAART therapy for HIV treatment has been in routine use since 1997. Symbols indicate mean number and range.

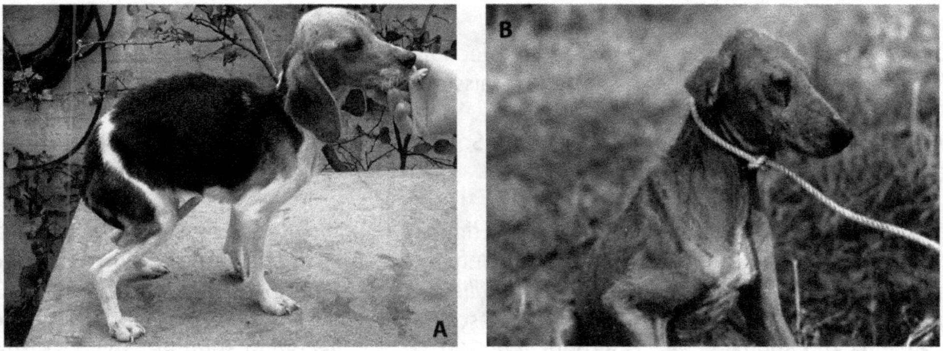

Figure 5. Symptomatic canine leishmaniasis. (A) severe weight loss and exfoliative dermatitis; (B) diffuse alopecia.

coast of Southern Italy. The great majority of parasite strains examined were identified as *L. infantum* zymodeme MON-1 (Gradoni 1999, Alvar *et al.* 2004). As with humans, asymptomatic infections may also be identified in dogs which represent either pre-patent phases of the disease before chronic progression, or resistant forms which eventually recover spontaneously. The time-course of the infection is very variable, but in progressive disease the diagnostic markers become positive in the following sequence: PCR-leishmanial DNA detection in bone marrow/peripheral blood → positive serology/parasite demonstration in tissues → development of clinical signs (Oliva *et al.* 2006). Infectiousness to phlebotomine vectors tends to be associated with disease progression: polysymptomatic animals can infect as many as 90% of sandflies feeding on them. However, asymptomatic dogs with positive serology can be as infectious as the symptomatic ones (Molina *et al.* 1994). Furthermore, sick dogs retain a high potential of infectiousness even if they are repeatedly treated with antileishmanial drugs (Gradoni *et al.* 1987).

Other animal hosts: Wild canids have been found infected with *L. infantum* zymodeme MON-1 in countries of the region, however their role as VL reservoirs was not elucidated. The red fox (*Vulpes vulpes*) was found infected at high rates in Portugal, and occasionally in France and Italy (Abranches *et al.* 1984). Infection records in wolves or jackals are anecdotic. No major reservoir role was identified in any rodent species, although a few natural infections by *L. infantum* MON-1 were reported from, and experimental phlebotomine xenodiagnosis was successful in *Rattus rattus* (Gradoni *et al.* 1983). In the past few years, there have been five confirmed cases of equine leishmaniasis in Europe, all of them being caused by *L. infantum* (Gramiccia and Gradoni 2005). The infections consisted of nodular or ulcerated skin lesions self-resolving without any therapy, suggesting that horses may represent an incidental host. Interestingly, cases were not only from traditionally endemic areas (Spain and Portugal), but also from a non-endemic area of Southern Germany, i.e. at border of the northern limit of leishmaniasis distribution in Southern Europe.

Recently, high infection prevalence rates were reported in domestic cats from France, Italy, Portugal and Spain (Gramiccia and Gradoni 2005). Titres of antileishmanial antibodies detected were usually lower than in CanL. A few isolates were identified as *L. infantum* zymodeme MON-1, and one symptomatic cat was infectious to phlebotomine vectors at xenodiagnosis (Maroli *et al.* 2007). However, the role of cats as VL reservoir is still controversial, as they are have a certain degree of resistance to experimental *Leishmania* infection, but naturally-infected cats do not recover spontaneously and the common co-infection with immunosuppressive viruses (FIV and FeLV) is supposed to facilitate the agent transmission.

Vectors: The proven vectors of zoonotic VL in Southern Europe are members of the subgenus *Phlebotomus* (*Larroussius*) (Figure 6).

Five species fulfilled the criteria for VL vector incrimination (Killick Kendrick 1990), which basically consist of the detection of natural *Leishmania* infections and parasite identification as an agent of human disease: *P. perniciosus*, *P. ariasi*, *P. perfiliewi*, *P. neglectus* and *P. tobbi* (Léger *et al.* 2000). Another *Larroussius* species, *P. langeroni* Nitzulescu, reported as a *L. infantum* vector in Northern Africa (Doha and Shehata 1992, Guerbouj *et al.* 2007) was collected in Southern Spain (Martinez Ortega *et al.* 1996) but its role in transmission was not ascertained locally. The vector species have different distribution in Southern Europe, with the Italian peninsula having a probable role of bridge, but also of barrier (e.g. due to Alps or the large Padana valley) between western and eastern species (Table 3).

Females are active during the warm months, usually from May through September-November depending on latitude and altitude above sea level. In southern latitudes (e.g. Sicily and Malta)

Figure 6. A female sandfly Phlebotomus perniciosus *taking a blood meal.*

Table 3. Distribution of the L. infantum vector species in some countries of Southern Europe, from west to east.

Country	*P. perniciosus*	*P. ariasi*	*P. perfiliewi*	*P. neglectus*	*P. tobbi*
Portugal	X	X			
Spain	X	X			
France	X	X			
Italy	X	X[a]	X	X	
Malta	X				
Croatia			X	X	X
Albania			X	X	X
Greece			X	X	X
Cyprus			X		X

[a]At the border with France.

usual collecting methods reveal two density peaks in June and August respectively, whereas in northern latitudes (e.g. sub-alpine continental Italy) there is only one peak in July. Blood-meal analyses of the most efficient VL vector *P. perniciosus* showed that this species does not exhibit specific host preferences, as it may feed on various warm-blooded animals depending on the host availability (De Colmenares *et al.* 1995, Bongiorno *et al.* 2003)

Evidence for northward spreading: Incidence of zoonotic VL has been increased in Southern Europe since the 1990s, with new foci detected within traditional boundaries of endemic transmission but also in northern regions previously regarded as non-endemic. This phenomenon was particularly investigated in Northern continental Italy, a traditional *Leishmania*-free territory at the border with France, Switzerland, Austria and Slovenia (Maroli *et al.* in press). From 1990 through 2005, over 200 human VL cases were recorded, of which 4 had no travel history. Seven endemic CanL foci were retrospectively identified starting from 1990, whereas 2003-2005 prospective investigations in dogs disclosed about 150 autochthonous clinical cases and asymptomatic seropositives among 5,500 animals examined from 16 sites of 6 northern regions of Italy. Comparisons with historical data showed that *P. perniciosus* and *P. neglectus* have increased in density and expanded their geographic range in the area. These findings demonstrated conclusively that Northern continental Italy became focally endemic for zoonotic VL after the 1990s.

Sporadic cutaneous leishmaniasis by L. infantum

Human disease: The term 'sporadic' is employed to distinguish this *L. infantum* CL entity from other Old World entities characterised by hyperendemic or epidemic patterns (i.e. *L. tropica* and *L. major* CLs). Lesions are usually single but can be highly polymorphic, varying from small nodules to large ulcerative lesions (Figure 7). Although being probably more diffuse than zoonotic VL, it is rarely reported due to its benign nature and therefore available notifications are unreliable. There is evidence that *L. infantum* is the only agent of endemic CL in most of the Southern European countries: Portugal, Spain, Southern France, Italy, Malta, Croatia, Albania and Cyprus (Gramiccia *et al.* 1995, Pratlong *et al.* 2004, Velo *et al.* 2003, Bosnić *et al.* 2006). The parasites are usually regarded as low-virulence genetic variants of viscerotropic *L. infantum*, although in several instances the parasites were typed as zymodeme MON-1. The so-called dermotropic zymodemes are differently

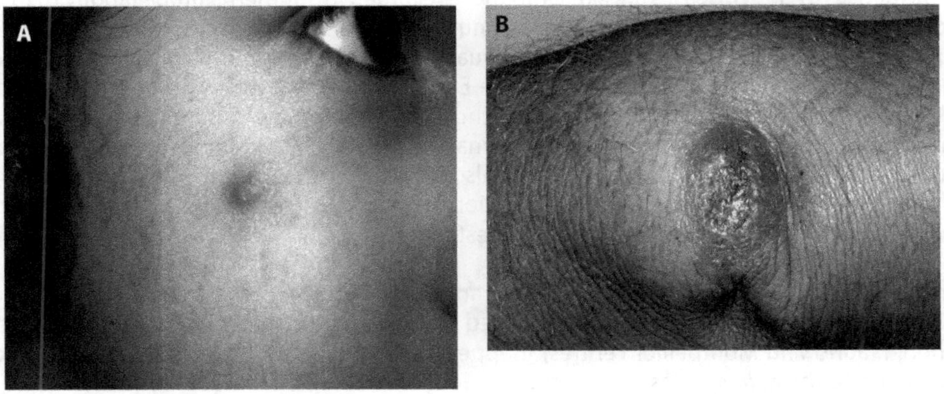

Figure 7. Some clinical aspects of sporadic cutaneous leishmaniasis caused by Leishmania infantum *in Southern Europe. (A) nodular lesion; (B) ulcerative inflamed lesion.*

distributed in the region, with 'western' zymodemes MON-11, MON-29 and MON-33 identified exclusively in CL cases from the Iberian peninsula and South-Western France (Pratlong *et al.* 2004), or the 'central' zymodeme MON-111 found exclusively in Italy, or the 'southern' zymodeme MON-78 found in Sicily and Malta only (Gramiccia *et al.* 1995, Gramiccia *et al.* 2004).

Reservoir hosts and vectors: Although dogs are strongly suspected to be also the main reservoir hosts for the dermotropic zymodemes of *L. infantum* different from MON-1, several authors have failed to demonstrate conclusively such hypothesis. On the other hand, various phlebotomine vector species have been incriminated because harbouring natural infections with dermotropic *L. infantum* zymodemes, and they are basically the same VL vectors: *P. perniciosus* and *P. ariasi* in the Iberian peninsula and Southern France (Pratlong *et al.* 2004; Campino *et al.* 2006) and *P. perfiliewi* (Maroli *et al.* 1987) and probably *P. perniciosus* in Italy.

Sporadic cutaneous leishmaniasis by L. tropica

As mentioned above, it appears that the western limit of the Asian *L. tropica* distribution extends to Southern Europe by including only inland and insular Greece. Here the disease is sporadic, and it is rarely reported (Frank *et al.* 1993). The suspected vector is *P. sergenti*, a species incriminated to be *L. tropica* vector in Turkey, Middle East and Morocco (Depaquit *et al.* 2002). This sandfly species was reported to extend its geographical range to southern territories of Europe, mainly in Spain, Portugal and Italy (Sicily) (D'Urso *et al.* 2004). The risk for *L. tropica* introduction through infected dogs or human patients in these territories has been recently emphasised (Gramiccia and Gradoni 2005).

The leishmaniases in countries bordering Southern Europe and disease importation

The burden of leishmaniases in endemic countries bordering Europe at south (Afro-Mediterranean region) and south-east (Turkey, Near and Middle East) is much higher than in Southern Europe. Three to four leishmaniasis entities affect most of these territories (Desjeux 1996, Gramiccia and Gradoni 2005). Zoonotic VL and sporadic CL due to *L. infantum* are found in Mediterranean biotopes of Morocco, Algeria, Tunisia, Egypt, Israel and West Bank, Lebanon and Turkey. Average low incidences of both diseases are similar to those recorded in Southern Europe (about 10 to 150 VL cases/year according to country). On the contrary, thousands cases of zoonotic CL caused by *L. major*, often with epidemic patterns, are annually recorded in semi-arid zones of all Northern African (including Libya) and Near and Middle East countries, but not in Turkey. Furthermore, anthroponotic CL caused by *L. tropica* affects thousands individuals in urban areas of Morocco (which suffered from a recent outbreak) (Guernaoui *et al.* 2005), all Middle East countries and South-East Turkey. Sporadic *L. tropica* cases are also reported in rural areas of Tunisia, Jordan, Israel and West Bank. Hence, the risk for the introduction of novel parasites into leishmaniasis-endemic zones of Southern Europe should not be underestimated.

From published literature (El Hajj *et al.* 2004, Gramiccia *et al.* 2004; Berger *et al.* 2006; Antinori *et al.* 2007) and unpublished information provided by European Leishmaniasis Reference Centres (Lisbon, Madrid and Montpellier centres) it appears that the importation of cases due to non-indigenous *Leishmania* species in countries of Southern Europe is indeed frequent. Ten to 70 patients infected by exogenous parasites were recorded by each centre, but it should be pointed out that species identification was made only in a small proportion of actual imported cases.

Patients consisted of immigrants from endemic extra-European countries, or nationals that travelled for tourism or work, as well as military staff. Cases were mainly represented by CL forms due to *L. tropica* or *L. major*, but also to neotropical parasites of the *L. braziliensis*, *L. amazonensis* or *L. guyanensis* complexes. Although for the latter parasites the adaptation to Old World vectors seems less probable, phlebotomine species susceptible to *L. tropica* (*P. sergenti*, see above) or *L. major* (*P. papatasi*) are recorded in several Southern European countries (Killick Kendrick 1990). On this respect, the recent report of hybrid *Leishmania* from Portugal that shares *L. infantum* and *L. major* genetic traits (Ravel *et al.* 2006) suggests that adaptation of novel parasites to Southern European vectors may indeed occur (Volf *et al.* 2007).

Global warming and risk of spreading or introducing leishmaniasis vectors in Europe

It is generally believed that leishmaniases and its vectors re-emerged in Europe after the 1940s-1960s DDT house spraying campaigns against malaria. If novel control strategies can be made effective (see below), the implementation of a successful control programme will depend on a detailed knowledge of the changing distribution of *L. infantum* transmission. Following local eradication of the disease, continual vigilance will be required. Some malaria studies have concluded that global warming is likely to cause disease spread into higher altitudes and temperate regions from adjacent endemic zones, especially where vectors abound without disease. Sandfly life-cycles in Europe depend on temperature, with typically 2-3 generations each year except at the temperate edges of their range (north to Paris, Alsace-Rhine Valley and Southern Hungary), and larval diapause permits overwintering (Ready and Croset, 1980). The persistence in Europe of endemic leishmaniases with relatively high transmission rates, despite the many post-war socio-economic changes in rural areas, indicates that sandfly-borne diseases may be less sensitive to human modifications of the landscape and changes in lifestyle. There is then a high probability that leishmaniases distribution in Europe will change with climate, especially along the temperate edge of its range.

Since 2004 the EU Integrated project 'Emerging Diseases in a changing European Environment (EDEN)', http://www.eden-fp6project.net (Accessed 4 September 2007) aims to identify and catalogue environmental conditions which can influence the spatial and temporal distribution and dynamics of human pathogenic *Leishmania*. Sandfly diversity and abundance in Europe are compared in defined bioclimate zones, at different altitudes and latitudes, and extensive datasets are produced from a variety of published and unpublished reports. These data are collated to define standard entomological measurements for each geo-referenced sampling site. New collections are made along transects crossing the distribution boundaries known for all species of European *Phlebotomus*. In general, sandflies are more abundant in Mediterranean than in temperate bioclimates (Rioux *et al.* 1997), and so most new transects cover a range of altitudes and environments where these bioclimates meet in Northern Portugal and Spain, Central France and Southern Germany, Northern Italy, and Turkey-Northern Greece-Hungary, which are regions where the vectors are likely to be expanding their ranges. The results are compared to other data from meteorological stations and environmental databases (e.g. human population density, agricultural practices), in order to quantify relationships between sandflies, climate and other environmental change, with the aim of extrapolating throughout the range of the more widespread vectors, e.g. *P. perniciosus* in Western Europe.

Small samples of the sandflies in the new collections are genetically characterised in order to identify their races and then to determine whether the races are associated with specific climatic and environmental conditions, or a specific parasite strains. Molecular phylogenetics and population genetics of sandflies have identified geographical races of the more important European vectors, and indicated their origins (Aransay *et al.* 2003, Depaquit *et al.* 2002, Di Muccio *et al.* 2000, Esseghir *et al.* 2000, Perrotey *et al.* 2005, Pesson *et al.* 2004). Mitochondrial and microsatellite DNA markers have provided support not only for the expected post-glacial spread of vectors from Mediterranean refuges, but also for the continued existence of isolated races that have not shown the same ability to spread northwards (e.g. races of *P. perniciosus* in South and Northeast Spain).

Perspectives for leishmaniases surveillance and control

International health agencies have increased efforts to improve methodologies for the surveillance and control of leishmaniasis entities characterised by a predominant anthroponotic transmission pattern. Better tools have recently been made available to developing countries for: (1) improved VL case detection, e.g. a direct agglutination test kit for laboratory serology, and a recombinant antigen (K39)-dipstick for field serology; (2) affordable VL treatment, e.g. the first oral antileishmanial drug, miltefosine; (3) a more efficient sand fly control for both anthroponotic VL and CL, e.g. long-lasting insecticide-impregnated bednets (Desjeux 2004).

In the field of zoonotic leishmaniases, new tools are under development for the surveillance and control of zoonotic VL, which are based on the control of the canine domestic reservoir. Because culling of infected dogs is not considered an acceptable measure for ethical reasons, control approaches are targeted to both immunological protection and prevention from sandfly bites. Several *Leishmania* vaccine candidates have been tested in the past 15 years (Gradoni 2001). New preparations are under field evaluation in Southern Europe: a fucose-mannose-ligand enriched *Leishmania* fraction already registered in Brazil for veterinary use against symptomatic canine leishmaniasis (Borja-Cabrera *et al.* 2002, 2004), and an excreted-secreted *Leishmania* antigen that confers long-lasting protection from infection (Lemesre *et al.* 2007).

A number of insecticide-based preparations have been specifically registered for dog protection against sandfly bites, which include deltamethrin impregnated collars, topical ('spot-on') permethrin and imidacloprid/permethrin. Laboratory and field studies have shown elevated efficacy of these preparations for both individual and mass protection. In Italy, the mass application of deltamethrin collars for two consecutive transmission seasons led to 86% reduction in canine leishmaniasis incidence in an endemic village as compared to untreated villages of the same area (Maroli *et al.* 2001). The efficacy of this control measure against human VL was shown in an endemic setting of Iran, where infantile VL incidences dropped by 50% in treated villages after the application of dog collars during one season (Mazloumi Gavgani *et al.* 2002).

Acknowledgments

We wish to thank our colleagues of the following Leishmaniasis Reference Centres: Prof. Lenea Campino, Instituto de Higiene e Medicina Tropical, Lisboa, Portugal; Dr Carmen Cañavate, Instituto de Salud Carlos III, Majadahonda, Madrid, Spain; and Prof. Jean-Pierre Dedet, CNRS-Université Montpellier1, France, for providing information on imported and HIV leishmaniasis cases. This work was partially supported by the EU FP6 project 'EDEN' CE N. 010284-2, Subproject EDEN-LEI.

References

Abranches P, Conceiçao-Silva FM and Silva-Pereira MC (1984) Kala-azar in Portugal. V. The sylvatic cycle in the enzootic endemic focus of Arrabida. J Trop Med Hyg 87: 197-200.

Alvar J, Yactayo S and Bern C (2006) Leishmaniasis and poverty. Trends Parasitol 22: 552-557.

Alvar J, Canavate C, Molina R, Moreno J and Nieto J (2004) Canine leishmaniasis. Adv Parasitol 57: 1-88.

Antinori S, Calattini S, Longhi E, Sestetti G, Piolini R, Magni C, Orlando G, Gramiccia M, Acquaviva V, Foschi A, Corvasce S, Colomba C, Titone L, Parravicini C, Cascio A and Corbellino M (2007) Clinical use of polymerase chain reaction on peripheral blood and bone marrow samples for the diagnosis and monitoring of visceral leishmaniasis in HIV-infected and HIV-uninfected patients: a single-center, 8 year experience in Italy and review of the literature. Clin Infect Dis 44: 1602-1610.

Antoniou M, Doulgerakis C, Pratlong F, Dedet JP and Tselentis Y (2004) Short report: Treatment failure due to mixed infection by different strains of the parasite *Leishmania infantum*. Am J Trop Med Hyg 71: 71-72.

Aransay AM, Ready PD and Morillas-Marquez F (2003) Population differentiation of *Phlebotomus perniciosus* in Spain following post-glacial dispersal. Heredity 90: 316-325.

Basset D, Faraut F, Marty P, Dereure J, Rosenthal E, Mary C, Pratlong F, Lachaud L, Bastien P and Dedet JP (2005) Visceral leishmaniasis in organ transplant recipients: 11 new cases and a review of the literature. Microbes Infect 7: 1370-1375.

Bassetti M, Pizzorni C, Gradoni L, Del Bono V, Cutolo M and Viscoli C (2006) Visceral leishmaniasis infection in a rheumatoid arthritis patient treated with adalimumab. Rheumatology (Oxford) 45: 1446-1448.

Berger F, Romary P, Brachet D, Rapp C, Imbert P, Garrabé E, Debord T and Spiegel A (2006) Outbreak of cutaneous leishmaniasis in military population coming back from French Guiana. Rev Epidemiol Sante Publique 54: 213-221.

Bongiorno G, Habluetzel A, Khoury C and Maroli M (2003) Host preferences of phlebotomine sand flies at a hypoendemic focus of canine leishmaniasis in central Italy. Acta Trop 88: 109-116.

Borja-Cabrera GP, Correia Pontes NN, da Silva VO, Paraguai de Souza E, Santos WR, Gomes EM, Luz KG, Palatnik M and Palatnik de Sousa CB (2002) Long lasting protection against canine kala-azar using the FML-QuilA saponin vaccine in an endemic area of Brazil (Sao Goncalo do Amarante, RN). Vaccine 20: 3277-3284.

Borja-Cabrera GP, Cruz Mendes A, Paraguai de Souza E, Hashimoto Okada LY, de A Trivellato FA, Kawasaki JK, Costa AC, Reis AB, Genaro O, Batista LM, Palatnik M and Palatnik-de-Sousa CB (2004) Effective immunotherapy against canine visceral leishmaniasis with the FML-vaccine. Vaccine 22: 2234-2243.

Bosnić S, Gradoni L, Khoury C and Maroli M (2006) A review of leishmaniasis in Dalmatia (Croatia) and results from recent surveys on phlebotomine sandflies in three southern counties. Acta Trop 99: 42-49.

Campino L, Pratlong F, Abranches P, Rioux JA, Santos-Gomes G, Alves-Pires C, Cortes S, Ramada J, Cristovāo JM, Afonso MO and Dedet JP (2006) Leishmaniasis in Portugal: enzyme polymorphism of *Leishmania infantum* based on the identification of 213 strains. Trop Med Int Health 11: 1708-1714.

Chicharro C, Jiménez MI and Alvar J (2003) Iso-enzymatic variability of *Leishmania infantum* in Spain. Ann Trop Med Parasitol 97: S57-S64.

Cruz I, Nieto J, Moreno J, Canavate C, Desjeux MCP and Alvar J (2006) *Leishmania*/HIV co-infections in the second decade. Indian J Med Res 123: 357-388.

Depaquit J, Ferté H, Léger N, Lefranc F, Alves-Pires C, Hanafi H, Maroli M, Morillas-Marquez F, Rioux JA, Svobodova M and Volf P (2002) ITS2 sequences heterogeneity in *Phlebotomus sergenti* and *Phlebotomus similis* (Diptera, Psychodidae): possible consequences in their ability to transmit *Leishmania tropica*. Int J Parasitol 32: 1123-1131.

Dereure J, Rioux JA, Gallego M, Perieres J, Pratlong F, Mahjou J and Saddiki H (1991) *Leishmania tropica* in Morocco: infection in dogs. Trans R Soc Trop Med Hyg 85: 595.

Dereure J, El-Safi SH, Bucheton B, Boni M, Kheir MM, Davoust B, Pratlong F, Feugier E, Lambert M, Dessein A and Dedet JP (2003) Visceral leishmnaisis in eastern Sudan: parasite identification in human and dogs; host-parasite relationships. Microbes Infect 5: 1103-1108.

Desjeux P (1996) Leishmaniasis. Public health aspects and control. Clin Dermatol 14: 417-423.

Desjeux P (2004) Leishmaniasis: current situation and new perspectives. Comp Immunol Microbiol Infect Dis 27: 305-318.

Desjeux P and Alvar J (2003) *Leishmania*/HIV co-infections: epidemiology in Europe. Ann Trop Med Parasitol 97: S3-S15.

De Colmenares M, Portus M, Botet J, Dobaño C, Gállego M, Wolff M and Seguí G (1995) Identification of blood meals of *Phlebotomus perniciosus* (Diptera: Psychodidae) in Spain by a competitive enzyme-linked immunosorbent assay biotin/avidin method. J Med Entomol 32: 229-233.

Den Boer M and Davidson RN (2006) Treatment options for visceral leishmaniasis. Expert Rev Anti Infect Ther 4: 187-197.

Di Muccio T, Marinucci M, Frusteri L, Maroli M, Pesson B and Gramiccia M (2000) Phylogenetic analysis of *Phlebotomus* species belonging to the subgenus *Larroussius* (Diptera, Psychodidae) by ITS2 rDNA sequences. Insect Biochem Mol Biol 30: 387-393.

Doha S and Shehata MG (1992) *Leishmania infantum* MON-98 isolated from naturally infected *Phlebotomus langeroni* (Diptera: Psychodidae) in El Agamy, Egypt. J Med Entomol 29: 891-893.

Duprey ZH, Steurer FJ, Rooney JA, Kirchhoff LV, Jackson JE, Rowton ED and Schantz PM (2006) Canine visceral leishmaniasis, United States and Canada, 2000-2003. Emerg Infect Dis 12: 440-446.

D'Urso V, Ruta F, Khoury C, Bianchi R, Depaquit J and Maroli M (2004) About the presence of *Phlebotomus sergenti* Parrot, 1917 (Diptera: Psychodidae) in Eastern Sicily, Italy. Parasite 11: 279-283.

El Hajj L, Thellier M, Carrière J, Bricaire F, Danis M and Caumes E (2004) Localized cutaneous leishmaniasis imported into Paris: a review of 39 cases. Int J Dermatol 43: 120-125.

Esseghir S, Ready PD and Ben-Ismail R (2000) Speciation of *Phlebotomus* sandflies of the subgenus *Larroussius* coincided with the late Miocene-Pliocene aridification of the Mediterranean subregion. Biol J Linn Soc Lond 70: 189-219.

Frank C, Hadziandoniou M, Pratlong F, Garifallou A and Rioux JA (1993) *Leishmania tropica* and *Leishmania infantum* responsible for cutaneous leishmaniasis in Greece: sixteen autochthnous cases. Trans R Soc Trop Med Hyg 87: 184-185.

Gállego M, Pratlong F, Fisa R, Riera C, Rioux JA, Dedet JP and Portús M (2001) The life-cycle of *Leishmania infantum* MON-77 in the Priorat (Catalonia, Spain) involves human, dogs and sandflies; also literature review of distribution and hosts of *L. infantum* zymodemes in the Old World. Trans R Soc Trop Med Hyg 95: 269-271.

Gebre-Michael T, Balkew M, Ali A, Ludovisi A and Gramiccia M (2004) The isolation of *Leishmania tropica* and *L. aethiopica* from *Phlebotomus* (*Paraphlebotomus*) species (Diptera: Psychodidae) in the Awash Valley, northeastern Ethiopia. Trans R Soc Trop Med Hyg 98: 64-70.

Gradoni L (1999) Epizootiology of canine leishmaniasis in southern Europe. In: R. Killick-Kendrick (ed.), Canine Leishmaniasis: An Update. Hoechst Roussel Vet, Wiesbaden, pp 32-39.

Gradoni L (2001) An update on antileishmanial vaccine candidates and prospects for a canine Leishmania vaccine. Vet Parasitol 100: 87-103.

Gradoni L and Gramiccia M (1994) *Leishmania infantum* tropism. Strain genotype or host immune status? Parasitol Today 10: 264-267.

Gradoni L, Gramiccia M and Scalone A (2003) Visceral leishmaniasis treatment, Italy. Emerg Infect Dis 9: 1617-1620.

Gradoni L, Maroli M, Gramiccia M and Mancianti F (1987) *Leishmania infantum* infection rates in *Phlebotomus perniciosus* fed on naturally infected dogs under antimonial treatment. Med Vet Entomol 1: 339-342.

Gradoni L, Pozio E, Gramiccia M, Maroli M and Bettini S (1983) Leishmaniasis in Tuscany (Italy): VII. Studies on the role of the black rat, *Rattus rattus*, in the epidemiology of visceral leishmaniasis. Trans R Soc Trop Med Hyg 77: 427-431.

Gramiccia M (2003) The Identification and variability of the parasites causing leishmaniasis in HIV-positive patients in Italy. Ann Trop Med Parasitol 97: S65-S73.

Gramiccia M and Gradoni L (2005) The current status of zoonotic leishmaniases and approaches to disease control. Int J Parasitol 35: 1169-1180.

Gramiccia M, Di Muccio T and Marinucci M (2004) Parasite identification in the surveillance of imported leishmaniasis cases in Italy. Parassitologia 44: 207-210.

Gramiccia M, Gradoni L and Troiani M (1995) Heterogeneity among zymodemes of *Leishmania infantum* from HIV-positive patients with visceral leishmaniasis in south Italy. FEMS Microbiol Lett 128:33-38.

Gramiccia M, Gradoni L, Di Martino L, Romano L and Ercolini D (1992) Two syntopic zymodemes of *Leishmania infantum* cause human and canine visceral leishmaniasis in the Naples area, Italy. Acta Trop 50: 357-359.

Grech V, Mizzi J, Mangion M and Vella C (2000) Visceral leishmaniasis in Malta – An 18 year paediatric, population based study. Arch Dis Child 82: 381-385.

Guerbouj S, Chemkhi J, Kaabi B, Rahali A, Ben Ismail R and Guizani I (2007) Natural infection of *Phlebotomus* (*Larroussius*) *langeroni* (Diptera: Psychodidae) with *Leishmania infantum* in Tunisia. Trans R Soc Trop Med Hyg 101: 372-377.

Guernaoui S, Boumezzough A, Pesson B and Pichon G (2005) Entomological investigations in Chichaoua: an emerging epidemic focus of cutaneous leishmaniasis in Morocco. J Med Entomol 42: 697-701.

Hailu A, Di Muccio T, Abebe T, Hunegnaw M, Kager PA and Gramiccia M (2006) Isolation of *Leishmania tropica* from an Ethiopian cutaneous leishmaniasis patient. Trans R Soc Trop Med Hyg 100: 53-58.

Harms G, Schönian G and Feldmeier H (2003) Leishmaniasis in Germany. Emerg Infect Dis 9: 872-875.

Jacobson RL, Eisenberger CL, Svobodova M, Baneth G, Sztern J, Cravalho J, Nasereddin A, El Fari M, Shalom U, Volf P, Votypka J, Dedet JP, Pratlong F, Schonian G, Schnur LF, Jaffe CL and Warburg A (2003) Outbreak of cutaneous leishmaniasis in northern Israel. J Infect Dis 188: 1065-1073.

Killick Kendrick R (1990) Phlebotomine vectors of the leishmaniases: a review. Med Vet Entomol 4: 1-24.

Killick Kendrick R (1999) The biology and control of phlebotomine sandfles. Clin Dermatol 17:279-289.

Koehler K, Stechele M, Hetzel U, Domingo M, Schönian G, Zahner H and Burkhardt E (2002) Cutaneous leishmaniosis in a horse in southern Germany caused by *Leishmania infantum*. Vet Parasitol 109: 9-17.

Kuhls K, Keilonat L, Ochsenreither S, Schaar M, Schweynoch C, Presber W and Schönian G (2007) Multilocus microsatellite typing (MLMT) reveals genetically isolated populations between and within the main endemic regions of visceral leishmaniasis. Microbes Infect 9: 334-343.

Le Fichoux Y, Quaranta JF, Aufeuvre JP, Lelievre A, Marty P, Suffia I, Rousseau D and Kubar J (1999) Occurrence of *Leishmania infantum* parasitemia in asymptomatic blood donors living in an area of endemicity in southern France. J Clin Microbiol 37: 1953-1957.

Legér N, Depaquit J, Ferté H, Rioux JA, Gantier JC, Gramiccia M, Ludovisi A, Michaelides A, Christophi N and Economides P (2000) Les phlébotomes (Diptera-Psychodidae) de l' île de Chypre. II – Presence de *Leishmania* (*Leishmania*) *infantum* Nicolle, 1908 (zymodeme MON1) chez *Phlebotomus* (*Larroussius*) *tobbi* Adler et Theodor, 1930. Parasite 7: 143-146.

Lemesre JL, Holzmuller P, Gonçalves RB, Boudoiseau G, Hugnet C, Cavaleyra M and Papierok G (2007) Long-lasting protection against canine visceral leishmaniasis using the LiESAp_MDP vaccine in endemic areas of France: double-blind randomised efficacy field trial. Vaccine 25: 4223-4234.

Lukeš J, Mauricio IL, Schönian G, Dujardin JC, Soteriadou K, Dedet JP, Kuhls K, Quispe Tintaya KW, Jirku M, Chocholovà E, Haralambous C, Pratlong F, Oborník M, Horàk A, Ayala FJ and Miles MM (2007) Evolutionary and geographical history of the *Leishmania donovani* complex with a revision of current taxonomy. Proc Natl Acad Sci USA 104: 9375-9380.

Maltezou HC, Siafas C, Mavrikou M, Spyridis P, Stavrinadis C, Karpathios T and Kafetzis DA (2000) Visceral leishmaniasis during childhood in southern Greece. Clin Infect Dis 31: 1139-1143.

Maroli M, Gramiccia M and Gradoni L (1987) Natural infection of *Phlebotomus perfiliewi* with *Leishmania infantum* in a cutaneous leishmaniasis focus of the Abruzzi Region (Italy). Trans R Soc Trop Med Hyg 81: 596-598.

Maroli M, Mizzoni V, Siragusa C, D' Orazi A and Gradoni L (2001) Evidence for an impact on the incidence of canine leishmaniasis by the mass use of deltamethrin-impregnated dog collars in southern Italy. Med Vet Entomol 15: 358-363.

Maroli M, Rossi L, Baldelli R, Capelli G, Ferroglio E, Genchi C, Gramiccia M, Mortarino M, Pietrobelli M and Gradoni L (2007) The northward spread of leishmaniasis in Italy: evidence from retrospective and ongoing studies on the canine reservoir and phlebotomine vectors. Trop Med Int Health, in press.

Maroli M, Pennisi MG, Di Muccio T, Khoury C, Gradoni L and Gramiccia M (2007) Infection of sand flies by a cat naturally infected with *Leishmania infantum*. Vet Parasitol 145: 357-360.

Martinez Ortega E, Conesa-Gallego and Romera-Lozano H (1996) *Phlebotomus* (*Larroussius*) *langeroni* Nitzulescu, 1930 (Diptera, Psychodidae), espèce nouvelle pour l'Espagne. Parasite 3: 77-80.

Mazloumi Gavgani AS, Hodjati MH, Morite H and Davies CR (2002) Effect of insecticide-impregnated dog collars on incidence of zoonotic visceral leishmaniasis in Iranian children: a matched-cluster randomised trial. Lancet 360: 374-379.

Mohebali M, Hajjaran H, Hamzavi Y, Mobedi I, Arshi S, Zarei Z, Akhoundi B, Naeini KM, Avizeh R and Fakhar M (2005) Epidemiological aspects of canine visceral leishmaniosis in the Islamic Republic of Iran. Vet. Parasitol 129: 243-251.

Molina R, Amela C, Nieto J, San-Andrés M, González F, Castello JA, Lucientes J and Alvar J (1994) Infectivity of dogs naturally infected with *Leishmania infantum* to colonized *Phlebotomus perniciosus*. Trans R Soc Trop Med Hyg 88: 491-493.

Oliva G, Scalone A, Foglia Manzillo V, Gramiccia M, Pagano A, Di Muccio T and Gradoni L (2006) Incidence and time course of *Leishmania infantum* infections examined by parasitological, serologic, and Nested-PCR techniques in a cohort of naïve dogs exposed to three consecutive transmission seasons. J Clin Microbiol 44: 1318-1322.

Ossandon A, Bompane D, Alessandri C, Marocchi E, Conti F and Valesini G (2006) *Leishmania* in SLE mimicking an exacerbation. Clin Exp Rheumatol 24: 186-190.

Oumeish OY (1999) Cutaneous leishmaniasis: a historical perspective. Clin Dermatol 17: 249-254.

Pagliano P, Carannante N, Rossi M, Gramiccia M, Gradoni L, Faella FS and Gaeta GB (2005) Visceral leishmaniasis in pregnancy: a case series and a systematic review of the literature. J Antimicrob Chemother 55: 229-233.

Pagliano P, Carannante N, Gramiccia M, Ascione T, Stornaiuolo G, Gradoni L, Faella FS and Gaeta GB (2007) Visceral leishmaniasis causes fever and decompensation in patients with cirrhosis. Gut 56: 893-894.

Perrotey S, Mahamdallie SS, Pesson B, Richardson KJ, Gállego M and Ready PD (2005) Postglacial dispersal of *Phlebotomus perniciosus* into France. Parasite 12: 283-291.

Pesson B, Ready JS, Benabdennbi I, Martín-Sánchez J, Esseghir S, Cadi-Soussi M, Morillas-Marquez F and Ready PD (2004) Sandflies of the *Phlebotomus perniciosus* complex: mitochondrial introgression and a new sibling species of *P. longicuspis* in the Moroccan Rif. Med Vet Entomol 18: 25-37.

Pratlong F, Dereure J, Deniau M, Marty P, Farrault-Gambarelli F and Dedet JP (2003) Enzymatic polymorphism during *Leishmania*/HIV co-infection: a study of 381 *Leishmania* strains received between 1986 and 2000 at the International cryobank in Montpellier, France. Ann Trop Med Parasitol 97: S47-S56.

Pratlong F, Rioux JA, Marty P, Farrault-Gambarelli F, Dereure J, Lanotte G and Dedet JP (2004) Isoenzymatic analysis of 712 strains of *Leishmania infantum* in the south of France and relationship of enzymatic polymorphism to clinical and epidemiological features. J Clin Microbiol 42: 4077-4082.

Ravel C, Cortes S, Pratlong F, Morio F, Dedet JP and Campino L (2006) First report of genetic hybrids between two very divergent *Leishmania* species: *Leishmania infantum* and *Leishmania major*. Int J Parasitol 36: 1383-1388.

Ready PD and Croset H (1980) Diapause and laboratory breeding of *Phlebotomus perniciosus* Newstead and *Phlebotomus ariasi* Tonnoir (Diptera: Psychodidae) from southern France. Bull Entomol Res 70: 511-523.

Rioux JA, Lanotte G, Serres E, Pratlong F, Bastien P and Perieres J (1990) Taxonomy of Leishmania. Use of isoenzymes. Suggestions for a new classification. Ann Parasitol Hum Comp 65: 111-125.

Rioux JA, Akalay O, Périères J, Dereure J, Mahjour J, Le Houérou HN, Léger N, Desjeux P, Gallego M, Saddiki A, Barkia A and Nachi H (1997) L'évaluation écoépidémiologique du 'risque leishmanien' au Sahara atlantique marocain. Intérêt heuristique de la relation 'Phlébotomes - bioclimats'. Ecol Mediterr 23: 73-92.

Rose K, Curtis J, Baldwin T, Mathis A, Kumar B, Sakthianandeswaren A, Spurck T, Low Choy J and Handman E (2004) Cutaneous leishmaniasis in red kangaroos: isolation and characterisation of the causative organisms. Int J Parasitol 34: 655-664.

Sacks DL and Perkins PV (1984) Identification of an infective stage of *Leishmania* promastigotes. Science 223: 1417-1419.

Sang DK, Njeru WK and Ashford RW (1994) A zoonotic focus of cutaneous leishmaniasis due to *Leishmania tropica* at Utut, Rift Valley Province, Kenya. Trans R Soc Trop Med Hyg 88: 35-37.

Schönian G, Mauricio I, Dedet JP, Gramiccia M, Canavate C, Abo-Shehada M, Boelaert M and Dujardin JC (2007) *Mare nostrum*, a model for the molecular epidemiology of leishmaniases. Trends Parasitol, in press.

Sudhakar S, Srinivas T, Palit A, Kar SK and Battacharya SK (2006) Mapping of risk prone areas of kala-azar (Visceral leishmaniasis) in parts of Bihar State, India: an RS and GIS approach. J Vector Borne Dis 43: 115-122.

Velo E, Bino S, Gradoni L and Maroli M (2003) Recrudescence of visceral leishmaniasis in Albania: retrospective analysis of cases during 1997 to 2001 and results of an entomological survey carried out during 2001 in some districts. Trans R Soc Trop Med Hyg 97: 288-290.

Volf P, Benkova I, Myskova J, Sadlova J, Campino L and Ravel C (2007) Increased transmission potential of *Leishmania major/Leishmania infantum* hybrids. Int J Parasitol 37: 589-593.

Bluetongue, West Nile, Usutu, Chikungunya and Dengue viruses

6. Modelling of vector-borne diseases and transmission of bluetongue virus in North-West Europe

Aline A. de Koeijer and Armin R.W. Elbers

Abstract

In 2006, Northwest Europe was surprised by an extreme heat wave in July and following that, the detection of blue tongue virus, an infection that was thought not to spread under the Northwest European climatic conditions. The clinical pattern and the mortality among sheep, one of the main hosts of the infection, also differed from the commonly accepted pattern of Blue tongue disease. We describe the diagnosis of this disease and the epidemic progress in 2006 with a lot of details on the clinical pattern of the disease. During 2006, there were few good measures to control spread of the infection. Retrospectively, only the regulations on animal transport were considered to be valuable in limiting the spread of the infection to other areas. Regulations to limit the contact between the vector and the host were considered less effective, partly because they were based on the wrong assumption that *Culicoides* species which transmit the infection are active in twilight and in the dark. At the time, there was also no good vaccine available. Thus, the options for control were limited at the time. With an increasing knowledge on the *Culicoides* and newly developed vaccines, there is more interest in analysing their efficacy and their efficiency in cost-benefit terms. This can best be done using models of the transmission of the infections, first to assess the 2006 epidemic and subsequently adjusting the parameters to assess various control measures. A model that uses the basic reproduction ratio as the main measure to determine the control and risk level of the epidemic is explained. Temperature is assumed to be one of the major drivers of the Blue Tongue Virus (BTV) epidemic, so this aspect is specifically addressed. Although little is known on the exact temperature effect on the various model parameters, some basic assumptions are applied, which are mainly based on the knowledge of BTV transmission by a tropical *Culicoides* species. The analysis shows that in the temperature range between 15 and 25 °C the reproduction ratio increases very fast, from a 'no risk' to a 'very high risk' situation. This is a good indication of the order of magnitude at which the temperature can affect the transmission, but should not be considered as exact results, given the nature of the underlying parameter estimates. The second extremely important driver of BTV epidemics is vector abundance. This is thought to depend heavily on temperature and humidity over a longer time period, but the life history parameters of the relevant *Culicoides* species are never quantified, which means that this abundance can only be quantified from insect traps, and extrapolation from those. When more is known on the vector, the present model can be extended with the population dynamics of the vector, which should lead to a much more complete picture of when and where BTV epidemics can be expected, and which control measures should be applied to prevent or restrict these.

Introduction

Bluetongue (BT) is an arthropod-borne viral non-contagious disease of domestic and wild ruminants, particularly affecting sheep with severe clinical disease including mortality. Until recently, bluetongue was known as an infection of the tropical regions. At present 24 different BT virus (BTV)-serotypes have been identified and the disease is transmitted by biting midges (*Culicoides*). BT has a worldwide distribution between approximate latitudes 35°S and 40°N, although in parts of western North America, China and in Kazakhstan BT may occasionally extend up to almost 50°N

(Lundervold *et al.* 2003). Since 1999, BTV has invaded Europe. Several BT incursions have been seen since, in Italy, Greece, Turkey, the French island of Corsica, the Spanish islands of Menorca and Mallorca and Portugal. Up to now, BT serotypes 1, 2, 4, 9 and 16 were involved in epidemics in the EU member-states (Mellor and Wittmann 2002). In some areas around the Mediterranean, virus activity has ended, in others virus activity is still present. Long term persistence appears to be difficult for this virus, and fade-out during winter often leads to regional extinction.

A recent large-scale epidemic of BTV in North-West Europe surprised the local and international veterinary community, since the infection was not expected to be so successful that far North. Recently there were warnings that local *Culicoides* species are also vector competent and that a high abundance of the tropical vector *Culicoides imicola* Kieffer is not necessary for BTV transmission (Mellor and Wittmann 2002). Since Europe is recently seeing warmer and more humid summers, this was considered to become a problem in future, but it was not yet taken that seriously on the short term, until BT outbreaks had actually started.

In this Chapter we describe the BTV epidemic in North-West Europe in 2006 and some of the basic modelling work that has been undertaken. We give some considerations in choosing a suitable model for BTV transmissions in the field, describe a measure of BTV risk and show some results on the influence of temperature on the risk of BTV epidemics. Finally some of the possible contributions of modelling for future control of BTV are presented.

BTV epidemic in North-West Europe in 2006

On 14 August 2006, a private veterinary practitioner in the southern province of Limburg in the Netherlands – with the help of veterinary specialists of the Animal Health Service in Deventer - notified the veterinary authorities of suspect clinical situations in 4 different holdings with sheep in which BT was suggested as a likely cause (van Wuijckhuise *et al.* 2006). Blood samples were taken and submitted to the Central Institute for Animal Disease Control (CIDC-Lelystad, Wageningen UR). Two days later, on 16 August, BTV infection was confirmed by reverse transcription real-time PCR-test in one of the 4 sheep flocks. As a result of the official announcement to the OIE and neighbouring countries, these countries came on full alert. The first laboratory confirmed BTV-outbreak in Belgium was announced on 19 August, in Germany on 21 August and in France on 30 August. This was the first observation of a major BTV epidemic that affected cattle and sheep holdings in the Netherlands, Belgium, Germany, France and Luxemburg. The affected area lies between latitudes 49°N and 52°N.

On 28 August 2006, the Community Reference Laboratory in Pirbright announced that BT virus serotype 8 (BTV-8) was causing the outbreaks. Typing of the virus is important in tracking where BTV-8 virus came from. BTV-8 is new to the European region. The gene sequence data is closely related to an earlier isolation in sub-Saharan Africa (Nigeria) in the beginning of the 1980's. However, from this region no regular isolations have been made in the last 20-25 years, and therefore the source of the BTV-8 introduction into North-Western Europe remains unclear. There have been earlier reports of BT outbreaks of serotype 8 in South-east Asia (Herniman *et al.* 1980, Hassan 1992, Prasad *et al.* 1992, Sreenivasulu *et al.* 2004), in the Caribbean (Mo *et al.* 1994), and several countries in Africa (Davies 1978, Herniman *et al.* 1983, Elfatih *et al.* 1987, Haresnape *et al.* 1988, Nevill *et al.* 1992, Barnard *et al.* 1998).

Since BT was previously absent in this part of Europe, farmers' and veterinarians' awareness of its clinical signs was likely to be low in the initial stage of the epidemic. Investigations in Belgium suggest that the very first clinical signs of BT appeared mid July in a cattle herd (Toussaint *et al.* 2007). In the first 10 BT outbreaks in the Netherlands, the owners indicated that the first clinical signs started approximately 12–17 days before a suspicion was reported to the veterinary authorities via a veterinary practitioner. These data indicate that in the first BT outbreaks, there was a period of at least 2–4 weeks between observation of the first clinical signs by the farmer and reporting of a suspicion by the farmer or veterinary practitioner.

Between the beginning of the epidemic in mid August 2006 and February 1, 2007, a total of 2,122 cases were reported (Table 1): 1111 in cattle, 951 in sheep, 13 in wild ruminants and 47 in mixed herds.

The overall trend of the epidemic curve (Figure 1) indicates a continuing increase in the total number of confirmed cases, with a peak in mid-October (week 42). From week 43 on, the number of new cases has continually decreased, except for another peak in week 45. The last clinical outbreaks were reported mid December 2006.

Both in Belgium and the Netherlands relatively more sheep flocks became infected compared to cattle herds, while in Germany the opposite was true. Since clinical signs of BT can resemble Foot and Mouth Disease (FMD), this disease and other vesicular diseases must be included in the differential diagnosis for cattle and sheep.

Other diseases that should be included for differentiation in cattle include Bovine Viral Diarrhoea (BVD), Infectious Bovine Rhinotracheitis (IBR), Malignant Catarrhal Fever (MCF), Epizootic Haemorrhagic Disease (EHD), photosensitisation, and miscellaneous causes of bovine stomatitis (Luedke and Jones 1984). Diseases that should be included for differentiation in sheep include Peste des Petits Ruminants (PPR), photosensitisation, polyarthritis, foot rot, white muscle disease, contagious ecthyma, sheep pox, necrobacillosis, heartwater and pulpy kidney disease (Verwoerd and Erasmus 2004).

Table 1. Number of Bluetongue serotype 8 outbreaks reported by country and by animal species from mid August 2006 until February 1, 2007.

Country	Animal species				Total
	Cattle	Sheep	Wild ruminants	Mixed[1]	
Germany	624	302	13	13	952
France	7	0	0	0	7
The Netherlands	176	250	0	34	460
Belgium	296	399	0	0	695
Luxemburg	8	0	0	0	8
Total	1,111	951	13	47	2,122

[1] Herds with two or more of the following species present: cattle, sheep, wild ruminants.

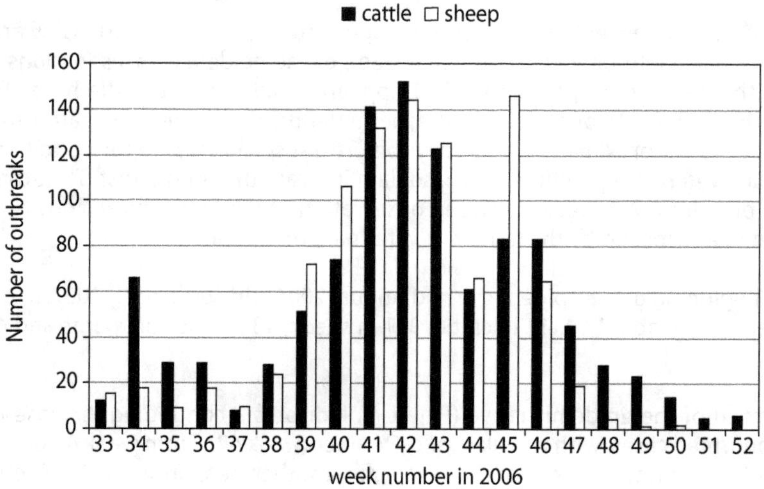

Figure 1. Distribution of the number of confirmed BTV-8 outbreaks by week in sheep and cattle holdings up to 31 December 2006 for the Netherlands, Belgium, Germany, France and Luxemburg together.

BTV may infect many different species of ruminants, but clinical disease signs are generally associated with sheep and consequently most descriptions of the disease apply to sheep (Erasmus 1975). Experienced farmers in BTV-endemic areas of South Africa may have observed from time to time mild clinical BT in their cattle, but the general impression is that BTV did not produce more than transient and mild, if any, clinical signs in cattle (Hourrigan and Klingsporn 1975).

Bluetongue in sheep has its most significant economic impact in temperate areas of the world. While infection of animals in the tropics and sub-tropics is common, clinical disease in indigenous species is unusual. The presence of endemic BTV activity is usually noticeable in the tropics and sub-tropics only when susceptible sheep are imported from temperate countries into the area (Gibbs and Greiner 1994).

One of the first reported incursions of BTV in Europe between 1956 and 1960 in Portugal and Spain killed approximately 200,000 sheep (Manso-Ribeiro *et al.* 1957). This catastrophic event was an important reason for sheep-rearing countries – strongly supported by the Netherlands - to carry a proposal in the mid-sixties to put BT on the list of notifiable diseases of the OIE.

During the 2006-epidemic, BTV-8 associated clinical signs were much more prominent in sheep than in cattle and relative frequency of specific clinical signs was different in cattle compared to sheep. In BTV-8 affected sheep, erosions of the oral mucosa, fever, salivation, facial and mandibular oedema, apathy and tiredness, oedema of the lips, lameness, and dysphagia were among the most frequent clinical signs recorded. The most prominent clinical signs in BTV-8 affected cattle were: crusts and lesions of the nasal mucosa, erosions of lips/crusts in or around the nostrils, erosions of the oral mucosa, salivation, fever, conjunctivitis, coronitis, muscle necrosis, and stiffness in limbs (Elbers *et al.* 2007a)

Almost 80% of the sheep flocks had a total of one or two sick sheep (BT-associated) per flock in the period up to 31 December 2006, in 0.4% of the sheep flocks no clinically sick sheep at all were present. Almost 68% of the cattle herds had a total of one or two sick cattle in the same period, in 14% of the cattle herds no clinically sick cattle at all were present (Elbers *et al.* 2007b).

Morbidity rate in sheep flocks and cattle herds in the Netherlands ranged between 0 and 20 per 100 sheep-month at risk and between 0 and 11 per 100 cattle-month at risk respectively. Mean morbidity rate was significantly ($p < 0.001$) higher in sheep flocks compared to cattle herds: 2.8 per 100 sheep-month at risk and 1.1 per 100 cattle-month at risk respectively. A total of 45% of the BT-affected sheep and 86% of the BT-affected cattle recovered.

Approximately 52% of BTV-8 infected sheep flocks had a total of 1-2 dead sheep (BT-associated) per flock in the period up to 31 December 2006, 38% of the infected sheep flocks showed no mortality at all. Fourteen percent of BTV-8 infected cattle herds had a total of 1-2 dead cattle (BT-associated) per herd in the same period, 85% of the infected cattle herds showed no mortality at all.

Mortality rate in sheep flocks and cattle herds in the Netherlands ranged between 0 and 20 per 100 sheep-month at risk of dying and between 0 and 5 per 100 cattle-month at risk of dying respectively. Mean mortality rate was significantly ($p < 0.001$) higher in sheep flocks compared to cattle herds: 1.5 per 100 sheep-month at risk of dying and 0.08 per 100 cattle-month at risk of dying respectively. Mean case fatality (proportion of sick animals that died) in sheep flocks was 51% and in cattle 11%.

In conclusion it can be stated that morbidity and mortality in cattle was very limited during the 2006-epidemic in North-Western Europe. In sheep, morbidity was fairly limited, with exceptions for a few flocks with high morbidity. However, almost 50% of the clinically sick sheep died.

Modelling of BTV transmission

General

The indirect transmission of BTV via a vector brings about some typical differences of the disease dynamics as compared to contagious viral infections. One of the most important aspects of infections with an insect as a vector is the strong influence of weather and climate on transmission. This is because climate has a major influence on vector abundance and their efficacy in replicating the infection (temperature influence). Various models have been developed to describe the transmission dynamics of vector-borne infections (e.g. Torina *et al.* 2004, Purse *et al.* 2005). Most of these models incorporate the climate effect. An underlying model that can predict vector abundance would also have been extremely useful. However, there is very little known about the life history parameters of *Culicoides* species in general, and even less on the differences between the many species that may contribute to BTV transmission. Therefore the modelling of the vector density as such is not (yet) possible. It has become clear that at least for *Culicoides imicola* a direct link to humidity and temperature can give good predictions of appearance of BTV epidemics (Purse *et al.* 2004a,b).

More recently models for vector-transmitted infections to describe the transmission of BTV in a herd of susceptible hosts have been developed. An overview can be found in basic mathematical epidemiology books like Diekmann and Heesterbeek (2000) and Anderson and May (1991). The first well known models for vector-borne infections were developed to analyse the transmission of malaria. At the time, there were no good tools available to control the disease and it was hoped that the models would show the options for better control of the infection. Later published models of BTV often use satellite weather information to generate risk maps, and these methods have proven to be effective for many regions. However, these models were generally developed for BTV epidemics which are transmitted (mainly) by the vector *Culicoides imicola*, and have not been applied much to BTV transmissions which are driven by other vectors, like *Culicoides dewulfi* Goetghebuer, the *Culicoides obsoletus* (Meigen) and *Culicoides pulicaris* (L.) groups which are considered to be possible vectors for BT in Northwest Europe.

Because there is not much experience with models for the specific situation of BTV in Europe, a basic analysis of the BTV dynamics is of primary importance. Once that has been assessed sufficiently, it may become possible to continue the work with satellite data at a later stage. In modelling infectious diseases, it is often practical to first focus on the basic reproduction ratio of the infection. There are several reasons to do so. Firstly, it easily distinguishes situations in which the host population is at risk and when an epidemic cannot occur, due to unfavourable conditions. Secondly, it is generally easy to derive an exact formulation of the basic reproduction ratio in terms of the relevant parameters of the infection, which makes it a powerful tool in an analysis of risk and control. The basic reproduction ratio is defined as the expected number of new infections, which are generated by one initial infective animal. That makes it a measure of the level of transmission per infectious generation. The practical use of this number lies mainly in the threshold behaviour when the reproduction ratio equals one. If the number is lower than one, an epidemic cannot occur, because the number of new infections decreases per generation. If the number is higher than one, the population is at risk of an epidemic, but it may not necessarily happen.

In the case of a vector-transmitted infection, the situation is slightly more complicated than for contagious, directly transmitted infections, because there are two main types of infectives, which may behave very differently. This is technically solved by analysing the two versions separately (see below).

Recently, the basic reproduction ratio has become one of the important measures that are used for risk assessment of infectious diseases, because it is one of the few measures that actually determine the difference between no risk and at risk. Thus it is a first step in assessing the risk of infectious diseases. A next step may be assessing other aspects, like the probability of introduction of the infection in the region or the probability that the temperature remains above 20 °C for an extended period of time.

Model

There are various aspects which are relevant for studying the BTV transmission dynamics and it can be studied on various scales of time and space. A good solution here is to start by observing the local dynamics within the active flying range of the midges (a few km^2) and on a short time scale (within one summer). On the long run, this needs to be extrapolated to national or EU spatial level and needs to include the yearly dynamics of BTV. The survival of the infection over winter is the main aspect to be added for the longer time scale. Vector dynamics and winter survival is a secondary aspect that may also turn out to be of strong influence. These aspects are as yet, still one step to far, given the paucity of information about winter survival and activity of the *Culicoides*, and about the frequency of extended infectiousness among various host species.

Basic information on the *Culicoides* indicates that they follow their hosts. They quickly migrate to an area where hosts are abundant, and thus, they hardly come to a situation where the host density is limiting the number of bites per insect. Thus we assume that they generally bite at a constant rate, which is not influenced by host density. Then we can describe a simple model, i.e. a model with few parameters.

Now let v denote the vector density and h denote the host density (sheep and cattle). Let a denote the (maximum) biting frequency of the vector. Furthermore p_1 denotes the probability of BTV transmission when an infectious host is bitten by a susceptible vector and p_2 denotes the probability of transmission when an infectious vector bites a susceptible host. The duration of the infectious period in the host is denoted by T and the latent period (incubation time) of the vector is denoted by τ. When the vector has become infectious, it is assumed to remain so for the rest of its lifespan, L. Assuming a constant death rate, μ, of the adult *Culicoides*, than the probability that an infected vector survives long enough to become infectious is given by $\exp[-\mu\tau]$. Under this assumption, the expected lifespan L of the vector after it has become infectious equals $1/\mu$.

We assume that in the relevant situation, the density of hosts is always sufficient for the vector to approach its maximal biting frequency. Otherwise, the hosts would be so rare that the infection will not spread. Therefore, the transmission from vector to host is not dependent on host density and the transmission from host to vector is proportionate to the density of the vector divided by the density of the host (because the more hosts there are, the lower the chance that an infectious host gets bitten by a specific vector).

Following Diekmann and Heesterbeek (2000) we can define the basic reproduction ratio, R_0, straight from the definitions of the parameters and use the following arguments. In a susceptible environment, one infected host will (on average) infect p_1Tav/h new vectors, and no hosts. One infected vector will (on average) infect no vectors directly and $p_2a\exp[\mu\tau]/\mu$ new hosts. Thus we find the BTV transmission matrix:

$$\begin{pmatrix} p_1Tav/h & 0 \\ 0 & p_2ae^{\mu\tau}/\mu \end{pmatrix}$$

In practice we may find some host to host transmission via veterinary activities, like vaccination and bleeding. Since this risk cannot be quantified and since it should be limited if the presence of the infection in the region is established, we ignore this aspect for the time being. From the transmission matrix we can derive the basic reproduction ratio, which is the dominant eigenvalue of the transmission matrix and we find

$$R_0^2 = (p_1p_2Tae^{\mu\tau}v) / (\mu h)$$

For this calculation it is not necessary to define the disease dynamics in more detail than the description given above. It follows these very general terms on transmission and is suitable for many deterministic and stochastic versions of the general model we describe above.

To get further from this initial analysis, we show an analysis of the effect of temperature on the basic reproduction ratio, and therefore also on the risk that an epidemic can spread and the speed at which it develops. This should be taken as an indication of what temperature can do in such a situation. The analysis is not nearly exact in real values, since very little is known on the exact values of the various parameters over an increasing temperature.

Parameters and temperature effects

To analyse the effect of climatic conditions we look at parameters which are strongly influenced by the temperature and other weather conditions, and which are not. Although we know that the vector density is influenced by weather, this is an effect of the underlying processes in the past like fecundity and survival of the various larval stages. Thus the effect is delayed and since very little is known on the relevant parameters at all, this is very difficult to analyse. Given that real data on vector density can be collected, we will assume that the observed number of vectors can be used as a known variable for a given situation. Therefore a temperature effect is not incorporated as such in the applied vector density. Since it is not possible at this moment to relate the countings of insect trappings to real densities, for the time being we analyse the BTV transmission using a reasonable guestimated density of midges under (warm) summer conditions.

Both h and p_1 are assumed not to be influenced by temperature. We copy a regularly applied assumption that p_2 equals 1; when the insect has passed the latent period in which the virus needs to replicate to sufficient numbers, any bite to a susceptible host should lead to viral transmission of BTV.

The other parameters, T, a and L are known to depend on temperature. Based on literature and expert opinion, we explore the effect of temperature on the basic reproduction ratio and the rate of spread of the infection, assuming the following temperature relations.

The biting frequency, a, increases with insect activity, i.e. with temperature. For tropical conditions it is known that there is an optimum temperature above which activity decreases again, but this is assumed not to be the case under West European conditions, where dryness is rarely an issue. The insects have a very low activity at low temperature, i.e. below 10 °C it is assumed that the biting frequency approaches 0. The maximum biting frequency is approached at about 25 to 30 °C, and equals one bite per 2 days (based on *C. imicola* data, see Figure 2).

The latent period in the vector, τ, is important because it determines the probability that the vector lives long enough to become infectious. Under low temperatures (<15 °C) it may take weeks before the virus replicates to sufficient levels, and then the probability that the vector survives long enough becomes rather low. High temperature (25 °C) leads to fast replication of the virus and under such conditions the vector may become infectious within a few days. Figure 3 shows the duration of the latent period in the vector as affected by temperature, based on the above arguments.

The lifespan of the infectious vector is the last relevant aspect that is strongly influenced by temperature. It probably has a rather low impact, because the temperature influence on the death rate of the adult vector is not that strong. Based on international data on various *Culicoides* species, we assume the lowest death rates at about 15 °C, very high death rates when the temperature decreases to 0 °C and slowly increasing death rates for temperatures above 15 °C, due to faster metabolism. Figure 4 shows the life expectancy which we applied in this analysis.

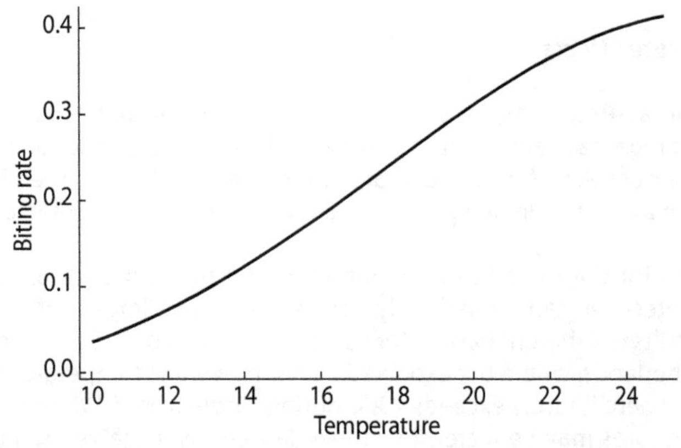

Figure 2. Biting frequency (per day per adult female insect) increases with temperature.

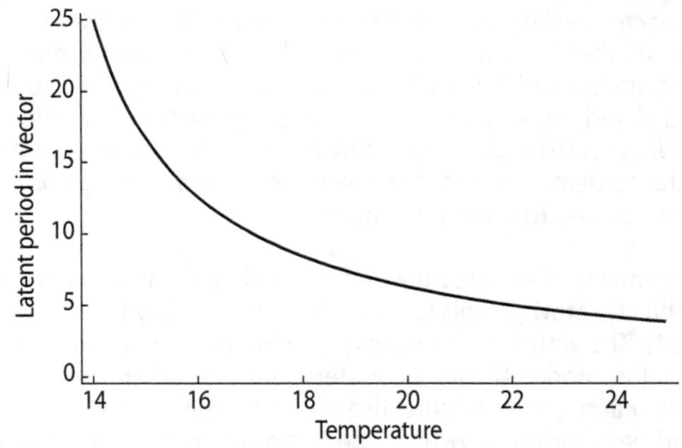

Figure 3. Decrease of the duration of the latent period (incubation period) in the infected vector due to increasing temperature.

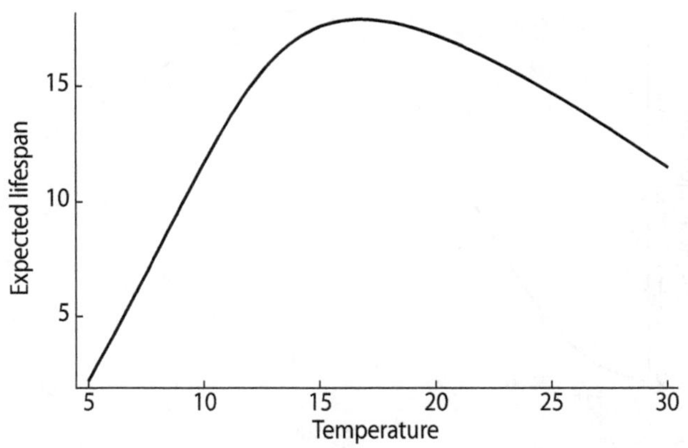

Figure 4. Expected lifespan of (infectious) vector over temperature.

Results of temperature effects

Based on the above assumptions we derive an estimate of the basic reproduction ratio as it depends on the average day temperature (Figure 5). The basic reproduction ratio is defined as the expected number of new infections that are induced by one typical infection. In the case of a vector this means an average over a typical infected host and a typical infected vector.

From the expression for the basic reproduction ratio and the influence of the temperature on some of the parameters, we can immediately derive the relative impact of temperature and the sensitivity of the results to different parameter values. The region between 15 and 20 °C is the most relevant for the Netherlands, where the expected 24 hours average temperature in July and august is around 18 °C and where it rarely exceeds 20 °C during single weeks. When we keep in mind that the average temperatures may be increasing a few degrees over the years, a temperature range up to 25 °C is relevant here. Figure 5 shows that there is a very strong impact on the transmission dynamics of BTV in this temperature range and a few degrees warmer can induce the switch from very little risk of an epidemic to almost certainty that if the infection gets introduced into an area with sufficient vector density an epidemic will develop. Thus, a few degrees higher summer temperatures may bring the Netherlands into the BTV risk area. This was proven to be true by the BTV epidemic in the summer and fall of 2006. A very warm period preceded the detection of the epidemic which had already spread widely over space by the time that the virus was detected. So far it is not clear whether the high temperature of July 2006 (day and night average 23 °C) was necessary to push the epidemic, or that it only enhanced the size of it, while a smaller epidemic would have developed during that summer anyway.

Based on Figure 5 we find that there are a few short periods every summer in which an epidemic of BTV can grow, and the rate at which this happens is determined by two aspects, the temperature and the vector density. The actual vector density is extremely variable over time and can not yet be fitted into a predictive model. The impact of temperature is therefore even stronger than the graph suggests, since warmer temperature also enhances the growth of the vector population. This effect is delayed because the growth of the vector population takes several weeks of high temperatures, since there are several larval stages to go through before the adult stage is reached at which they can transmit the infection. Actually the weather history since the last winter is

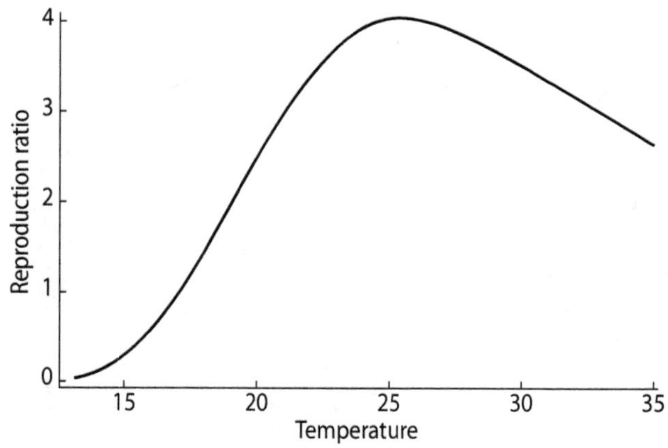

Figure 5. Basic Reproduction Ratio for a temperature range from 15 to 35 °C.

expected to have a major influence on vector abundance, but little is known on the precise conditions that favour the vector density.

The assumed temperature effect on the rate of the virus replication has an extremely high influence on the steepness and the height of the R_0-curve. Also the survival function for the adult *Culicoides* has a major influence. In this study we assume a constant death rate, independent of age, as has been found to fit for many insects. The actual applied value for survival is estimated from data from *C. imicola*.

There is much uncertainty about the exact value of the various parameters and the resulting R_0. However, it is also clear that the shape of the curve as given in Figure 5 gives a good description of the temperature effect in the relevant range.

Conclusions temperature effect

There is a strong influence of temperature on the reproduction ratio of BTV, especially in the range slightly above the normal 24 hours summer temperatures in the Netherlands (15-25 °C). This means that, especially in the Netherlands, the climatic conditions are very close to the level where the BTV can spread. When the climate in general becomes warmer over the years, we should expect that BTV is an infection that will become a structural threat to our ruminant population. So far it seems, however, that a normal Dutch climate (summers average at about 16 °C) will not allow successful spread of the virus if it gets introduced. This does not mean that eradication of an existing epidemic is equally simple. The 2006 epidemic took very long to stop, because there was no removal of infected and infectious hosts, which meant that the fading out of the epidemic continued for a very long time, even though the basic reproduction ratio decreased below one at the end of the summer. The warm winter of 2006-7, with very few frost nights did not contribute to ending the epidemic.

This model does not include an analysis of the vector dynamics, which is influenced by the temperature, but also by humidity. Observed numbers for vector densities should be applied. With a temperate winter, a humid spring, followed by a very warm summer, we obtained ideal conditions for BTV in 2006. The winter and spring led to favourable conditions for the *Culicoides* survival and growth, whereas the warm summer leads to fast replication of the virus in the vector. When the summer temperature is relatively low (<16 °C), we expect that the virus cannot spread widely, while harsh winter conditions will restrict survival of the virus through fade out after an epidemic and also reduces the density of the vector in the next spring. This would contradict the fast spread of the virus in the summer of 2007 despite the rather normal summer temperature. However, the warm winter and humid spring and summer of 2007 again offered favourable conditions for *Culicoides*, which appeared in much higher densities than in 2006. This allows for viral spread at lower temperatures.

There is a high level of uncertainty in the parameter estimates, since they are often based on limited data and tend to be extrapolated from data on other *Culicoides* species and tropical conditions. Furthermore there is some difference between the various BTV strains that can influence transmission.

Discussion

During the 2006 BTV epidemic a large area in North-West Europe became infected (Figure 6). The epidemic surprised many, because of the high number of animals and herds that became infected, because of the large area covered by the infection and because of the high mortality in sheep. Many reasons underlie the wide spread and the surprisingly high morbidity of the infection in cattle, which is not common for BTV. The question is than, whether these aspects result from a new evolving virus that adapts to different climatic conditions, or that it results only from the extreme weather conditions, in combination with the North-West European situation, where we should think of issues like high humidity but also large-scale transportation of hosts over distances in the order of 100 km (as compared to tropical areas). The EFSA report on the BTV serotype 8 epidemic in North-West Europe (2007) gives a good overview of the facts of this epidemic and of the aspects that may have contributed to the scale of the epidemic.

The attention was drawn to the epidemic due to morbidity and mortality in sheep, and diagnosis was difficult, because the infection was not expected to be found this high North. The nearest location of BTV-infections was at least 1000 km further south. However, once the diagnosis was made in sheep, it also explained some earlier clinical disease in cattle, which were not properly diagnosed so far. This shows that awareness for exotic diseases should be kept high, and because the clinical pattern may differ from the one in the standard textbooks, extra vigilance is important.

Obviously it is always difficult to properly prepare for emerging diseases, since they were not considered to be a risk factor before. General surveillance is therefore essential and regular laboratory testing of animals with unexplained diseases is important for the early diagnosis. In the case of the BTV serotype 8 epidemic in 2006, the options for controlling the epidemic once it had been diagnosed were limited. However, since then, the knowledge of the abundance of the vector in North-West Europe is increasing quickly, and the development of vaccines has been enhanced. These two factors will probably be essential for future control of BTV.

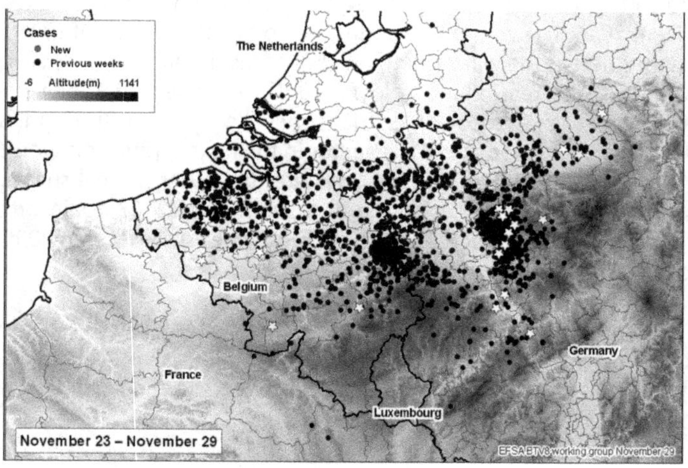

Figure 6. An overview of the spatial spread of BTV in summer and fall 2006. Late October, still a lot of new positive herds were detected. Figure taken from the EFSA report (2007).

The origin of the infection and the route of introduction could not be established. The first is probably due to the erratic testing and reporting of such viruses in tropical areas where they frequently occur. The latter is due to the fact that there are so many possibilities. The EFSA report (2007) gives a good overview of the work done on this issue and the results found.

During 2006, there were very few good measures to control spread of the infection available. Retrospectively, only the regulations on animal transport were considered to be valuable in limiting the spread of the infection to other areas. These regulations have been maintained during 2007. The regulations that were aiming at a decrease of the contacts between the vector and the host were considered less effective, partly because they were based on the wrong assumption that *Culicoides* species which transmit the infection are active in twilight and in the dark. At the time, there was no good vaccine available, and it was also wrongly assumed that the infection would not spread once the temperature would drop. However, the fade-out of the epidemic took a long time and detection of herds with a low level of infection was possibly long after the infection was introduced. Thus, a lot of new cases were found still in November and December of 2006 (see Table 1 and Figure 6).

To analyse whether control measures are effective and efficient, modelling is an essential tool. The model included in this paper is a first step along the process of assessing the effect of various factors on the transmission of the infection. It clearly shows the temperature effect. Unfortunately, the temperature is not something that can easily be controlled. However, this model can also be used to assess the efficacy of a vaccination programme, once the effect of the vaccine on the replication and transmission of the virus in the hosts can be quantified. Furthermore this model can be used to assess the maximum vector density, below which an epidemic is not expected. (See Annex 4b of the EFSA report, 2007). When more knowledge is generated on the life history and dynamics of the vector population, such information may become useful in assessing the options for vector control or to limit the contacts between vectors and susceptible hosts. For example, if we can find some animals which are not susceptible to the virus but are still very attractive for the vector, they could be used to reduce the number of bites per susceptible host and thus, reduce the risk of BTV epidemics.

References

Anderson RM and May RM (1991) Infectious Diseases of Humans. Dynamcis and Control. Oxford University Press, Oxford, UK.

Barnard BJH, Gerdes GH and Meiswinkel R (1998) Some epidemiological and economic aspects of a bluetongue-like disease in cattle in South Africa – 1995/1996 and 1997. Onderstepoort J Vet Res 65: 145-151.

Davies FG (1978) Bluetongue studies with sentinel cattle in Kenya. J Hyg Camb 80: 197-204.

Diekmann O and Heesterbeek JAP (2000) Mathematical epidemiology of infectious diseases: Model building, analysis and interpretation. John Wiley & Sons, UK.

European Food Safety Authority (EFSA) (2007) Epidemiological analysis of the 2006 Bluetongue virus serotype 8 epidemic in north-western Europe. EFSA, Parma, Italy. 42 pp. with 9 additional appendices. www.efsa.europe.eu (Accessed 9 September 2007).

Elbers ARW, Mintiens K, Staubach C, Gerbier G, Meiswinkel R, Hendrickx G, Backx A, Conraths FJ, Meroc E, Ducheyne E, Gethmann J, Heesterbeek JAP, De Clercq K, Unger F, Stegeman JA (2007a) Bluetongue virus serotype 8 epidemic in North-Western Europe in 2006: preliminary findings. Proc Soc Vet Epidemiol Prev Med 28-30 March 2007, Helsinki, Finland: 231-245.

Elbers ARW, Backx A, Meroc E, Gerbier G, Staubach C, Hendrickx G, van der Spek A, Mintiens K (2007b) Nature and severity of disease in sheep and cattle herds during the Bluetongue virus serotype 8 epidemic in 2006 in north-western Europe. Prev Vet Med, in press.

Elfatih M, Mohammed H and Taylor WP (1987) Infection with bluetongue and related orbiviruses in the Sudan detected by the study of sentinel calf herds. Epidem Infect 99: 533-545.

Erasmus BJ (1975) Bluetongue in sheep and goats. Austr Vet J 51: 165-170.

Gibbs EPJ and Greiner EC (1994) The epidemiology of Bluetongue. Comp Immun Microbiol Infect Dis 17: 207-220.

Haresnape JM, Taylor WP and Lungu SAM (1988) The epidemiology of bluetongue in Malawi. Epidem Infect 100: 493-499.

Hassan A (1992) Epidemiology of bluetongue virus infection in Malaysia. In: Walton TE, Osburn BI (eds.), Bluetongue, African horse sickness and Related Orbiviruses, Proceedings 2nd Int. Symp., Boca Raton, CRC Press, pp. 155-161.

Herniman KAJ, Gumm ID, Owen L, Taylor WP and Sellers RF (1980) Distribution of Bluetongue viruses and antibodies in some countries of the eastern hemisphere. Bull Off Int Epiz 92: 581-586.

Herniman KAJ, Boorman JPT and Taylor WP (1983) Bluetongue virus in a Nigerian dairy cattle herd. J Hyg Camb 90: 177-193.

Hourrigan JL and Klingsporn AL (1975) Bluetongue: the disease in cattle. Austr Vet J 51: 170-174.

Luedke AJ and Jones RH (1984) Bluetongue: diagnosis and significance in the bovine animal. The Bovine Practitioner 19: 79-86.

Lundervold M, Milner-Gulland EJ, O'Callaghan CJ and Hamblin C (2003) First evidence of bluetongue virus in Kazakhstan. Vet Microbiol 92: 281-287.

Manso-Ribeiro J, Rosa-Azevedo JA, Noronha FO, Braco-Forte MC, Grave-Pereira C and Vasco-Fernandez M (1957) Fievre catarrhale du mouton (bluetongue). Bull Off Int Epiz 48: 350-367.

Mellor PS and Wittmann EJ (2002) Bluetongue virus in the Mediterranean Basin 1998-2001. Vet J 164: 20-37.

Mo CL, Thompson LH, Homan EJ, Oviedo M-T, Greiner EC, Gonzalez J and Saenz MR (1994) Bluetongue virus isolations from vectors and ruminants in central America and the Caribbean. Am J Vet Res 55, 211-215.

Nevill EM, Erasmus BJ and Venter GJ (1992) A six year survey of viruses associated with culicoides biting midges throughout South Africa (Dipitera: Ceratopogonidae). In: Walton TE, Osburn BI (eds.), Bluetongue, African horse sickness and Related Orbiviruses, Proceedings 2nd Int. Symp., Boca Raton, CRC Press, pp. 314-319.

Prasad G, Jain NC and Gupta Y (1992) Bluetongue virus infection in India: a review. Rev Sci Tech Off Int Epiz 11: 699-711.

Purse BV, Tatem AJ, Caracappa S, Rogers DJ, Mellor PS, Bayliss M and Torina A (2004a) Modelling the distributions of *Culicoides* bluetongue vrus vectors in Sicily in relation to satellite-derived climate variables. Med Vet Entomol **18**: 90-101.

Purse BV, Bayliss M, Tatem AJ, Rogers DJ, Mellor PS, van Ham M, Chizov-Ginzburg A and Braverman Y (2004b) Predicting the risk of bluetongue through time: climate models of temporal patterns of outbreaks in Israel. Rev Sci Tech Off Int Epiz 23: 761-775.

Purse BV, Mellor PS, Rogers DJ, Samuel AR, Mertens PPC and Baylis M (2005) Climate change and the recent emergence of bluetongue in Europe. Nature Rev Microbiol 3: 171-181.

Sreenivasulu D, Subba Rao MV, Reddy YN and Gard GP (2004) Overview of bluetongue disease, viruses, vectors, surveillance and unique features: the Indian sub-continent and adjacent regions. Vet Ital 40, 73-77.

Torina A, Caracappa S, Mellor PS, Bayliss M and Purse BV (2004). Spatial distribution of bluetongue virus and its *Culicoides* vectors in Sicily. Med Vet Entomol 18: 1-9.

Toussaint J-F, Sailleau C, Mast J, Houdart P, Czaplicki G, Demeestere L, VandenBussche F, van Dessel W, Goris N, Breard E, Bounaadja L, Thiry E, Zientara S and De Clercq K (2007) Bluetongue in Belgium, 2006. Emerg Infect Dis 13: 614-616.

Van Wuijckhuise L, Dercksen D, Muskens J, De Bruijn J, Scheepens M and Vrouenraets R (2006) Bluetongue in the Netherlands: description of the first clinical cases and differential diagnosis 9/14 – common symptoms just a little bit different and in too many herds. Tijdschr Diergeneeskd 131: 649-654.

Verwoerd DW and Erasmus BJ (2004) Bluetongue. In: Coetzer JAW, Tustin RC (eds.), Infectious Diseases of Livestock, Vol. 2, Oxford University Press, pp. 1201-1220.

7. Bluetongue: an emerging vector-borne disease outbreak in North-western Europe

Willem Takken, Eugene M.A. van Rooij, Niels O. Verhulst, Frans Jacobs, Silvie Huijben, Jacob Beeuwkes, Nikée Groot, Veronique C.A. Vos, Jeroen Spitzen, Rene C.G. Heutink and Piet A. van Rijn

Abstract

The animal vector-borne disease bluetongue (BT) was discovered in the Netherlands and neighbouring areas of Belgium, France, Germany and Luxemburg for the first time in August 2006. The BT virus was identified as serotype 8. This serotype had never before been found in Europe. Vector surveillance in the outbreak area revealed nine different potential vector species, with >90% belonging to the *Culicoides obsoletus* and *C. pulicaris* group of midges. BT virus was found in *C. dewulfi*, a member of the *C. obsoletus* group. Following a winter period free of reported cases in the affected areas, renewed outbreaks were reported on 17 July 2007, indicating that the virus had survived undetected throughout the winter of 2006/2007. It is argued that the introduced BT virus may have benefited from the unusually high temperatures in July 2006 in northwestern Europe, and that climate, vectors and virus conspired to cause a major epidemic in farm animals. Aspects of this outbreak are discussed, with relevance to the origin of the serotype, the vector ecology and competence, the climate, and the route of introduction.

Keywords: Bluetongue virus, North-Western Europe, *Culicoides*, outbreak, control

Introduction

Climate change and increasing international trade have been blamed for the dissemination of vector-borne diseases at an unprecedented speed in recent years. Dengue fever, a disease previously limited to S.E. Asia, has become endemic in the Caribbean and Central- and South America; West Nile fever has arrived in the continental USA and Chikungunya fever has hit hard on the island of Réunion (Jetten and Focks 1997, Morse 2003, Paganin *et al.* 2006). In addition, the incidence of several tick-borne diseases is rising in the northern hemisphere, e.g. Lyme disease, granular ehrlichiosis (anaplasma) and tick-borne encephalitis (Parola 2004, Piesman and Gern 2004). Now, another infectious disease can be added to this list: bluetongue (BT) was confirmed in sheep in the Netherlands on 15 August 2006, the first case ever reported north of the Alps.

BT is endemic in South Africa, most of southern Asia, Australia and the Americas, but since 1998 it has also been present at one time or another in many parts of Mediterranean Europe. Cyprus, certain Greek Islands and the Iberian peninsula have previously suffered occasional outbreaks of BT probably caused by the wind-borne spread of vectors infected by bluetongue virus (BTV) from Israel and Turkey, or North Africa (Purse *et al.* 2005). These early incursions were short lived and involved only a single serotype on each occasion. However, this picture has changed since 1998, with new introductions each year except 2003, involving BTV strains of five distinct serotypes (1, 2, 4, 9 and 16) killing many thousands of sheep. In southern Europe, the competent vector was found to be the well-known vector of BT, *Culicoides imicola* Kieffer, a blood feeding midge with a distribution from South Africa to southern Europe. In Europe, however, *C. imicola* is a relatively recent arrival. Since 2000, Italy has witnessed yearly epidemics of BT, which are mostly controlled by vaccination. In northern Italy and the Balkans, the disease is most likely transmitted by *C.*

obsoletus (Meigen), *C. scoticus* Downes and Kettle, and *C. pulicaris* (Linnaeus), midge species that tolerate a cooler and wetter climate than *C. imicola* (De Liberato *et al.* 2005, Gomulski *et al.* 2006). *Culicoides obsoletus* and *C. scoticus* belong to the subgenus Avaritia Fox and *C. pulicaris* to the subgenus *Culicoides* Latreille, respectively. Recent epidemiological models predicted that the distribution of BTV in Europe would not reach beyond the Balkan countries or southern France (Purse *et al.* 2005).

First outbreaks of BTV

The index case of BT in the Netherlands occurred in Kerkrade, in the far south of the country, on the border with Germany and Belgium. That same day, five other flocks of sheep were recorded as suspect (Figure 1). By 17 August, three farms had been confirmed positive for BTV, with eight others suspected. All cases were confirmed by PCR in combination with antibody detection. By 19 and 21 August, several cases were also recorded from neighbouring areas in Belgium and Germany, respectively. On 26 August, the isolate was typed by the European Community Reference Laboratory at Pirbright in the UK as a BT serotype 8 using RT-PCR and sequence analysis of genome segment 2. The virus is derived from a sub-Saharan African lineage but is distinct from the reference and attenuated vaccine strains of BTV-8. It is the first time that this serotype has been isolated from animals in Europe. Moreover, this strain not only affected sheep but also cattle with clinical symptoms.

By late August 2006, 28 farms in the Netherlands had been confirmed positive with BTV, all within an area of 400 km^2 in Limburg province. Three cows were found positive 150 km north of the outbreak area, at an animal exportation centre near Utrecht. The three animals had been brought there from the infected area shortly before confirmation of the BT introduction. A further 46 and 21 farms were infected in Belgium and Germany, respectively. Shortly thereafter, numerous cases were reported to the authorities in Belgium, Germany, Luxemburg and the Netherlands, and there were a few cases reported in northern France (Mehlhorn *et al.* 2007, Toussaint *et al.* 2007) (http://www.efsa.europa.eu/en/in_focus/bluetongue/bluetongue_report_s8.html). By the end of 2006, more than 450 infected farms had been reported in the Netherlands, with many more in

Figure 1. Sheep on a farm in the affected area in Limburg province, August 2006.

Belgium, Germany and Luxemburg (Figure 2). The weekly number of new infections is shown in Figure 3, with a clear peak of cases by the end of October. Only in France the number of infected farms remained low (n=6) and confined to the Northeast. Adhering to contingency plans adopted by the European Community, a series of regulatory steps were issued to the farming community with regard to animal movement and other aspects of their business. According to these rules, by 17 November 2006 about 50% of the Netherlands was declared to lie within the infected area (Figure 4). Livestock owners in the infected area were also required to treat animals with a pour-on insecticide (deltamethrin or permethrin or pyrethrin/piperonylbutoxide) and treat walls of the

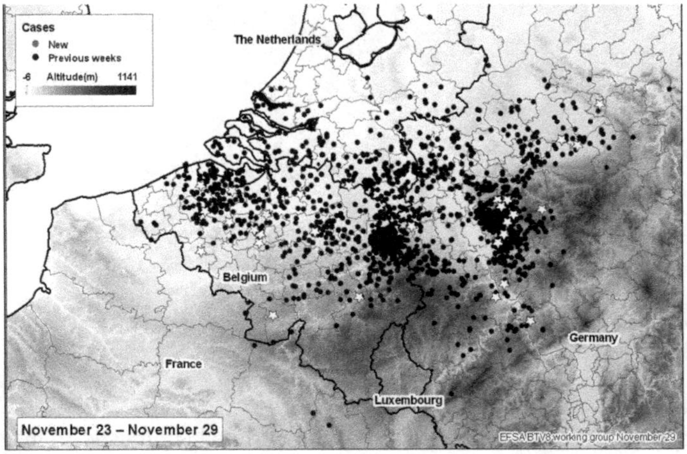

Figure 2. Overview of cases of BTV 8 recorded in North-western Europe between 17 August 2006 and 27 February 2007 (source: DEFRA, UK).

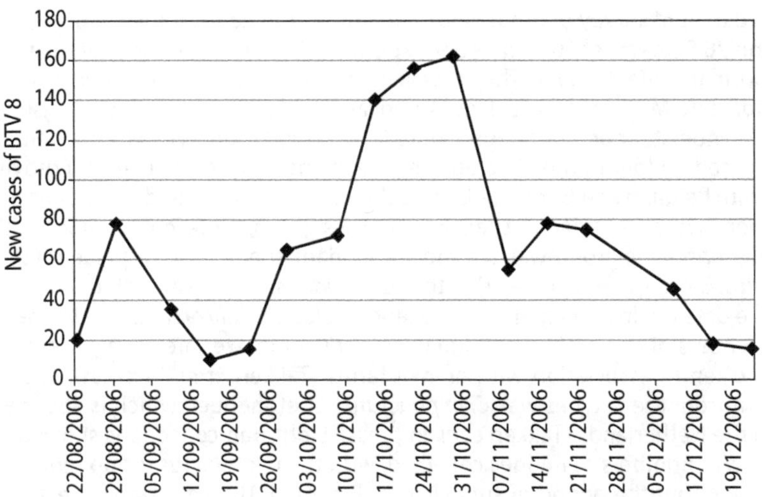

Figure 3. Weekly number of BTV cases during the bluetongue outbreak in North-western Europe in 2006 (source: EFSA, Parma, Italy).

Figure 4. Restricted zone for live ruminant transport in the Netherlands, November 2006 (modified after Dutch Ministry of Agriculture, Nature and Food Quality).

stables with this insecticide. At the early start of the outbreak, domestic animals were required to be kept indoors from one hour before sunset to one hour after sunrise (http://www.minlnv.nl/portal/page?_pageid=116,1640524&_dad=portal&_schema=portal).

Vector surveillance

We conducted a rapid survey of putative BTV vectors in the heart of the outbreak zone between 22 August and 28 September 2006, in order to obtain information on the presence and density of *Culicoides* spp. in the affected area. This surveillance was conducted with Liberty plus counterflow CO_2 traps (Mosquito Magnet® Liberty Plus, American Biophysics Corp., North Kingstown, RI 02852, USA) and with Onderstepoort light traps (Figure 5) (Kline 2002, Meiswinkel 1998, Venter *et al.* 1996). The study focused on four of the affected livestock farms in southern Limburg province, close to the borders with Belgium and Germany. During this survey, we collected 6,441 *Culicoides*. Of these, 3,902 specimens were identified to species or species group, representing at least nine different species. The *C. obsoletus* group was the most abundant species group present (88%), followed by the *C. pulicaris* group (9%) (Table 1). Other *Culicoides* species were relatively rare among the catches. These data are in agreement with recent studies on *Culicoides* in the Netherlands, which show that members of the *C. obsoletus* and *C. pulicaris* complexes are widely present throughout the country, often in association with animal farms (Takken *et al.* in press). The only notable species absent from the catches was *C. impunctatus* Guetghebuer, which is abundantly present elsewhere in the Netherlands (Takken *et al.* in press). Light trap collections showed considerable daily variations in numbers of midges collected, with the *C. obsoletus* group remaining the most abundant throughout the period of surveillance (Figure 6). The light trap collected much larger numbers of *Culicoides*, but the CO_2 trap was in continuous operation and therefore could also

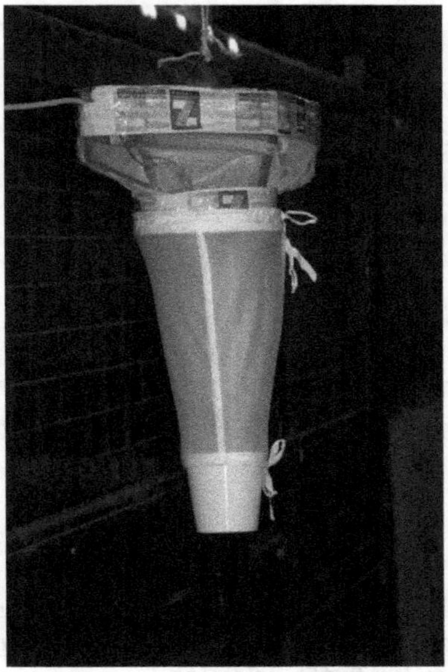

Figure 5. Onderstepoort light trap in one of the sites of the Culicoides survey in September 2006.

Table 1. Total catch of Culicoides *spp. in Onderstepoort light traps during the survey in Limburg province, 22 Aug – 28 Sept 2006.*

Species	Total number collected	Percentage of total
C. obsoletus group[1]	3,449	88%
C. pulicaris group	351	9%
C. punctatus	28	1%
C. newsteadi	6	0%
C. festivipennis	35	1%
C. circumscriptus	5	0%
C. stigma	2	0%
C. nubeculosus	2	0%
Unidentified *Culicoides* spp.	24	1%
Totaal	3,902	100%

[1]The *C. obsoletus* group includes *C. obsoletus, C. scoticus, C. chiopterus* and *C. dewulfi.*

collect midges during daylight hours, a time when in northwestern Europe midges can be very active (Service, 1971). Interestingly, the daily numbers of midges in the catches were relatively low (1–50 *Culicoides* spp. per light trap per night) unlike *Culicoides* catches in southern Europe, where often hundreds of midges are collected during one night (De Liberato *et al.* 2003).

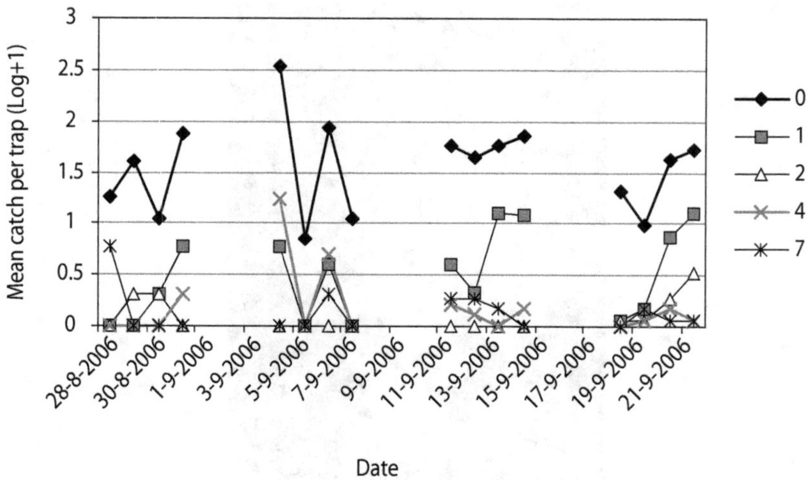

Figure 6. Mean number of Culicoides spp. collected in Onderstepoort light traps between 28 August 28 2006 and 23 September 2006 in the outbreak area in the Netherlands. 0 – C. obsoletus group; 1 – C. pulicaris group; 2 – C. punctatus; 4 – C. festivipennis; 7 – unidentified Culicoides spp.

Overwintering and new outbreaks

During the winter of 2006/2007 no new confirmed infections of BTV were reported, whereas newly found BT positive animals may have resulted from infections in late 2006. This suggests that active viral transmission had halted due to lower temperatures leading to the death of adult (infected) midges. Although no information about overwintering of adult biting midges in Europe is available, we frequently caught adult midges throughout the winter in an outdoor trap on a livestock farm near Afferden, the Netherlands (51°52′52″N, 5°38′21″E, 5 m above sea level), where in May 2006 large numbers of *C. obsoletus* had been caught. Numbers of *Culicoides* spp. in the trap, however, were much lower between December 2006 and March 2007 compared to summer and autumn populations (W. Takken, unpublished results). The activity of adult *Culicoides* throughout the winter was also reported by Losson *et al.* (2007), who frequently caught adult midges in a light trap in a calving unit near Liège in Belgium between November 25, 2006 and March 9, 2007. Although catches were initially high, they dropped to <10 *Culicoides* per night after December 10, 2006.

On 17 July 2007 the first new case of BT was reported from Germany, ostensibly from an animal taken sick on 6 July 2007. This case was followed by several cases in Belgium and the Netherlands and one case in France (Anonymous 2007a,b). By 1 August 2007, BT had been recorded in the Netherlands from more than 22 farms, all located within the 2006 affected areas. The southern part of the Netherlands was, once more, declared a BTV infected area, requiring restrictions for transport of live ruminants (Figure 4)[2]. By 22 August 2007 more than 336 farms were reported affected with BT, with the disease having also spread into the northern provinces of the Netherlands.

[2] On August 3, 2007, the British authorities reported a case of foot and mouth disease on a livestock farm. As a preventive measure, the Dutch government on August 4, 2007, issued a complete ban on all ruminant transport throughout the country. The ban was lifted when it was clear that the FMD outbreak had been contained.

Discussion

This BT outbreak in North-western Europe in 2006 and 2007, caused by a novel BT serotype for Europe, raises several important questions that need further attention in the future. Although southern Europe has experienced recurrent BT outbreaks in recent years, some involving *C. obsoletus* and *C. pulicaris*, until 2006 the virus had limited its distribution to the Iberian peninsula, France (Corsica), Italy, the Balkan countries, European Turkey and Greece (Calistri *et al.* 2003). Recent epidemiological models, incorporating climatic factors, did not predict an outbreak in the Netherlands, not even under future climate scenarios (Purse *et al.* 2005, Tatem *et al.* 2003). Although members of both the *C. obsoletus* and *C. pulicaris* complexes have been incriminated as vectors in the Balkans and northern Italy, and both species complexes are widely distributed in central and northern Europe, it was thought that temperature would be the limiting factor for establishment of the virus and indeed, the 2006 record high temperatures recorded in some parts of central and northern Europe may have been a factor in the observed virus spread. Thus, while competent vectors are present in the Netherlands, we should question how the virus has arrived this far north, leaping across a gap of hundreds of kilometres. A second aspect is that this outbreak was caused by a virus isolate that is new to Europe. Bluetongue virus serotype 8 is found in Africa and the Americas, and suspected on the Indian subcontinent, but had, prior to 2006, not been recorded from Europe[3]. Sheep and cattle, although widely traded across Europe, are rarely imported from outside the continent. Therefore, accidental importation through an exotic animal should be considered. A third, curious, aspect of this epidemic is the initial patchy distribution. Within Limburg province, BTV had been found on 28 farms by 29 August 2006. Many of these farms are not adjacent to each other, but are often separated by several km in a highly diverse terrain. In addition to commercial livestock holders, there are numerous smallholder or hobby farms, with <15 sheep. Midges are found throughout the area, mostly concentrated in river valleys, where they find shade and moisture. A first analysis of the epidemic showed that not all farms were affected. However, as BTV is vector-borne, one might expect a rapid spread affecting all livestock holdings indiscriminately. Clearly this farm to farm variation requires urgent investigation.

Given that the epidemic was present across borders, affecting at least five countries, control of the epidemic proved difficult. A ban on animal movement was implemented within an area of 20 km around the affected farms, with further restrictions within a 150 km radius. This zone reached across Luxemburg into northern France and far into Germany, covering all of Belgium and about 50% of the Netherlands. The fact that infected animals were not slaughtered and hence may have served as infectious reservoirs for midges, may have resulted in many more animals becoming infected and an increased risk of further spreading of the disease. It seems more likely, however, that viral distribution is accomplished by the *Culicoides* vectors, which can disperse over several hundred m per day, or be rapidly distributed over a large distance by wind (Bishop *et al.* 2000). Whereas infected ruminants are thought to lose infectious BTV after 60 days, the infection can persist the entire life of adult *Culicoides*, which may vary from two weeks to several months. Hence containment of infectious midges, if possible, would provide an effective strategy for BTV control. Livestock holders were required to spray barns and stables with insecticides, and treat their animals with insecticide as well. However, there is no proven evidence demonstrating the efficacy of these measures and in the 2006 outbreak they failed to halt BTV dispersal. It is reported

[3] There are reports of sheep positive with BT serotype 8 from Bulgaria, but no virus has been isolated from these animals.

from Italy and the USA that treatment of farm animals with insecticides failed to stop the spread of BTV to adjacent herds (Mullens *et al.* 2001, Satta *et al.* 2004).

The month of July 2006 was the warmest on record in the Netherlands (http://145.23.254.254/VinkCMS/news_detail.jsp?id=33586) with an average monthly temperature >22.3 °C (the historic mean temperature for July is 17.4 °C). These high temperatures may have favoured the development of BTV in *Culicoides* vectors and its transmission to susceptible animals, given the presence of an infected animal imported into the region. Climate change has already been blamed for changes in the distribution of vector-borne diseases (Epstein 2005), which are widely considered to affect the health of humans and animals. It was assumed that the epidemic would fade out in the winter of 2006/2007, but recent studies in the Netherlands and Belgium demonstrated that at least a very small proportion of *Culicoides* survived as adults, thus presenting a possible reservoir of BTV (Losson *et al.* 2007; Takken *et al.*, unpublished results). The reported outbreaks in Belgium, Germany, France and the Netherlands in late July 2007 support the hypothesis that BTV remained active in the affected areas. Although at the time of writing it is too early to speculate about its 2007 distribution, it is likely that a larger area than in 2006 may become affected. The month of July 2007 was, in stark contrast to that of 2006, relatively cool and very wet. This suggests that high summer temperatures are not required for BTV distribution, and that the indigenous *Culicoides* species are competent vectors. Considering the size of the infected areas and the large number of farms, vector control seems not feasible. One could consider to focus vector control on the overwintering adult vectors, which are likely to be concentrated in animal stables and barns. Here, a focussed campaign with insecticides targeted to kill the overwintering adult midges might be effective and is expected to result in a drastic reduction of the vector reservoir. Alternatively, vector control could be started when the population of the responsible *Culicoides* spp. starts to grow, therewith delaying the spread of BT. Extensive knowledge of the competent vector(s) and communication with, for instance, weather forecasters and farmers is needed to effectively reduce the growth of the vector population. Ultimately, vaccination seems to be the only way to eradicate BT.

Although at present we have little knowledge about the origin of the current BTV outbreak, its vectors or epidemiology, and the possible effect of climate change, the fact that the disease occurred this far north and its ability to spread rapidly are reasons for serious concern about the risk of similar emerging diseases and our preparedness to diagnose and control them.

Acknowledgements

We appreciate the support and encouragement for these studies provided by Wim Pelgrim and Paul Leijs. We are grateful to Philip Mellor and Simon Carpenter for having made available Onderstepoort Light traps. We thank Rudy Meiswinkel and Maria Goffredo for their useful advice on the taxonomy of *Culicoides*.

References

Anonymous (2007a) Bluetongue in Belgium. Vet Rec 161: 77.

Anonymous (2007b) Bluetongue in France. Vet Rec 161: 148.

Bishop AL, McKenzie HJ, Barchia IM and Spohr LJ (2000) Moon phase and other factors affecting light-trap catches of *Culicoides brevitarsis* Kieffer (Diptera: Ceratopogonidae). Austr J Entomol 39: 29-32.

Calistri P, Goffredo M, Caporale V and Meiswinkel R (2003) The distribution of *Culicoides imicola* in Italy: Application and evaluation of current Mediterranean models based on climate. J Vet Med Ser B-Infect Dis Vet Pub Health 50: 132-138.

De Liberato C, Purse BV, Goffredo M, Scholl F and Scaramozzino P (2003) Geographical and seasonal distribution of the bluetongue virus vector, *Culicoides imicola*, in central Italy. Med Vet Entomol 17: 388-394.

De Liberato C, Scavia G, Lorenzetti R, Scaramozzino P, Amaddeo D, Cardeti G, Scicluna M, Ferrari G and Autorino GL (2005) Identification of *Culicoides obsoletus* (Diptera: Ceratopogonidae) as a vector of bluetongue virus in central Italy. Vet Rec 156: 301-304.

Epstein PR (2005) Climate change and human health. N Engl J Med 353: 1433-1436.

Gomulski LM, Meiswinkel R, Delecolle JC, Goffredo M and Gasperi G (2006) Phylogeny of the subgenus Culicoides and related species in Italy, inferred from internal transcribed spacer 2 ribosomal DNA sequences. Med Vet Entomol 20: 229-238.

Jetten TH and Focks DA (1997) Potential changes in the distribution of dengue transmission under climate warming. Am J Trop Med Hyg 57: 285-297.

Kline DL (2002) Evaluation of various models of propane-powered mosquito traps. J Vector Ecol 27: 1-7.

Losson B, Mignon B, Paternostre J, Madder M, De Deken R, De Deken G, Deblauwe I, Fassotte C, Cors R, Defrance T, Delecolle JC, Baldet T, Haubruge E, Frederic F, Bortels J and Simonon G (2007) Biting midges overwintering in Belgium. Vet Rec 160: 451-452.

Mehlhorn H, Walldorf V, Klimpel S, Jahn B, Jaeger F, Eschweiler J, Hoffmann B and Beer M (2007) First occurrence of *Culicoides obsoletus*-transmitted Bluetongue virus epidemic in Central Europe. Parasitol Res 101: 219-228.

Meiswinkel R (1998) The 1996 outbreak of African horse sickness in South Africa--the entomological perspective. Arch Virol Suppl 14: 69-83.

Morse DL (2003) West Nile virus - Not a passing phenomenon. N Engl J Med 348: 2173-2174.

Mullens BA, Gerry AC and Velten RK (2001) Failure of a permethrin treatment regime to protect cattle against bluetongue virus. J Med Entomol 38: 760-762.

Paganin F, Borgherini G, Staikowsky F, Arvin-Berod C and Poubeau P (2006) Chikungunya on Réunion Island: chronicle of an epidemic foretold. Presse Med 35: 641-646.

Parola P (2004) Tick-borne rickettsial diseases: emerging risks in Europe. Comp Immunol Microbiol Infect Dis 27: 297-304.

Piesman J and Gern L (2004) Lyme borreliosis in Europe and North America. Parasitol 129 Suppl.: S191-220.

Purse BV, Mellor PS, Rogers DJ, Samuel AR, Mertens PPC and Baylis M (2005) Climate change and the recent emergence of bluetongue in Europe. Nature Rev Microbiol 3: 171-181.

Satta G, Goffredo M, Sanna S, Vento L, Cubeddu GP and Mascherpa E (2004). Field disinfestation trials against *Culicoides* in north-west Sardinia. Vet Ital 40: 329-335.

Service MW (1971) Adult flight activities of some British Culicoides species. J Med Entomol 8: 605-609.

Takken W, Verhulst N, Scholte E-J, Jacobs AFG, Jongema Y and Van Lammeren R, in press. Longitudinal sampling of *Culicoides* spp. to determine the risk of bluetongue outbreaks in the Netherlands. Prev Vet Med, in press.

Tatem AJ, Baylis M, Mellor PS, Purse BV, Capela R, Pena I and Rogers DJ (2003) Prediction of bluetongue vector distribution in Europe and north Africa using satellite imagery. Vet Microbiol 97: 13-29.

Toussaint JF, Sailleau C, Mast J, Houdart P, Czaplicki G, Demeestere L, VandenBussche F, van Dessel W, Goris N, Breard E, Bounaadja L, Etienne T, Zientara S and De Clercq K (2007) Bluetongue in Belgium, 2006. Emerg Infect Dis 13: 614-616.

Venter GJ, Meiswinkel R, Nevill EM and Edwardes M (1996) Culicoides (Diptera: Ceratopogonidae) associated with livestock in the Onderstepoort area, Gauteng, South Africa as determined by light-trap collections. Onderstepoort J Vet Res 63: 315-325.

8. West Nile virus in Europe: waiting for the start of the epidemic?

Marion Koopmans, Byron Martina, Chantal Reusken and Kees van Maanen

Abstract

The recent changes in West Nile virus (WNV) epidemiology, with incursions into the New World and reporting of increased neuroinvasive disease, have raised concerns that this might happen in Europe as well. Recent studies suggest that a single viral mutation may have resulted in increased virulence specifically for American crows, raising viraemic levels above the critical threshold for mosquitoes present in North America. This illustrates the complexity of predicting by extrapolation what could happen in Europe, a conclusion that has led us to undertake a systematic review of available information from a surveillance perspective. Surveillance studies undertaken since the late 1990s have shown that WNV is present in parts of Europe, particularly but not exclusively in wetland areas. Continued reporting of human or animal cases in subsequent years suggests that in some areas the virus is enzootic, but the picture is far from complete. The fact that additional cases are often found in these areas when active surveillance is done indicates that there is significant underreporting. Whether or not birds, arriving from Central Europe, could establish a sylvatic cycle resulting in an outbreak of WNV in Western Europe will, among other factors, depend on mosquito species present, their activity, and feeding behaviour in late autumn. Consequently, climatic changes that shift the balance of ecological factors towards one favouring the start and maintenance of the transmission cycle may increase the risk of WNV outbreaks in the future, although at present we consider this risk to be low. In addition, several other, largely unknown, factors play a role in the epidemiology of WNV as well. Therefore, monitoring of WNV presence in areas with occasional illness reports should continue but a more coordinated effort is needed to provide a reliable risk map of WNV for Europe that can be used for public health preparedness.

Keywords: West Nile virus, flavivirus, Viraemia, epidemiology, vector-borne disease, arbovirus, mosquito, birds

Introduction

West Nile virus (WNV) is an arthropod-borne (arbo) flavivirus that is endemic in Africa, Europe, and East Asia (Dauphin *et al.* 2004). Its impact was limited to occasional outbreaks in Africa and Europe, but the picture changed dramatically in 1999. In that year, the virus was introduced into North America which resulted in a rapid dissemination across the USA and into Canada, Central and South America where the virus now is endemic (Briese *et al.* 1999). This rapid emergence and spread of an arbovirus that had not been detected in the New World before 1999 sparked discussions about the potential for its emergence in Europe. In January 2003, the European commission organised a workshop to discuss the current state of knowledge on arboviral infections, including WNV, and their impact on public health in Europe[4]. The meeting concluded that the existing data gaps about WNV incidence and prevalence made predictions about the situation in Europe virtually impossible. The scientific committee on veterinary measures relating to public health in April 2004 published an opinion in which they recommended that all countries in Europe should start passive surveillance of unexplained encephalitis in horses and humans, and countries with documented

[4] Presentations held at this workshop are available at http://ec.europa.eu/health/ph_threats/com/west_nile/ev_20030120_en.htm (Accessed 16 September 2007).

West Nile activity should have an active surveillance strategy for horses (http://ec.europa.eu/food/fs/sc/scv/out67_en.pdf)(Accessed 16 September 2007). At present, four years later, the anxiety level has gone down because West Nile did not emerge in Europe, although many of the same questions are still outstanding. Here, we review factors that have been considered and studies that have been undertaken since then to provide data needed for a more evidence- based assessment for the potential of WNV expansion in Europe.

A short history of WNV disease

West Nile virus was first isolated in 1937 from a woman with fever in Uganda, in the West Nile Province. From studies in the region it was established that native birds are the natural reservoir in which the virus can replicate to sufficient levels to allow transmission to other birds via mosquitoes (Figure 1) (Davis *et al.* 2006). As the viruses could also be isolated from healthy children in Egypt, with high levels of seropositivity in inhabitants of the same region, they were not seen as a serious health threat because humans (and animals) were considered to be dead end hosts (Melnick *et al.* 1951). Occasionally, epidemics of illness in humans were reported, first in Israel in a small region near Tel Aviv, where cases kept recurring for some years during the 1950s (Zeller and Schuffenecker 2004). Here, in addition to fever, occasionally meningo-encephalitis was seen. In the 1960s, the virus emerged in southern France in the Rhone delta (Camargue), to disappear again. In the 1970s a huge outbreak in South Africa was attributed to WNV, and smaller outbreaks again in the mid 1990s in Tunisia and Algeria (Table 1).

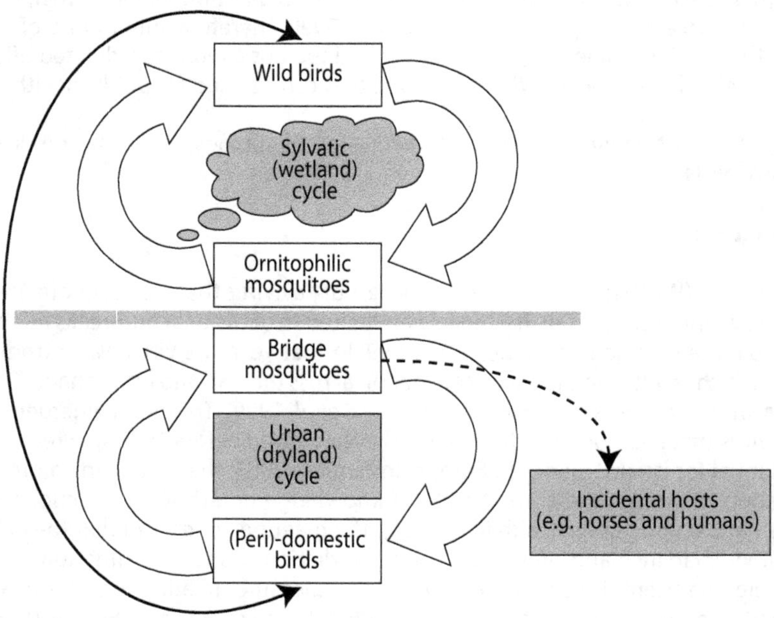

Figure 1. West Nile virus transmission cycle.The sylvatic cycle is sustained by wild birds (migratory and resident) and ornithophilic mosquitoes. Bridge vectors present in both wetland and urban areas can transmit the virus to horses and humans (dashed line). Peri-domestic bird species such as the house sparrow (Passer domesticus) can function as introductory and amplifying hosts in urban settings (solid line). An urban cycle may be established and maintained independently of the sylvatic cycle.

Table 1. Reports of human illness caused by WNV in Europe and the Mediterranean basin.

Country	Reference	Year	Nr cases	Method
France	Panthier *et al.* 1968	1968	?	
Algeria	Zeller and Schuffenecker 2004	1994	50	Serology
Tunisia	Zeller and Schuffenecker 2004	1997	173	Serology
Romania	Tsai *et al.* 1998, Han *et al.* 1999, Ceianu *et al.* 2001, Campbell *et al.* 2001	1996	393	Serology
Romania	Cernescu *et al.* 2000, Ceianu *et al.* 2001	1997	15	Serology
Romania	Ceianu *et al.* 2001	1998	5	Serology
Romania	Ceianu *et al.* 2001	1999	7	Serology
Romania	Ceianu *et al.* 2001	2000	13	Serology
Russia- Volga delta	Platonov *et al.* 2001, Zeller and Schuffenecker 2004	1999	826	RT-PCR and serology
Israel	Giladi *et al.* 2001	1999	2	RT-PCR and serology
Hungary	Bakonyi *et al.* 2006	2003	14	RT-PCR and serology
Israel	Zeller and Schuffenecker 2004	2000	417	Serology
Russia	Zeller and Schuffenecker 2004	2000	56	Serology
Russia	Zeller and Schuffenecker 2004	2001	64	Serology
France	Mailles *et al.* 2003, Del Giudice *et al.* 2004	2003	4	Serology
Volga delta	Kovtunov *et al.* 2003	2002	33	Serology
Spain	Kaptoul *et al.* 2007	2004	1	Serology
Russia- Novosibirsk	Ternovoi 2007	2004	3	RT-PCR and serology
Portugal	Connell *et al.* 2004	2004	2	Serology

For Europe, the first major outbreak of human WNV that was recognised occurred in 1996 in Romania, in a region around the capital city of Bucharest, bordering the Black Sea and flanked by the River Danube (Tsai *et al.* 1998). In total, 393 persons had laboratory-confirmed WNV neuroinvasive disease, but hundreds more with similar symptoms were hospitalised. By extrapolation of data from serosurveys an estimated 74,000-126,000 people had become infected, with a similar rate of seropositivity across all age groups. Case-control studies showed that transmission was peri-domestic, because people living in houses which had mosquitoes and flooded basements were at increased risk for acquiring infection and for developing clinical disease (Han *et al.* 1999). The flooding was caused by sewage contaminated water leaking from poorly maintained plumbing, resulting in optimal breeding conditions for mosquitoes. By matching this data with results from surveys, it was concluded that the city-dwelling mosquito species *Culex pipiens pipiens* L. was the most probable epidemic vector (Han *et al.* 1999). In the years following this outbreak, enhanced surveillance in Bucharest and South-eastern Romania identified some cases each year between 1997 and 2000 but there were no major outbreaks (Table 1) (Ceianu *et al.* 2001).

While all the above examples fit the picture that WNV is endemic in (parts of) Africa, Asia and Europe with occasional incursions and local outbreaks in areas with favourable conditions, a very different pattern was seen when WNV was introduced in the USA. This was recognised in 1999 when the New York City department of health announced a cluster of cases with neurological illness in Queens (Asnis *et al.* 1999). Based on serological testing, St Louis encephalitis (SLE) was suspected, since these related viruses occasionally cause infections on the East coast of the USA.

However, since simultaneously unusual mortality of birds was observed – both in town and at the Bronx Zoo – other possible etiologies were explored because SLE is not a cause of death in birds. Eventually, WNV was detected in tissues from birds and from humans (Asnis *et al.* 1999, Briese *et al.* 1999). Phylogenetic analysis on the basis of the gene sequence of the E and M protein (proteins associated with the virus envelope), showed that the viruses circulating in the USA were all highly related, suggesting a common source, with direct links to the viruses circulating in the Mediterranean (Lanciotti *et al.* 1999, Giladi *et al.* 2001). In contrast with the familiar picture from the old world, after two years with localised activity, WNV subsequently spread across the continent with unprecedented bird mortality, and infections in humans and horses[5]. Throughout the course of the epidemic it became clear that a range of other animals could become infected, a.o. dogs, cats, bats, chipmunks, skunks, squirrels, rabbits, and llamas[6]. Very recently, a codon change in a virus-encoded protein, the helicase, was linked to increased virulence particularly for American crows, which may have contributed to the unprecedented spread of the infection after introduction in the USA (Brault *et al.* 2007).

Risk assessment for WNV emergence in Europe

West Nile virus is a member of the family Flaviviridae which includes other arboviruses that may be transmitted by mosquitoes (dengue virus, yellow fever virus, Japanese encephalitis virus, JE) or ticks (Tick-borne encephalitis; TBE). Based on antigenic cross reactivities, WNV is grouped with the JE serocomplex, that includes viruses endemic in Europe (Usutu), South America (Cacipacore), North America (St Louis encephalitis), Africa (Koutango, Yaounde), Asia (JE) and Australia (Murray Valley, Kunjin). All flaviviruses are closely related, and extensive cross-reactivity may be seen in serological assays. This explains the initial misdiagnoses in New York, when SLE antibodies were detected in the early WNV cases. It also implies that results of serosurveys determining the level of circulation of these viruses need to be interpreted with care, including the studies listed later in this chapter (Petersen and Roehrig 2001). Since WNV is rare in Europe, and the majority of cases with meningitis, encephalitis, or meningo-encephalitis goes without diagnosis, recognition of the clinical disease by physicians *if* cases occur is unlikely, unless outbreaks occur (Rockx *et al.* 2006). Diagnosis of WNV infection is typically based on serology, but is complicated by the presence of related viruses endemic in Europe, such as the tick-borne viruses TBE and louping ill (Grard *et al.* 2007). Even though genetically these viruses are quite distinct from WNV, extensive serological cross reactivity has been documented, particularly for IgG antibody assays (Niedrig *et al.* 2007).

The mosquito-borne flaviviruses are maintained in an enzootic cycle, which for WNV involves birds as reservoir hosts (Figure 1). Viruses are primarily transmitted between these avian hosts and mosquito vectors, thus the presence of birds and mosquitoes that can produce sufficiently high levels of virus for efficient transmission is essential. While Culicidae are regarded to be preferred vectors, WNV has been identified in a wide range of arthropods, including ticks and biting midges (Higgs *et al.* 2004). Similarly, birds may differ greatly in their susceptibility to WNV and the levels of viraemia that are produced (Komar 2003). Phylogenetic studies have shown two main lineages of WNV strains relevant for Europe at present (although at present five lineages have been described; Bondre *et al.* 2007). Strains from lineage I have recently emerged in Europe, the Mediterranean basin and the USA, and strains from lineage II have been reported predominantly in sub-Saharan

[5] For detailed maps of the distribution of WNV over time see http://www.cdc.gov/ncidod/dvbid/westnile/surv&control.htm#surveillance (Accessed 16 September 2007).

[6] http://www.cdc.gov/ncidod/dvbid/westnile/Mapsactivity/surv&control04Maps.htm (Accessed 16 Sept 2007).

Africa although there is recent evidence that both lineages circulate in Europe (Bakonyi *et al.* 2006, Zeller and Schuffenecker 2004). Questions that arose when WNV emerged in the USA were: (1) do bird populations in Europe have the same level of susceptibility to WNV as American bird populations or could repeated incursions of WNV from Africa have resulted in partial immunity? (2) What effect do the mutations in the helicase of many lineage I viruses have on viraemia levels in European susceptible bird species? (3) Which vector species are relevant, are those present in Europe, and do the European lineages have the same behaviour and physiology as their New World counterparts? (4) Could WNV emerge and become endemic in Europe? Available data will be reviewed in the following paragraphs.

West Nile virus in humans

Lessons from outbreaks in Europe and the US

WNV infections rarely result in clinical disease. In early reports, illness was described as rather mild, with fever, headache, myalgia and fatigue, sometimes with other symptoms like rash, lymphadenopathy, vomiting and diarrhoea but rarely neurological symptoms (Mostashari *et al.* 2001). Studies from the USA and Romania have shown that for every reported clinical case between 140 and 250 unreported infections have taken place (Mostashari *et al.* 2001, Han *et al.* 1999). The New York survey estimated that approximately 20% of the persons with serological evidence for infection had an episode with symptoms consistent with WNV infection. People of all age groups can be infected, but the proportion of infections with clinical symptoms increases with age (Han *et al.* 1999). Most patients that are symptomatic develop the syndrome called West Nile fever, described in the early studies. Less than 1% of infected individuals develop severe neuroinvasive disease, which can be classified either as WNV meningitis, meningo-encephalitis, encephalitis or acute flaccid paralysis (AFP) (Kramer *et al.* 2007). Recovery of WNV fever typically is within a few days to months. The prognosis for encephalitis is poor and high mortality rates are seen when this complication develops, typically in elderly patients. For the other forms of neuroinvasive disease the outcome varies considerably (Kramer *et al.* 2007, Davis *et al.* 2006).

The introduction of WNV into the Americas and the resulting epidemic has revealed properties of the infection that had not been recognised previously. In addition to mosquito-borne transmission, infection can spread through blood transfusion, transplantation, breast-feeding and intra-uterine transmission. As a result, blood banks in the USA have established routine screening of blood donations for WNV, resulting in withdrawal of hundreds of samples from viraemic donors (818 in 2003; 224 in 2004; 417 in 2005 and 361 in 2006). For Europe, donor screening was not considered feasible by representatives of a research network named BOTIA (Blood and Organ Transmissible Infectious Agents)[7]; as long as disease outbreaks occur sporadically. The group did recommend a deferral policy for people travelling to the USA and Canada (Lefrère *et al.* 2003).

Data from surveillance studies

A recommendation of the 2004 report of the European Union scientific committee on veterinary measures relating to public health was that all countries in Europe should start passive surveillance of unexplained encephalitis in horses and humans, and countries with documented West Nile

[7] http://ec.europa.eu/research/fp6/ssp/botia_en.htm (Accessed 16 September 2007).

activity should have an active surveillance strategy[8]. Following this, several surveillance studies were initiated (Table 2). A problem in trying to aggregate this data is that there is no agreed protocol. Therefore, studies differ substantially in the case definitions used, the antibody tests used, and levels of confirmation. In addition, many studies have been published in national journals that are not accessible for non-native speakers. Despite these limitations, the following observations can be made: there is a significant percentage of serologically positive persons in different surveys in regions with reported WNV activity (e.g. Romania, Spain, Volga delta), as well as occasional cases identified through enhanced surveillance. A study from Germany with screening for WNV of blood donors reported four WNV positive persons (confirmed by neutralisation tests), which is remarkable because no endemic WNV activity has been recorded in Germany (Pfleiderer *et al.* 2006). The authors do not mention possible travel history of these donors or otherwise comment on their findings. Surveillance studies targeting illness in humans and horses in the UK and The Netherlands did not find any evidence for WNV activity, despite a report from the UK suggesting endemic circulation of a WNV like virus (Rockx *et al.* 2006, Morgan 2006, Buckley *et al.* 2006). Overall, the studies conducted so far suggest that there is some circulation of WNV in southern and eastern regions of Europe, but data are too incomplete to draw a conclusion whether this results from sustained enzootic presence or repeated introductions.

Trend reports from the Centers for Disease Control and Prevention (CDC) surveillance programme in the USA clearly illustrate an important issue, namely that this type of surveillance is biased towards reporting of most severe cases (Figure 2). With increasing coverage of reporting and more knowledge about WNV of physicians, a decrease of the reported overall case fatality rate was observed because WNV illness was recognised also in non neurological patients. The ratio of the number of deaths and the number of neuroinvasive diseases remained more or less at the same level. Rate of illness, severity of symptoms, and prognosis of neuroinvasive disease increase steeply with age. For this reason, surveillance of neurological illness in humans in Europe has been targeted at persons over 50 years of age (Rockx *et al.* 2006, Morgan 2006). On the basis of phylogenetic analyses of genes encoding structural proteins, WNV isolates can be grouped in five distinct lineages, with most of the recent outbreaks associated with lineage I (Bondre *et al.* 2007). Viruses belonging to this lineage are neurotropic in many animal species (Ceccaldi *et al.* 2004). Experimental data show that WNV viruses differ substantially and that increased neurovirulence may be more common in lineage I viruses, suggesting that the observed differences in epidemiology result from emergence of this lineage (Beasley *et al.* 2002). In addition, the viruses have evolved following incursion into the USA, and the observed increase in the intensity of WNV transmission beginning in 2002 was associated with the emergence of a distinct phylogenetic variant. This genotype seems to have an advantage over previously recognised WNV strains because the extrinsic incubation period was shorter (Ebel *et al.* 2004). A recent study showed that the expanding viruses from lineage I have a distinct mutation. Viruses with this mutation replicate to much higher levels in American crows than other (older) WNV strains, which actually may have led to passing of a critical threshold for infection of mosquitoes (Brault *et al.* 2007). What effect this mutation has on levels of viraemia in European resident birds is presently not known but is crucial information for risk assessment.

[8] http://ec.europa.eu/health/ph_threats/com/west_nile/wnv_case_def_en.pdf (Accessed 16 September 2007).

Table 2. WNV surveillance studies conducted in the old world.

Country	Reference	Year	Surveillance target	N tested	Incidence/ prevalence	Remarks
Romania	Cernescu *et al.*2000	1997-8	Human illness	NR	15 cases	
Romania	Ceianu *et al.* 2001	1997-2000	Human illness	NR	0.95/million/year	
Czech Republic	Hubalek *et al.* 2000	1997	Human illness	NR	2.1% hospitalised	Flooding
Spain	Lozano and Filipe 1998	1997	Population 10 towns	NR	8%	Ebro delta
Russia-Volga delta	Platonov *et al.* 2001	1998	Human illness	NR	35 cases	Retrospective study
Russia-Volga delta	L'vov *et al.* 2000	1999	Human illness	NR	81 cases	
United Kingdom	Morgan 2006	2003-06	Equine illness	???	None	Syndromic surveillance
Italy	Autorino *et al.* 2002	1998	Horses	159	14 cases	
Austria	Weissenböck 2003	2000-02	Horses	350	0%	4 imported
France	Durand *et al.* 2002	2000	Horses	432	8.5% (of which 42% also IgM positive)	
France	Charrel *et al.* 2001	2000?	Human illness		None	
Russia	Vasil'ev *et al.* 2005	2001-04	Horses, cattle, camels, pigs, sheep	2,884	9.8%, 6.4%, 5.2%, 3.1%, 2.2%	Astrakhan, in Russian
France	Charrel *et al.* 2001	2000?	Blood donors		None	
Spain	Bofill 2006	2001	Population survey	992	4%	Ebro delta
Russia-Volga delta	L'vov 2004	2000-02	Mosquitos, mammals, birds		Endemic, new variant?	Russian article
Czech Republic	Hubalek *et al.* 2004	2002	Residents, flooding		1.2%, TBE?	in Szech
Netherlands	Rockx *et al.* 2006	1999-2004	Humans, horses	781 human 71 equine	None	Syndromic surveillance
Netherlands	Koppelman 2006	2004	Pooled donor blood		None	
Germany, Austria	Pfleiderer *et al.* 2006	2004	Donor blood	15365	4 cases	High % IgG
Turkey	Ozkul *et al.* 2006	2004?	Systematic sampling humans and animals	259 (equine) 88 human	13.5% horses 20.4% humans	

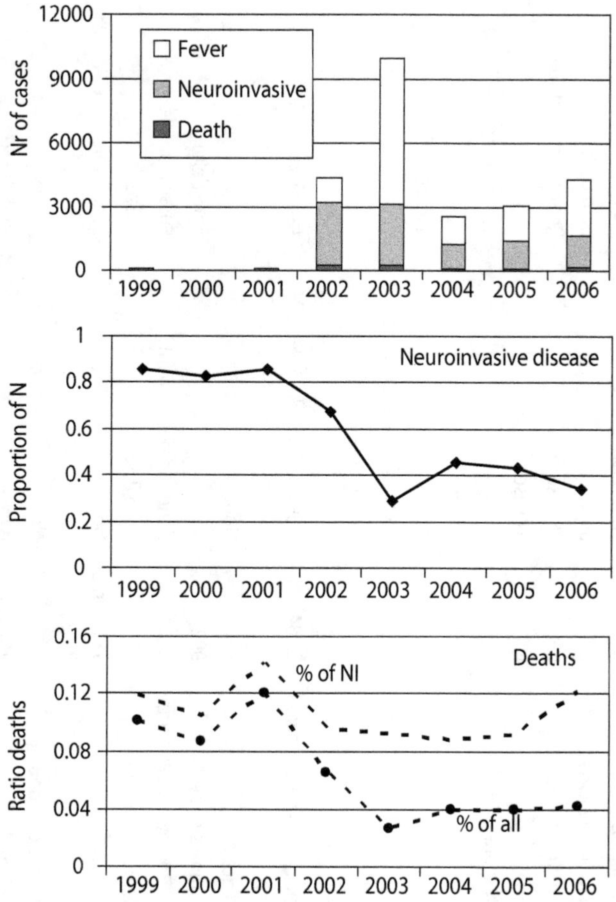

Figure 2. WNV epidemic in the USA, with (top) number of reported cases with fever, neuroinvasive illness, and death; (middle) proportion of reported cases with neuroinvasive disease and (bottom) ratio of deaths compared to all reported cases (dotted line with symbols) and of cases with neuroinvasive disease (dotted line).

West Nile virus in horses

WNV has been reported previously as a disease of horses with small outbreaks in Africa, The Middle East, Asia, and several European countries (Joubert *et al.* 1971, Cantile *et al.* 2000, Trock *et al.* 2001, Murgue *et al.* 2001a). The largest outbreak of equine neurologic disease caused by WNV started in 1999 in New York City. Since that time, more than 25,000 cases of equine West Nile encephalomyelitis have been reported in horses in the USA. The estimated case-fatality rate of 30 to 40% was also very high (Long 2007). As in humans, the detailed studies following incursion of WNV into the USA led to greatly increased understanding about the clinical symptoms, pathogenesis and diagnosis of WNV in horses.

Pathogenesis and clinical findings in horses

As in other hosts, most WNV infections in horses are asymptomatic. However, the outbreaks of WNV in the last decennium saw an increased proportion of neurological disease in both humans and horses (Marfin *et al.* 2001, Castillo-Olivares and Wood 2004). Neurological disease in horses consists of changes in mental status, signs consistent with spinal cord abnormalities, and defects in cranial nerves of the hindbrain. Pathology induced in the neurons of thalamus, medulla and pons are held responsible for changes in behaviour, ranging from aggression to somnolence and even coma. Predominant symptoms are weakness and ataxia. Lesions in the spinal cord, consisting of perivascular cuffing and gliosis, tend to increase in severity caudally, resulting in lower motor neuron disease. Involuntary skin and muscle fasciculations, tremors and hyperesthesia are also often observed in equine WNV disease, and can be explained by loss of fine motor control after neuron destruction in the basal ganglia. WNV reaches the equine central nervous system either by the vascular route or the neuronal route (through transaxonal transmission). Evidence indicates, that toll-like receptors are crucial for entry of WNV into the central nervous system (Tesh *et al.* 2005, Wang *et al.* 2004).

The most commonly reported presenting signs include ataxia, abnormal gait, weakness, muscle fasciculations, depression and recumbency (Cantile *et al.* 2000, Murgue *et al.* 2001b, Porter *et al.* 2003, Salazar *et al.* 2004, Ward *et al.* 2004, 2006). Also signs like circling, hypersensitivity to noise and hyperexcitability have been reported (Morales *et al.* 2006). The reported case-fatality rate for horses with neurological illness varies between 20 and 40%, but is about 70% for recumbent horses (Porter *et al.* 2003, Salazar *et al.* 2004). In a cross-sectional study in Nebraska and Colorado, 80% of the horses that survived WNV infection recovered fully with a mean duration of disease of 22 days (Salazar *et al.* 2004). In 2002/2003 both inactivated and recombinant canarypox WNV vaccines became available in the USA, which reduced morbidity and mortality significantly (Schuler *et al.* 2004, Ward *et al.* 2006, Gardner *et al.* 2007).

Diagnosis in horses including pathologic findings

Clinical symptoms are not specific for WNV infection and a full diagnostic investigation should be pursued. Depending on the geographical and regional situation differential diagnoses include alphavirus encephalitis, rabies, equine protozoal myelo-encephalitis (EPM), equine herpesvirus type 1, botulism, besides non-infectious causes. Since viraemia in horses is low and short-lived, confirmation of WNV infection is primarily based on detection of antibodies (Kleiboeker *et al.* 2004). In case of neurological disease, the preferred test is an IgM-capture ELISA (MAC-ELISA) with an estimated sensitivity and specificity of 91 and 99%, respectively (Long *et al.* 2006). On exposure to WNV, horses develop a significant IgM response that lasts approximately six weeks. Since increases in IgM hardly occur after vaccination, this test is also feasible in endemic situations where vaccination is practised. Balasuriya *et al.* (2006) described a microsphere immunoassay incorporating recombinant WNV envelope proteins (rE) and recombinant non-structural proteins (rNS1, rNS3, and rNS5). The rE immunoassay had a high diagnostic sensitivity and specificity, but could not differentiate between vaccinated and infected horses. The rNS1 based immunoassay could differentiate between horses vaccinated with a recombinant WNV vaccine on one hand and infected horses or horses vaccinated with a whole virus inactivated vaccine on the other hand. The diagnostic sensitivity of the latter assay was poor, however. Cross-reactions with closely related flaviviruses have been observed. For detection of IgG or total Ig also indirect, competitive and blocking ELISAs have been described (Blitvich *et al.* 2003, Choi *et al.* 2007). The virus neutralisation test or plaque reduction neutralisation test (PRNT) is considered the gold standard for detection of WNV specific antibodies.

Pathologic features consist of mild to severe polioencephalomyelitis with constant involvement of the ventral horns of the thoracic and lumbar spinal cord. Perivascular cuffing, consisting predominantly of T-lymphocytes, and multifocal gliosis are commonly observed, whereas neuronophagia is occasionally seen. Perivascular haemorrhages can be observed in about 50% of the horses, even macroscopically sometimes. WNV infection in horses is characterised by lesions with little associated antigen (as demonstrated by immunohistochemistry) when compared with WNV infection in birds and some fatal human cases (Cantile *et al.* 2000, 2001). Although the diagnostic sensitivity of RT-PCR is low on whole blood collected from clinically affected horses (Kleiboeker *et al.* 2004, Garcia-Tapia *et al.* 2006), PCR is a good diagnostic method for confirmation of WNV in either fresh or formalin-fixed neural tissues with the medulla as a predilection site for sampling (Kleiboeker *et al.* 2004).

Surveillance in horses

Horses are considered accidental end-hosts for WNV strains belonging to lineage I and II. From 1996 onwards several human and equine outbreaks of encephalitis have occurred in Europe, North-Africa and the Americas, all caused by lineage I strains (Tables 1 and 3). In southern Africa Lineage II strains, however, are not associated with equine neurologic disease (Guthrie *et al.*

Table 3. Case/outbreak reports of WNV in horses in Europe and the Mediterranean basin.

Country	Reference	Year	N	Method	Remarks
France	Lapras *et al.* 1968	1968	??	Serology	
Morocco	Murgue *et al.* 2001	1996	94	Serology	CFR 45%
Italy	Cantile *et al.* 2000, 2001, Murgue *et al.* 2001	1998	14		CFR 43%
Israel	Murgue *et al.* 2000, Steinman *et al.* 2002	1998	76	Serology/virus isolation	9/76 horses with neurological disorders laboratory confirmed
Austria	Weissenböck *et al.* 2003	2000-02	12	PCR/IHC	Unresolved encephalomyelitis, all cases WNV negative
France	Murgue *et al.* 2001	2000	131	Serology	71/131 horses with neurological disorders laboratory confirmed
Morocco	Schuffenecker *et al.* 2005	2003	9	PCR/virus isolation	5/9 (56%) euthanised

2003). Also for lineage I strains, clinical expression seems to be variable and dependent on yet unknown virulence factors and host factors. Also, after experimental challenge with neurotropic WNV lineage I strains, most horses did not develop neurological disease. (Bunning *et al.* 2002, Siger *et al.* 2006, Bowen and Nemeth 2007). In the last decade results of syndromic surveillance in several countries have been reported, e.g. Israel, USA, Canada, UK, France, Italy, Romania, Austria, Tunisia, Morocco and Argentina (Murgue *et al.* 2001a, Weissenböck *et al.* 2003, Abutarbush *et al.* 2004, Schuffenecker *et al.* 2005, Morales *et al.* 2006, Morgan 2006, Corrigan *et al.* 2006, Rockx *et al.* 2006, Gardner *et al.* 2007) (Tables 2 and 3).

Syndromic surveillance is an efficient tool only under the presumption that a WNV introduction in an immunologically naïve equine population will induce neurological symptoms in a significant proportion of infected individuals. Gardner *et al.* (2007) demonstrated in a prospective cohort study, that 5.4% of a cohort of unvaccinated ponies developed clinical neurologic disease, whereas 68% developed antibodies. Results of a serosurvey in France (Durand *et al.* 2002, 2005) showed, after an outbreak with 76 equine clinical cases in 2000, that 8.5% (n=432) of horses were seropositive within a 10-km radius of laboratory-confirmed cases. In another study in 2003 only four equine clinical cases were found in another department, whereas within a 25-km radius 328/906 horses (36%) were seropositive for WNV indicating widespread virus circulation. Also a wide range of seroprevalences (1-20%) were reported recently in Croatia, Romania, Russia, Turkey, Latin America and the Caribbean (Lorono-Pino *et al.* 2003, Vasil'ev *et al.* 2005, Lefrancois *et al.* 2005, Pupo *et al.* 2006, Komar and Clark 2006, Ozkul *et al.* 2006, Morales-Betoulle *et al.* 2006) accompanied by no or sparse reports of equine disease. These studies illustrate the need for standardisation of serological methods and their interpretation. The low rate of clinically apparent infections shows the need for systematic surveillance that includes population-based serological screening to get information on true prevalence of infection. Surveillance should include efforts to identify the exact viral strains involved, to help understand the epidemiology of WNV at the molecular level.

WNV vectors

Ticks and mosquitoes

In nature WNV has been isolated from a wide range of arthropod vectors including 11 genera of mosquitoes (Diptera: Culicidae) (Reusken 2006), six genera of ticks (Acari) (Higgs *et al.* 2004) and one biting midge species (Diptera: Ceratopogonidae) (Naugle *et al.* 2004) (Table 4). WNV has been isolated from both Ixodid and Argasid tick species in Europe, Asia and Africa. Laboratory studies have demonstrated WNV transmission by a number of tick species (Lawrie *et al.* 2004). However, the role of ticks as vector in WNV epidemiology is not yet clear. Mosquitoes are considered the principal vectors for WNV transmission and until now the presence of WNV in the field has been demonstrated in mosquito species covering North America, Africa, Europe and Asia. In Europe WNV has been detected in six different mosquito species in France, Portugal, Israel, Moldavia, Romania, Slovenia, Tsjech republic, Ukraine and Belarus. Identified vectors are *Culex (Cx). pipiens*, *Cx. modestus* Ficalbi, *Cx. univittatus* Theobald, *Coquillettidia richiardii* (Ficalbi), *Ochlerotatus cantans* (Meigen) and *Anopheles maculipennis* Meigen (Hubalek and Halouzka 1996, Hannoun *et al.* 1964, Savage *et al.* 1999, Labuda *et al.* 1974, Filipe 1972). The ornithophilic species *Cx. pipiens* is the most commonly identified vector species for WNV both in Europe and the USA. In Africa *Cx. univittatus* seems to be the most important bridge vector. The basic pattern of transmission is that of sustained cycles between reservoir hosts and vector species. For the discussion about potential endemic transmission

Table 4. WNV isolations from vectors in field studies in Europe. Tick species are indicated in bold, mosquito species in regular font.

Country	Reference	Year	Vector species
France	Hannoun *et al.* 1964	1964	*Culex modestus*
Russia-Volga delta	Shalunova *et al.* 1968	1965	***Hyalomma plumbeum plumbeum***
Israel	Nir *et al.* 1968	1968	*Anopheles coustani, Culex molestus, Culex univittatus*
Russia Volga delta	Butenko *et al.* 1968	1968	***Hyalomma marginatum***
Russia	Chumakov *et al.* 1968	1968	***Hyalomma plumbeum plumbeum***
France	Mouchet *et al.* 1970	1970	*Culex modestus*
Russia Volga-delta	Berezin *et al.* 1971	1971	*Culex pipiens, Culex modestus*
Russia Volga delta	Berezin *et al.* 1972	1972	*Culex univittatus, C. vishnui, C. quinquefasciatus, Coquillettidia richiardii*
Portugal	Filipe 1972	1972	*Anopheles maculipennis*
Slovenia	Labuda *et al.* 1974	1974	*Ochlerotatus cantans*
Moldavia	Chumakov *et al.* 1974	1974	Tick spp.
Ukraine	Vinograd and Obukhova 1974	1974	*Ochlerotatus caspius, Ochlerotatus excrusians, Ochlerotatus cantans, Anopheles maculipennis, Aedes excrusians*
Bulgaria	Katsarov *et al.* 1980	1980	*Culex pipiens, Coquillettidia richiardii*
Belarus	Samoilova *et al.* 2003	1985-1999	*Anopheles sp., Aedes sp.*
Israel	Samina *et al.* 1986	1986	*Culex perexiguus, Culex pipiens*
Romania	Savage *et al.* 1999, Tsai *et al.* 1998	1996	*Culex pipiens*
Czech Republic	Hubalek *et al.* 1998, Hubalek *et al.* 2000	1997-1998	*Culex pipiens*
Russia-Volga delta	L'vov *et al.* 2004	2000-02	Mosquito and tick spp.?
Russia-Volga delta	L'vov *et al.* 2002	2001	***Hyalomma marginatum***
Russia-Volga delta	Fyodorova *et al.* 2006	2003	*Culex modestus, Culex pipiens*
Portugal	Estevez *et al.* 2005	2004	*Culex pipiens, Culex univittatus*
Israel	Mumcuoglu *et al.* 2005	2005	***Argas arboreus, Dermanyssus gallinae, Ornithonyssus sylvarium, Rhipicephalus turanicus***
Armenia	Manukian *et al.* 2006	?	*Anopheles maculipennis*, ***Dermacentor marginatus***

of WNV in Europe with its moderate climate zones, other modes of transmission may be necessary. Vertical and nonviraemic transmission of WNV has been demonstrated for several mosquito species. However the role of these types of transmission in WNV epidemiology is unclear (Baqar *et al.* 1993, Goddard *et al.* 2003, Mishra and Mourya 2001, Miller *et al.* 2000, Higgs *et al.* 2005).

Mosquitoes and WNV establishment

The main route of introduction of WNV in the regions of Europe with a temperate climate is probably by infected migrating birds from Africa (Hubalek and Halouzka 1999, Rappole *et al.* 2000). For efficient local transmission, mosquito species are needed that are capable of sustaining and transmitting the virus among the indigenous bird and wildlife population and humans (Higgs *et al.*

2004). The role of a specific mosquito species in WNV transmission in a specific area is influenced by several factors that will be discussed below:

1. *Vector competence (mosquito genetics)*. WNV has been isolated in the field from more than 100 different mosquito species, including species that are widely distributed in Europe (e.g. twelve potential species in the UK and nine in The Netherlands (Medlock *et al.* 2005, Higgs *et al.* 2004, Reusken 2006). Therefore, the presence of 'the right' mosquito species is generally not considered to be a limiting factor for virus establishment in Europe. However, vector competence is determined by intrinsic (genetic) mosquito factors, which may vary not only between mosquito species but also between different strains (isolates from different areas) (Hayes *et al.* 1984). Therefore data are needed to determine the competence and capacity of mosquito species isolated from different areas in Europe, even if these species have been implicated in WNV transmission elsewhere (USA) and even if WNV has been isolated in field-collected individuals in the area of interest (Hardy 1988). These studies should include both lineage I and II strains of WNV since there is evidence that both lineages circulate in Europe (Bakonyi *et al.* 2006) and both lineages might respond differently.
2. *Mosquito population dynamics*. The abundance and distribution of competent vector species are crucial determinants for successful spread of the virus upon introduction. This information is sparse in Europe, in particular longitudinal data sets are seldomly reported.
3 *Climate*. Arboviruses and their vectors are very sensitive to climate (Rogers and Randolph 2003). It has been predicted that the observed climate change ('global warming') will increase the incidence and intensity of arbovirus transmission (Cook 1992, Martens and Moser 2001, IPCC 2001). The average global temperature is predicted to increase between 1.1 and 6.4 °C until 2100, in comparison to 1990. Effects are expected to be most pronounced for night-time and winter temperatures, which are likely to enable mosquitoes to transmit arboviruses for longer periods and to expand their activities to new areas. The introduction and spread of exotic mosquito species like *Aedes albopictus* (Skuse) and *Ochlerotatus japonica* (Theobald) in Europe illustrate the potential for the succesful establishement of exotic species in Europe which will largely be temperature-dependent (Gratz 2004, Schaffner *et al.* 2003).
   The average annual precipitation and number of rain days are expected to increase in northern Europe and decrease in southern Europe. Intensified rainfall and floods could trigger disease outbreaks by creating vector breeding grounds (IPCC 2007). WNV amplifications are thought to occur under the climate conditions of warm winters followed by hot, dry summers (Paz 2006). Meteorological conditions limit mosquito activity to seasons and regions where temperatures stay above certain minimum threshold levels. Within their survivable range of temperatures, mosquitoes proliferate faster and have an increasing biting activity when temperature increases. Furthermore, the viral replication rates increase with increasing temperatures (Epstein 2001, Sutherst 2004). Therefore, the extrinsic incubation period (EIP, time required for the virus to replicate and disseminate to the salivary glands upon intake by the vector) is temperature dependent. For WNV lineages I and II the thermal minimum for females to become infected and eventually transmit WNV is 14 °C for a sufficient amount of time: the EIP is estimated to require 109 degree-days (sum of degrees exceeding 14 °C per day for a given period of time) (Reisen *et al.* 2006, Cornel *et al.* 1993). Several studies suggest that the influence of environmental temperature might differ between lineages I and II WNV (Lanciotti *et al.* 2002, Reisen *et al.* 2006). Reisen *et al.* (2006) showed that WNV lineage I strain NY-99 required warmer incubation periods within a mosquito host for efficient replication and transmission than lineage II strain WNV-SA.
4. *Feeding behaviour*. Ornithophilic mosquito species (feeding preferentially on avian species) are important for the maintenance of the virus in nature through the enzootic bird-mosquito-

bird cycle (Figure 1). For transmission to humans, vectors are needed that have a more opportunistic feeding behaviour and therefore can serve as bridge vectors. Intrinsic (genetic) mosquito factors influence mosquito feeding behaviour and strains from different areas differ genetically as was illustrated for the *Cx. pipiens* complex. In Europe (genetic) hybrids between a human-biter form and an bird-biter form are practically non-existent (only at a very low level in southernmost Europe) while in the USA a large proportion of individuals of this complex represent such hybrids. These hybrids are considered efficient bridge vectors that readily transmit the virus both in the sylvatic and urban cycle (Fonseca *et al.* 2004). However, under local circumstances the actual feeding preferences will be influenced by the availability of the various hosts. Kilpatrick *et al.* (2006) showed that preferential feeding of *Cx. pipiens* and *Cx. tarsalis* Coquillet shifted from birds to humans in late summer/early fall. These shifts coincided with the dispersal of their preferred avian hosts and triggered the rise in human WNV infections in the USA. As WNV in Europe is most likely introduced by bird migration, ornithophilic mosquito species will probably be the most relevant species for WNV establishment in an enzootic cycle in Europe (Higgs *et al.* 2004, Rappole *et al.* 2000). Fonseca *et al.* (2004) suggested that the ubiquitous presence of *Cx. pipiens* hybrids between human-biters and bird-biters in the USA might have contributed to the unprecedented severity of the WNV epidemic in North America and that the arrival of such hybrids in northern Europe might drastically change the WNV dynamics in Europe.

Surveillance

In known infected areas, mosquito-based surveillance enables monitoring of virus presence and especially of virus infection rates. In an intensive surveillance programme the detection of WNV in mosquitoes can precede the detection of virus activity by surveillance based on human and bird illness reports[9]. This is particularly true for Europe, where outbreaks of WNV have rarely resulted in bird mortality.

Geographical Information System (GIS) analyses have been used in Canada and the USA for real-time monitoring of WNV activity and to identify regions at high risk for WNV introduction, outbreaks and enzootic circulation (Theophilides *et al.* 2003, Gosselin *et al.* 2005, Tachiiri *et al.* 2006). In the UK GIS-based monitoring has been used to simulate the potential life cycle of new vector species (*Ae. albopictus*) in specific areas to dentifying the target regions for surveillance for introduction of new vector species (Medlock *et al.* 2006).

Role of birds in the epidemiology and ecology of WNV

Lessons from Africa and the Middle-East

Studies of WNV in birds have historically been linked closely to reports of illness in humans. Following the first isolation of West Nile viruses from humans in Egypt, the role of birds in their ecology became apparent from serological studies showing the presence of antibodies in blood of house sparrows (*Passer domesticus*), hooded crows (*Corvus corone*), buff-backed herons (*Bubulcus ibis*), palm doves (*Streptopelia senegalensis*), kestrels (*Falco tinnunculus*), pigeons (*Columbia livia*), chickens (*Gallus domesticus*), ducks (*Anas platyrhynchos*), and geese (*Anser anser*) (Hurlbut *et al.* 1956, Work *et al.* 1953, 1955) collected mainly in the upper Nile Delta area (Table 5). No epizootics

[9] http://www.cdc.gov/ncidod/dvbid/westnile/resources/wnv-guidelines-aug-2003.pdf (Accessed 16 September 2007).

Table 5. Bird species in the Old World and Europe for which evidence of WNV infection exist. Species from which WNV was isolated under experimental conditions are indicated in bold. For details see text.

Family	Common name	Latin name
Accipitridae	Black kite	Milvus migrans
Accipitridae	Common buzzard	Buteo buteo
Anatidae	Grey leg goose	Anser anser
Anatidae	Eurasian wigeon	Anas penelope
Anatidae	Mallard	Anas platyrhynchos
Anatidae	Gargeney	Anas querquedula
Ardeidae	Night heron	Nycticorax nycticorax
Ardeidae	Cattle egret	Bubulcus ibis
Ardeidae	Little egret	Egretta garzetta
Ardeidae	Grey heron	Ardea cinerea
Ardeidae	Purple heron	Ardea purpurea
Charadriidae	Kentish plover	Charadrius alexandrinus
Charadriidae	Lapwing	Vanellus vanellus
Columbidae	Collared dove	Streptopelia decaocto
Columbidae	Turtle dove	Streptopelia turtur
Columbidae	Rock dove	Columba livia
Corvidae	Common magpie	Pica pica
Corvidae	Eurasian jay	Garrulus glandarius
Corvidae	Carrion crow	Corvus corone
Emberizidae	Reed bunting	Emberiza schoeniculus
Falconidae	Common kestrel	Falco tinnunculus
Falconidae	Red-footed falcon	Falco vespertinus
Fringillidae	Goldfinch	Carduelis carduelis
Hirundinidae	Sand martin	Riparia riparia
Hirundinidae	Barn swallow	Hirundo rustica
Hirundinidae	House martin	Delichon urbica
Laridae	Yellow-legged gull	Larus cachinnans
Laridae	Little gull	Larus minutus
Laridae	Black-headed gull	Larus ridibundus
Motacillidae	Meadow pipit	Anthus pratensis
Motacillidae	Pied wagtail	Motacilla alba
Paridae	Great tit	Parus major
Paridae	Blue tit	Parus caeruleus
Passeridae	Tree sparrow	Passer montanus
Passeridae	House sparrow	Passer domesticus
Phalacrocoracidae	Cormorant	Phalacrocorax carbo
Picidae	Green woodpecker	Picus viridis
Podicipedidae	Black-necked grebe	Podiceps nigricollis
Podicipedidae	Great crested grebe	Podiceps cristatus
Rallidae	Moorhen	Gallinula chloropus
Rallidae	Common coot	Fulica atra
Recurvirostridae	Black-winged stilt	Himantopus himantopus
Recurvirostridae	Pied avocet	Recurvirostra avosetta
Remizidae	Penduline tit	Remiz pendulinus

Table 5. Continued.

Family	Common name	Latin name
Scolopacidae	Curlew sandpiper	Calidris ferruginea
Scolopacidae	Common snipe	Gallinago gallinago
Scolopacidae	Ruff	Philomachus pugnax
Sternidae	Common tern	Sterna hirundo
Sternidae	Little tern	Sterna albifrons
Sternidae	Black tern	Chlidonias nigra
Sturnidae	Common starling	Sturnus vulgaris
Sylviidae	Reed warbler	Acrocephalus scirpaceus
Sylviidae	Sedge warbler	Acrocephalus schoenobaenus
Sylviidae	Garden warbler	Sylvia borin
Sylviidae	Blackcap	Sylvia atricapilla
Timaliidae	Bearded tit	Panurus biarnicus
Turdidae	Common redstart	Phoenicurus phoenicurus
Turdidae	Eurasian robin	Erithacus rubecula
Upupidae	Hoopee	Upupa epops

were reported during the surveillance period but virus was isolated from blood of naturally infected rock pigeons and hooded-crows, indicating local transmission of the virus. Experimental infections of hooded crows, house sparrows and kestrels, using infected mosquitoes, resulted in high levels of viraemia in especially crows and house sparrows, suggesting that these bird species could serve as amplifying hosts in Egypt, producing virus titers high enough to infect *Culex* species.

The first outbreaks of WNV in humans were reported in Israel in 1951 and 1952 (Bernkopf *et al.* 1953), followed by reports of severe human cases in 1957 and 1962, all without reports of WNV epizootics among birds. Serological studies performed between 1959 and 1960 revealed high sero-prevalence of anti-flavivirus antibodies in wild birds in Israel, especially turtle doves (*Streptopelia turtur)*, but the methods used in these studies lacked specificity (Akov and Goldwasser 1966, Nir *et al.* 1969). The isolation of WNV from turtle doves in the same region indicated that WNV circulated among this bird species in Israel, but their role in the enzootic cycle remained obscure.

In 1958 the first human cases of WNV were documented in South Africa (Kokernot and McIntosh 1959). The role of birds in this WNV transmission was studied some years later between 1961 and 1965. Fifty-one species of wild birds, comprising 17 families, were investigated in this study, which showed high seroprevalence in birds belonging to the family *Ploceidae* (weavers and sparrows) (McIntosh *et al.* 1969b). Subsequent experimental infections of several members of the families *Columbidae* and *Ploceidae* confirmed their susceptibility to WNV infection (McIntosh *et al.* 1969a). The next and largest human epidemic in South Africa occurred in 1974, with thousands of confirmed clinical cases recorded, but without significant bird mortality. Taken together, the serological studies performed in Egypt and South-Africa suggested that wild birds were central in the transmission cycle of WNV and that infection with WNV was not fatal in the majority of avian species. The role of birds as amplifying hosts was elucidated by the experimental infections and virus isolation studies, showing that some species in the orders Passeriformes (crow and sparrow) and Columbiformes (pigeons) sustained significant viraemia titers.

In 1997 the first WNV epizootic among flocks of domestic geese was reported in Israel. Affected animals displayed signs of acute encephalitis resulting in high mortality rates. Epizootics among several flocks of geese were also observed in 1998, 1999 and 2000 (Bin *et al.* 2001, Swayne *et al.* 2001, Weinberger *et al.* 2001). Interestingly, in the same period of the epizootics, WNV was isolated from white storks (*Ciconia ciconia*), which were probably displaced from their normal migration route by unusual weather conditions (Malkinson *et al.* 2002), leading to the hypothesis that a more virulent strain had emerged or was introduced into the region. Sequence analyses of the isolates indeed showed that the virus isolated from storks were similar to what had caused the epizootics but different than what was isolated prior to 1997, indicating that storks may have introduced this new WNV variant in Israel. This finding suggested for the first time that migratory birds could carry WNV into previously unaffected geographical areas and play a role as amplifying hosts. Serological analyses of wild birds following the epizootics in Israel suggested that crows were not an important species in the transmission cycle, whereas storks and pigeons may have played a prominent role. In addition, the high viraemia levels obtained in experimentally infected geese suggested their possible role as amplifying hosts in the Israeli urban cycle.

Lessons from the USA

In the summer of 1999, WNV unexpectedly emerged in New York, killing thousands of native bird species pertaining to the following orders (Komar 2003): (1) Anseriformes, (2) Apodiformes, (3) Caprimulgiformes, (4) Ciconiiformes, (5) Charadriiformes, (6) Columbiformes, (7) Cuculiformes, (8) Coraciiformes, (9) Falconiformes, (10) Galliformes, (11) Gaviiformes, (12) Gruiformes, (13) Passeriformes, (14) Pelecaniformes, (15) Piciformes, (16) Podicipediformes, and (17) Strigiformes. Especially members of the family *Corvidae*, order Passeriformes (crows, jays and magpies) developed severe fatal infections. Diagnoses of most fatal infections in birds were based on positive laboratory tests conducted between 1999 and 2002. The susceptibility of most birds species for developing nonfatal WNV-associated disease as a result of natural infection is, however, not well-documented. In experimental settings members of the orders Passeriformes, Charadriiformes, Strigiformes, Columbiformes, and Falconiformes sustain high virus replication, with fatality rates ranging from 40%-100%, indicating a potential involvement of these species in the North American WNV transmission cycle. In contrast, members of the orders, Paciformes and Anseriformes did not sustain sufficiently high viraemia titers, questioning their potential to infect *Culex* mosquitoes (Komar 2003). Komar *et al.* (2003) looked at the competence index of several bird species, which was calculated as the product of bird susceptibility to infection, duration of bird infectiousness, and the proportion of mosquitoes that became infectious per day. The authors concluded that passeriform and charadriiform birds were more likely to function as amplifying hosts, as these species sustained viraemia titers greater than 10^5 PFU/ml. The threshold viraemia level that was used by Komar *et al.* (2003) was based on studies showing that 10^5 PFU/ml was infectious for 50% or more (ID_{50}) of *Cx. pipiens*, *Cx. quinquefasciatus* Say and *Cx. tarsalis* tested in laboratory studies. It is important to realise that if certain host viraemia levels results in rates of infection in mosquitoes higher than the minimum infection rate (MIR) needed to produce epizootics, then such a bird species may still be competent. To date, the MIR values for competent mosquito vectors are not known. Interestingly, in 2000 relatively low infection rate values were reported in *Cx. restuans* Theobald, *Cx. pipiens*, and *Cx. salinarius* Coquillet in Connecticut, which were still associated with intense bird epizootics, but no human outbreaks (Hadler *et al.* 2001), indicating that infection rates in mosquitoes lower than the MIR may still be sufficient to ensue and maintain WNV transmission cycle among birds. Therefore, in absence of field-derived data, the inference of host-competence potential from experimental infection must be done with caution.

The difference in susceptibility to WNV infection and disease development among several bird species in North America was shown to be age-dependent and likely to be the result of genetic differences. Also experiments conducted in Egypt and South Africa indicated that the mortality rates among young birds were significantly higher than for adult birds (Malkinson and Banet 2002, Murgue *et al.* 2002). It is intriguing to note that some North American bird species are more susceptible to fatal WNV infection than their counterparts in Africa, the Middle East and Europe. For instance, American carrion crows (*Corvus corone*) are highly susceptible to WNV infection and high fatality rates are reported, whereas no WNV-associated mortalities have been observed in Europe (France, Italy, Hungary). One explanation for this difference in susceptibility to fatal infection is that WNV strains with different levels of virulence circulate, as indicated by Brault *et al.* (2007). However, how WNV was introduced in New York City in 1999 remains a mystery. Phylogenetic analyses of the New York WNV strain indicated high sequence identity with a WNV strain isolated in Israel in 1998 (Giladi *et al.* 2001), suggesting a possible introduction from the Middle East. A possible role for migratory birds in the ecology of WNV is discussed below.

Lessons from Europe

Although several outbreaks of WNV in horses and/or humans occurred, the role of birds in these outbreaks has not consistently been investigated in Europe. The first reported epizootics of WNV in domestic horses occurred in the Camargue region in France in 1962 and 1965, and WNV strains were isolated from *Cx. modestus* as well as from the blood of individuals working in the field as early as 1964, it took much longer before ecological studies in birds were done (Hannoun *et al.* 1964). During the outbreak of severe WNV infection in Romania in 1996 (Table 6) neutralising antibodies were detected in domestic fowls (ducks, chickens, turkeys), while passeriform birds (house sparrow, wood warbler, willow warbler and song thrush) were seronegative (Savage *et al.* 1999). The limited wild bird surveillance data collected during 1997-2000 demonstrated seroprevalence rates of up to 8%, suggesting that WNV either persisted locally for several years in a poorly defined enzootic cycle with sporadic spillover to humans, or the virus was introduced yearly into Romania.

A WNV outbreak in humans was reported in 1999 and 2000 in the Volgograd region in Russia, but no mortalities were seen among horses or birds (Hubalek and Halouzka 1999, L'vov *et al.* 2000, Platonov *et al.* 2001). Furthermore, no serological surveys in birds were conducted during or after this outbreak, leaving the bird species involved in WNV transmission in Russia speculative.

In 1994 it was shown that several wild bird species were susceptible to experimental infection without showing clinical signs (Malkinson and Banet 2002). In 2000, when another epizootic occurred among horses in France, this time outside the Camargue region, serological studies in birds were initiated. These investigations showed the presence of neutralising antibodies in mallards (*Anas platyrhynchos*), yellow-legged gulls (*Larus cachinnans*) and common magpies (*Pica pica*). No mortality was detected among these bird species.

In the summer of 2003 an outbreak of WNV encephalitis was reported in a flock of domestic geese (*Anser domesticus*) in Hungary (Bakonyi *et al.* 2006, Glavits *et al.* 2005). This was the first report of

Table 6. Summary table of studies of WNV presence in birds in Europe and the Mediterranean basin.

Country	Reference	Year	Strain	Prevalence	Remarks
Czechoslovakia	Malkinson *et al.* 2002	1984-1990		5.3 %	
Belarus	Samoilova *et al.* 2003	1985-1999	Plural	6.5-16.7%	
Israel	Malkinson *et al.* 2002	1965-1966		4-23%	Data should be interpreted with caution given the small sample sizes used.
Israel	Murgue *et al.* 2002, Deubel *et al.* 2001	1996-1997		2-29%	Data should be interpreted with caution given the small sample sizes used.
Israel	Malkinson *et al.* 2002	1998			isolation from migratory storks suggests introduction by migratory birds
Romania	Savage *et al.* 1999	1996		41 (27-67) % seroprevalence in domestic fowl	(no difference between urban and rural sites)
Romania	Ceianu *et al.* 2001	1997-2000		7.8-29% of domestic fowl	
Russia (West Siberia)	Ternovoi *et al.* 2004	2002			
Russia- far east	Ternovoi *et al.* 2006	2002-4			
Russia (baraba)	Kononova *et al.* 2006	2003-4			WNV introduction. No data on circulation of WNV in resident birds
Russia-Volga delta	Lvov *et al.* 2004	2000-02	I	Present	
UK	Buckley *et al.* 2003	2001		???	
Austria	Weissenböck *et al.* 2003	?	None		This study concluded that WNV did not circulate in Austria
UK	Buckley *et al.* 2006	2005		10-30%	Sentinel chickens (n=20), sequential study from D0
Hungary	Clavitz *et al.* 2005	2004-5?	L1, L2, new strain		
Germany	Hlinak *et al.* 2006	2005?		No isolations	
Turkey	Ozkul *et al.* 2006	2005?			
Poland	Murgue *et al.* 2002, Malkinson *et al.* 2002	1995-1996		3-12%	

a WNV epizootic in birds in Europe, coinciding with human cases. Phylogenetic analysis showed that the Hungarian 2003 strain was closely related to strains isolated in Israel in 1998 and in New York in 1999, both places where epizootics in birds also occurred, suggesting that the increased virulence properties of this strain for American crows may also apply for other bird species (Brault *et al.* 2007). In 2004 WNV was detected in a dead goshawk (*Accipiter gentiles*) in a national park located in South-East Hungary. The same WNV strain re-emerged in 2005, suggesting that the virus has become established in the region. Phylogenetic analysis of the 2004 strain indicated that these strains belonged to lineage II of WNV, suggesting that a virulent lineage II WNV is emerging in Europe. It is noteworthy that a putative new WNV lineage (Rabensburg) has also been isolated from *Cx. pipiens* mosquitoes, which were collected on the border of the Czech Republic and Austria, only a few hundred kilometers away from where the Hungarian-2003 epizootic took place, suggesting that several strains of WNV may be circulating in the region. More studies among resident and migratory birds are needed to increase our understanding of the risks that migratory birds bring in introducing new viruses to Europe.

Role of migratory birds in WNV emergence

It is noteworthy that WNV outbreaks are often focused in or near wetlands where high concentrations of birds come into contact with large numbers of mosquitoes. However, outbreaks have also been observed in dryland areas, indicating the existence of at least two transmission cycles that can exist independently. The principal cycle appears to be sylvatic (wetland), where WNV circulates and is amplified in wild birds that share their habitat with ornithophilic mosquitoes (Figure 1). The urban (dryland) cycle of WNV is probably maintained by domestic birds. However, bird species with a peridomestic habitat (e.g. house sparrows) may carry the virus from a close wetland area into an urban area, allowing transmission of the virus to susceptible hosts by bridge vectors. It is worth noting that the house sparrow has been suspected to play a role in the transmission cycle of WNV in Egypt, but also in the epidemiology of Saint Louis encephalitis virus in North America (Gruwell *et al.* 2000, McLean *et al.* 1983). The importance of a particular bird species for maintaining a local transmission cycle is likely to vary geographically depending on the diversity of the bird, mosquito, virus populations, as well has human habits.

Migratory birds have been suspected as introductory hosts of WNV in Israel, France, North America, and Hungary. The role of migratory birds in infectious disease is exemplified by the fact that Western Equine Encephalitis (WEE) and Eastern Equine Encephalitis (EEE), ecological relatives of WNV, as well as influenza A viruses, have been isolated from several species of migratory birds (Duan *et al.* 2007, Gerhold *et al.* 2007, Hubalek 2004, Jourdain *et al.* 2007), a scenario that could explain the EEE virus epidemic in Jamaica in 1962 (Hart *et al.* 1964) and the H7N7 influenza epizootics in poultry in the Netherlands (Fouchier *et al.* 2004). Wild birds which maintain high and long-term viraemia could allow transmission of WNV to competent mosquitoes and spring migrations could be essential for the introduction of WNV into new locations. Paradoxically, experimental infections in several bird species indicated that viraemic levels remain high only for a short duration (maximum 7 days) making introduction of WNV by long distance migrators an unlikely event. It cannot be excluded, however, that host-virus interaction might be modified because of stress associated with migration, thereby lengthening the duration of viraemia. It is believed for instance that migrating storks introduced a virulent WNV strain in Israel in 1998 (Malkinson *et al.* 2002, 2001, Zeller and Schuffenecker 2004). Follow-up studies indicated that storks had probably hatched in Central Europe during their autumn migration southwards, where they were probably exposed to WNV. The wetlands of southeastern Hungary are breeding habitats for storks and many other wild bird species and may represent a complex of biotopes that enables establishment of the sylvatic cycle. Still, for migrating storks to introduce the virus into Israel, the virus had to pass to other hosts along their migration route, thereby assuming that the storks were sufficiently viraemic to infect mosquitoes. It must be emphasised that although antibodies to WNV have been found in the blood of many migratory bird species, these studies must be interpreted with caution, since cross-reactivity with other flaviviruses cannot be ruled out and most studies were based on small sample sizes that made reliable detection of low prevalence sero-positivity impossible. More systematic studies are needed to elucidate the sources of WNV incursion into Europe, including the role of (migratory) birds.

Conclusions

The recent changes in WNV epidemiology with incursions into the USA, Canada and South America have raised concerns that this might happen in Europe as well. Surveillance studies done since the late 1990s have shown that WNV is present in several regions of Europe, particularly, but not exclusively, in wetland areas. Whether or not birds arriving from Central Europe can establish a sylvatic cycle resulting in an outbreak in Western Europe will depend on a complex interplay between birds, mosquitoes and viruses, as well as local climate and changes thereof. Therefore, monitoring of WNV presence in areas with occasional illness reports should continue but a more coordinated effort is needed to provide a reliable risk map of WNV for Europe that can be used for public health preparedness. This includes investment in research capacity to study vector competence and feeding behaviour in the European setting.

References

Abutarbush SM, O'Connor BP, Clark C, Sampieri F and Naylor JM (2004) Clinical West Nile virus infection in 2 horses in western Canada. Can Vet J 45: 315-317.

Akov Y and Goldwasser R (1966) Prevalence of antibodies to arboviruses in various animals in Israel. Bull World Health Organ 34: 901-9.

Asnis D, Conetta R, Waldman Q *et al.* (1999) Outbreak of West Nile-like viral encephalitis: New York. MMWR Morb Mortal Wkly Rep 48: 845–849.

Autorino GL, Battisti A, Deubel V, Ferrari G, Forletta R, Giovannini A, Lelli R, Murri S, Scicluna MT (2002) West Nile virus epidemic in horses, Tuscany region, Italy. Emerg Infect Dis 8: 1372-1378.

Baqar S, Hayes CG, Murphy JR and Watts DM (1993) Vertical transmission of West Nile virus by *Culex* and *Aedes* species mosquitoes. Am J Trop Med Hyg 48: 757-762.

Bakonyi T, Ivanics E, Erdelyi K, Ursu K, Ferenczi E, Weissenböck H and Nowotny N (2006) Lineage 1 and 2 strains of encephalitic West Nile virus, central Europe. Emerg Infect Dis 12: 618-23.

Balasuriya UB, Shi PY, Wong SJ, Demarest VL, Gardner IA, Hullinger PJ, Ferraro GL, Boone JD, De Cino CL, Glaser AL, Renshaw RW, Ledizet M, Koski RA and MacLachlan NJ (2006) Detection of antibodies to West Nile virus in equine sera using microsphere immunoassay. J Vet Diagn Invest 18: 392-395.

Beasley D, Li L, Suderman M and Barrett A (2002) Mouse Neuroinvasive Phenotype of West Nile Virus Strains Varies Depending upon Virus Genotype. Virology 296: 17-23.

Berezin VV, Chumakov MP, Semenov BF, Reshetnikov IA, Hannoun C, Corniu B and Mouchet J (1971) Investigation on the ecology of mosquito-borne viruses using sentinel animals in the Volga delta. Vopr Virusol 16: 739-745.

Berezin VV, Semenov BF, Reshetnikov IA and Bashkirtsev VN (1972) Importance of birds in natural cycle of mosquito-borne viruses in the Volga delta. In: Cherepanov AI (ed.): Transcontinental Connections of Migratory Birds and their Role in the Distribution of Arboviruses. Nauka, Novosibirsk: 310-313.

Bernkopf H, Levine S and Nerson R (1953) Isolation of West Nile virus in Israel. J Infect Dis 93: 207-218.

Bin H, Grossman Z, Pokamunski S, Malkinson M, Weiss L, Duvdevani P, Banet C, Weisman Y, Annis E, Gandaku D, Yahalom V, Hindyieh M, Shulman L and Mendelson E (2001) West Nile fever in Israel 1999-2000: from geese to humans. Ann N Y Acad Sci 951: 127-142.

Blitvich BJ, Bowen RA, Marlenee NL, Hall RA, Bunning ML and Beaty BJ (2003) Epitope-blocking enzyme-linked immunosorbent assays for detection of west nile virus antibodies in domestic mammals. J Clin Microbiol 41: 2676-2679.

Bofill D, Domingo C, Cardenosa N, Zaragoza J, de Ory F, Minguell S, Sanchez-Seco MP, Dominguez A and Tenorio A (2006) Human West Nile virus infection, Catalonia, Spain. Emerg Infect Dis 12: 1163-1164.

Bondre V, Jadi R, Mishra A, Yergolkar A and Arankalle V (2007) West Nile virus isolates from India: evidence for a distinct genetic lineage. J Gen Virol 88: 875-884.

Bowen RA and Nemeth NM (2007) Experimental infections with West Nile virus. Curr Opin Infect Dis 20: 293-297.

Brault AC, Huang CY, Langevin SA, Kinney RM, Bowen RA, Ramey WN, Panella NA, Holmes EC, Powers AM and Miller BR (2007) A single positively selected West Nile viral mutation confers increased virogenesis in American crows. Nat Genet 39:1162-1166.

Briese T, Jia X, Huang C, Grady L and Lipkin WL (1999) Identification of a Kunjin/West Nile-like flavivirus in brains of patients with New York encephalitis. Lancet 354: 1261-1262.

Buckley A, Dawson A, Moss S, Hinsley S, Bellamy P and Gould E (2003) Serological evidence of West Nile virus, Usutu virus and Sindbis virus infection of birds in the UK. J Gen Virol 84: 2807-2817.

Buckley A, Dawson A and Gould EA (2006) Detection of seroconversion to West Nile virus, Usutu virus and Sindbis virus in UK sentinel chickens. Virol J 3: 71.

Bunning ML, Bowen RA, Cropp CB, Sullivan KG, Davis BS, Komar N, Godsey MS, Baker D, Hettler DL, Holmes DA, Biggerstaff BJ and Mitchell CJ (2002) Experimental infection of horses with West Nile virus. Emerg Infect Dis 8: 380-386.

Butenko AM, Chumanov MP, Bashkirtsev VN, Tkachenko EA, Rubin SG and Stolbov DN (1968) New investigations of west Nile virus infections in the U.S.S.R.-Astrakhan region. MaterXV Nauch Sess Inst Poliom Virus Encefal (Moskva) 3: 175-176.

Campbell GL, Ceianu CS and Savage HM (2001) Epidemic West Nile encephalitis in Romania: waiting for history to repeat itself. Ann NY Acad Sci 951: 94-101

Cantile C, Di Guardo G, Eleni C and Arispici M (2000). Clinical and neuropathological features of West Nile virus equine encephalomyelitis in Italy. Equine Vet J 32: 31-35.

Cantile C, Del Piero F, Di Guardo G and Arispici M (2001) Pathologic and immunohistochemical findings in naturally occurring West Nile virus infection in horses. Vet Pathol 38: 414-421.

Castillo-Olivares J and Wood J (2004) West Nile virus infection of horses. Vet Res 35: 467-483.

Ceccaldi P, Lucas M and Despres P (2004) New insights on the neuropathogenicity of West Nile virus. FEMS Microbiol Lett 233:1-6.

Ceianu C, Ungureanu A, Nicolescu G, Cernescu C, Nitescu L, Tardei G, Petrescu A, Pitigoi D, Martin D, Ciulacu-Purcarea V, Vladimirescu A and Savage HM (2001) West nile virus surveillance in Romania: 1997-2000. Viral Immunol 14: 251-262.

Cernescu C, Ruta SM, Tardei G, Grancea C, Moldoveanu L, Spulbar E and Tsai T (1997) A high number of severe neurologic clinical forms during an epidemic of West Nile virus infection. Rom J Virol 48: 13-25.

Cernescu C, Nedelcu NI, Tardei G, Ruta S, Tsai TF (2000) Continued transmission of West Nile virus to humans in southeastern Romania, 1997-1998. J Infect Dis 181: 710-712.

Charrel RN, Durand JP, de Lamballerie X and Drancourt M (2001) Absence d'argument serologique pour l'implication du virus West Nile dans les meningo-encephalites du sud-est de la France. Presse Med 30: 534.

Choi KS, Ko YJ, Nah JJ, Kim YJ, Kang SY, Yoon KJ and Joo YS (2007) Monoclonal antibody-based competitive enzyme-linked immunosorbent assay for detecting and quantifying West Nile virus-neutralizing antibodies in horse sera. Clin Vaccine Immunol 14: 134-138.

Chumakov MP, Belyaeva AN, Butenko AM and Martyanova LI (1968) West Nile virus in the USSR.1. Isolation of a strain of west Nile virus from *Hyalomma plumbeum plumbeum* Panz. ticks. Trudy Inst Poliom Virus Encef (Moscow) 12: 365-373.

Clavits R, Ferenczi E, Ivanics E, Bakonyi T, Mato T, Zarka P and Palya V (2005) Co-occurrence of West Nile Fever and circovirus infection in a goose flock in Hungary. Avian Pathol 34: 408-414.

Chumakov MP, Bashkirtsev VN and Golger EI (1974) Isolation and identification of Crimean Hemorrhagic fever and West Nile fever viruses from ticks collected in Moldavia. Med Virusol 22: 45-49.

Connell J, McKeown P, Garvey P, Cotter S, Conway A, O'Flanagan DP, O'Herlihy B, Morgan D, Nicoll A and Lloyd G (2004) Two linked cases of West Nile virus (WNV) acquired by Irish tourists in the Algarve, Portugal Eurosurveill Weekly 8 (32): 040805.

Cook GC (1992) Effect of global warming on the distribution of parasitic and other infectious diseases: a review. J R Soc Med 85: 688-691.

Cornel AJ, Jupp PG and Blackburn NK (1993) Environmental temperature on the vector competence of Culex univittatus (Diptera: Culicidae) for West Nile virus. J Med Entomol 30: 449-456.

Corrigan RL, Waldner C, Epp T, Wright J, Whitehead SM, Bangura H, Young E and Townsend HG (2006) Prediction of human cases of West Nile virus by equine cases, Saskatchewan, Canada, 2003. Prev Vet Med 76: 263-272.

Dauphin G, Zientara S, Zeller H and Murgue B (2005) West Nile: worldwide current situation in animals and humans. Comp Immunol Microbiol Infect Dis 28: 249-250.

Davis L, DeBiasi R, Goade D, Haaland K, Harrington J, Harnar J, Pergam S, King M, DeMasters B and Tyler K (2006) West Nile virus neuroinvasive disease. Ann Neurol 60: 286-300.

Del Giudice P, Schuffenecker I, Vandenbos F, Counillon E and Zeller H (2004) Human West Nile virus, France. Emerg Infect Dis 10: 1885-1886.

Deubel V, Fiette L, Gounon P, Drouet MT, Khun H, Huerre M, Banet C, Malkinson M and Desprès P (2001) Variations in biological features of West Nile viruses. Ann NY Acad Sci 951: 195-206.

Dohm DJ, Sardelis MR and Turell MJ (2002) Experimental vertical transmission of West Nile virus by Culex pipiens (Diptera: Culicidae). J Med Entomol 39: 640-644.

Duan L, Campitelli L, Fan XH, Leung YH, Vijaykrishna D, Zhang JX, Donatelli I, Delogu M, Li KS, Foni E, Chiapponi C, Wu WL, Kai H, Webster RG, Shortridge KF, Peiris JS, Smith GJ, Chen H and Guan Y (2007) Characterization of low pathogenic H5 subtype influenza viruses from Eurasia: Implications for the origin of highly pathogenic H5N1 viruses. J Virol 81: 7529-7539.

Durand B, Chevalier V, Pouillot R, Labie J, Marendat I, Murgue B, Zeller H and Zientara S (2002) West Nile virus outbreak in horses, southern France, 2000: results of a serosurvey. Emerg Infect Dis 8: 777-782.

Durand B, Dauphin G, Zeller H, Labie J, Schuffenecker I, Murri S, Moutou F and Zientara S (2005) Serosurvey for West Nile virus in horses in southern France. Vet Rec 157: 711-713.

Ebel GD, Carricaburu J, Young D, Bernard KA, Kramer LD (2004) Genetic and phenotypic variation of West Nile virus in New York, 2000-2003. Am J Trop Med Hyg 71:493-500

Epstein PR (2001) West Nile virus and the climate. J Urban Health 78: 367-371.

Esteves A, Almeida AP, Galao RP, Parreira R, Piedade J, Rodriguez JC, Sousa CA, Novo MT (2005) West Nile virus in Southern Portugal, 2004. Vector Borne Zoon Dis 5: 410-413.

Filipe AR (1972) Isolation in Portugal of West Nile virus from *Anopheles maculipennis* mosquitoes. Acta Virol 16: 361.

Fonseca DM, Keyghobadi N, Malcolm CA, Mehmet C, Schaffner F, Mogi M, Fleischer RC and Wilkerson RC (2004) Emerging vectors in the *Culex pipiens* complex. Science 303:1535-1538.

Fouchier RA, Schneeberger PM, Rozendaal FW, Broekman JM, Kemink SA, Munster V, Kuiken T, Rimmelzwaan GF, Schutten M, Van Doornum GJ, Koch G, Bosman A, Koopmans M and Osterhaus AD (2004) Avian influenza A virus (H7N7) associated with human conjunctivitis and a fatal case of acute respiratory distress syndrome. Proc Natl Acad Sci USA 101: 1356-61.

Fyodorova MV, Savage HM, Lopatina JV, Bulgakova TA, Ivanitsky AV, Platonova OV and Platonov AE (2006) Evaluation of potential West Nile virus vectors in Volgograd region, Russia 2003 (Diptera:Culicidae)'species composition, bloodmeal host utilization, and virus infection rates of mosquitoes. J Med Entomol 43: 552-563.

Garcia-Tapia D, Loiacono CM, Kleiboeker SB (2006) Replication of West Nile virus in equine peripheral blood mononuclear cells. Vet Immunol Immunopathol 110: 229-244.

Gardner IA, Wong SJ, Ferraro GL, Balasuriya UB, Hullinger PJ, Wilson WD, Shi PY and MacLachlan NJ (2007) Incidence and effects of West Nile virus infection in vaccinated and unvaccinated horses in California. Vet Res 38: 109-116.

Gerhold RW, Tate CM, Gibbs SE, Mead DG, Allison AB and Fischer JR (2007) Necropsy findings and arbovirus surveillance in mourning doves from the southeastern United States. J Wildl Dis 43: 129-35.

Giladi M, Metzkor-Cotter E, Martin DA, Siegman-Igra Y, Korczyn AD, Rosso R, Berger SA, Campbell GL and Lanciotti RS (2001) West Nile encephalitis in Israel, 1999: the New York connection. Emerg Infect Dis 7:659-661.

Glavits R, Ferenczi E, Ivanics E, Bakonyi T, Mato T, Zarka P and Palya V (2005). Co-occurrence of West Nile Fever and circovirus infection in a goose flock in Hungary. Avian Pathol 34: 408-414.

Goddard LB, Roth AE, Reisen WK and Scott TW (2003) Vertical transmission of West Nile Virus by three California *Culex* (Diptera: Culicidae) species. J Med Entomol 40: 743-746.

Gosselin P, Lebel G, Rivest S and Douville-Fradet M (2005) The integrated system for Public Health Monitoring of West Nile virus (ISPHM-WNV): a real-time GIS for surveillance and decision-making. Int J Health Geogr 4: 21.

Grard G, Moureau G, Charrel RN, Lemasson JJ, Gonzalez JP, Gallian P, Gritsun TS, Holmes EC, Gould EA and de Lamballerie X (2007) Genetic characterization of tick-borne flaviviruses: new insights into evolution, pathogenetic determinants and taxonomy. Virology 361: 80-92.

Gratz NG (2004) Critical review of the vector status of *Aedes albopictus*. Med Vet Entomol 18: 215-27.

Gruwell JA, Fogarty CL, Bennett SG, Challet GL, Vanderpool KS, Jozan M and Webb JP Jr (2000) Role of peridomestic birds in the transmission of St. Louis encephalitis virus in southern California. J Wildl Dis 36: 13-34.

Guthrie AJ, Howell PG, Gardner IA, Swanepoel RE, Nurton JP, Harper CK, Pardini A, Groenewald D, Visage CW, Hedges JF, Balasuriya UB, Cornel AJ and MacLachlan NJ (2003) West Nile virus infection of Thoroughbred horses in South Africa (2000-2001). Equine Vet J 35: 601-605.

Hadler J, Nelson R, McCarthy T, Andreadis T, Lis MJ, French R, Beckwith W, Mayo D, Archambault G and Cartter M (2001) West Nile virus surveillance in Connecticut in 2000: an intense epizootic without high risk for severe human disease. Emerg Infect Dis 7: 636-642.

Han L, Popovici F, Alexander J, Laurentia V, Tengelsen L, Cernescu C, Gary H, Ion-Nedelcu N, Campbell G, and Tsai T (1999) Risk factors for West Nile virus infection and meningoencephalitis, Romania, 1996. J Infect Dis 179:230-233.

Hannoun C, Panthier R, Mouchet J and Ezouan J (1964) Isolation in france of the WNV from patients and from the vector *Culex modestus* Ficalbi. C R Hebd Seances Acad Sci 259: 4170-4172.

Hardy JL (1988) Susceptibility and resistance of vector mosquitoes. In: Monath T (ed.) Arboviruses: epidemiology and ecology, vol. I, pp. 87-126.

Hart KL, Keen D and Belle EA (1964) An Outbreak of Eastern Equine Encephalomyelitis in Jamaica, West Indies. I. Description of Human Cases. Am J Trop Med Hyg 13: 331-334.

Hayes CG, Baker RH, Baqar S and Ahmed T (1984) Genetic variation for West Nile virus susceptibility in *Culex tritaeniorhynchus*. Am J Trop Med Hyg 33: 715-724.

Higgs S, Snow K and Gould E (2004) The potential for West Nile virus to establish outside of its natural range: a consideration of potential mosquito vectors in the United Kingdom. Trans R Soc Trop Med Hyg 98: 82-87.

Higgs S, Schneider BS, Vanlandingham DL, Klingler KA and Gould EA (2005) Nonviremic transmission of West Nile virus. Proc Natl Acad Sci USA 102: 8871-8874.

Hlinak A, Muhle R, Werner O, Globig A, Starick E, Schirrmeier H, Hoffmann B, Engelhardt A, Hubner D, Conraths F, Wallschlager D, Kruckenberg H and Muller T (2006) A virological survey in migrating waders and other waterfowl in one of the most important resting sites of Germany. J Vet Med B Infect Dis Vet Public Health 53: 105-110.

Hubalek Z (2004) An annotated checklist of pathogenic microorganisms associated with migratory birds. J Wildl Dis 40: 639-659.

Hubalek Z and Halouzka J (1996) Arthropod-borne viruses of vertebrates in Europe. Acta Sci Nat Brno 30: 1-95.

Hubalek Z, Halouzka J, Juricova Z and Sebesta O (1998) First isolation of mosquito-borne West Nile virus in the Czech republic. Acta Virol 42: 119-120.

Hubalek Z and Halouzka J (1999) West Nile fever-a reemerging mosquito-borne viral disease in Europe. Emerg Infect Dis 5: 643-650.

Hubalek Z, Savage HM, Halouzka J, Juricova Z, Sanogo YO and Lusk S (2000) West Nile virus investigations in South Moravia, Czechland. Viral Immunol 13: 427-433.

Hubalek Z, Zeman P, Halouzka J, Juricova Z, St'ovickova E, Balkova H, Sikutova S and Rudolf I (2004) Antibodies against mosquito-borne viruses in human population of an area of Central Bohemia affected by the flood of 2002. Epidemiol Mikrobiol Imunol 53: 112-20. [article in Czech]

Hurlbut HS, Rizk F, Taylor RM and Work TH (1956) A study of the ecology of West Nile virus in Egypt. Am J Trop Med Hyg 5: 579-620.

IPCC (2001) Climate change 2001: Synthesis report. A contribution of working groups I, II and III to the Third assessment Report of the Intergovernmental Panel on Climate Change (eds. Watson *et al.*).

IPCC (2007) Climate change 2007: Synthesis report. A contribution of working groups I, II and III to the fourth assessment Report of the Intergovernmental Panel on Climate Change (eds. Watson *et al.*).

Joubert L, Oudar J, Hannoun C and Chippaux M (1971) [Experimental reproduction of meningo-encephalomyelitis of horses with West Nile arbovirus. 3. Relations between virology, serology, and anatomo-clinical evolution. Epidemiological and prophylactic consequences]. Bull Acad Vet Fr 44: 159-167.

Jourdain E, Gauthier-Clerc M, Bicout DJ and Sabatier P (2007) Bird migration routes and risk for pathogen dispersion into western Mediterranean wetlands. Emerg Infect Dis 13: 365-72.

Kaptoul D, Viladrich PF, Domingo C, Niubo J, Martinez-Yelamos S, De-Ory F and Tenorio A (2007) West Nile virus in Spain: report of the first diagnosed case (in Spain) in a human with aseptic meningitis. Scand J Infect Dis 39: 70-71.

Katsarov G, Vasilenko S, Vargin V, Butenko S and Tkachenko E (1980) Serological studies on the distribution of some arboviruses in Bulgaria. Problems Infect Parasit Dis 8: 32-35.

Kilpatrick AM, Kramer LD, Jones MJ, Marra PP and Daszak P (2006) West Nile virus epidemics in North America are driven by shifts in mosquito feeding behavior. PLoS Biol 4: e82.

Kleiboeker SB, Loiacono CM, Rottinghaus A, Pue HL and Johnson GC (2004) Diagnosis of West Nile virus infection in horses. J Vet Diagn Invest 16: 2-10.

Kokernot RH and McIntosh BM (1959) Isolation of West Nile virus from a naturally infected human being and from a bird, Sylvietta rufescens (Vieillot). S Afr Med J 33: 987-989.

Komar N and Clark GG (2006) West Nile virus activity in Latin America and the Caribbean. Rev Panam Salud Publica 19: 112-117.

Komar N (2003) West Nile virus: epidemiology and ecology in North America. Adv Virus Res 61: 185-234.

Komar N, Langevin S, Hinten S, Nemeth N, Edwards E, Hettler D, Davis B, Bowen R and Bunning M (2003) Experimental infection of North American birds with the New York 1999 strain of West Nile virus. Emerg Infect Dis 9: 311-322.

Kononova IuV, Ternovoi V, Shchelkanov M, Protopopova E, Zolotykh S, Iurlov A, Druziaka A, Slavskii A, Shestopalov A, L'vov D and Loktev V (2006) West Nile virus genotyping among wild birds belonging to ground and tree-brush bird populations on the territories of the Baraba forest-steppe and Kulunda steppe (2003-2004). Vopr Virusol 51: 19-23.

Koppelman MH, Sjerps MS, de Waal M, Reesink HW, Cuypers HT (2006) No evidence of West Nile virus infection in Dutch blood donors. Vox Sang 90: 166-169.

Kovtunov AI, Kolobukhina LV, Moskvina TM, Shishkina EO, Dzharkenov AF, Kisteneva LB, L'vov DN, Shchelkanov MI, Aristova VA, Larichev VF, Zlobina LV, Grishanova AP, Grenkova EP, Arshba TE and Oganesian IV (2003) West Nile fever among residents of the Astrakhan region in 2002. Vopr Virusol 48: 9-11 [in Russian].

Kramer L, Li J and Shi P (2007) West Nile virus. Lancet Neurol 6:171-181.

Labuda MO, Kozuch O and Gresikova M (1974). Isolation of West Nile virus from *Aedes cantans* mosquitoes in West Slovakia. Acta Virol 18: 429-433.

Lanciotti R, Roehrig J, Deubel V, Smith J, Parker M, Steele K, Crise B, Volpe K, Crabtree M, Scherret J, Hall R, MacKenzie J, Cropp C, Panigrahy B, Ostlund E, Schmitt B, Malkinson M, Banet C, Weissman J, Komar N, Savage H, Stone W and McNamara T (1999) Origin of the West Nile virus responsible for an outbreak of encephalitis in the northeastern United States. Science 286: 2333-2337.

Lanciotti RS, Ebel GD, Deubel V, Kerst AJ, Murri S, Meyer R, Bowen M, McKinney N, Morrill WE, Crabtree MB, Kramer LD and Roehrig JT (2002) Complete genome sequences and phylogenetic analysis of West Nile virus strains isolated from the United States, Europe, and the Middle East. Virology 298: 96-105.

Lapras M, Florio R, Joubert L, Oudar J, Guillon JC and Hannoun C (1968) Electroclinical study of meningo-encephalo-myelitis in horses due to equino-human (West Nile) arbovirus isolated in Camargue. J Med Lyon 49: 1423-1441.

Lawrie CH, Uzcategui NY, Gould EA and Nuttall PA (2004) Ixodid and argasid tick species and west nile virus. Emerg Infect Dis 10: 653-657.

Lefrancois T, Blitvich BJ, Pradel J, Molia S, Vachiery N, Pallavicini G, Marlenee NL, Zientara S, Petitclerc M and Martinez D (2005) West Nile virus surveillance, Guadeloupe, 2003-2004. Emerg Infect Dis 11: 1100-1103.

Lefrere JJ, Allain JP, Prati D, Sauleda S and Reesink H (2003) West Nile virus and blood donors. Lancet 361: 2083-2084.

Long MT, Jeter W, Hernandez J, Sellon DC, Gosche D, Gillis K, Bille E and Gibbs EP (2006) Diagnostic performance of the equine IgM capture ELISA for serodiagnosis of West Nile virus infection. J Vet Intern Med 20: 608-613.

Lorono-Pino MA, Blitvich BJ, Farfan-Ale JA, Puerto FI, Blanco JM, Marlenee NL, Rosado-Paredes EP, Garcia-Rejon JE, Gubler DJ, Calisher CH and Beaty BJ (2003) Serologic evidence of West Nile virus infection in horses, Yucatan State, Mexico. Emerg Infect Dis 9: 857-859.

Lozano A and Filipe AR (1998) Anticuerpos frente a virus West Nile y otros virus transmitidos por artropodos en la poblacion del Delta Del Ebro. 1998. Rev Esp Salud Publ 72: 245-50. [in Spanish]

L'vov DK, Butenko AM, GromashevskyVL, Larichev VP, Gaidamovich SY, Vyshemirsky OI, Zhukov AN, Lazorenko VV, Salko VN, Kovtunov AI, Galimzyanov KM, Platonov AE, Morozova TN, Khutoretskaya NV, Shishkina EO and Skvortsova TM (2000) Isolation of two strains of West Nile virus during an outbreak in southern Russia, 1999. Emerg Infect Dis 6: 373-376.

L'vov DK, Dzakenov AF, L'vov DN, Aristova VA, Kovtunov AI, Gromahevskii VL, Vyshemirskii OI, Galkina IV, Al'khovskii SV, Samokhvalov EI, Prilipov AG, Deriabin PG, Odolevskii EI and Ibragimov RM (2002) Isolation of the West Nile fever virus from the great cormorant *Phalacrocotax carbo*, the crow *Corvus corone* and *Hyalomma marginatum* ticks associated with them in natural and synanthroic biocenosis in the Volga delta (Astrakhan region, 2001). Vopr Virusol 47: 7-12. [in Russian]

L'vov DK, Butenko AM, Gromashevsky VL, Kovtunov AI, Prilipov AG, Kiney R, Aristova VA, Dzharkenov AF, Samokhvalov EI, Savage HM, Shchelkanov MY, Galkina IV, Deryabin PG, Gubler DJ, Kulikova LN, Alkhovsky SK, Moskvina TM, Zlobina LV, Sadykova GK, Shalatov AG, L'vov DN, Usachev VE and Voronina AG (2004) West Nile virus and other zoonotic viruses in Russia: examples of emerging-reemerging situations. Arch Virol Suppl. 18: 85-96.

Mailles A, Dellamonica P, Zeller H, Durand JP, Zientara S, Goffette R, *et al.* (2003) Human and equine West Nile virus infections in France, August – September 2003. Eurosurveill Weekly 7: 43.

Malkinson M and Banet C (2002) The role of birds in the ecology of West Nile virus in Europe and Africa. Curr Top Microbiol Immunol 267: 309-322.

Malkinson M, Banet C, Weisman Y, Pokamunski S, King R, Drouet MT and Deubel V (2002) Introduction of West Nile virus in the Middle East by migrating white storks. Emerg Infect Dis 8: 392-397.

Malkinson M, Weisman Y, Pokamonski S, King R and Deubel V (2001) Intercontinental transmission of West Nile virus by migrating white storks. Emerg Infect Dis 7: 540.

Manukian DV, Oganesian AS, Shakhnazarian SA, Aleksanian IT (2006). The species composition of mosquitoes and ticks in Armenia. Med Parazitol (Mosk.) 1:31-33.

Marfin AA, Petersen LR, Eidson M, Miller J, Hadler J, Farello C, Werner B, Campbell GL, Layton M, Smith P, Bresnitz E, Cartter M, Scaletta J, Obiri G, Bunning M, Craven RC, Roehrig JT, Julian KG, Hinten SR and Gubler DJ (2001) Widespread West Nile virus activity, eastern United States, 2000. Emerg Infect Dis 7: 730-735.

Martens P and Moser SC (2001) Health impacts of climate change. Science 292: 1065-1066.

McIntosh BM, Dickinson DB and McGillivray GM (1969a) Ecological studies on Sindbis and West Nile viruses in South Africa. V. The response of birds to inoculation of virus. S Afr J Med Sci 34: 77-82.

McIntosh BM, Madsen W and Dickinson DB (1969b) Ecological studies on Sindbis and West Nile viruses in South Africa. VI. The antibody response of wild birds. S Afr J Med Sci 34: 83-91.

McLean RG, Mullenix J, Kerschner J and Hamm J (1983) The house sparrow (Passer domesticus) as a sentinel for St. Louis encephalitis virus. Am J Trop Med Hyg 32: 1120-1129.

Medlock JM, Snow KR and Leach S (2005) Potential transmission of West Nile virus in the British Isles: an ecological review of candidate mosquito bridge vectors. Med Vet Entomol 19: 2-21.

Medlock JM, Avenell D, Barrass I and Leach S (2006) Analysis of the potential for survival and seasonal activity of *Aedes albopictus* (Diptera: Culicidae) in the United Kingdom. J Vector Ecol 31: 292-304.

Melnick J, Paul J, Riordan J, Barnett V, Goldblum N and Zabin E (1951). Isolation from human sera in Egypt of a virus apparently identical to West Nile virus. Proc Soc Exp Biol Med 77: 661-665.

Miller BR, Nasci RS, Godsey MS, Savage HM, Lutwama JJ, Lanciotti RS and Peters CJ (2000) First field evidence for natural vertical transmission of West Nile virus in *Culex univittatus* complex mosquitoes from Rift Valley province, Kenya. Am J Trop Med Hyg 62: 240-246.

Mishra AC and Mourya DT (2001) Transovarial transmission of West Nile virus in Culex vishnui mosquito. Indian J Med Res 114: 212-214.

Morales MA, Barrandeguy M, Fabbri C, Garcia JB, Vissani A, Trono K, Gutierrez G, Pigretti S, Menchaca H, Garrido N, Taylor N, Fernandez F, Levis S and Enria D (2006) West Nile virus isolation from equines in Argentina, 2006. Emerg Infect Dis 12: 1559-1561.

Morales-Betoulle ME, Morales H, Blitvich BJ, Powers AM, Davis EA, Klein R, Cordon-Rosales C (2006) West Nile virus in horses, Guatemala. Emerg Infect Dis 12: 1038-1039.

Morgan D (2006) Control of arbovirus infections by a coordinated response: West Nile Virus in England and Wales. FEMS Immunol Med Microbiol 48:305-312.

Mostashari F, Bunning M, Kitsutani P, Singer D, Nash D, Cooper M, Katz N, Liljebjelke K, Biggerstaff B, Fine A, Layton M, Mullin S, Johnson A, Martin D, Hayes E and Campbell G. (2001) Epidemic West Nile encephalitis, New York, 1999: results of a household-based seroepidemiological survey. Lancet 358: 254-255.

Mouchet, J, Rageau J, Laumond C, Hannoun C, Beytout D, Oudar J, Corniou B and Chippaux A (1970) Epidemiology of West Nile virus : study of a focus in Camargue. V. The vector: *Culex modestus* Ficalbi (Diptera: Culicidae). Ann Inst Pasteur (Paris): 118: 839-855.

Mumcuoglu KY, Banet-Noach C, Malkinson M, Shalom U and Galun R (2005) Argasid ticks as possible vectors of West Nile virus in Israel. Vector Borne Zoon Dis 5: 65-71.

Murgue B, Murri S, Zientara S, Durand B, Durand JP and Zeller H (2001b). West Nile outbreak in horses in southern France, 2000: the return after 35 years. Emerg Infect Dis 7: 692-696.

Murgue B, Murri S, Triki H, Deubel V and Zeller HG (2001) West Nile in the Mediterranean basin: 1950-2000. Ann NY Acad Sci 951: 117-126.

Murgue B, Zeller H and Deubel V (2002) The ecology and epidemiology of West Nile virus in Africa, Europe and Asia. Curr Top Microbiol Immunol 267: 195-221.

Naugle DE, Aldridge CL and Walker BL (2004) West Nile virus: pending crisis for greater sage-grouse. Ecol lett 7: 704-813

Niedrig M, Sonnenberg K, Steinhagen K, Paweska JT (2007) Comparison of ELISA and immunoassays for measurement of IgG and IgM antibody to West Nile virus in human sera against virus neutralisation. J Virol Methods 139:103-105.

Nir Y, Lasowski Y, Avivi A and Cgoldwasser R (1969). Survey for antibodies to arboviruses in the serum of various animals in Israel during 1965-1966. Am J Trop Med Hyg 18: 416-422.

Ozkul A, Yildirim Y, Pinar D, Akcali A, Yilmaz V and Colak D (2006) Serological evidence of West Nile Virus (WNV) in mammalian species in Turkey. Epidemiol Infect 134: 826-829

Panthier R, Hannoun C, Beytout D, Mouchet J (1968) Epidémiologie du virus West Nile: étude d'un foyer en Camargue. III. Les maladies humaines. Ann Inst Pasteur (Paris) 115: 435-445. [in French]

Paz S (2006) The West Nile Virus outbreak in Israel (2000) from a new perspective: the regional impact of climate change. Int J Environ Health Res 16: 1-13.

Petersen L and Roehrig J (2001) West Nile virus: a reemerging global pathogen. Emerg Infect Dis 4:611-614.

Pfleiderer C, Konig C, Chudy M, Schmidt M, Roth WK, Nübling CM (2006) Molecular epidemiology of West Nile Virus in humans. Dev Biol (Basel) 126: 197-201, discussion 326-327.

Platonov AE, Shipulin GA, Shipulina OY, Tyutyunnik EN, Frolochkina TI, Lanciotti RS, Yazyshina S, Platonova OV, Obukhov IL, Zhukov AN, Vengerov YY and Pokrovskii VI (2001) Outbreak of West Nile virus infection, Volgograd Region, Russia, 1999. Emerg Infect Dis 7: 128-132.

Porter MB, Long MT, Getman LM, Giguere S, MacKay RJ, Lester GD, Alleman AR Wamsley HL, Franklin RP, Jacks S, Buergelt CD and Detrisac CJ (2003) West Nile virus encephalomyelitis in horses: 46 cases (2001). J Am Vet Med Assoc 222: 1241-1247.

Pupo M, Guzman MG, Fernandez R, Llop A, Dickinson FO, Perez D, Cruz R, Gonzalez T, Estevez G, Gonzalez H, Santos P, Kouri G, Andonova M, Lindsay R, Artsob H, Drebot M (2006) West Nile Virus infection in humans and horses, Cuba. Emerg Infect Dis 12: 1022-1024.

Rappole JH, Derrickson SR and Hubalek Z (2000) Migratory birds and spread of West Nile virus in the Western Hemisphere. Emerg Infect Dis 6: 319-328.

Reisen WK, Fang Y and Martinez VM (2006) Effects of temperature on the transmission of west nile virus by *Culex tarsalis* (Diptera: Culicidae). J Med Entomol 43: 309-317.

Reusken CBEM (2006) Samenstelling vectorenbestand van West Nile virus, een literatuuroverzicht [Vectors for West Nile virus transmission, a review]. RIVM briefrapport 164/06 MGB CR.

Rockx B, van Asten L, van den Wijngaard C, Godeke G, Goehring L, Vennema H, van der Avoort A, van Pelt W and Koopmans M (2006) Syndromic surveillance in The Netherlands for the early detection of West Nile virus epidemics. Vector Borne Zoon Dis 6: 161-169.

Rogers DJ and Randolph SE (2003) Studying the global distribution of infectious diseases using GIS and RS. Nat Rev Microbiol 1: 231-237.

Salazar P, Traub-Dargatz JL, Morley PS, Wilmot DD, Steffen DJ, Cunningham WE and Salman MD (2004) Outcome of equids with clinical signs of West Nile virus infection and factors associated with death. J Am Vet Med Assoc 225: 267-274.

Samina I, Margalit J and Peleg J (1986) Isolation of viruses from mosquitoes of the Negev, Israel. Trans R Soc Trop Med Hyg 80: 471-472.

Samoilova TI, Votiakov VI and Titov LP (2003) Virologic and serologic invetsigations of West Nile virus circulation in Belarus. Cent Eur J Public Health 11: 55-62.

Savage HM, Ceianu C, Nicolescu G, Karabatsos N, Lanciotti R, Vladimirescu A, Laiv L, Ungureanu A, Romanca C and Tsai TF (1999) Entomologic and avian investigations of an epidemic of West Nile fever in Romania in 1996, with serologic and molecular characterization of a virus isolate from mosquitoes. Am J Trop Med Hyg 61: 600-611.

Schaffner F, Chouin S and Guilloteau J (2003) First record of *Ochlerotatus (Finlaya) japonicus japonicus* (Theobald, 1901) in metropolitan France. J Am Mosq Control Assoc 19: 1-5.

Schuffenecker I, Peyrefitte CN, el Harrak M, Murri S, Leblond A and Zeller HG (2005) West Nile virus in Morocco, 2003. Emerg Infect Dis 11: 306-309.

Schuler LA, Khaitsa ML, Dyer NW and Stoltenow CL (2004) Evaluation of an outbreak of West Nile virus infection in horses: 569 cases (2002). J Am Vet Med Assoc 225: 1084-1089.

Shalunova NV, Zgurskaya GN and Berezin VV (1968) [West Nile virus in the USSR. 3. Isolation of another strain of West Nile virus from *Hyalomma plumbeum plumbeum* Panz in Astrakhan Oblast in 1965]. Trudy Inst Poliom Virus Encef (Moscow) 12: 389-393.

Siger L, Bowen R, Karaca K, Murray M, Jagannatha S, Echols B, Nordgren R and Minke JM (2006) Evaluation of the efficacy provided by a recombinant canarypox-vectored equine West Nile virus vaccine against an experimental West Nile Virus intrathecal challenge in horses. Vet Ther 7: 249-256.

Steinman A, Banet C, Sutton GA, Yadin H, Hadar S and Brill A (2002) Clinical signs of West Nile virus encephalomyelitis in horses during the outbreak in Israel in 2000. Vet Rec 151: 47-49.

Sutherst RW (2004) Global change and human vulnerability to vector-borne diseases. Clin Microbiol Rev 17: 136-173.

Swayne DE, Beck JR, Smith CS, Shieh WJ and Zaki SR (2001) Fatal encephalitis and myocarditis in young domestic geese (Anser anser domesticus) caused by West Nile virus. Emerg Infect Dis 7: 751-753.

Tachiiri K, Klinkenberg B, Mak S and Kazmi J (2006) Predicting outbreaks: a spatial risk assessment of West Nile virus in Britisch Columbia. Int J Health Geogr 5: 21.

Ternovoi VA, Shchelkanov M, Shestopalov A, Aristova V, Protopopova E, Gromashevskii V, Druziaka A, Slavskii A, Zolotykh S, Loktev V and L'vov D (2004) Detection of West Nile virus in birds in the territories of Baraba and Kulunda lowlands (West Siberian migration way) during summer-autumn of 2002. Vopr Virusol 49: 52-56.

Ternovoi VA, Protopopova E, Surmach S, Gazetdinov M, Zolotykh S, Shestopalov A, Pavlenko E, Leonova G and Loktev V (2006) The genotyping of the West Nile virus in birds in the far eastern region of Russia in 2002-2004. Mol Gen Mikrobiol Virusol 4: 30-35.

Ternovoi VA, Protopopova EV, Kononova IuV, Ol'khovikova EA, Spiridonova EA, Akopov GD, Shestopalov AM, Loktev VB (2007) Cases of West Nile fever in Novosibirsk region in 2004, and the genotyping of its viral pathogen. Vestn Ross Akad Med Nauk 1: 21-26 [in Russian]

Tesh RB, Siirin M, Guzman H, Travassos da Rosa AP, Wu X, Duan T, Lei H, Nunes MR and Xiao SY (2005) Persistent West Nile virus infection in the golden hamster: studies on its mechanism and possible implications for other flavivirus infections. J Infect Dis 192: 287-295.

Theophilides CN, Ahearn SC, Grady S and Merlino M (2003) Identifying West Nile virus risk areas: the Dynamic Continuous Area Space-Time system. Am J Epidemiol 157: 843-854.

Tsai TF, Popovici F, Cernescu C, Campbell GL and Nedelcu NI (1998) West Nile encephalitis epidemic in southeastern Romania. Lancet 352: 767-771.

Trock SC, Meade BJ, Glaser AL, Ostlund EN, Lanciotti RS, Cropp BC, Kulasekera V, Kramer LD and Komar N (2001) West Nile virus outbreak among horses in New York State, 1999 and 2000. Emerg Infect Dis 7: 745-747.

Vasil'ev AV, Shchelkanov MI, Dzharkenov AF, Aristova VA, Galkina IV, L'vov DN, Morozova TN, Kovtunov AI, Grenkova EP, Zhernovoi AV, Shatilova VP, Slavskii AA, Petrenko MS, Chirkizov PF, Dybal'VD, Leont'ev EA, Gabbasov FB, Odolevskii EA, Ibragimov RM, Idrisova RZ, Sokolova NN, Artiukh NP, Andreeva NI, Bondarev AD, Deriabin PG, Gromashevskii VL, Nepoklonov EA, Aliper TI and L'vov DK (2005) [West Nile virus infection of agricultural animals in the Astrakhan region, as evidenced by the 2001-2004 serological surveys]. Vopr Virusol 50: 36-41.

Vinograd IA and Obukhova VR (1974) [Isolation of arboviruses from birds in western Ukraine.] Sbornik Trudov Inst Virusol (Moskva) 3: 84-87.

Wang T, Town T, Alexopoulou L, Anderson JF, Fikrig E and Flavell RA (2004) Toll-like receptor 3 mediates West Nile virus entry into the brain causing lethal encephalitis. Nat Med 10: 1366-1373.

Ward MP, Levy M, Thacker HL, Ash M, Norman SK, Moore GE and Webb PW (2004) Investigation of an outbreak of encephalomyelitis caused by West Nile virus in 136 horses. J Am Vet Med Assoc 225: 84-89.

Ward MP, Schuermann JA, Highfield LD and Murray KO (2006) Characteristics of an outbreak of West Nile virus encephalomyelitis in a previously uninfected population of horses. Vet Microbiol 118: 255-259.

Weinberger M, Pitlik SD, Gandacu D, Lang R, Nassar F, Ben David D, Rubinstein E, Izthaki A, Mishal J, Kitzes R, Siegman-Igra Y, Giladi M, Pick N, Mendelson E, Bin H and Shohat T (2001) West Nile fever outbreak, Israel, 2000: epidemiologic aspects. Emerg Infect Dis 7: 686-691.

Weissenböck H, Hubalek Z, Halouzka J, Pichlmair A, Maderner A, Fragner K, Kolodziejek J, Loupal G, Kolbl S and Nowotny N (2003) Screening for West Nile virus infections of susceptible animal species in Austria. Epidemiol Infect 131: 1023-1027.

Work TH, Hurlbut HS and Taylor RM (1953) Isolation of West Nile virus from hooded crow and rock pigeon in the Nile delta. Proc Soc Exp Biol Med 84: 719-722.

Work TH, Hurlbut HS and Taylor RM (1955). Indigenous wild birds of the Nile Delta as potential West Nile virus circulating reservoirs. Am J Trop Med Hyg 4: 872-888.

Zeller HG and Schuffenecker I (2004) West Nile virus: an overview of its spread in Europe and the Mediterranean basin in contrast to its spread in the Americas. Eur J Clin Microbiol Infect Dis 23: 147-156.

9. Emergence of Usutu virus in Central Europe: diagnosis, surveillance and epizootiology

Herbert Weissenböck, Sonja Chvala-Mannsberger, Tamás Bakonyi and Norbert Nowotny

Abstract

Usutu virus (USUV), a member of the mosquito-borne clade within the Flaviviridae family, emerged in the summer 2001 in Austria and caused fatalities in several species of wild birds. This was the first time that USUV appeared outside Africa, where it was considered to be unimportant in terms of pathogenicity. USUV continued to kill birds in Austria, especially blackbirds (*Turdus merula* L), during the summer periods of consecutive years. Since the emergence of USUV provided a unique opportunity to study the further development of the activity of a mosquito-borne virus in a previously unexposed ecosystem and a naive host population, a surveillance system was established and maintained for a period of three consecutive years (2003-2005). The survey identified the geographical distribution of infected birds and the most affected bird species, and revealed marked differences of USUV-associated bird mortality in the three seasons investigated. Different methods for detection of USUV in organs were established (immunohistochemistry, reverse transcription - polymerase chain reaction, and in situ hybridisation) demonstrating viral antigens and nucleic acids. Mosquito surveillance showed the presence of USUV nucleic acid in *Culex* spp. and some other species. Serological screenings of wild birds were performed in the years 2003-2006, showing a remarkable increase of antibody response to USUV beginning in 2005, which indicated the establishment of herd immunity. Analysis of human blood samples for antibodies and viral RNA revealed a considerable percentage of seropositives and in one case viral nucleic acid was detected. This indicates that infections of humans do occur, however without causing notable clinical signs. While USUV-associated bird losses significantly declined in Austria, the first cases of avian mortality due to USUV occurred in Hungary, Switzerland and Italy starting in the year 2005. The currently available data indicate that USUV, a pathogen initially restricted to the African continent, has become a resident infectious agent in Europe with a tendency of further spread within Central Europe.

Keywords: blackbird, bird mortality, flavivirus, mosquito, vector-borne disease, Usutu virus

Historical background

Much of the information concerning isolations of Usutu virus (USUV) in Africa is difficult to access, because (1) it is hidden in tables or a few lines of papers primarily focusing on other viruses, (2) it has been published in reports of different research organisations which are not easily available in most libraries and (3) it is mentioned only in books without citing an original source. In this context special credit has to be given to the website authored by Adam and Digoutte (2005), which is provided by the Pasteur Institute and turned out to be an excellent source of valuable information on African flaviviruses in general.

There is general agreement that the first isolation of USUV was accomplished by McIntosh in 1959 from a mosquito pool of *Culex univittatus* Theobald collected in Ndumu, Natal, South Africa. The isolate was registered under the reference number SA Ar 1776. It was found to be closely related to West Nile virus (WNV) but antigenically distinct, and the obviously new virus was named Usutu virus, according to a river located close to the area where the mosquitoes were captured (Williams *et al.* 1964, Theiler and Downs 1973). In the following years USUV was also detected in other mosquito species, like *Coquillettidia (Mansonia) aurites* Theobald (Woodall 1964, Williams *et al.* 1964), *Culex* spp. (Henderson *et al.* 1971), *Culex perfuscus* Edwards, *Aedes minutus* Theobald and *Mansonia africana* Theobald (Adam and Digoutte 2005). These additional isolates originated from different areas of Africa, such as Uganda, Senegal, and the Central African Republic. In addition to SA Ar 1776, which is considered the prototype strain of USUV and thus has been distributed to a number of laboratories worldwide, several other isolates have received official reference numbers and seem to be deposited in virus repositories. Such strains are MP1626, MP8174, Ar D19848, and Ar D101291, all isolated from mosquitoes in Uganda and Senegal, respectively.

Besides mosquitoes USUV was also isolated from at least 3 sedentary African bird species: a kurrichane thrush (*Turdus libonyanus* Smith) from Nigeria (Odelola and Fabiyi 1976), a forest canopy hornbill (*Bycanistes fistulator sharpii* Elliot) from the Central African Republic (Sureau and Germain 1978) and a little greenbul (*Andropadus virens* Cassin) (Adam and Digoutte 2005). The only isolate ever derived from a mammal was reported from an African soft-furred rat (*Praomys* spp.). USUV seems to have been found in a human only once: in a man who suffered from febrile illness and rash in the Central African Republic (Adam and Digoutte 2005). Simpson (1966) was unable to demonstrate significant viraemia in Abyssinian grass rats (*Arvicanthis abyssinicus* Rüppell) experimentally inoculated with USUV. During an investigation of flavivirus infections of cattle in Zimbabwe the antigenic cross-reactivity of USUV and other flaviviruses (Wesselsbron, Spondweni, Banzi, West Nile, and yellow fever viruses) was studied in antisera prepared in guinea pigs using different serological assays (Blackburn and Swanepoel 1980). Strong cross-reactivity was observed in the haemagglutination inhibition test, moderate cross-reactivity in the complement fixation test, and practically no cross-reactivity in the microneutralisation test. Antibody testing of field sera of cattle revealed no antibodies to USUV. Thus the authors concluded that there was no evidence for circulation of USUV in Zimbabwe.

Taken together, prior to the emergence of USUV in Europe information on this virus was fragmentary and difficult to access, because USUV was largely scientifically ignored. This is most likely due to the fact that there was no indication of pathogenicity for any host of the virus strains circulating in Africa. The only paper providing a partial sequence of USUV was published by Kuno *et al.* (1998). In this study on the molecular phylogeny of 71 flaviviruses, the authors sequenced approximately 1 kb of a part of the NS5 gene of these viruses, including USUV. This partial sequence information was later very useful for rapid identification of USUV as causative agent of avian mortality in Austria.

First recognition of Usutu virus-associated bird mortality in Austria

From the beginning of August through mid-September 2001, a considerable die-off of blackbirds was observed in and around Vienna, Austria (Weissenböck *et al.* 2002). In addition, within five days in mid-August, five great grey owls (*Strix nebulosa* Forster) died in the Vienna Zoo (Tiergarten Schönbrunn). This unusual mortality of different species of birds was similar to the beginning of the WNV epidemic in the United States. Immediate pathological and virological investigations of this episode of avian deaths in Austria determined USUV as the causative agent. This was a surprising finding because it was for the first time that USUV had emerged outside Africa and had caused recognised fatalities in warm-blooded hosts.

In the consecutive year, ornithologists and the general public were alerted by newspaper- and other mass media reports that bird die-offs could occur again (Weissenböck *et al.* 2003a). People who found dead birds were requested to submit them for investigation. In this year out of 72 investigated birds 28 blackbirds, one blue tit (*Parus caeruleus* L) and one house sparrow (*Passer domesticus* L) were tested USUV-positive. The positive birds were found in Vienna and in districts of the federal state of Lower Austria (south, east and west, close to Vienna). USUV-associated bird mortality was restricted to the time period of mid-July to mid-September (Weissenböck *et al.* 2002, 2003a). In parallel, ornithologists observed focally dramatic population decreases especially of blackbirds in urban areas of Vienna and Lower Austria (Straka 2005, Steiner and Holzer 2007)

This observation had proved that USUV had managed to overwinter and had obviously been able to establish an efficient local bird-mosquito transmission cycle (Weissenböck *et al.* 2003b). Because dead bird surveillance had already been successfully applied for monitoring of WNV activity in North America (Eidson *et al.* 2001, Guptill *et al.* 2003, Glaser 2004), a surveillance system for USUV was established and maintained for the period of the three following years (2003-2005), giving a unique opportunity to study the further development of the activity of a mosquito-borne virus in a previously unexposed ecosystem and a naive host population (Chvala *et al.* 2007).

Genome analysis and characterisation of the Austrian USUV strain

The complete genome sequence of the USUV strain that emerged in Austria in 2001 (Vienna 2001-blackbird) was determined using direct sequencing of overlapping RT-PCR amplification products, which covered the genome of the virus. The genome sequence was compared to the South African USUV reference strain SA Ar 1776. The isolates exhibited 97% nucleotide, and 99% amino acid similarities, and particularly variable regions were not identified (Bakonyi *et al.* 2004). When comparing USUV with other flaviviruses, the highest nucleotide sequence similarities were found with representatives of the Japanese encephalitis virus (JEV) antigenic group, i.e. Murray Valley encephalitis virus (73%), JEV (71%) and WNV (68%). The probable genetic relatedness between USUV and other flaviviruses is illustrated in a phylogenetic tree (Figure 1) (Bakonyi *et al.* 2004).

Diagnostic methods

Necropsy and histology

An easy and efficient method for substantiating an initial suspicion of USUV infection is necropsy and consecutive histological examination of dead birds. Macroscopically, the disease is characterised by spleno- and hepatomegaly (Figure 2), empty stomachs and seromucous enteritis. Typical histological findings are multifocal neuronal necrosis and glial nodules in the brain, myocardial degeneration and nonsuppurative myocarditis, as well as necrotising splenitis and hepatitis (Figure 3) (Chvala *et al.* 2004).

Methods of virus detection

Methods for detecting USUV in organ-, blood- and mosquito samples as well as in paraffin-wax embedded tissues were developed in order to unequivocally diagnose the infection and to provide information on the distribution of this relatively unknown virus in different organs. The established methods included virus isolation in cell cultures, immunohistochemistry (IHC) for detection of USUV-viral antigen in tissues, in situ hybridisation (ISH) with an USUV-specific

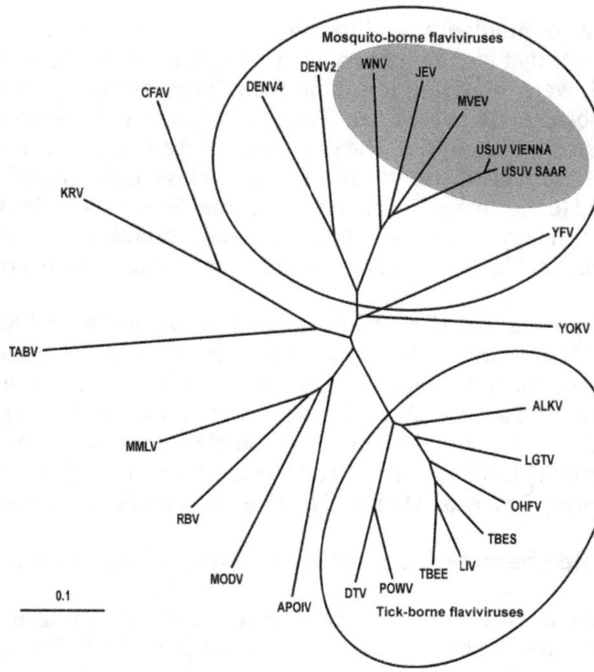

Figure 1. Phylogenetic tree illustrating the genetic relationship between flaviviruses, based on their complete genome sequences. (Abbreviations: ALKV - Alkhurma virus, APOIV - Apoi virus, CFAV - Cell Fusing Agent virus, DTV - Deer Tick virus, DENV2 - Dengue virus 2, DENV4 - Dengue virus 4, JEV - Japanese Encephalitis virus, KRV - Kamiti River virus, LGTV - Langat virus, LIV - Louping Ill virus, MODV - Modoc virus, MMLV - Montana myotis leukoencephalitis virus, MVEV - Murray Valley Encephalitis virus, OHFV - Omsk Hemorrhagic Fever virus, POWV - Powassan virus, RBV - Rio Bravo virus, TABV - Tamana Bat virus, TBEE - Tick Borne Encephalitis virus Europe, TBES - Tick Borne Encephalitis virus Siberia, USUV SAAR - Usutu virus strain SA Ar 1776, USUV VIENNA - Usutu virus strain Vienna 2001-blackbird, WNV - West Nile virus, YFV - Yellow Fever virus, YOKV - Yokose virus.) The JEV group viruses are marked with grey underlay. Bar on the left demonstrates the genetic distance.

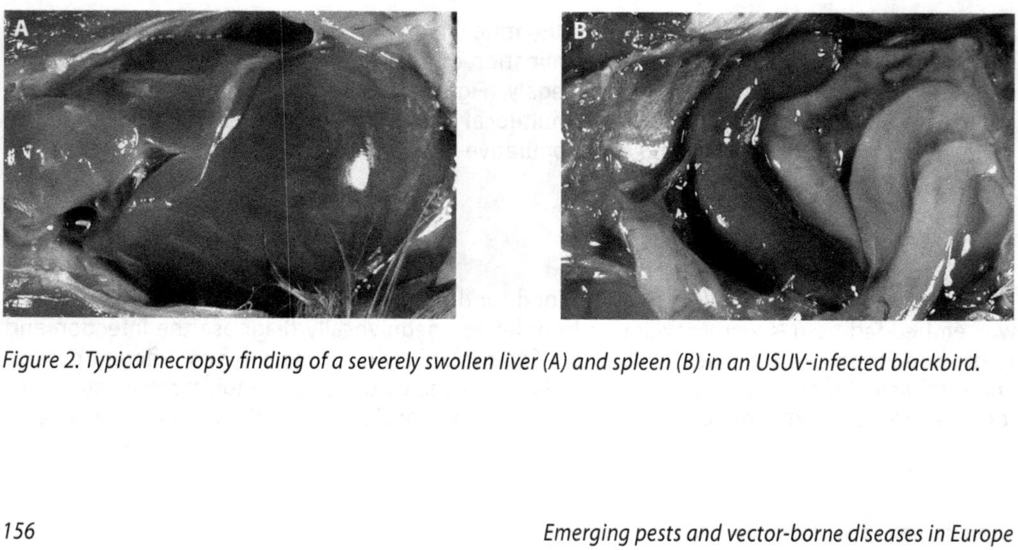

Figure 2. Typical necropsy finding of a severely swollen liver (A) and spleen (B) in an USUV-infected blackbird.

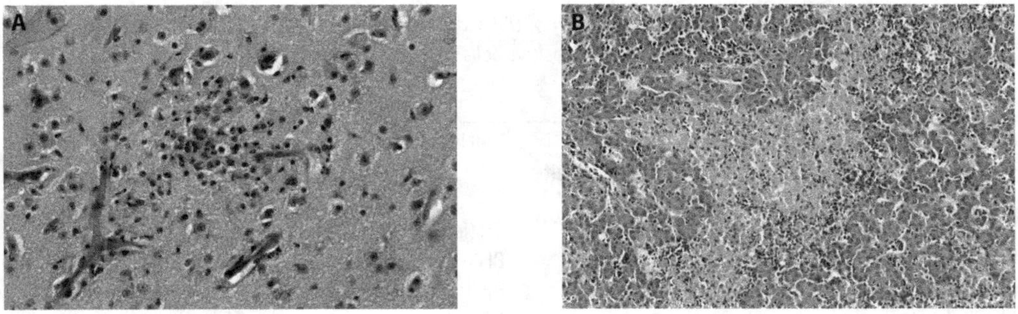

Figure 3. (A) Glial nodule in the cerebral cortex of a blackbird; in addition swelling of endothelial cells of adjacent capillaries and mild migration of inflammatory cells is also present. (B) Focal parenchymal coagulation necrosis in the liver of a blackbird.

oligonucleotide probe demonstrating viral nucleic acid, reverse transcription-polymerase chain reaction (RT-PCR) and real time RT-PCR.

Virus isolation and cell susceptibility studies

The virus was isolated from brain homogenates of blackbirds in Vero cell line. Electron microscopy revealed typical flaviviral morphology (Figure 4).

Subsequently, the susceptibility of 13 permanent cell lines, and three primary cell cultures to USUV infection was investigated (Bakonyi *et al.* 2005a). Although viral multiplication was detected in all tested mammalian cell types by IHC (Table 1), only Vero, PK-15, and goose embryo fibroblast cells developed cell rounding, which was the most typical cytopathic effect (CPE) of USUV infection. A similar study by Pfeffer *et al.* (2003) led to partially contradictory results: the virus did not replicate in human (Hela, Hep-2, 293) and equine (Edmin) cell lines, but did grow to moderate titres in PK-15

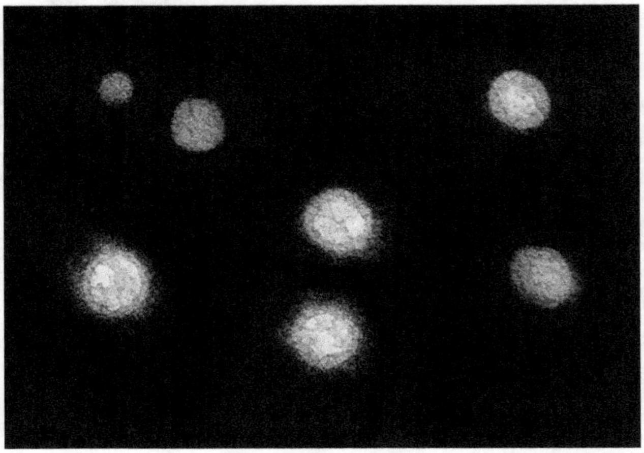

Figure 4. USUV virus particles shown by negative staining electron microscopy in tissue culture supernatant of infected Vero cells.

Table 1. Semiquantitative evaluation of the number of USUV antigen-positive cells (three days p.i., IHC staining). Scoring criteria: (+): 1-5% positive cells; +: 6-25% positive cells; ++: 26-50% positive cells. Primary cell cultures are indicated in italics.

Cell line	IHC result	Cell line	IHC result	Cell line	IHC result	Cell line / culture	IHC result
HeLa (human)	++	PK-15 (porcine)	++	CR (feline)	+	TH1 (turtle)	++
Vero (simian)	++	RK-13 (lapin)	++	BHK-21 (hamster)	+	*EqK* (equine)	++
ED (equine)	++	MDCK (canine)	++	BF (hamster)	+	*CEF* (chicken)	-
MDBK (bovine)	+	DK (canine)	(+)	C6 (rat)	+	*GEF* (goose)	++

(porcine) and Nie-168 (murine) cells. The virus was also isolated by inoculation into the allantoic sac of embryonated goose eggs, and into the brain of suckling mice (Weissenböck *et al.* 2004, Chvala *et al.* 2006). Interestingly, chicken embryo fibroblast cells and chicken embryos were found resistant to experimental USUV infections.

Immunohistochemistry

Using USUV-specific or cross-reactive WNV-specific polyclonal antibodies viral antigen could be found in the cerebral cortices (Figure 5), thalamus and the metencephalon, including the cerebellum of naturally infected birds. In many cases viral antigen was also detected in the heart (myofibres and walls of large blood vessels, Figure 5), spleen capsule (in fibrocytes and smooth muscle cells), kidney (glomeruli, tubular epithelial cells), proventriculus and gizzard, small intestine (crypt epithelial cells), lung and liver (Chvala *et al.* 2004, 2007). Infrequently pancreas, ovaries, testes, adrenals and bursa of Fabricius were examined for USUV-antigen, mostly with a positive result.

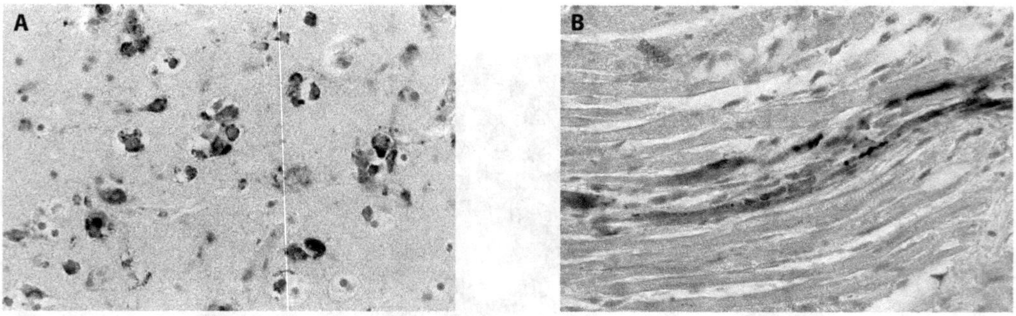

Figure 5. (A) Cerebral cortex of a blackbird, showing numerous neurons contain flavivirus antigen within their cytoplasm by IHC. (B) Heart of a blackbird showing flavivirus antigen within myofibres by IHC.

In situ hybridisation

This technique was performed with an USUV-specific digoxigenin-labeled oligonucleotide probe. In general, the presence of viral RNA matched the presence of viral antigen (Figure 6) (Chvala *et al.* 2004). However, in epithelial cells of intestinal crypts and proventricular glands viral RNA was more abundant than viral proteins (Figure 6), whereas in certain cell populations of the spleen, kidney, lung and intestine the situation was reversed. Reasons for RNA preponderance might be restricted infection implying some translation inhibition in the cell types involved. Alternatively, the affected cells may have been in an early stage of viral infection, in which viral protein synthesis had not become fully operative.

The reason for viral protein preponderance is not obvious. Although it is possible that RNA degradation plays a role, it seems doubtful that some tissues - particularly those not rich in RNAse - are unusually prone to RNA degradation. However, both methods, ISH and IHC, are reliable and efficient diagnostic tools for detecting USUV infection in organs.

Reverse transcription – polymerase chain reaction (RT-PCR)

Nucleic acid of the USUV strain, which emerged in Austria, was first detected in RT-PCR assays using universal JEV-group specific primer pairs (Weissenböck *et al.* 2002). After the determination of the complete genome sequence of the virus, further specific, and highly sensitive, one-step RT-PCR assays were developed for the detection of USUV-specific nucleic acid in brain or parenchymal tissue samples, swab samples, peripheral blood mononuclear cells, plasma and in mosquito homogenates. The specificity of the assays was validated by sequencing of the amplification products. False positive results due to contamination were precluded using non-nested, one-step RT-PCR systems. The viral nucleic acid was consequently detected in USUV-infected samples, even if the sample material was partially autolysed (i.e. dead bird tissues), if the nucleic acid was partially denatured (i.e. formalin-fixed, paraffin-embedded tissue sections), or if the sample contained low amounts of virus (i.e. mosquito homogenates). In 99% of the cases of USUV-infected birds, the results of IHC and RT-PCR were concordant, indicating an excellent correlation and a high specificity and sensitivity of both techniques (Chvala *et al.* 2004, Chvala *et al.* 2007). Sequence comparisons of RT-PCR amplification products targeting the major part of the E protein coding region of USUV strains were used to demonstrate the diversity of the genotypes detected within five consecutive years in Austria (Chvala *et al.* 2007).

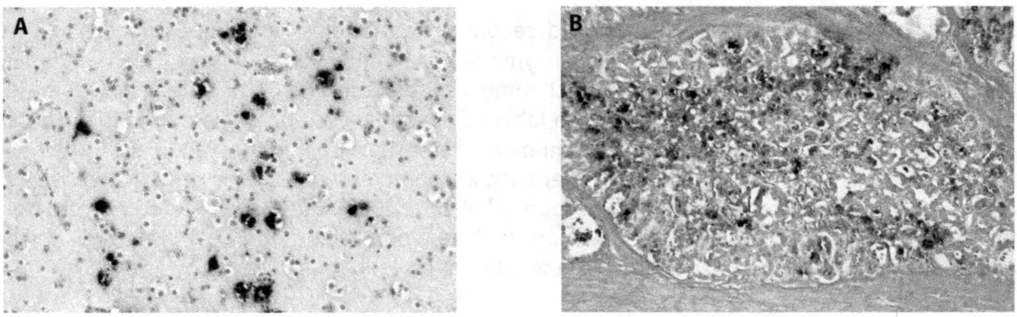

Figure 6. (A) Cerebral cortex of a blackbird, showing USUV nucleic acid in the cytoplasm of numerous neurons by ISH. (B) Proventriculus of a blackbird showing abundant specific signal within cells of a proventricular gland by ISH.

Real-time RT-PCR

TaqMan-based, real-time one-step RT-PCR assays were developed for the quick and sensitive detection and quantification of USUV genomic RNA in cultivated virus strains, as well as organ homogenates of birds and mosquitoes. Oligonucleotide primers and fluorescently labelled TaqMan probes were designed, targeting the putative NS3 and NS5 protein coding regions of the virus genome. The sensitivity of the assays was determined using cultivated and titrated USUV strains. The real-time RT-PCR assays detected the viral nucleic acid in cell culture supernatants containing 0.006 to 0.01 $TCID_{50}$ /ml infectious particles. The specificity of the assay was tested on USUV-related flaviviruses (WNV, DENV 1-4, JEV, Saint Louis encephalitis virus [SLEV], YFV and TBEV), and non-infected host nucleic acid samples. No amplification signal was detected in these reactions. The real-time RT-PCR assays could provide quantitative results within four hours after a sample was received in the laboratory, and reliably detected the virus in partially autolysed samples, paraffin-embedded tissue sections, or in mosquito pools.

Serology

The epizootiological data of the USUV outbreak in Austria indicated that many more potential hosts were exposed to virus infection than those which showed symptoms of the disease and died. Experimental animal infections also revealed that USUV infections are often subclinical. The direct demonstration of the virus infection in these cases is rarely successful, because the virus multiplication and shedding is transient. Therefore, for a comprehensive active surveillance programme, indirect (serological) methods were needed to detect previous infections and seroconversion. The serology of flaviviruses is difficult because the different virus species share common surface antigens which are cross-reactive in several assays. In Austria, the European tick-borne encephalitis virus (TBEV), a tick-transmitted flavivirus is widespread in small mammals and ticks. However, birds are mainly occasional hosts of TBEV and some of these (i.e. anatid and gallinaceous birds) can be amplifying hosts as well (Hubálek 1994). TBEV was isolated from several bird species including *Turdus merula* (Brummer-Korvenkontio *et al.* 1973) and from ticks collected on them (Ernek *et al.* 1968). Another flavivirus, which is antigenically and genetically closely related to WNV, Rabensburg virus (RABV), was isolated from mosquitoes at the Austrian-Czech border in 1997 and in 1999 (Hubálek *et al.* 1998, 2000, Bakonyi *et al.* 2005b). Ecologically and epizootiologically this area overlaps with the USUV-infected areas in Austria. Therefore, the probable host populations of USUV are possibly exposed to at least two further flaviviruses. During the development of serological assays, they were tested for cross-reactivity for TBEV and WNV.

Several serological assays employ labelled secondary antibodies, which are specific to the immunoglobulins of the host species (i.e. enzyme-linked immunosorbant assay [ELISA], indirect immunofluorescence assay [IFA] and indirect immunoperoxidase assay). The application of these tests is restricted to host species for which labelled secondary antibodies are available. These conjugates are commercially produced for humans, domestic and laboratory animals, but rarely for wild birds. Anti-wild bird horseradish peroxidase-conjugated antibodies were developed to detect IgG in species from several avian orders (Ebel *et al.* 2002), but these antibodies were not tested for Central European species. Therefore such serological methods were preferred, which are independent from the host origin of the sera under investigation.

PRNT

The plaque-reduction neutralisation test (PRNT) is the 'gold standard' of flavivirus serology. This test is considered to be the most specific and does not employ secondary antibodies. The main drawback of the assay is that it is time-consuming and laborious. USUV induces CPE just in certain cell types, and after subsequent passages, the intensity of the CPE is decreasing. With routine cultivation techniques USUV forms diffuse CPE, therefore the plaque formation is facilitated by overlaying the cells with semi-liquid maintenance medium (i.e. agar or carboxymethylcellulose-containing medium). Some cells are sensitive to such agents and may develop cell degeneration, which may disturb the assay. More frequently sera contain toxic components, especially if they were collected during necropsy or were haemolysed.

The PRNT technique described by De Madrid and Porterfield (1969) and modified by Hubálek *et al.* (1979) for USUV serology. After inactivation and an initial 1:10 dilution, twofold serial dilutions of the sera were incubated with approximately 15 plaque forming units (PFUs) of USUV. After incubation, susceptible cells were added to the serum-virus mixtures, dispensed on microneutralisation plates and covered with 1.5% carboxymethylcellulose-containing medium. Plates were incubated for three to four days, and when plaques had developed in the control wells, the cells were stained with naphthol blue, and plaques were counted. The assays were evaluated on 50% and 90% reduction level criteria.

Cross-reactivity studies between USUV, TBEV and WNV indicated that the antibody titres to the homologous virus antigens were 4 to 8 times higher than to the related viruses both on 50% and 90% plaque reduction level. Neutralising antibodies to USUV were detected in the sera samples of wild birds from Austria, experimentally infected chicken and geese, as well as in Austrian human serum samples (Weissenböck and Nowotny 2006, Meister *et al.* 2007).

HI

USUV haemagglutinates goose red blood cells at pH 6.2 to 6.4. Consequently, a haemagglutination-inhibition (HI) test was developed for the detection of antibodies raised against the haemagglutinating antigens of the virus. The original method of Clarke and Casals (1958) was adapted to USUV (Chvala *et al.* 2005). The main advantages of the HI method compared to PRNT are that HI is quicker and simpler, susceptible cells are not required, and the evaluation of the assay is less influenced by serum abnormalities. Nevertheless, in general, HI is considered to be less specific than PRNT. Cross-reactivity tests using positive control sera were performed on USUV, WNV and TBEV in order to determine the specificity of the assay. The HI titres to the homologous viruses were 4 to 16 times higher than to the related viruses. Austrian wild bird sera showed HI activity to USUV up to a titre of 1:2560, and HI antibodies were demonstrated in experimentally infected animals and in Austrian human serum samples, too.

Human and animal serum samples were further investigated for the presence of USUV and related flavivirus antibodies with commercially available WNV indirect IgG ELISA, WNV capture IgM ELISA, flavivirus IFA, and in-house developed WNV IFA (Ferenczi *et al.* 2005) techniques. The reliability and reproducibility of these test systems were lower than that of the PRNT and HI tests. In conclusion, PRNT and HI are appropriate methods for the indirect diagnosis of USUV infections; however, due to the possible cross-reactions, the results should be validated by simultaneous titrations of the tested sera with related flavivirus antigens. Therefore the processing of high numbers of serum samples is time-consuming and expensive.

Monitoring of USUV activity by dead bird surveillance 2003-2005

Chvala *et al.* (2007) described a surveillance programme and its results based on the participation of the general public which was requested by mass media advises to collect and submit dead birds to local collection centres usually installed in the headquarters of the district veterinarians, during the summer period 2003-2005.

In total 107 birds (of 504) were found USUV positive by RT-PCR and/or by IHC. The most affected bird species were blackbirds. In addition two great tits, one nuthatch (*Sitta europaea* L), one song thrush (*Turdus philomelos* Brehm), and one robin (*Erithacus rubecula* L) were found positive for USUV. The majority of the birds were found dead, but in some cases clinical symptoms like apathy, ruffled plumage, uncoordinated movements, inability to fly and seizures with consecutive death within 2–24 hours were reported. In the summer of the year 2003 a peak of 92 USUV-positive birds was reached, which corresponded to 52% of the total number of birds investigated this year, whereas in 2004 USUV-associated bird mortality decreased to 11 blackbirds (5% of the total number of submissions) and 2005 only 4 dead blackbirds (4% of the total) were USUV-positive. Another observation was that the proportion of birds with low viral titres in their tissues increased continuously. At the beginning of the USUV epidemic, practically all affected birds showed severe tissue lesions and abundant viral signals within a broad range of tissues which indicated that the affected birds had definitely died as a result of the virus infection. Beginning with the third season of USUV activity there were positive birds which did not display lesions and only showed very limited amounts of viral signal within the tissue. These birds might have transiently harboured the virus and the infection most likely was not the cause of death. This phenomenon may be explained by decreased virulence of the virus or more efficient defence strategies of the host, e.g. by immunity. However, such birds were not consistently seropositive; thus, at least the involvement of neutralising antibodies cannot be proved. Within the surveillance period the virus had spread from its initial area of emergence and circulation, the surroundings of Vienna, to large areas of the federal states of Lower Austria, Burgenland and Styria.

The surveillance programme also included serological examinations of bird sera, available from birds captured in mist nets or other trapping devices, further from sick or injured birds brought to the avian clinic (University of Veterinary Medicine Vienna), or from an owl and raptor rehabilitation centre and finally from dead birds submitted for necropsy (Meister *et al.* 2007). In total 442 sera were examined. In this serological survey the HI was used for screening and the PRNT for confirmation. The data showed a proportion of seropositives below 10% in the years 2003-2004. In 2005 and 2006 the percentage of seroreactors increased to more than 50%. The overall trend of declining numbers of USUV-associated bird deaths, together with an increase of seropositives in the wild bird population has been interpreted as developing herd immunity. The phenomenon of reduced mortality might also be explained by reduced virulence of the circulating virus strain, an assumption which currently cannot be formally excluded. Data from the related WNV suggest that especially mutations in certain E and NS1 protein genes resulting in loss of glycosylation were responsible for reduced virulence or neuroinvasion (Samuel and Diamond 2006). For USUV, currently no complete sequences or experimental data of virus strains isolated in different years are available. However, sequencing of 88% of the E coding region of 12 USUV isolates from three consecutive years (2003-2005) revealed only single random mutations, all of which except one did not result in amino acid changes (Chvala *et al.* 2007). These limited data do not provide evidence for alterations in virulence over time.

In addition to serosurveillance Meister *et al.* (2007) also performed a longitudinal study in captive birds of prey in order to recognise seroconversions and titre dynamics during one transmission season. Sera taken before, during and after the transmission season were serologically examined. The data indicate a continuing USUV activity with several seroconversions or significant titre rises despite a high level of pre-existing herd immunity in the birds of this enclosure.

The genetic relatedness of 12 USUVs from different geographic origins and from three consecutive years showed 99.7 to 100% identity to each other and to the Vienna 2001 strain (Chvala *et al.* 2007). This observation clearly implies that all the USUV strains present in Austria originate from a single introduction around the year 2000.

Identification of mosquitoes as vectors for USUV

During the years 2002-2003 mosquitoes from different areas with proven USUV activity were collected, taxonomically identified and tested by RT-PCR for USUV specific nucleic acid. USUV nucleic acid was present in pools containing *Culex pipiens* L, *Cx. hortensis* Ficalbi, *Cx. territans* Walker, *Culiseta annulata* Schrank, *Aedes vexans* Meigen and *Ae. rossicus* Dolbeskin, Gorickaja & Mitrofanova (Nowotny and Weissenböck 2004). In mosquito collections from 2004-2005 USUV was detected in 4% of overwintering *Culex* species, an observation which revealed where USUV might hide during periods of non-activity. (Weissenböck and Nowotny 2006).

Pfeffer (2003) performed preliminary vector competence studies on intrathoracally inoculated female *Cx. pipiens* originating from the area of the Austrian USUV outbreak. Whole mosquitoes were individually tested for virus in a plaque assay one week later. Viral titres were found to reach from 1.5×10^2 to 6.3×10^3, corresponding to a maximum 200-fold increase. This moderate level of viral replication, however, underlines the likely role of *Cx. pipiens* as the main competent vector of USUV in Europe.

Spread of USUV to other European countries 2005-2006

In August 2005 increased dead blackbird sightings were reported in Budapest, Hungary. One blackbird was received for investigation and tested USUV positive. In the consecutive year nine dead blackbirds were received from the Budapest area for avian influenza monitoring. Six of these were also tested for USUV by RT-PCR and all six proved positive (Bakonyi *et al.* 2008).

In the summer of 2006 unusual peracute mortality in wild and captive birds (a total of 105 birds of the order Passeriformes and Strigiformes) was observed in and around Zürich Zoo, Switzerland. In the majority of these an USUV infection was diagnosed and this virus was considered the causative agent (Steinmetz *et al.* 2007). In the same period a great grey owl originating from a northern Italian breeding collection of many different Strigiformes, where unusually high bird losses had occurred, was investigated. By IHC large amounts of USUV antigen were found in many different organs. This finding was confirmed by PCR and sequencing of amplification products (Dorrestein *et al.* 2007). The partial sequences of the Hungarian, Swiss and Italian USUV strains were compared to each other and demonstrated to be 99% identical. Furthermore they proved to be 99% identical to the Vienna 2001-blackbird strain (Nowotny *et al.* 2007).

Already in 2003 Buckley *et al.* published serological data suggesting presence of USUV-specific antibodies in wild birds in the UK. Later the same authors (Buckley *et al.* 2006) also reported USUV-seropositive sentinel chickens. These authors used the lower stringency $PRNT_{50}$ endpoint, which should allow detection of antibodies of low titres and low avidity. Intriguingly, almost 54% of all sera of different wild bird species investigated with this method yielded a positive result, suggesting circulation of USUV in the UK. However, there are no reports on USUV-associated bird mortality from the UK and also, USUV has never been isolated in that country. In case these serological data represent a genuine immune response to USUV, the virus strain present in the UK has to be considered unrelated to the Central European strain, because its virulence seems to be significantly lower. A final answer to the nature of the agent inducing neutralising antibodies to USUV in wild birds in the UK cannot be given before successful isolation and characterisation of the virus.

Experimental infections of mice, chickens and geese

Except for wild birds in Central Europe, clinical disease due to USUV infection has not been reported from any other animal species. Practically all members of the Flavivirus and Alphavirus genera are neuropathogenic for mice, and experimental infections with some of these viruses serve as models for the study of apoptosis, inflammation or demyelination induced by neurotropic viruses. This fact encouraged a study on the neuropathogenicity of USUV for mice which had never been systematically investigated before (Weissenböck *et al.* 2004). Following intraperitoneal inoculation 1-week-old suckling mice showed clinical signs like depression, paraplegia, paralysis and coma 6 to 11 days post infection. Histopathological examination revealed widespread neuronal apoptosis but also clear signs of demyelination, especially in the brain stem, associated with viral signals. Only in newborn mice and mice of one week of age USUV was neuroinvasive. Older mice were resistant to peripheral USUV infection: neither peripheral viral replication nor neuroinvasion were observed. The age dependence of neuroinvasiveness is well known from other flavi- and alphaviruses. The age until which neuroinvasion by peripheral inoculations can be achieved ranges from 12 days (Semliki forest virus, SFV) (Fazakerley 2002), over 6 weeks (WNV) (Eldadah *et al.* 1967), and 4 months (SLEV) (Andersen and Hanson 1974, Monath *et al.* 1974) to even adult age (Banzi virus) (Bhatt and Jacoby 1976; Jacoby and Bhatt 1976). Thus it seems that the age-dependent neuroinvasiveness of the Austrian USUV strain is the lowest ever been tested among several mosquito-borne viruses.

Domestic chickens usually do not develop clinical disease following flavivirus infections, but they mount a serological response. This is frequently taken advantage of by using chickens as sentinels in arbovirus monitoring and surveillance programmes. The potential pathogenicity of USUV in chickens was studied by Chvala *et al.* (2005). The established data suggest that domestic chickens are not at risk of developing clinical disease following USUV infection and that chickens are unlikely to be useful for sentinel purposes in USUV surveillance programmes.

The closely related WNV has been shown to be responsible for mortality in domestic geese (Austin *et al.* 2004). Also, USUV showed to efficiently replicate in goose embryonic fibroblast cell cultures and in goose embryos (Bakonyi *et al.* 2005a, Chvala *et al.* 2006), which encouraged an experimental study to determine whether USUV can cause disease in domestic geese (Chvala *et al.* 2006). This experiment showed that USUV is able to replicate in geese, but does not induce clinical disease or mortality, and only infrequently leads to viraemia or virus shedding.

USUV-infections in humans

203 human blood samples from individuals exposed to mosquitoes in USUV-endemic areas were examined by RT-PCR for detection of USUV-specific nucleic acid and 208 human blood samples were investigated by HI test and by PRNT for the presence of antibodies to USUV. Many of the subjects included in this study exhibited rash of unknown origin, and they tested negative for Lyme-Borreliosis. In this pre-selected group, 83 individuals showed antibody titres to USUV between 1:20 and >1:160 in HI test, and 52 of them could be confirmed by PRNT, which results in an overall seroprevalence of 25% in this pre-selected group. In one 'fresh' case USUV nucleic acid was demonstrated by RT-PCR. These data indicate that USUV is able to infect humans without inducing severe disease; in particular, no association between neurological disease and USUV infection could be established in human beings. Transient rash, however, seems to be a clinical symptom associated with USUV infection in human beings (N. Nowotny, unpublished data, Weissenböck and Nowotny 2006).

Conclusions

The current distribution of USUV in Europe is not completely known. Except for Austria no attempts at detecting the virus in mosquitoes have been made. Although avian - especially blackbird - mortality is an excellent indicator for USUV activity, it may be missed, especially in rural areas where dead birds are not easily spotted. In addition, as clearly demonstrated in Austria, several seasons of USUV activity within a defined geographic range tend to lead to a significant decline of USUV-associated bird deaths, a phenomenon which is at least partially attributable to herd immunity. Thus it is not unlikely that introduction of USUV into a new area was not recognised because initial increased avian mortality was not registered, ignored, or not correctly interpreted. After several seasons the infection may have become 'silent' and is only detectable by serological examinations of the natural hosts - wild birds, or by surveillance of the mosquito vectors. Consequently, the actual distribution range of USUV may well extend beyond the areas identified by virus recovery in dead birds. This may explain the obvious gaps between the different areas where so far outbreaks have been noticed, i.e. eastern Austria, Budapest, Zürich and northern Italy. Alternatively, however, as all known outbreaks so far occurred within or in close proximity to bird-keeping facilities (zoos and a private breeding station) the introduction of USUV by trade of latently infected birds cannot be excluded completely.

In any case, it has to be expected that USUV continues to spread in Europe. It is also most likely that it will cause unusual episodes of avian mortalities which may especially be problematic for zoos and breeding facilities keeping certain species of owls.

The origin of the USUV strain present in Europe is unknown. At present, molecular phylogenetic studies are not helpful, because only the prototype strain isolated in 1959 in southern Africa and kept in laboratories since and one Austrian and one Hungarian USUV strain have been fully sequenced. It would probably be elucidating to sequence all USUV strains which are deposited in different virus repositories worldwide (and probably others which are not mentioned in the literature).

It is also currently entirely unknown why USUV is highly pathogenic for blackbirds and certain owl species and why many other species of birds are totally resistant. Such differential vulnerability is also well established for the related WNV (Komar *et al.* 2003) but seems to be more pronounced for USUV.

Once incidentally introduced into Austria, the 'tropical' USUV easily established a local mosquito-bird transmission cycle without stringent climatic requirements. The virus found competent vectors in local mosquito species, vulnerable hosts in the local wild bird population and is able to 'survive' the winter months in overwintering female *Culex* mosquitoes. Although – fortunately enough – only of very low pathogenicity for human beings, this introduction of a 'tropical' flavivirus into temperate climate zones deserves continuing scientific attention, as it could serve as a valuable model how other, probably more devastating vector-borne viruses may behave in a new ecosystem with a naive vector- and host population. Finally, seroepidemological surveys focusing on WNV and related flaviviral pathogen will need to be aware of the presence of USUV in certain areas of Europe and to consider the fact that antibodies against this virus will be able to interfere with WNV and other flaviviral test systems.

References

Adam F and Diguette JP (2005) Virus d'Afrique (base de donnes). Centre Collaborateur OMS de référence et de recherche pour les arbovirus et les virus de fièvres hémorrhagiques (CRORA). Institut Pasteur de Dakar. http://www.pasteur.fr/recherche/banques/CRORA (Accessed 23 August 2007).

Andersen AA and Hanson RP (1974) Influence of sex and age on natural resistance to St. Louis encephalitis virus infection in mice. Infect Immun 9: 1123-1125.

Austin RJ, Whiting TL, Anderson RA and Drebot MA (2004) An outbreak of West Nile virus-associated disease in domestic geese (*Anser anser domesticus*) upon initial introduction to a geographic region, with evidence of bird to bird transmission. Can Vet J 45: 117-123.

Bakonyi T, Gould EA, Kolodziejek J, Weissenböck H and Nowotny N (2004) Complete genome analysis and molecular characterization of Usutu virus that emerged in Austria in 2001 – comparison with the South African strain SA AR-1776 and other flaviviruses. Virology 328: 301-310.

Bakonyi T, Lussy H, Weissenböck H, Hornyák A and Nowotny N (2005a) In vitro host-cell susceptibility to Usutu virus. Emerg Infect Dis 11: 298-301.

Bakonyi T, Erdélyi K, Ursu K, Ferenczi E, Csörgő T, Lussy H, Chvala S, Bukovsky C, Meister T, Weissenböck H and Nowotny N (2008) Emergence of Usutu virus in Hungary. J Clin Microbiol (in press).

Bakonyi T, Hubálek Z, Rudolf I and Nowotny N (2005b) Novel flavivirus or new lineage of West Nile virus, central Europe. Emerg Infect Dis 11: 225-231.

Bhatt PN and Jacoby RO (1976) Genetic resistance to lethal flavivirus encephalitis. II. Effect of immunosuppression. J Infect Dis 134: 166-173.

Blackburn NK and Swanepoel R (1980) An investigation of flaviviruses infections of cattle in Zimbabwe Rhodesia with particular reference to Wesselsbron virus. J Hyg 85: 1-33.

Brummer-Korvenkontio M, Saikku P, Korhonen P and Oker-Blom N (1973) Arboviruses in Finland. I. Isolation of tick-borne encephalitis (TBE) virus from arthropods, vertebrates, and patients. Am J Trop Med Hyg 22: 382-389.

Buckley A, Dawson A and Gould EA (2006) Detection of seroconversion to West Nile virus, Usutu virus and Sindbis virus in UK sentinel chickens. Virol J 4: 71.

Buckley A, Dawson A, Moss SR, Hinsley SA, Bellamy PE and Gould EA (2003) Serological evidence of West Nile virus, Usutu virus and Sindbis virus infection of birds in the UK. J Gen Virol 84: 2807-2817.

Chvala S, Kolodziejek J, Nowotny N and Weissenböck H (2004) Pathology and viral distribution in fatal Usutu virus infections of birds from the 2001 and 2002 outbreaks in Austria. J Comp Path 131: 176-185.

Chvala S, Bakonyi T, Bukovsky C, Meister T, Brugger K, Rubel F, Nowotny N and Weissenböck H (2007) Monitoring of Usutu virus activity and spread by using dead bird surveillance in Austria, 2003-2005. Vet Microbiol 122: 237-245.

Chvala S, Bakonyi T, Hackl R, Hess M, Nowotny N and Weissenböck H (2005) Limited pathogenicity of Usutu virus for the domestic chicken (*Gallus domesticus*). Avian Pathol 34: 392-395.

Chvala S, Bakonyi T, Hackl R, Hess M, Nowotny N and Weissenböck H (2006) Limited pathogenicity of Usutu virus for the domestic goose (*Anser anser f. domestica*) following experimental inoculation. J Vet Med B 53: 171-175.

Clarke DH and Casals J (1958) Techniques for hemagglutination and hemagglutination-inhibition with arthropod-borne viruses. Am J Trop Med Hyg 7: 561-573.

De Madrid AT and Porterfield JS (1969) A simple micro-culture method for the study of group B arboviruses. Bull World Health Org 40: 113-121.

Dorrestein GM, Crosta L, Steinmetz HW, Bakonyi T, Nowotny N and Weissenböck H (2007) Usutu virus activity is spreading in Europe. In: Proceedings 9th European AAV Conference, Zürich, p. 7.

Ebel GD, Dupuis AP 2nd, Nicholas D, Young D, Maffei J and Kramer LD (2002) Detection by enzyme-linked immunosorbent assay of antibodies to West Nile virus in birds. Emerg Infect Dis 8: 979-982.

Eidson M, Komar N, Sorhage F, Nelson R, Talbot T, Mostashari F, McLean R and the West Nile virus avian mortality surveillance group (2001) Crow death as a sentinel surveillance system for West Nile virus in the northeastern United States, 1999. Emerg Infect Dis 7, 615-620.

Eldadah AH, Nathanson N and Sarsitis R (1967) Pathogenesis of West Nile virus encephalitis in mice and rats. 1. Influence of age and species on mortality and infection. Am J Epidemiol 86: 765-775.

Ernek E, Kozuch O, Lichard M and Nosek J (1968) The role of birds in the circulation of tick-borne encephalitis virus in the Tribec region. Acta Virol 12: 468-470.

Fazakerley JK (2002) Pathogenesis of Semliki Forest virus encephalitis. J Neurovirol 8 Suppl 2: 66-74.

Ferenczi E, Rácz G, Faludi G, Czeglédi A, Mezey I and Berencsi G (2005) Natural foci of classical and emerging viral zoonoses in Hungary. In: G. Berencsi, A.S. Khan & J. Halouzka (eds.), Emerging biological threat. IOS Press, Amsterdam, pp. 43-49.

Glaser A (2004) West Nile virus and North America: an unfolding story. Rev Sci Tech Off Int Epiz 23: 557-568.

Guptill SC, Julian KG, Campbell GL, Price SD and Marfin AA (2003) Early-season avian deaths from West Nile virus as warnings of human infection. Emerg Infect Dis 9: 483-484.

Henderson BE, McCrae AWR, Kirya BG, Ssenkubuge Y and Sempala SDK (1972) Arbovirus epizootics involving man, mosquitoes and vertebrates at Lunyo, Uganda 1968. Ann Trop Med Parasitol 66: 343-355.

Hubálek Z (1994) Pathogenic microorganisms associated with free-living birds (a review). Acta Sci Nat Brno, 28: 1-74.

Hubálek Z, Chanas AC, Johnson BK and Simpson DI (1979) Cross-neutralization study of seven California group (Bunyaviridae) strains in homoiothermous (PS) and poikilothermous (XTC-2) vertebrate cells. J Gen Virol 42: 357-362.

Hubálek Z, Savage HM, Halouzka J, Juricova Z, Sanogo YO and Lusk S (2000) West Nile virus investigations in South Moravia, Czechland. Viral Immunol 13: 427-433.

Hubálek Z, Halouzka J, Juricova Z and Sebesta O (1998) First isolation of mosquito-borne West Nile virus in the Czech Republic. Acta Virol 42:119-120.

Jacoby RO and Bhatt N (1976) Genetic resistance to lethal flavivirus encephalitis. I. Infection of congenic mice with Banzi virus. J Infect Dis 134: 158-165.

Komar N, Langevin S, Hinten S, Nemeth N, Edwards E, Hettler D, Davis B, Bowen R and Bunning M (2003) Experimental infection of North American birds with the New York 1999 strain of West Nile virus. Emerg Infect Dis 9: 311-322.

Kuno G, Chang GJ, Tsuchiya KR, Karabatsos N and Cropp CB (1998) Phylogeny of the genus Flavivirus. J Virol 72: 73-83.

Meister T, Lussy H, Bakonyi T, Šikutová S, Rudolf I, Vogl W, Winkler H, Frey H, Hubálek Z, Nowotny N and Weissenböck H (2007) Serological evidence of continuing high Usutu virus (*Flaviviridae*) activity and establishment of herd immunity in wild birds in Austria. Vet Microbiol (in press).

Monath TP, Cropp CB and Harrison AK (1983) Mode of entry of a neurotropic arbovirus into the central nervous system. Reinvestigation of an old controversy. Lab Invest 48: 399-410.

Nowotny N and Weissenböck H (2004) Auftreten von Usutu Virus in Österreich – Erforschung des Usutu Virus und seiner Bedeutung für Mensch und Tier in Österreich. Project report, Bundesministerium für Gesundheit und Frauen.

Nowotny N, Bakonyi T, Gould EA, Steinmetz HW, Dorrestein GM, Kolodziejek J and Weissenböck H (2007) Genomic comparison of Usutu virus that emerged in Austria in 2001 with a South African strain, isolated in 1958, and Usutu viruses that emerged in 2005/2006 in Hungary, Switzerland and Italy. In: Proceedings International Meeting on Emerging Diseases and Surveillance, Vienna, p. 99.

Odelola HA and Fabiyi A (1976) Antigenic relationships among Nigerian strains of West Nile virus by complement fixation and agar gel precipitation techniques. Trans Roy Soc Trop Med Hyg 70: 138-140.

Pfeffer M (2003) Usutu virus in Austria. In: Proceedings, 12th Meeting of the European Network for Diagnostics of „Imported' Viral Diseases (ENIVD), Helsinki. http://www.enivd.de/MEMBERS/MP03.PDF

Pfeffer M, Kuhn R, Weissenböck H, Seidel B and Nowotny N (2003) Preliminary characterization of an Usutu virus isolated from a Blue Tit (*Parus caeruleus*) In: Proceedings Jahrestagung der Gesellschaft für Virologie (GfV), Berlin, p. 400.

Samuel MA and Diamond MS (2006) Pathogenesis of West Nile virus infection: a balance between virulence, innate and adaptive immunity, and viral evasion. J Virol 80: 9349-9360.

Simpson DIH (1966) The susceptibility of *Arvicanthis abyssinicus* (Rüppell) to infection with various arboviruses. Trans Roy Soc Trop Med Hyg 60: 248-254.

Steiner HM and Holzer T (2007) Kleinräumige Unterschiede im Zeitraum und Ausmaß des Bestandseinbruchs bei Wiener Amseln (*Turdus merula* L.) nach Auftreten des Usutu Virus. Vogelwarte, submitted.

Steinmetz HW, Bakonyi T, Chvala S, Weissenböck H, Eulenberger U, Hatt JM, Robert N, Hoop R and Nowotny N (2007) Emergence of Usutu virus in Switzerland. In: Proceedings 43rd International Symposium on Diseases of Zoo and Wild Animals, Edinburgh, pp. 129-131.

Straka U (2005) Mehrjährige Beobachtungen an Amseln *Turdus merula* in der Stadt Stockerau (Niederösterreich) vor und nach dem Auftreten des Usutu-Virus. Vogelkundl Nachr Ostösterr 16, 8-11.

Sureau P and Germain M (1978) The role of birds in the ecology of arboviruses in central Africa. In: A.I. Cherepanov (ed.), Transcontinental connections of migratory birds and their role in distribution of arboviruses. Nauka, Novosibirsk, pp. 147-150.

Theiler M and Downs WG (1973) The arthropod-borne viruses of vertebrates. Yale University Press, New Haven and London, pp. 147-408.

Weissenböck H and Nowotny N (2006b) Kontinuierliche Überwachung der Usutu Virus Infektion in Österreich – Geographische Ausbreitung, empfängliche Spezies und Erkennung von Fällen beim Menschen. Project report, Bundesministerium für Gesundheit und Frauen.

Weissenböck H, Kolodziejek J, Url A, Lussy H, Rebel-Bauder B and Nowotny N (2002) Emergence of Usutu virus, an African mosquito-borne Flavivirus of the Japanese encephalitis virus group, central Europe. Emerg Infect Dis 8: 652-656.

Weissenböck H, Kolodziejek J, Fragner K, Kuhn R, Pfeffer M and Nowotny N (2003a) Usutu virus activity in Austria, 2001-2002. Microbes Infect 5: 1132-1136.

Weissenböck H, Kolodziejek J, Chvala S, Bakonyi T, Seidel B, Kuhn R, Pfeffer M and Nowotny N (2003b) Usutu virus activity in Austria: an update. In: Proceedings 4th International Conference on Emerging Zoonoses, Ames, Iowa, p. 106.

Weissenböck H, Bakonyi T, Chvala S and Nowotny N (2004) Experimental Usutu virus infection of suckling mice causes neuronal and glial cell apoptosis and demyelination. Acta Neuropathol 108: 453-460.

Williams MC, Simpson DIH, Haddow AJ and Knight EM (1964) The isolation of West Nile virus from man and of Usutu virus from the bird-biting mosquito *Mansonia aurites* (Theobald) in the Entebbe area of Uganda. Ann Trop Med Parasitol 56: 367-374.

Woodall JP (1964) The viruses isolated from arthropods at the East African Virus Research Institute in the 26 years ending December 1963. Proc E Afr Acad II: 141-146.

10. Should we expect Chikungunya and Dengue in Southern Europe?

Didier Fontenille, Anna Bella Failloux and Roberto Romi

Abstract

Chikungunya and dengue arboviral diseases are in expansion throughout the world. *Aedes aegypti*, the principal vector of dengue and Yellow Fever has disappeared in Europe. In contrast, populations of *Ae. albopictus* established in Europe recently and are spreading to most southern regions. In the laboratory, this species has proven competent for transmitting various viruses. At the same time, an increase of virus importation is observed due to increased international travel. A limited Chikungunya outbreak which occurred in August - September 2007 in Italy confirmed that Europe is definitely at risk for *Aedes*-borne diseases. The risk of local transmission of Chikungunya and/or dengue in southern Europe depends on many historical, social, economic, environmental, climatic and entomological factors that we are presenting and discussing.

Keywords: Chikungunya, dengue, Europe, risk, mosquitoes

Introduction

Establishment and spread of the Asian tiger mosquito *Aedes albopictus* in Europe, as well as the steady increase of imported cases of *Aedes*-borne viruses such as Chikungunya and Dengue, raise the question of risk of these tropical diseases becoming established and spreading in Europe.

What are Chikungunya and dengue?

Epidemiology and symptoms

Chikungunya (CHIK) and dengue (DEN) are diseases caused by arboviruses (arthropod-borne viruses) transmitted to vertebrates by mosquitoes. Vectors responsible for transmission to humans are *Aedes aegypti* (which is also the vector of the Yellow Fever virus), and *Ae. albopictus*. Both diseases are in expansion throughout the world. A major outbreak of CHIK virus occurred in the Indian Ocean between 2004 and 2007, causing more than 3 million cases in Kenya, the West Indian Ocean islands, India, Indonesia, etc. (Pialoux *et al.* 2007). DEN has a worldwide distribution and is present in almost all tropical countries (Gubler 2006).

Chikungunya virus (CHIKV) is an alphavirus of the family Togaviridae, which was isolated for the first time in 1952 in Tanzania (Ross 1956). Its natural vectors are African forest *Aedes* mosquitoes of the subgenera *Diceromyia*, *Stegomyia*, and *Aedimorphus* that feed preferentially on wild primates, which represent the natural hosts of the virus (Diallo *et al.* 1999). Autochthonous human cases and outbreaks have been observed in Africa, East African islands and Asia. Three major phylogenetic groups of virus have been described: West African, Central-East-South African and Asian genotypes (Powers *et al.* 2000). The 'domestic' form of *Ae. aegypti* is closely associated with human habitations, readily enters houses, feeds almost exclusively on human beings, and is ubiquitous throughout the tropics. *Ae. albopictus,* the species implicated in the La Réunion outbreak, is sometimes present far from human habitation, and feeds readily on many species of mammals and birds (Reiter *et al.* 2006). CHIKV is responsible for an acute infection, characterised by high fever, arthralgia, myalgia, headache, and rash which last about 10 days. The typical clinical sign of the disease is arthralgia, which may persist for several months (Jupp and McIntosh 1988, Powers and Logue 2007). Despite being considered as slightly pathogenic, some deaths have been directly attributed to the virus, and abnormally high death rates have been observed during the recent outbreak in La Réunion. (Pialoux *et al.* 2007).

Dengue virus (DENV) is a group of 4 flaviviruses of the family Flaviviridae (DEN1, DEN2, DEN3, DEN4). DEN1 serotype was first isolated in 1943 by Hotta and Kimura, and the other serotypes were isolated between 1944 and 1957. The viruses probably originated from Asian non-human primates (see Gubler 1997 for references on dengue), where sylvatic cycles involving forest mosquitoes exist, as is also the case in Africa (Rudnick 1986). However, virus strains which are currently circulating within human populations are distinct from forest strains (Wang *et al.* 2000). Dengue is found in almost all tropical and sub-tropical regions around the world, predominantly in urban and semi-urban areas. Symptoms of dengue fever vary from a non-symptomatic infection to severe haemorrhagic fever. Normally it involves a few days of flu-like illness with high fever, severe headache, muscle and joint pains, and rash. DENV are responsible for 50-100 million annual infections, 500.000 cases of dengue haemorrhagic fever and more than 20.000 deaths every year, mainly children (Gubler 2006). There has been a dramatic geographical expansion of the disease over the last 50 years, with severe forms like dengue haemorrhagic fever becoming more frequent. Like CHIKV, DENV is transmitted to humans by *Ae. aegypti,* which is the unique and major vector in South America, the Caribbean, and Asia, and by *Ae. albopictus*, which has been implicated in several outbreaks, like in La Réunion in 1978 (Coulanges *et al.* 1979).

Historical distribution of Aedes-borne diseases in temperate and tropical regions

Dengue: It is difficult to distinguish DEN and CHIK diseases based only on descriptions of clinical symptoms in old reports. The first description of dengue was provided in a Chinese encyclopaedia published in 992 (Gubler 1997). DEN-like outbreaks were recorded in the Caribbean in 1635, in Panama in 1699, in Egypt and Indonesia in 1779. It seems that DEN was first observed in Europe in the Mediterranean area in 1778, particularly in Cadiz and Sevilla in Spain (Angolotti 1980) as a consequence of the global expansion of ship movements and urbanisation. Outbreaks occurred again in Cadiz in 1860-1868 and in the Southern USA between 1870–1873, where the virus remained established. Dengue was identified in Athens, in Greece, in 1883–1886, and in 1927–1928 when a major outbreak occurred with more than 1000 deaths and one million cases (Copanaris 1928, Gubler 1997).

The current distribution of DEN covers all tropical countries, with regular and major outbreaks occurring in South East Asia, the Caribbean and South America. No indigenous cases have been recorded in European temperate countries in recent years. The last outbreak in Japan was reported during World War II, the vector being *Ae. albopictus*. Sporadic cases occur in sub-tropical Southern USA in Texas and in Queensland in Northern Australia, where *Ae. aegypti* and *Ae. albopictus* are both present.

Chikungunya: Diagnosis of CHIK disease, between outbreak periods, is rarely achieved, and distribution of the disease may be underestimated. Based on historical description of clinical signs, Carey (1971) suggested that CHIK outbreaks may have occurred as early as 1779 in Jakarta (formerly Batavia, in Indonesia) and in Cairo (Egypt), and later on, in Zanzibar (The Republic of Tanzania) in 1823.

Using more specific diagnostic methods, it has been shown that the virus circulates in intertropical Africa from Senegal to Kenya (Thonnon *et al.* 1999, Chretien *et al.* 2007) and in tropical Asia (Pialoux *et al.* 2007). The recent 2005-2007 outbreak occurred in all Indian Ocean islands and surrounding countries (Kenya, Comoros, Madagascar, Mauritius, Seychelles, La Réunion, Maldives, Sri Lanka, India, Indonesia, and Malaysia). For the first time in August 2007, a Chikungunya outbreak has been declared in Europe with more than 150 human cases in two small towns close to Ravenna in the Emilia Romagna province, on the Adriatic coast of Italy. The vector was *Ae. albopictus* which was very abundant in the area. The virus was detected by PCR (polymerase chain reaction) in several *Ae. albopictus* pools. The virus was very probably introduced by a traveller who returned from South-West India where the virus was circulating.

Yellow Fever: Yellow Fever (YF) is mainly transmitted to humans by *Ae. aegypti* (and *Haemagogus* spp. in the Americas). Outbreaks regularly occur in Africa despite a very efficient vaccine, and to a lesser extent, in South America. Paradoxically, while originating from Africa, YF outbreaks were first reported in the Caribbean after the introduction of its vector via the slave trade from West African slave harbours. The first Caribbean outbreaks occurred in Guadeloupe in 1635 and in 1647 in Barbados. Yellow Fever and *Ae. aegypti* were introduced as early as 1701 in Europe, particularly in areas surrounding harbours of Portugal, Spain, France and Italy. Outbreaks occurred in Gibraltar, Sevilla, Cadiz, Malaga, Lisbon, Porto, Barcelona, Marseille, Brest, Saint- Nazaire, Rochefort, Bordeaux, and Livorno during the 19th and beginning of the 20th Century (Bres 1986). More than 5000 people died in Barcelona between 1821 and 1824, and more than 6000 in Lisbon in 1857. Yellow Fever cases remained restricted to main harbour cities. No outbreak was observed in inland cities. Currently, YF occurs mainly in tropical West and Central Africa, and in South America. No case has ever been observed in Asia despite the presence of competent *Ae. aegypti* populations.

Historical and current distribution of potential vectors in Europe

According to current knowledge on dengue and Chikungunya transmission to man, the two potential vectors in Europe are *Ae. aegypti* and *Ae. albopictus*.

Aedes aegypti, described by Linnaeus in 1762, takes its name from specimens collected in Cairo, Egypt. Despite being recorded in mosquito checklists from several European countries such as Albania, Croatia, France, Greece, Italy, Portugal, Spain (Schaffner *et al.* 2001), it seems that this species is no longer present in Europe as well as in neighbouring countries. However, *Ae. aegypti* was very abundant in southern Europe and western Asia at the beginning of 20th century, particularly in Syria, Lebanon, Turkey, Greece, former Yugoslavia, Italy, Corsica (France) and Spain (Curtin 1967, Aitken 1954). *Aedes aegypti* was considered common in Spain until the 1950's (Rico-Avelló y Rico 1953). In Italy, the species was very common up to World War II but most of the official records stated the presence in harbour cities. (Capra 1944, La Face and Raffaele 1928, Piras 1917). The species was active in summer and was probably re-introduced regularly in these cities. The outbreak which occurred in Livorno in 1804 (a harbour city of the central West Coast of Italy, Leghorn in English) lasted about four months, from August to November, raising the question of mosquito overwintering (Levrè 2002). The last record of *Ae. aegypti* in Italy occurred in Desenzano del Garda (Brescia province, northern Italy) in 1971 (Callot and Delecolle 1972). Two major causes have been suggested for installation and spreading of this species since the 15th century: (1) increase of international exchanges and importation of exotic insect species from Africa, the Carribean and Central-South America, and (2) dramatic development of urbanisation and increase of domestic larval development sites. This period was the golden age of *Ae. aegypti* in Europe. Vector populations disappeared after the 1950's for two major reasons. Firstly the development of sanitation and management of urban water collections, which led to almost complete disappearance of domestic water storage practices, and secondly anti-malaria vector control with dichloro-diphenyl-trichloroethane (DDT), which had a collateral effect on other mosquito species.

Aedes albopictus, originates from Asia, was first described by Skuse in 1894, and has recently spread to all continents. It was recorded for the first time in Europe in Albania in 1979 (Adhami and Reiter 1998), then in Italy (Sabatini *et al.* 1990, Dalla Pozza and Majori 1992). It is now present in all European countries around the Mediterranean Sea (see Chapter 14). It is particularly abundant in Italy and is becoming a major nuisance in South-East France, close to Italian border, since 2005. *Aedes albopictus* was introduced in Italy through shipments of used tyres from the USA (Dalla Pozza *et al.* 1994). In the USA, the species was imported at the beginning of the 1980's from South-East Asia and/or, more probably from the northern regions of Japan (Moore *et al.* 1988). Despite a specific federal law (the so called 'Tyre act'), that imposed a certificate of disinfestation to tyre containers entering the USA (but not leaving) from endemic *Ae. albopictus* areas, the species was reported across most of the USA within a few years (Moore and Mitchell 1997). In Italy, the species has quickly spread across the country, since the first record of a breeding population in the outskirts of Padova (Veneto Region) in 1991, showing a great ability to adapt to different ecological situations. At present, scattered foci of the species are reported in all the regions of Italy, from coastal plains to low inland hills (up to about 600 m of altitude) with the exception of Valle d'Aosta (North-West Italy). *Aedes albopictus* density depends on rainfall, relative humidity and available breeding sites. In Rome, *Ae. albopictus* has encountered particularly favourable conditions. Since the first record in the capital city in 1997, the species has colonised the entire metropolitan area, despite efforts to block its spread (Romi *et al.* 1999, Romi 2001). In Europe and other temperate regions, *Ae. albopictus* mainly colonises man-made breeding sites such as flower plates and pots, particularly in cemeteries and sometimes tree-holes and rock pools. In Italy, after seventeen years of surveillance, *Ae. albopictus* is confirmed to be a strict 'container breeding mosquito' never found in 'natural' breeding sites, with the exception of tree holes (*Platanus spp., Robinia pseudoacacia).*

How to assess Dengue and Chikungunya risk in southern Europe?

The risk of local transmission of DEN, CHIK or other mosquito-borne viruses, is the result of simultaneous presence of the virus, competent mosquitoes, susceptible human hosts, and contact between these three entities. DEN and CHIK viruses are not yet endemic in Europe, all reported cases are imported, with the remarkable exception of the August 2007 Ravenna Province outbreak in Italy. *Aedes aegypti* has disappeared, but *Ae. albopictus* is currently invading southern Europe, particularly in urban areas where human-mosquito contact is increasing. Assessing the risk of transmission in Europe requires evaluation of the frequency of virus importation, spreading of *Ae. albopictus*, human-vector contact, and vector competence of anthropophilic mosquito species. A simple approach for assessing such a risk is to evaluate the components of the basic reproductive rate, R_0, which can be summarised by the following model, first developed by MacDonald (1957) for malaria. If the virus is introduced by a viremic host in a location, it will generate secondary cases if R_0 is > 1.

$$R_0 = (m \cdot a^2 \cdot p^n / -\ln p) \cdot b \cdot c \cdot 1/r$$

m: Ratio of mosquitoes to humans

a: Human biting rate (number bites on a human, per mosquito, per day)
p: Daily survival rate of the mosquitoes
n: Incubation period (days); the extrinsic incubation cycle of the virus
b: Transmission efficiency from an infected mosquito to a human
c: Transmission efficiency from an infected human to a mosquito
1/r: Human infectious period (days)

Virus importation

There is no common institutional registration database for detected arbovirus cases in Europe. Sources of information are statistics of Ministries of Health, tropical disease networks such as TropNetEurop (www.tropnet.net; Accessed 28 August 2007), scientific articles, and reports. Regular reports based on cases observed in European hospitals, or from serological surveys in specific populations (troops, travellers, etc.) show that DENV and to a lesser extent, CHIKV, are regularly introduced to Europe (Jelinek 2000, Badiaga *et al.* 2003). Viruses are imported by tourists and professionals visiting tropical countries, European citizens or soldiers living in endemic regions and returning to Europe, and disease-endemic country citizens visiting or immigrating to Europe. Most DEN cases acquired between 1999 and 2002 by 483 travellers living in Europe came from South East Asia, Central America / the Caribbean and Brazil (Jelinek *et al.* 2002, Wichmann *et al.* 2003). Most imported CHIK cases reported by several European countries (Belgium, Czech Republic, France, Germany, Italy, Norway, Switzerland, UK[10], Spain and Italy) are related to the 2005-2006 Indian Ocean outbreak (Depoortere *et al.* 2006). CHIK and DEN fever are generally not easily recognised in European countries, which has led to a huge underestimation of cases. During the 2007 Italian outbreak, first declared cases were initially suspected as Phlebotomus fever. Moreover, asymptomatic or mild cases are not recorded. Notification has been implemented only in 2006 in metropolitan France (i.e. excluding French Caribbean, Indian Ocean and Pacific Ocean overseas departments). A total of 898 CHIK cases were reported between April 2005 and January 2007 in metropolitan France from confirmed laboratory data (Krastinova *et al.* 2006, see www.invs.sante.fr/(Accessed 28 August 2007). Almost all these cases originated from the Indian Ocean, particularly from La Réunion island where more than 266,000 cases were estimated to have occurred. Most imported cases were observed in Paris and in the South East region where *Ae. albopictus* is now established (in the Alpes Maritimes and Corsica departments). There is regular importation of DEN virus in metropolitan France, particularly from overseas French tropical departments. A total of 209 imported DEN cases have been recorded from laboratory data, between January and November 2006 [11]. The real number, including asymptomatic and mild cases, is probably much higher. Dengue is now endemic, with epidemic episodes of the different serotypes, in French Guyana, Martinique, Guadeloupe, Tahiti and New Caledonia. A limited outbreak of DEN1 occurred in La Réunion and Mayotte in 2004. The number of imported cases is lower in other European countries than in France. For example, in Italy the National reference center on arboviruses at Istituto Superiore di Sanita (ISS) confirmed an average of 40 DEN cases per year over the last five years (Nicoletti and Ciufolini, personal communication). A total of 23 CHIK cases have been detected and confirmed in 2006-2007, which were nearly all contracted by tourists visiting Mauritius (Fusco *et al.* 2006, Beltrame *et al.* 2007). The recent CHIKV outbreak near Ravenna also demonstrated that the virus could be introduced from India, and possibly other parts of Asia.

[10] See www.hpa.org.uk/cdr/archives/2006/cdr5006.pdf (Accessed 29 August 2007).

[11] See www.invs.sante.fr/presse/2007/le_point_sur/dengue_cas_importes_250107/dengue_cas_importes_250107.pdf (Accessed 29 August 2007).

Importation of infected mosquitoes seems negligible compared to travel of infected humans. Most aircrafts arriving from tropical regions are disinsectised and *Ae. aegypti* has rarely been found on aircraft (Gratz *et al.* 2000). During the last two decades, airport malaria cases have sometimes been noted, but never DEN or CHIK cases (Tatem *et al.* 2006). Similarly importation of DEN or CHIK infected *Aedes* eggs (through vertical transmission) by ships must be an exceptional event.

Vector competence of European mosquito populations

Vector competence describes the ability of a mosquito to serve as a disease vector. It is governed by genetic factors that influence the ability of a vector to transmit a pathogen (Hardy *et al.* 1983). It is assessed in the laboratory by experimental infections where infection rate corresponds to the proportion of female mosquitoes capable of transmitting the pathogen by bite after a period of virus replication following an infected blood meal. Vector competence experiments are conducted in secure laboratories using standardised protocols. Batches of mosquitoes of the same age from the same population or colony are allowed to feed on blood containing the virus at a given titer. The proportion of infected mosquitoes is calculated after two weeks by indirect immunofluorescence on head squashes (Kubersky and Rosen 1977) or by RT-PCR (reverse transcription Polymerase Chain Reaction) (Lanciotti *et al.* 1992).

Aedes aegypti

Important variations in vector competence have been observed between *Ae. aegypti* populations (Gubler *et al.* 1979). Two major morphological and behavioural forms have been recognised. *Aedes aegypti formosus* originates from tropical African forests, with both domestic and sylvatic populations, the more urban *Aedes aegypti aegypti is* present in the New World, in Asia and the Pacific area. By investigating the susceptibility of *Ae. aegypti* populations to infection with DEN virus, Failloux *et al.* (2002) confirmed recently that the *formosus* form had lower infection rates than the *aegypti* form. The genetic basis of these differences is not yet well understood. Several quantitative trait loci (QTLs) on chromosomes X, II and III were found to affect midgut susceptibility to dengue infection (Bennett *et al.* 2005, Gomez-Machorro *et al.* 2004). *Aedes aegypti* is also the major CHIK vector to humans in West and East Africa, Comoros, India, etc. A wide variation in CHIKV susceptibility to experimental infection has been demonstrated among different strains of *Ae. aegypti* (Banerjee *et al.* 1988). In the same way, the different strains of CHIKV behave differently in *Ae. aegypti* (Mourya *et al.* 1987). Mourya *et al.* (1994) showed that variation in susceptibility to CHIKV was due to genetic factors related to a gene located on chromosome III. Refractoriness is due to a mesenteron barrier since *Ae. aegypti* strains can support viral multiplication after intrathoracic inoculation (Mourya *et al.* 1998). The mesenteron barrier controlled by CHIKV-specific receptors could be one of the important factors determining susceptibility of mosquitoes to the virus. However, other interacting factors may affect this trait.

Aedes albopictus

Aedes albopictus has been incriminated as a DEN vector in Japan, Indonesia, the Seychelles, Thailand, Malaysia, Indian Ocean islands (Hawley 1988, Coulanges *et al.* 1979) and more recently, in 2001, in Hawaii (Hayes *et al.* 2006), and China (Xu *et al.* 2007). Similar to *Ae. aegypti*, variations among geographic strains of *Ae. albopictus* have been observed in susceptibility to infection with DENV (Gubler and Rosen 1976). In the Americas, *Ae. albopictus* has never been implicated in a DEN outbreak but it has been found naturally infected with DENV in Mexico and in Brazil

(Ibanez-Bernal *et al.* 1997). *Aedes albopictus* from North America has been shown to be an efficient experimental vector of DEN viruses and the YF virus (Mitchell *et al.* 1987, Boromisa *et al.* 1987). Two Italian populations of *Ae. albopictus*, from Padova and Rome respectively, were infected with DEN2 virus using an Emotec system. More than 30% of females were positive after two weeks (R Romi and MG Ciufolini, unpublished data). Regardless of their geographic origin, strains of *Ae. albopictus* are more susceptible to infection with CHIKV than are the strains of *Ae. aegypti* (Tesh *et al.* 1976, Turell *et al.* 1992, Konishi and Yamanishi 1986). *Aedes albopictus* was reported to be more susceptible to an Asian than an African CHIKV strain (Tesh *et al.* 1976). Like in *Ae. aegypti*, different strains of CHIKV replicate differently in *Ae. albopictus* (Vazeille, unpublished data). In areas of recent colonisation like in the Americas, *Ae. albopictus* may be implicated in the transmission of different viruses like West Nile virus (WNV) (Turell *et al.* 2001), viruses from the *Bunyamwera* group of Bunyaviridae (Francy *et al.* 1990), Potosi virus, Cache valley virus, La Crosse virus (Mitchell *et al.* 1998) and Eastern equine encephalitis virus (Niebylski *et al.* 1992). Under experimental conditions, *Ae. albopictus* is a competent vector of at least 22 arboviruses including DEN viruses, YF, CHIKV and Ross River virus (Gratz 2004). Experimental infections showed that *Ae. albopictus* from the South of France is also highly susceptible to CHIKV. At a viral titer of 10^7 pfu/mL in the artificial blood meal, 56-77% of females were able to transmit after two weeks of incubation at 28°C (Failloux and Vazeille, unpublished data). Finally, the 2007 CHIK outbreak in Italy suggests that Italian *Ae. albopictus* are also competent under field conditions.

Other mosquito species

Several other French mosquito species like *Aedes vittatus* and *Ochlerotatus caspius* are also able to experimentally transmit CHIKV whereas *Ae. vexans* and *Oc. detritus* are less susceptible (Failloux and Vazeille, unpublished data).

Vertical transmission

Several arboviruses might be maintained in vector populations through vertical transmission from the female to her progeny. *Aedes aegypti* has been shown capable of transmitting all four DEN serotypes vertically (Khin and Than 1983, Hull *et al.* 1984, Mourya *et al.* 2001, Joshi *et al.* 1997, Joshi *et al.* 2002), as well as many other viruses such as YF (Fontenille *et al.* 1997), WNV (Baqar *et al.* 1993), etc. CHIKV vertical transmission has not been demonstrated yet for *Ae. aegypti* (Mourya 1987). Vertical transmission of the four DEN serotypes has also been observed in the field and experimentally in *Ae. albopictus* (Mitchell and Miller 1990, Shroyer 1990, Serufo *et al.* 1993, Ibáñez-Bernal *et al.* 1997). Following experimental assays, Mourya (1987) concluded the absence of CHIKV vertical transmission in *Ae. albopictus*. However, the virus was isolated (and PCR detected) in two pools of larvae, out of more than 500 pools tested, during the 2006 CHIKV outbreak in La Réunion, which demonstrates that vertical transmission can occur in the field. Moreover, experimental transmission studies conducted with *Ae. albopictus* colonies from La Réunion showed that ovaries of females were infected with CHIKV only six days after the initial infection (Failloux, unpublished data). Finally, vertical transmission of WNV by *Ae. albopictus* has been experimentally demonstrated (Baqar *et al.* 1993). Moreover, Ross River and Sindbis viruses, which are genetically close to CHIKV, have been isolated from field-collected immature stages of *Ae. camptorhynchus* in Australia, indicating the existence of vertical transmission as a mechanism for arbovirus maintenance in mosquito populations in temperate regions (Dhileepan *et al.* 1996). Epidemiological consequences of such vertical transmission are often difficult to assess, and vary depending on epidemiological context.

Vectorial capacity of **Aedes albopictus**

Very few studies have been conducted on vectorial capacity components (i.e. density, trophic preference, longevity, duration of gonotrophic cycles) of European populations of *Ae. albopictus* (with *Ae. aegypti* having disappeared). *Aedes albopictus* is generally considered to have a wide range of hosts for blood feeding, from birds to humans. Like *Ae. aegypti* (Scott *et al.* 2000), *Ae. albopictus* females probably take several blood meals within the same gonotrophic cycle (Do Si Hien 1976). Eggs from the same batch might be laid over several days and be dispersed over several breeding sites. Gonotrophic cycles estimated in the field and in the laboratory last between 3–5 days (Mori and Wada 1977). Its longevity depends on climate, environment, habitat, and populations. It is widely assumed that longevity is around 4–6 weeks (Hawley, 1988).

In Italy, *Ae. albopictus* represents the major human biting pest throughout much of its range. The species has found a peculiar breeding site represented by the manholes of the rainwater collection system and has easily colonised man-made breeding sites in the peridomestic environment (Knudsen *et al.* 1996). In central Italy, diapausing winter eggs hatch during the second half of February and the first half of March. Larvae are present up to November and some females are active until December. The mean pre-imaginal duration of egg hatching to pupation lasts 6-7 days at 25 °C, which is usual for this species (Hawley, 1988). Adult abundance peaks in late August-September. It has recently been observed in Rome that some females remained active during winter and 30% of a set of 650 ovitraps were positive (Figure 1).

This may suggest a polymorphism in an adaptive physiological mechanism to temperate climate conditions and a selection for populations active all year long (Romi *et al.* 2006). While most females are diurnal and exophagic, mosquitoes might be captured biting at night and/or indoors, even at the higher floors of buildings, where breeding sites for larvae are indoor containers (flower pots, etc.).

Entomological risk related to global changes

Lessons from history teach us that Europe may have major outbreaks of *Aedes*- borne arboviruses. The establishment and spreading of a new competent DEN and CHIKV vector in southern

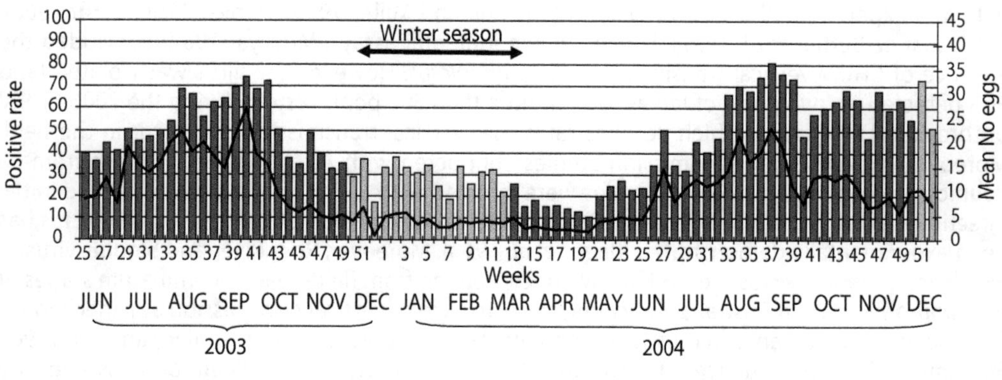

Figure 1. Weekly density of Aedes albopictus in Rome. Percentage of positive traps out of 650 traps (bars, left axis), and mean number of eggs by positive traps (black line, right axis).

Europe creates conditions for establishment of indigenous DEN, CHIK and even YF (witch can be experimentally transmitted by *Ae. albopictus*). Absence of such an outbreak since the time *Ae. albopictus* became established in Europe until 2007 is not an argument for considering this risk negligible. Despite an abundant and very competent vector, the tropical island La Réunion also didn't experience DEN or CHIK outbreaks for more than 36 years, since the last major DEN outbreak in 1978. Indigenous transmission occurred on at least eight occasions between 1980 and 2005 in the USA, following the introduction of the virus by travellers (Gubler 2006). Occurrence of the CHIK outbreak in Italy in August 2007 dramatically demonstrates that Europe must be definitively considered at risk for *Ae. albopictus* borne arboviruses.

Climate

Although temperate climate is less favourable to arbovirus transmission, DEN and YF which are transmitted by *Ae. aegypti* occurred in very temperate regions of Europe (i.e. Glasgow), USA (i.e. Boston), and in South America for example in Argentina: Concordia (18.9 °C; 12.3 °C for average yearly temperature and average coldest month temperature, respectively), Parana (18.2 °C, 11.2 °C) and even in Buenos Aires (16 °C, 10 °C) which experienced several YF outbreaks in the second part of the 19th century (Otero *et al.* 2006). In comparison, several cities in southern Europe have similar temperatures as Buenos Aires; Nice in France (15.6 °C, 9.1 °C), Rome in Italy 17.6 °C, 9.4 °C) and Athens in Greece (18.5 °C, 9.4 °C).

Environment

It is very unlikely that *Ae. aegypti* may re-colonise Europe. Its major breeding sites at the beginning of 20th century, the domestic water storage containers, have mostly disappeared and other ecological niches tend to be occupied by *Ae. albopictus* by which it is out-competed (Juliano *et al.* 2004). Distribution models predict that *Ae. albopictus* will continue to expand (Benedict *et al.* 2007, Knudsen *et al.* 1996, Medlock *et al.* 2006) depending on transport, environment and climate changes. Global warming may possibly contribute to increase the vectorial capacity (i.e. shortening virus extrinsic incubation cycle and mosquito pre-imaginal development time), but it is only a small component of the model. To date, most modelling studies focus on seasonality (i.e. correlation between annual variations of rainfall or temperature, and variation of DEN incidence), rather than on medium to long-term variations in climate (Chadee *et al.* 2007).

Contact between humans and *Ae. albopictus* contact will depend on the nature and location of breeding sites, alternative hosts, besides mosquito and human behaviour. In Italy, as well as in some French locations, *Ae. albopictus* is highly abundant in urbanised and populated areas and is now a major nuisance, demonstrating that human biting rates can be high.

It is difficult to estimate R_0 in Europe if viremic patients are bitten by mosquitoes. However, taking the following plausible values (based on published and non published data, see Hawley 1988) for R_0 parameters: m = 20, a =0.25, p = 0.85, n = 5 (for CHIKV), 1/r = 4, and considering that b and c are 0.7 each, one obtains R_0 = 0.96, close to 1, the threshold for having secondary cases. These values are rough estimations. Survival, duration of gonotrophic cycle, human biting rate and the extrinsic incubation period of the virus, may vary a lot depending on human and mosquito populations, virus strains and seasons. R_0 may therefore be much higher at certain times or in certain places. That means that secondary cases may occur if CHIK or DEN viruses are introduced in a given region where *Ae. albopictus* is abundant and anthropophilic during a warm period.

There is experimental evidence that other indigenous European vectors are sufficiently competent for transmission of DEN or CHIKV. However, human-vector contact and mosquito longevity (i.e. parameters of vectorial capacity) of these rural species must be carefully assessed to evaluate their real (and not theoretical) role as vectors under natural conditions.

Globalisation

During the last few years, imported DEN cases increased in Europe, in relation to expansion of DEN in the world and increase of international travel. In 2005, the total number of passengers transported by air in the European Union (EU) rose by 8.5% compared to 2004, to more than 700 million. Passenger numbers rose by 8.8% in 2004 and by 4.9% in 2003. Of these passengers, 35% were carried on external-EU flights (Figure 2) (De la Fuente Layos 2006).

Eurostat (a European Union agency) estimated that in 2004, a total of 1,474,218 people travelled from Madagascar (153,766), Mauritius (657,312), Mayotte (63,372), La Réunion (498,388) and the Seychelles (101,380) to the European mainland. More than 60 million passengers travelled to Europe from tropical countries in 2006, among them 2.5 million travelled from India, including about 120,000 to Italy. Several million passengers arrived in metropolitan France in 2005 from regions where CHIK or DEN were endemic or epidemic (Table 1).

Mosquito control

Mosquito control both at the community and household level, may reduce population density and contact with humans. Nevertheless, such control efforts have proven difficult to implement against *Ae. albopictus*, as observed in Italy and the USA (see Chapter 14).The establishment of *Ae. albopictus* in Italy has caused great concern among public health authorities (Knudsen *et al.* 1996) and a national plan of surveillance was implemented and funded for three years. Since 1991, a reference center (at ISS) is in charge of monitoring *Ae. albopictus,* by studying mosquito

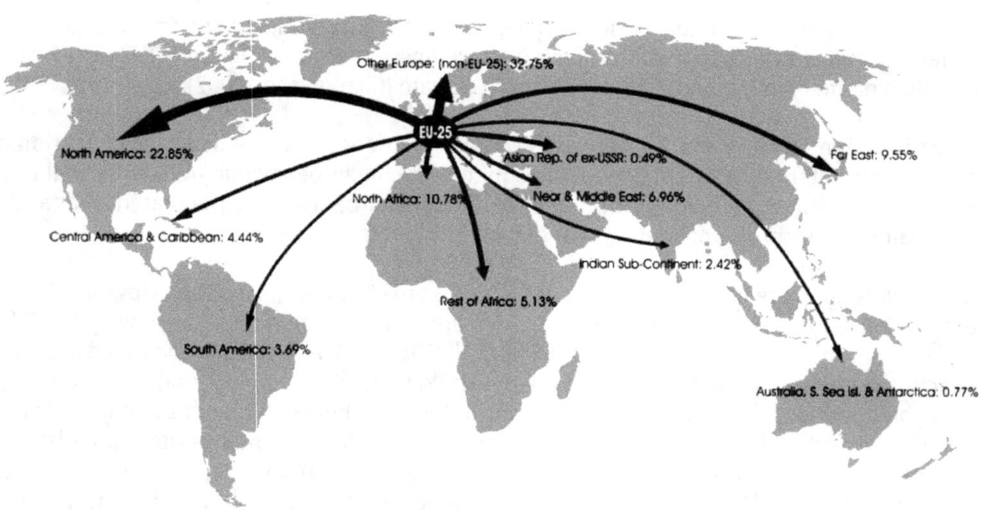

Figure 2. Extra-EU-25 transport of passengers: share in world regions in % of total extra-EU-25 transport in 2005. From De la Fuente Layos (2006).

Table 1. Example of passenger traffic in metropolitan France in 2005 (from Chikungunya or Dengue-endemic regions). From Ministère Français de l'Équipement des Transports, de l'Aménagement du territoire du Tourisme et de la Mer (2006).

Region, country	Number of passengers arriving in France in 2005
Africa, Cameroon	98,364
Africa, Gabon	56,310
Africa, Mauritius	222,215
Africa, Kenya	29,335
Asia, India	278,453
Asia, Malaysia	72,408
Asia, Thailand	189,861
Asia, Vietnam	133,678
South America, Brazil	419,944
South America, French Guyana	95,342
All French overseas departments	1,538,294

bionomics and producing guidelines for vector control. As, up to August 2007, despite its potential role as a powerful vector of arboviruses, *Ae. albopictus* in Italy has represented only a source of nuisance biting, the problem of its presence has been shifted from a health problem to an environmental issue. However, the recent Italian outbreak has led the health authorities to reconsider its vectorial status. Monitoring and control are currently carried out by municipal governments, by following the guidelines provided by the ISS. Control activities mainly consist of source reduction through community education and larvicidal treatment of breeding sites located in public areas, with temephos up to 2006, and with diflubenzuron, pyriproxyfen and *Bacillus thuringiensis israeliensis* formulations at present (Romi *et al.* 2003, Toma *et al.* 2003). Focal adulticide space spraying is used in case of local emergences due to massive biting activity of the species (usually carried out with mixtures of fast knock-down synergised pyrethrins and second-third generation pyrethoids). A plan of action in case of an outbreak had recently been implemented by the Ministry of Health. It involved all the local health structures (ASL, Azienda Sanitaria Locale), the departments of infectious diseases of the hospitals of reference, and the ISS (both MIPI Dept-the national centre of reference for aboviruses – and the Unit of vector control for entomological surveillance) but without specific financial support yet. This plan was activated in August 2007. Vector control consisted of five cycles of early morning outdoor spraying with pyrethroid insecticides immediately after the outbreak was detected. Larvicides (Diflubenzuron, an insect growth regulator, and *Bacillus thuringiensis israeliensis*) have also been used both in the public domain and in all private houses after a door to door inspection.

In metropolitan France, a surveillance system in high-risk areas for *Ae. albopictus* has been implemented since 1999, and successfully detected several *Ae. albopictus* importations, which have been controlled. However, in 2005 despite the French Ministry of Health monitoring system, newly established *Ae. albopictus* populations were observed in South-East France. Attempts to control these populations failed and distribution of the species expanded in 2006 to neighbouring regions and Corsica. Following CHIK outbreaks in French overseas departments, a new plan has been developed in 2006 which includes reinforcement of epidemiological surveillance, CHIK and

DEN case declaration, *Ae. albopictus* population monitoring with a dense network of ovitraps, implementation of mosquito control and a strong public communication plan.

At the European level, the European Centre of Disease Prevention and Control (ECDC) has organised a special meeting (ECDC 2006, Depoortere *et al.* 2006) and international consultations for assessing the risk of CHIK in Europe, after the Indian Ocean and Italian outbreaks. The main recommendations were the improvement of information by airline companies to passengers, by National public health authorities and by ECDC itself; the constitution of a group of experts and European reference centers; the development of research on vector biology; the improvement of measures for monitoring *Ae. albopictus* distribution, and for preventing its importation and spreading, and the improvement of information among health services and general physicians on Chikungunya disease and preventive measures.

Conclusions

The answer to the question 'Should we expect Chikungunya and dengue in Southern Europe?' is 'Yes, definitely'. Current data including the recent Italian outbreak clearly show that receptivity and infectivity of mosquito populations (i.e. vectorial capacity and competence), as well as vulnerability of southern Europe (i.e. virus importation) make indigenous transmission of CHIK and DEN possible. Based on current knowledge, it is impossible to predict the time of occurrence and level of a hypothetic outbreak, from localised secondary cases (as it was probably the case in the Italian province of Ravenna) to huge outbreaks similar to historical DEN or YF in Greece and Spain. A rebound of CHIK transmission had been predicted in La Réunion in 2007 after the 2006 episode yet only few cases have been observed without any clear explanation. Vigilance must, however, be maintained as the risk of an outbreak is definitely present, particularly in regions with well-established *Ae. albopictus* populations and mobile human populations. Unfortunately, ongoing global changes will render many urban areas in southern Europe prone to such conditions within the next few years (Sutherst 2004). Europe should consider establishing permanent surveillance and an adequate response strategy, should a new arbo-virus outbreak be discovered.

Acknowledgements

Fabrice Legros, for information on passenger traffic, Eurostat for data on transport, and Francesco Severini, ISS, for detailed data on *Ae. albopictus* bionomics in Rome.

References

Adhami J and Reiter P (1998) Introduction and establishment of *Aedes* (*Stegomyia*) *albopictus* Skuse (Diptera: Culicidae) in Albania. J Am Mosq Control Assoc 14: 340-343.

Aitken THG (1954) The Culicidae of Sardinia and Corsica (Diptera). Bull Entomol Res 45: 437-494.

Angolotti E (1980) La fiebre amarilla. Historia y situación actual. La fiebre amarilla en la Barcelona de 1821. Rev San Hig Púb 54: 89-102.

Badiaga S, Barrau K, Brouqui P, Durant J, Malva D, Janbon F, Bonnet E, Bosseray A, Sotto A, Peyramont D, Dydymski S, Cazorla C, Tolou H, Durant JP, Delmont J and Infectio-Sud Group (2003) Imported dengue in French university hospitals: a 6-year survey. J Travel Med 10: 286-289.

Banerjee K, Mourya DT and Malunjkar AS (1988) Susceptibility and transmissibility of different geographical strains of *Aedes aegypti* mosquitoes to Chikungunya virus. Indian J Med Res 87: 134-138.

Baqar S, Hayes CG, Murphy JR and Watts DM (1993) Vertical transmission of West Nile virus by *Culex* and *Aedes* species mosquitoes. Am J Trop Med Hyg 48: 757-762.

Beltrame A, Angheben A, Bisoffi Z, Monteiro G, Marocco S, Calleri G, Lipani F, Gobbi F, Canta F, Castelli F, Gulletta M, Bigoni S, Del Punta V, Iacovazzi T, Romi R, Nicoletti L, Ciufolini MG, Rorato G, Negri C and Viale P (2007) Imported Chikungunya infection in Italy. Report of 17 consecutive cases in Travelers. Emerg Infect Dis 13: 1264-1266.

Benedict MQ, Levine RS, Hawley WA and Lounibos LP (2007) Spread of the tiger: global risk of invasion by the mosquito *Aedes albopictus*. Vector Borne Zoon Dis 7: 76-85.

Bennett KE, Flick D, Fleming KH, Jochim R, Beaty BJ and Black WC 4th (2005) Quantitative trait loci that control dengue-2 virus dissemination in the mosquito *Aedes aegypti*. Genetics 170: 185-194.

Boromisa RD, Rai KS and Grimstad PR (1987) Variation in the vector competence of geographic strains of *Aedes albopictus* for dengue 1 virus. J Am Mosq Control Assoc 3: 378-386.

Bres PLJ (1986) A century of progress in combating yellow fever. Bull World Health Organ 64: 775-786.

Callot J and Delecolle JC (1972) Notes d'entomologie. - vi) localisation septentrionale d'*Aedes aegypti*. Ann Parasitol Hum Comp 47: 665.

Capra F (1944) Sulla presenza della *Aedes (Stegomya) aegypti* L. a Genova. 944. Mem Soc Entom Ital 23: 80-81.

Carey DE (1971) Chikungunya and dengue: a case of mistaken identity? J Hist Med Allied Sci 26: 243-262.

Chadee DD, Shivnauth B, Rawlins SC and Chen AA (2007). Climate, mosquito indices and the epidemiology of dengue fever in Trinidad (2002-2004). Ann Trop Med Parasitol 101: 69-77.

Chretien JP, Anyamba A, Bedno SA, Breiman RF, Sang R, Sergon K, Powers AM, Onyango CO, Small J, Tucker CJ and Linthicum KJ (2007) Drought-associated chikungunya emergence along coastal East Africa. Am J Trop Med Hyg 76: 405-407.

Copanaris P (1928) L'épidémie de Dengue en Grèce au cours de l'été 1928. Office International d'Hygiène Publique 20: 1590-1601.

Coulanges P, Clercy Y, Jousset FX, Rodhain F and Hannoun C (1979) Dengue à La Réunion: Isolement d'une souche à l'Institut Pasteur de Madagascar. Bull Soc Pathol Exot Filiales 72: 205-209.

Curtin TJ (1967) Status of *Aedes aegypti* in the Eastern Mediterranean. J Med Entomol 4: 48-50.

Dalla Pozza GL and Majori G (1992) First record of *Aedes albopictus* establishment in Italy. J Am Mosq Control Assoc 8: 318-320.

Dalla Pozza GL, Romi R and Severini C (1994) Source and spread of *Aedes albopictus* in the Veneto region of Italy. J Am Mosq Control Assoc 10: 589-592.

De la Fuente Layos L (2006) Air transport in Europe in 2005. Statistics in focus, Transport. Eurostat publications KS-SF-07-008-EN-N.

Depoortere E and Coulombier D and ECDC Chikungunya risk assessment group (2006) Chikungunya risk assessment for Europe: recommendations for action. Euro Surveill 11: E060511.2. www.eurosurveillance.org/ew/2006/060511.asp#2 (Accessed 29 August 2007)

Dhileepan K, Azuolas JK and Gibson CA (1996) Evidence of vertical transmission of Ross River and Sindbis viruses (Togaviridae: Alphavirus) by mosquitoes (Diptera: Culicidae) in southeastern Australia. J Med Entomol 33: 180-182.

Diallo M, Thonnon J, Traore-Lamizana M and Fontenille D (1999) Vectors of Chikungunya virus in Senegal: current data and transmission cycles. Am J Trop Med Hyg 60: 281-6.

Do Si Hien (1976) Biology of *Aedes aegypti* (L. 1762) and *Aedes albopictus* (Skuse, 1995) (*Diptera, Culicidae*). IV. The feeding of females. Acta Parasitol Pol 34: 27-35.

ECDC (2006) Consultation on Chikungunya risk assessment for Europe. Stockholm, 30 March 2006. meeting report. 10p

Failloux AB, Vazeille M and Rodhain F (2002). Geographic genetic variation in populations of the dengue virus vector *Aedes aegypti*. J Mol Evol 55: 653-663.

Fontenille D, Diallo M, Mondo M, Ndiaye M and Thonnon J (1997) First evidence of natural vertical transmission of yellow fever virus in *Aedes aegypti*, its epidemic vector. Trans R Soc Trop Med Hyg 91: 533-535.

Francy DB, Karabatsos N, Wesson DM, Moore CG Jr, Lazuick JS, Niebylski ML, Tsai TF and Craig GB Jr (1990) A new arbovirus from *Aedes albopictus*, an Asian mosquito established in the United States. Science 250: 1738-1740.

Fusco MF, Puro V, Di Caro A, Nicastri E, Carannante N, Faella FS, Barzon L, Di Cesare S, Palu G, Capobianchi MR and Ippolito G (2006) Casi di febbre Chikungunya in Italia in viaggiatori di ritorno dall'Oceano indiano e rischio di introduzione nel territorio Italiano. Infez Med 4: 238-245.

Gomez-Machorro C, Bennett KE, del Lourdes Munoz M and Black WC 4th (2004) Quantitative trait loci affecting dengue midgut infection barriers in an advanced intercross line of *Aedes aegypti*. Insect Mol Biol 13: 637-648.

Gratz NG (2004) Critical review of the vector status of *Aedes albopictus*. Med Vet Entomol 18: 215-227.

Gratz NG, Steffen R and Cocksedge W (2000) Why aircraft disinsection? Bull World Health Organ 78: 995-1004.

Gubler DJ (1997) Dengue and dengue hemorrhagic fever: its history and resurgence as a global public health problem. In: Gubler DJ, Kuno G, eds. Dengue and dengue hemorrhagic fever [Chapter 1]. Wallingford, United Kingdom: CABI International, pp. 1-22.

Gubler DJ (2006) Dengue/Dengue Haemorrhagic Fever: History and current status in new treatment strategies for dengue and other Flaviviral diseases, pp. 3-22. Novartis Foundation symposia, Gregory Bock, Jamie Goode editors.

Gubler DJ and Rosen L (1976) Variation among geographic strains of *Aedes albopictus* in susceptibility to infection with dengue viruses. Am J Trop Med Hyg 25: 318-825.

Gubler DJ, Nalim S, Tan R, Saipan H and Sulianti Saroso J (1979) Variation in susceptibility to oral infection with dengue viruses among geographic strains of *Aedes aegypti*. Am J Trop Med Hyg 28: 1045-1052.

Hardy JL, Houk EJ, Kramer LD and Reeves WC (1983) Intrinsic factors affecting vector competence of mosquitoes for arboviruses. Ann Rev Entomol 28: 229-262.

Hawley AH (1988) The biology of *Aedes albopictus*. J Am Mosq Control Assoc 4: 2-39.

Hayes JM, Rigau-Perez JG, Reiter P, Effler PV, Pang L, Vorndam V, Hinten SR, Mark KE, Myers MF, Street K, Bergau L, Meyer C, Amador M, Napier M, Clark GG, Biggerstaff BJ and Gubler DJ (2006) Risk factors for infection during a dengue-1 outbreak in Maui, Hawaii, 2001. Trans R Soc Trop Med Hyg 100: 559-566.

Hull B, Tikasingh E, Desouza M and Martinez R (1984) Natural trans-ovarial transmission of dengue-4 virus in *Aedes aegypti* in Trinidad. Am J Trop Med Hyg. 33: 1248-1250.

Ibanez-Bernal S, Briceño B, Mutebi JP, Argot E, Rodríguez G, Martinez-Campos C, Paz R, de la Fuente-San Roman P, Tapia-Conyer R and Flisser A (1997) First record in America of Aedes albopictus naturally infected with dengue virus during the 1995 outbreak at Reynosa, Mexico. Med Vet Entomol 11: 305-309.

Jelinek T (2000) Dengue Fever in International Travelers. Clin Infect Dis 31: 144–147.

Jelinek T, Mühlberger N, Harms G, Corachán M, Grobusch MP, Knobloch J, Bronner U, Laferl H, Kapaun A, Bisoffi Z, Clerinx J, Puente S, Fry G, Schulze M, Hellgren U, Gjørup I, Chalupa P, Hatz C, Matteelli A, Schmid M, Nielsen LN, da Cunha S, Atouguia J, Myrvang B, Fleischer K and European Network on Surveillance of Imported Infectious Diseases (2002) Epidemiology and Clinical Features of Imported Dengue Fever in Europe: Sentinel Surveillance Data from TropNetEurop. Clin Infect Dis 35: 1047–1052.

Joshi V, Singhi M and Chaudhary RC (1996) Transovarial transmission of dengue 3 virus by *Aedes aegypti*. Trans R Soc Trop Med Hyg 90: 643-644.

Joshi V, Mourya DT and Sharma RC (2002) Persistence of dengue-3 virus through transovarial transmission passage in successive generations of *Aedes aegypti* mosquitoes. Am J Trop Med Hyg 67: 158-161.

Juliano SA, Lounibos LP and O'Meara GF (2004) A field test for competitive effects of Aedes albopictus on A. aegypti in South Florida: differences between sites of coexistence and exclusion? Oecologia 139: 583-593.

Jupp PG and McIntosh BM (1988) Chikungunya virus disease. In: Monath TP, editor. The Arboviruses: Epidemiology and ecology. Boca Raton (Florida): CRC Press. pp. 137–157.

Khin MM and Than KA (1983) Trans-ovarial transmission of dengue-2 virus by *Aedes aegypti* in nature. Am J Trop Med Hyg 32: 590-594.

Knudsen AB, Romi R and Majori G (1996) Occurrence and spread in Italy of *Aedes albopictus*, with implications for its introduction into other parts of Europe. J Am Mosq Control Assoc 12: 177-183.

Konishi E and Yamanishi H (1986) Titer distribution analysis of chikungunya virus in *Aedes albopictus* (Diptera: Culicidae). J Med Entomol 23: 92-98.

Krastinova E, Quatresous I and Tarantola A (2006) Imported cases of chikungunya in metropolitan France: update to June 2006. Euro Surveill 11: E060824 1

Kubersky TT and Rosen L (1977) A simple technique for the detection of dengue antigen in mosquitoes by immunofluorescence. Am J Trop Med Hyg 26: 533-537.

La Face L and Raffaele G (1928) Sulla preenza della Stegomya fasciaa nell'Italia meridionale e nella Sicilia Il Policlinico, sezione pratica. XXXV (43) 2095.

Lanciotti RS, Calisher CH, Gubler DJ, Chang GJ and Vorndam AV (1992) Rapid detection and typing of dengue viruses from clinical samples by using reverse transcriptase-polymerase chain reaction. J Clin Microbiol 30: 545-551.

Levrè E (2002) The yellow fever outbreak of 1804 in Leghorn. Annali D'igiene 14 (suppl.1): 153-157.

MacDonald G (1957) The epidemiology and control of malaria, London, Oxford University Press. 201 pp.

Medlock JM, Avenell D, Barrass I and Leach S (2006) Analysis of the potential for survival and seasonal activity of *Aedes albopictus* (Diptera: Culicidae) in the United Kingdom. J Vector Ecol 31: 292-304.

Ministère Français de l'Équipement des Transports, de l'Aménagement du territoire du Tourisme et de la Mer, division de l'Observation du marché et des Etudes économiques (2006) Bulletin Statistique, Trafic Commercial, Année 2005: 89 pp.

Mitchell CJ and Miller BR (1990) Vertical transmission of dengue viruses by strains of *Aedes albopictus* recently introduced into Brazil. J Am Mosq Control Assoc 6: 251-253.

Mitchell CJ, Haramis LD, Karabatsos N, Smith GC and Starwalt VJ (1998) Isolation of La Crosse, Cache Valley, and Potosi viruses from *Aedes* mosquitoes (Diptera: Culicidae) collected at used-tire sites in Illinois during 1994-95. J Med Entomol 35: 573-577.

Mitchell CJ, Miller BR and Gubler DJ (1987) Vector competence of *Aedes albopictus* from Houston, Texas, for dengue serotypes 1 to 4, yellow fever and Ross River viruses. J Am Mosq Control Assoc 3: 460-465.

Moore CG, Francy DB, Eliason DA and Monath TP (1988) *Aedes albopictus* in the United States: rapid spread of a potential disease vector. Am Mosq Control Assoc 4: 360-361.

Moore CG and Mitchell CJ (1997) *Aedes albopictus* in the United States: ten-year presence and public health implications. Emerg Infect Dis 3: 329-334.

Mori A and Wada Y (1977) The gonotrophic cycle of *Aedes albopictus* in the field. Trop Med 19: 141-146.

Mourya DT (1987) Absence of transovarial transmission of Chikungunya virus in *Aedes aegypti* & *Ae. albopictus* mosquitoes. Indian J Med Res 85: 593-595.

Mourya DT, Gokhale MD, Malunjkar AS, Bhat HR and Banerjee K (1994) Inheritance of oral susceptibility of *Aedes aegypti* to Chikungunya virus. Am J Trop Med Hyg 51: 295-300.

Mourya DT, Gokhale MD, Basu A, Barde PV, Sapkal GN, Padbidri VS and Gore MM (2001) Horizontal and vertical transmission of dengue virus type 2 in highly and lowly susceptible strains of *Aedes aegypti* mosquitoes. Acta Virol 45: 67-71.

Mourya DT, Malunjkar AS and Banerjee K (1987) Susceptibility & transmissibility of *Aedes aegypti* to four strains of chikungunya virus. Indian J Med Res 86: 185-190.

Mourya DT, Ranadive SN, Gokhale MD, Barde PV, Padbidri VS and Banerjee K (1998) Putative chikungunya virus-specific receptor proteins on the midgut brush border membrane of *Aedes aegypti* mosquito. Indian J Med Res 107: 10-14.

Niebylski ML, Mutebi JP, Craig GB Jr, Mulrennann JA and Hopkins RS (1992) Eastern equine encephalitis virus associated with *Aedes albopictus* Florida 1991. MMWR Morbidity and Mortality Weekly Report 41: 115, 121.

Otero M, Solari HG and Schweigmann N (2006) A stochastic population dynamics model for *Aedes aegypti*: formulation and application to a city with temperate climate. Bull Math Biol 68: 1945-74.

Pialoux G, Gauzere BA, Jaureguiberry S and Strobel M (2007) Chikungunya, an epidemic arbovirosis. Lancet Infect Dis 7: 319-27.

Piras L (1917) Sulla presenza della *Stegomya calopus* (Blanchard, 1907) nell'ambito del porto di Genova. L'igiene moderna, X (1): 1-2

Powers AM and Logue CH (2007) Changing patterns of chikungunya virus: re-emergence of a zoonotic arbovirus. J Gen Virol 88: 2363-2377.

Powers AM, Brault AC, Tesh RB and Weaver SC (2000) Re-emergence of chikungunya and o'nyong-nyong viruses: evidence for distinct geographical lineages and distant evolutionary relationships. J Gen Virol 81: 471-479.

Reiter P, Fontenille D and Paupy C (2006) *Aedes albopictus* as an epidemic vector of chikungunya virus: another emerging problem? Lancet Infect Dis 6: 463-464.

Rico-Avello Y and Rico C (1953) Fiebre Amarilla en España (Epidemiología histórica). Revista de Sanidad e Higiene Pública, Año XXVII, núm 1-2: 29-87

Romi R (1996) Linee guida per la sorveglianza ed il controllo di Aedes albopictus in Italia. Rapporti ISTISAN. 96/4, 52p.

Romi R (2001) *Aedes albopictus* in Italia: un problema sanitario sottovalutato. Ann. Ist Super Sanità 37: 241-247.

Romi R, di Luca M and Majori G (1999) Current status of *Aedes albopictus* and *Ae. atropalpus* in Italy. J Am Mosq Control Assoc 15: 425-427.

Romi R, Severini F and Toma L (2006) Cold acclimatation and overwintering of female *Aedes albopictus* in Roma. J Am Mosq Control Assoc 22: 153-155.

Romi R, Toma L, Severini F and di Luca M (2003) Susceptibility of *Aedes albopictus* Italian populations to temephos and to other insecticides. J Am Mosq Control Assoc 19: 419-423.

Ross RW (1956) The Newala epidemic. III. The virus: Isolation, pathogenic properties and relationship to the epidemic. J Hyg 54: 177–191.

Rudnick A (1986) Dengue Virus Ecology in Malaysia. In Dengue fever studies in Malaysia, Bulletin num 23. Institute for medical research, Kuala Lumpur, Malaysia. Rudnick and Lim ed. P. 51-153.

Sabatini A, Raineri V, Trovato G and Coluzzi M (1990) *Aedes albopictus* in Italia e possibile diffusione della specie nell'area Mediterranea. Parassitologia 32: 301-304.

Schaffner F, Angel G, Geoffroy B, Hervy JP, Rhaim A and Brunhes J (2001) The mosquitoes of Europe. CD Rom IRD Editions.

Scott TW, Morrison AC, Lorenz LH, Clark GG, Strickman D, Kittayapong P, Zhou H and Edman JD (2000) Longitudinal studies of *Aedes aegypti* (Diptera: Culicidae) in Thailand and Puerto Rico: population dynamics. J Med Entomol 37: 77-88.

Serufo JC, de Oca HM, Tavares VA, Souza AM, Rosa RV, Jamal MC, Lemos JR, Oliveira MA, Nogueira RM and Schatzmayr HG (1993) Isolation of dengue virus type 1 from larvae of *Aedes albopictus* in Campos Altos city, State of Minas Gerais, Brazil. Mem Inst Oswaldo Cruz 88: 503-504.

Shroyer DA (1990) Vertical maintenance of dengue-1 virus in sequential generations of *Aedes albopictus*. J Am Mosq Control Assoc 6: 312-314.

Sutherst RW (2004) Global change and human vulnerability to vector-borne diseases. Clin Microbiol Rev 17: 136-173.

Tatem AJ, Rogers DJ and Hay S (2006) Estimating the malaria risk of African mosquito movement by air travel. Malar J. 5: 57.

Tesh RB, Gubler DJ and Rosen L (1976) Variation of geographic strains of *Aedes albopictus* in susceptibility in infection with chikungunya virus. Am J Trop Med Hyg 25: 326-335.

Thonnon J, Spiegel A, Diallo M, Diallo A and Fontenille D (1999) Epidémies à virus Chikungunya en 1996 et 1997 au Sénégal. Bull Soc Pathol Exot 92: 79-82.

Toma L, Severini F, Bella A and Romi R (2003) A semifield evaluation of VECTOBAC® DT (ABG-6499), a new formulation of *Bacillus thuringiensis israeliensis* for control of *Aedes albopictus*. J Am Mosq Control Assoc 19: 424-429.

Turell MJ, Beaman JR and Tammariello RF (1992) Susceptibility of selected strains of *Aedes aegypti* and *Aedes albopictus* (Diptera: Culicidae) to chikungunya virus. J Med Entomol 29: 49-53.

Turell MJ, O'Guinn ML, Dohm DL and Jones JW (2001) Vector competence of North American mosquitoes (Diptera: Culicidae) for West Nile virus. J Med Entomol 38, 130-134.

Wang E, Ni H, Xu R, Barrett AD, Watowich SJ, Gubler DJ and Weaver SC (2000) Evolutionary relationships of endemic/epidemic and sylvatic dengue viruses. J Virol 74: 3227–3234.

Wichmann O, Muhlberger N and Jelinek T (2003) Dengue - the underestimated risk in travellers. Dengue Bull 27: 126-137.

Xu G, Dong H, Shi N, Liu S, Zhou A, Cheng Z, Chen G, Liu J, Fang T, Zhang H, Gu C, Tan X, Ye J, Xie S and Cao G (2007) An outbreak of Dengue virus serotype 1 infection in Cixi, Ningbo, People's Republic of China, 2004, associated with a traveller from Thailand and high density of *Aedes albopictus*. Am J Trop Med Hyg 76: 1182-1188.

Tick-borne encephalitis and Lyme disease

11. Tick-borne encephalitis in Europe: dynamics of changing risk

Sarah E. Randolph and Dana Šumilo

Abstract

Across Europe, tick-borne encephalitis (TBE) shows significant spatio-temporal heterogeneity in changing incidence over the past 3-4 decades, and offers a perfect case study of the diverse impact of humans, both direct and indirect, on zoonotic risk. Attention is focused on the two sides of the zoonotic coin, transmission potential via enzootic cycles and exposure of humans to infection. With few exceptions, patterns of change are unrelentingly upwards, nowhere more dramatically so than in the Eastern European countries coincident with independence from Soviet rule. Also, where endemic risk areas have historically been more patchily distributed in western Europe, new foci have appeared since 2000. The environmental conditions in new and older foci can be compared using new generation satellite imagery in predictive risk mapping. When TBE epidemiology and climate were analysed at sufficiently fine spatial and temporal resolution in the Baltic States, it was clear that, although a step increase in spring temperatures occurred in 1989 that could promote the transmission of TBE virus between co-feeding ticks, the changes in climate were too uniform over this geographical region to explain the spatio-temporal heterogeneity in TBE upsurges. Data support the notion that other environmental changes associated with political upheaval could also have promoted increasingly robust enzootic cycles of TBE virus, while socio-economic transition would have created conditions likely to increase human exposure to ticks to a greater or lesser extent. Demographic and socio-economic factors associated with individual risk are consistent with this. Furthermore, in many parts of the Baltic countries, there were spatially variable decreases in incidence since 1999 that were too great to be accounted for by the recorded rates of vaccination, but significantly correlated with previous incidence. This suggests human responses to risk perception, quite possibly involving changes in behaviour associated with visits to tick-infested forests, as have indeed been reported in health surveys.

Keywords: epidemiology, new foci, vaccination, risk factors, environmental change, socio-economic transition

Recent changes in tick-borne encephalitis epidemiology

The most significant vector-borne diseases throughout central and northern Europe are those caused by pathogens transmitted by ticks. The two principal vector species of medical importance are *Ixodes ricinus* L., widely distributed from Ireland to the Urals and from Sweden to Tunisia, and *I. persulcatus* Schulze, occupying a broad northerly band from the Baltic countries to the Far East. The bacterial infection Lyme borreliosis (LB) is very widespread and prevalent throughout the ticks' ranges, but although it can be clinically very serious, it is virtually never fatal. The flavivirus that causes tick-borne encephalitis (TBE) is more virulent, with infection by the Western virus strain in Europe resulting in a c. 1% case fatality rate, long recovery processes, neurological sequelae and severely decreased quality of life (Dumpis *et al.* 1999, Charrel *et al.* 2004).

Whereas LB has been recorded non-systematically only since the mid 1980s, cases of TBE have been registered systematically for the past 30-50 years. Patterns of changing incidence of TBE over this period show significant spatio-temporal heterogeneity, thereby offering a perfect case study of the diverse impact of humans, both direct and indirect, on zoonotic risk. Many aspects of the transmission dynamics of tick-borne pathogens are sensitive to natural and human-induced variation in environmental factors of all sorts: climate affects the development and survival rates of ticks and the probability of contact between ticks of each stage and competent transmission hosts (Randolph and Storey 1999); landscape structure affects habitat suitability for ticks and hosts (Daniel *et al.* 1998, Kitron 1998, Estrada-Pena 2003); and abundance of wildlife and domestic animals affects the availability of blood-meal hosts for ticks and transmission hosts for the pathogens. The abundance of infected ticks is thus variable. Furthermore, changing human use of natural environments affects the degree of exposure to the risk of infection by these ticks.

Changing TBE incidence

The general pattern throughout Europe has been sudden or gradual increases in TBE incidence in each country over the past two decades (Figure 1), although there are several notable exceptions. In Austria, home of one of the safe and effective vaccines against TBE (produced by Baxter AG, Vienna), a very successful vaccination campaign since 1979 has culminated in nearly 90% protection of the population (Kunz 2003), matched by an equal reduction in TBE incidence (Šumilo, unpublished data). The step decrease in 1997 in Hungary (and Croatia) coincided with a change in funding arrangements for Hungarian public health practices, including TBE diagnosis, resulting in a decrease in the number of laboratory tests (Csepregi Lajos, personal communication). Even before this, however, there were shifts in the distribution of cases within Hungary, with fewer cases in the southwest of the country and new foci appearing progressively eastwards, both up to and after 1997 (unpublished analysis from the EU-F6 EDEN project, see http://www.eden-fp6project.net/ Accessed 10 September 2007). This not only suggests causes other than national public health changes, but also emphasises the importance of analysing epidemiological data at sufficiently fine spatial and temporal scales to allow progress from apparent coincidences to correctly identified causal relationships. Finally, the massive upsurge in TBE incidence that occurred in Eastern Europe, especially the Baltic countries, Poland and Slovakia, coincident with the end of Soviet rule (discussed below), has been followed by a decrease in Latvia since 1999. Not only must all these variable national patterns be accommodated within a consistent multi-factorial explanation, but the variable patterns within each country (as in Hungary, and see below) must also fit any explanation for both increases and decreases in incidence. This is the goal of on-going work, not yet fully achieved.

The zoonotic iceberg

An obvious analogy for zoonotic infections is the iceberg (Figure 2A). Human cases are only the visible tip of the iceberg. The wildlife transmission cycles upon which persistence of the pathogen depends are largely hidden beneath the surface, with exposure to humans varying in time and space. Incidence in humans depends on two interrelated features, the bulk of the iceberg (equivalent to the natural transmission potential) and the relative exposure (equivalent to the degree of human contact with infection and susceptibility to infection). The differential distribution and prevalence of Lyme borreliosis and TBE in Europe are due not to differences in exposure, as both are transmitted principally by *Ixodes ricinus* nymphs and both appear to be equally infective to humans, but to the very much greater transmission potential of *Borrelia burgdorferi s.l.* spirochaetes due to their long infective period in a wide range of competent transmission hosts (Randolph *et al.* 1996, Piesman and Gern 2004). Within any one disease system, the search for causes of changing incidence is helped by distinguishing between changing biotic and abiotic factors that may have altered the bulk of the iceberg, and changing human factors that may have altered its relative exposure without affecting the bulk.

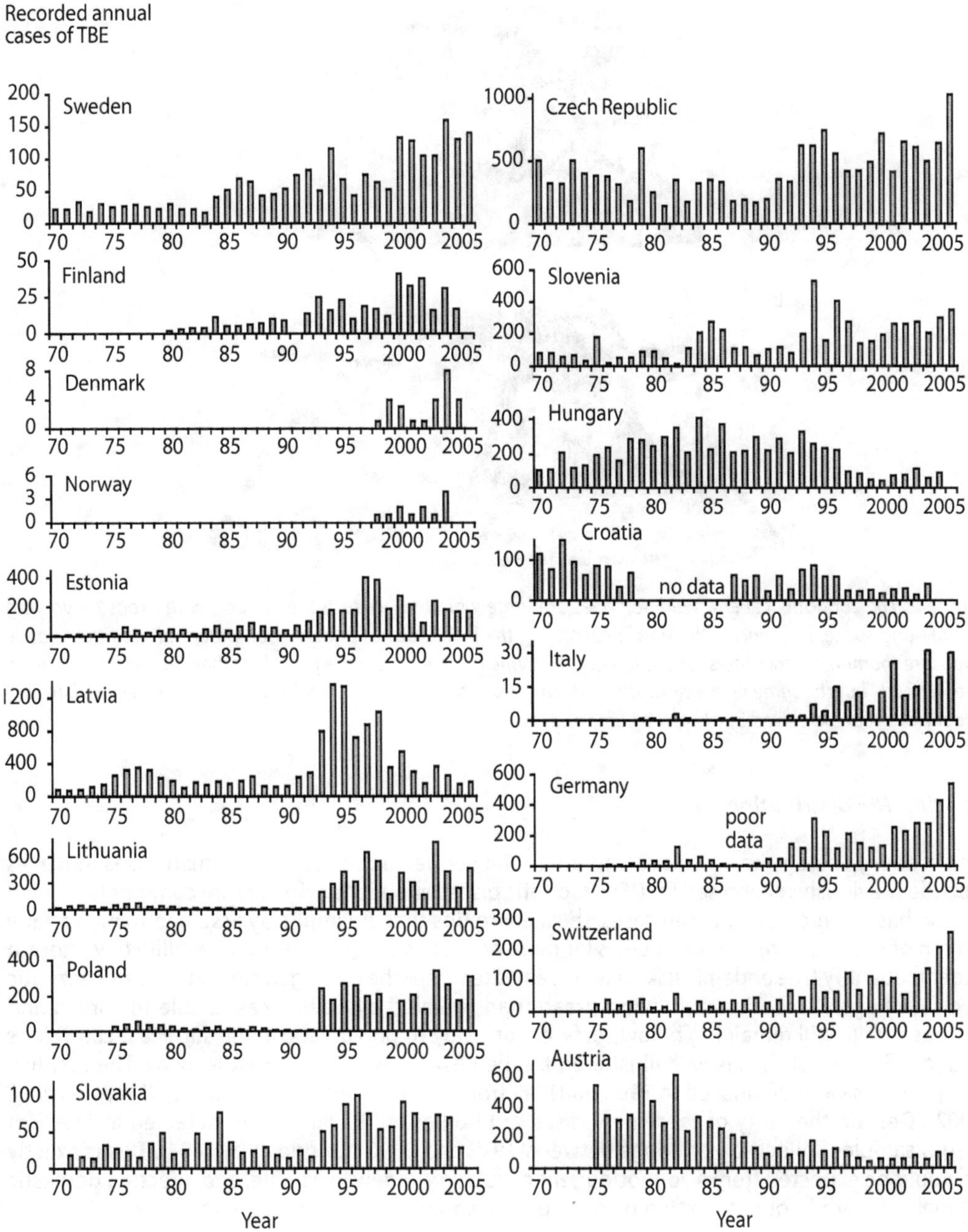

Figure 1. The recorded annual number of cases of tick-borne encephalitis in different countries in Europe from 1970 to 2006.

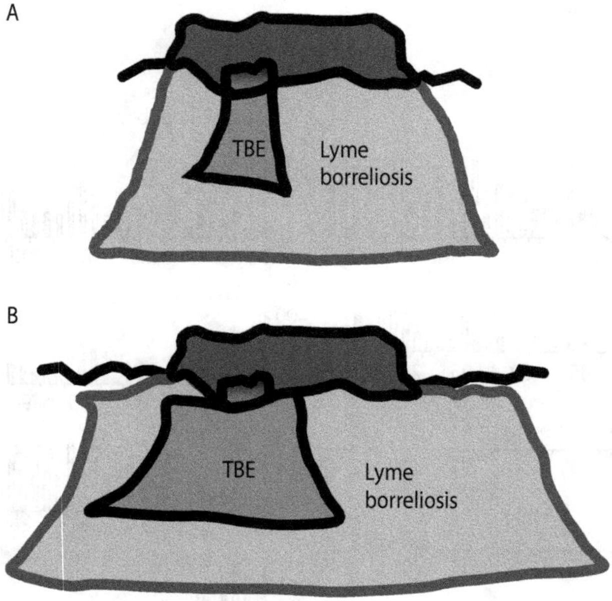

Figure 2. The zoonotic iceberg. Wildlife cycles are largely hidden beneath the surface, with exposure varying in time and space. Incidence in humans depends on the bulk (natural transmission potential) and the relative exposure (human contact rates) of the iceberg. (A) Lyme borreliosis is more prevalent than TBE because it has a greater bulk, but the same relative exposure. (B) True enzootic cycles are probably much more extensive beneath the surface than perceived by observed human cases.

Shifting TBE distribution

This iceberg analogy can also accommodate the sudden appearance of human cases in places outside previously recognised foci. For zoonotic diseases, endemic risk foci are commonly defined on the basis of recorded human cases, the iceberg tips that are implicitly assumed to map neatly on top of the submerged bulk. It takes far more effort to survey infections in wildlife or vectors in order to identify true endemic risk areas, i.e. any site where the pathogen circulates. These enzootic cycles are likely to be far more widespread than revealed by human cases, able to spread and increase while still remaining below surface, but ready to pop up above the surface if conditions change (Figure 2B). This is well illustrated by the new focus in South-East Norway. The first five human cases were diagnosed in 1998-2001, all from Tromøya, Aust-Agder county (Skarpass *et al.* 2002). Despite the rarity of these new cases, antibodies to TBE virus were detected in 16.4% of serum samples collected from dogs between 1992 and 2000 (Csango *et al.* 2004). This is exactly as would be expected from a zoonotic agent circulating widely in wildlife, and infecting domestic animals that are frequently bitten by ticks, before appearing in the human population.

For TBE, such events have occurred recently more or less simultaneously in many parts of Europe (Table 1), assuming that new recorded foci are not artefacts of increased surveillance/recording effort. It appears that new foci are appearing more commonly in Western than in Eastern Europe, but this may merely reflect the greater patchiness of foci here historically, leaving more vacant areas to be filled in. It seems highly unlikely that such a pattern of emergent foci is due to

Table 1. Examples of new foci of recorded human cases of tick-borne encephalitis in Western Europe.

Country	Regions	Date of first record	Source of information
Austria	Voralberg	2000	www.tbe-info.com/tbe.aspx
Germany	Amberg	2000	www.rki.de
	Sigmaringen	2002	www.rki.de
	Marburg	2000	www.rki.de
	Hildburg-hausen	2004	www.rki.de
	Mecklenburg-Stelitz	2004	(Hemmer *et al.* 2005)
Italy	Friuli-Venezia Giulia	1998	A. Rizzoli (pers. comm.)
France	Munster Valley	2002-03	(Hansmann *et al.* 2006)
	Grenobles & Faverges	?	www.baxtervaccines.com/?print/430
	Nancy	?	www.tbe-info.com/tbe.aspx
Switzerland	Vaud, northern Bern, Solothurn, Uri, Nidwalden, Obwalden, Schwyz, St Gallen, Glarus	2000 or later	www.bag.admin.ch
Norway	Aust-Agder county	1998	(Skarpass *et al.* 2002)
Sweden	Skåne	2000	(Falt *et al.* 2006)

independent introductions, carried by birds for example, as has been inferred for *B. burgdorferi s.l.* (Comstedt *et al.* 2006) and *Ehrlichia* (now *Anaplasma*) spp. (Bjoersdorff *et al.* 2001). Birds are not the ideal transporters of TBE virus; they are not known to be competent transmission hosts, so can only carry feeding ticks that were already infected before they attached, which are likely to be very few in number. If such introductions are nevertheless common enough to account for the apparently geographical mosaic of distantly related genotypes (Jääskeläinen *et al.* 2006, O. Valapahti, personal communication) and also this sudden rash of new foci, the question arises, why now? Alternatively, if such introductions are rather rare events, the question is why so many and why now? Answers to both questions involve the notion of altered conditions to increase the likelihood of either introduction or establishment, or both. By identifying the conditions associated with human cases, and investigating whether these also apply to enzootic cycles, we may begin to understand the underlying causes, and thereby predict the probability of shifting risk foci by observing changes in these limiting conditions. While some factors, such as natural environmental conditions, change relatively slowly and may therefore offer the potential for advance warning of future emergence, others, especially those involving human activities, change much more rapidly and unpredictably.

From descriptive to explanatory and predictive epidemiology

Descriptions of the biological mechanism of TBE virus transmission from tick to tick via rodents on the one hand, and of the well-mapped foci of TBE across Europe on the other hand, offer twin starting points for explaining how and therefore where TBE virus can persist under natural conditions. TBE virus is transmitted via short-lived, non-viraemic infections in its rodent transmission hosts, principally *Apodemus* spp. (Labuda *et al.* 1993b). Transmission therefore requires that infectible recipient ticks feed together with the infected donor ticks (so-called co-feeding) (Labuda *et al.* 1993a, 1997), which they have been observed to do disproportionately in the wild within

TBE foci in Slovakia (Randolph *et al.* 1999). Furthermore, because ticks feed only once per life stage, as larvae, nymphs and adults, and because only larvae and nymphs habitually feed on rodents, natural amplification of TBE virus to allow persistent cycles depends on transmission from the relatively few infected nymphs to the much more abundant infectible larvae (which then moult to infected, host-seeking nymphs up to one year later) (Randolph 2004a). Although other transmission routes are biologically possible (transovarially infected larva to other larvae, transstadially infected nymph to other nymphs, etc), only the transstadially infected nymph to larvae transmission route is quantitatively able to account for TBE virus persistence. This demands that larvae and nymphs show synchronous periods of seasonal feeding activity, which occurs to different degrees in different places according to the temperature-driven patterns of seasonal population dynamics of each tick stage (Randolph *et al.* 2000).

If we can define the climatic conditions under which these biological requirements for TBE virus transmission are met, and then identify where they occur, we can explain the current distribution of TBE foci. We can then go on to predict presence or absence in unmonitored places on the basis of the prevailing climate and be alert to any likely shift in distribution in response to changing climate. The most versatile approach is to construct full biological process-driven models incorporating all the dynamics of tick population ecology and virus transmission. Notice the emphasis on *full* and *all* in this ideal. Anything short of complete quantification of all relationships between the various biotic components and with the abiotic drivers will yield unreliable predictions. Fortunately, the indirect causal biological linkage between environmental drivers, rates of biological processes and resultant epidemiological patterns allows an alternative, complementary approach based on the direct correlation between environment and epidemiological pattern. Statistical, pattern-matching models are less demanding of detailed quantification, and although less versatile in their predictions, may themselves be used to identify the crucial environmental drivers. This is based on the reasonable assumption that the most significant environmental correlates of the observed distributional patterns are likely to be the most significant causal drivers. This assumption may then be tested by exploring the biological mechanisms underpinning the purported causality.

TBE in space: predictive mapping of foci

The first continental-scale predictive risk map for a vector-borne disease, TBE, was based on correlations between mapped human cases recorded up to 1997 and mean satellite-derived environmental conditions for 1982-93 (Randolph 2000). Remotely sensed surrogate measures of climatic conditions are used for such risk mapping (Hay *et al.* 2006a) because they provide blanket cover of data over continental areas at fine spatial resolution, unlike ground meteorological stations. Polar-orbitting satellites return raw images of the entire Earth's surface twice a day. Even after compositing over a number of days to remove cloud contamination, these also allow sufficiently fine temporal resolution to capture seasonality, an essential component of most vector-borne disease systems. The MODIS (MODerate Resolution Imaging Spectroradiometer) sensor on board the Terra satellite, launched in December 1999, now offers better spectral, spatial and temporal resolution than the earlier NOAA-AVHRR (National Oceanographic and Atmospheric Administration Advanced Very High Resolution Radiometer) that originally gave images at 8 x 8 km resolution and later at 1 x 1 km (Hay *et al.* 2006b). Nevertheless, the mean 1982-93 AVHRR imagery at 8 x 8 km pixel size yielded a predictive map of TBE presence with 85% correct fit to the observations, and, more importantly, identified specific features of summer moisture and seasonal temperature as the most significant limiting factors. The former factor (measured as the annual amplitude or maximum NDVI - Normalised Difference Vegetation Index) simply indicates the general suitability of the habitat for *I. ricinus*; this tick species requires high humidity at ground level for long-term survival and active host-seeking, which is typically found in deciduous woods with high maximum NDVI values (Daniel *et al.* 1998). This is clearly necessary but not sufficient, otherwise TBE would be found wherever there are tick populations, as with LB. The additional limiting factor involves seasonal temperature (measured as LST - Land Surface Temperature); detailed investigation identified a relatively high rate of cooling in the autumn, corrected for mid-summer maximum temperatures, as by far the most significant correlate of the presence of TBE and also of the synchronous seasonal activity of larval and nymphal ticks (Randolph *et al.* 2000).

This conclusion concerning autumnal cooling was statistically correct and not contradicted by anything known at the time. Later, however, it no longer matched the new quantitative framework for the tick's seasonal population dynamics derived from detailed field data (Randolph *et al.* 2002), when the rate of warming in the spring seemed intuitively to be biologically more important than autumn conditions. Nymphal ticks require a lower threshold temperature for activity (weekly mean daily maximum of at least 7 °C) (Macleod 1935, Perret *et al.* 2000) than do larvae (10 °C) (Randolph 2004a). The faster the rising temperatures pass through the interval between 7 and 10 °C, the shorter the period between the onset of activity by nymphs and larvae, thereby potentially reducing any wasted bites by infected nymphs before larvae are also feeding and infectible. This can only increase the transmission potential of TBE virus. Recently, the solution to this paradox arose from the (embarrassingly) simple expedient of correcting the rate of spring warming (from February to April) by January minimum temperatures (rather than annual maximum temperatures); then spring warming assumed its (apparently) correct position fitting both statistical patterns and biological explanations (Figure 3A). A principal components analysis (Figure 3B) confirmed the distinctly higher rate of spring warming in TBE-presence pixels than in TBE-absence pixels, t = 25.23, $P < 0.001$, on a grid of 1,992 regularly spaced pixels across Europe. There are now one fewer loose ends in the explanation for TBE focal distributions.

The time is now ripe to re-visit the original predictive risk map for TBE, equipped with the improved satellite imagery available for recent years, to explore the environmental correlates of new foci. We can ask a number of questions whose answers may help to explain the temporal and spatial dynamics of TBE risk. For example, do new foci fall within the false positive predictions of the original map, suggesting that conditions for virus circulation were appropriate before and the tip of the iceberg is now exposed for one reason or another? In which case, what are those reasons? Alternatively, have conditions changed to make sites of new foci more like those of old established foci? These investigations are just starting. Unfortunately, while satellite imagery can give us excellent comparisons in space (see above), they cannot give us comparisons over time because changes in the sensor platform, and in the precise timing and position of each satellite orbit, introduce shift and drift that are likely to be at least as large, if not larger, than any true changes in climatic conditions under investigation. It will be necessary to revert to ground-based instrumental meteorological records, even though periodic satellite-driven predictive maps will continue to provide important complementary insight.

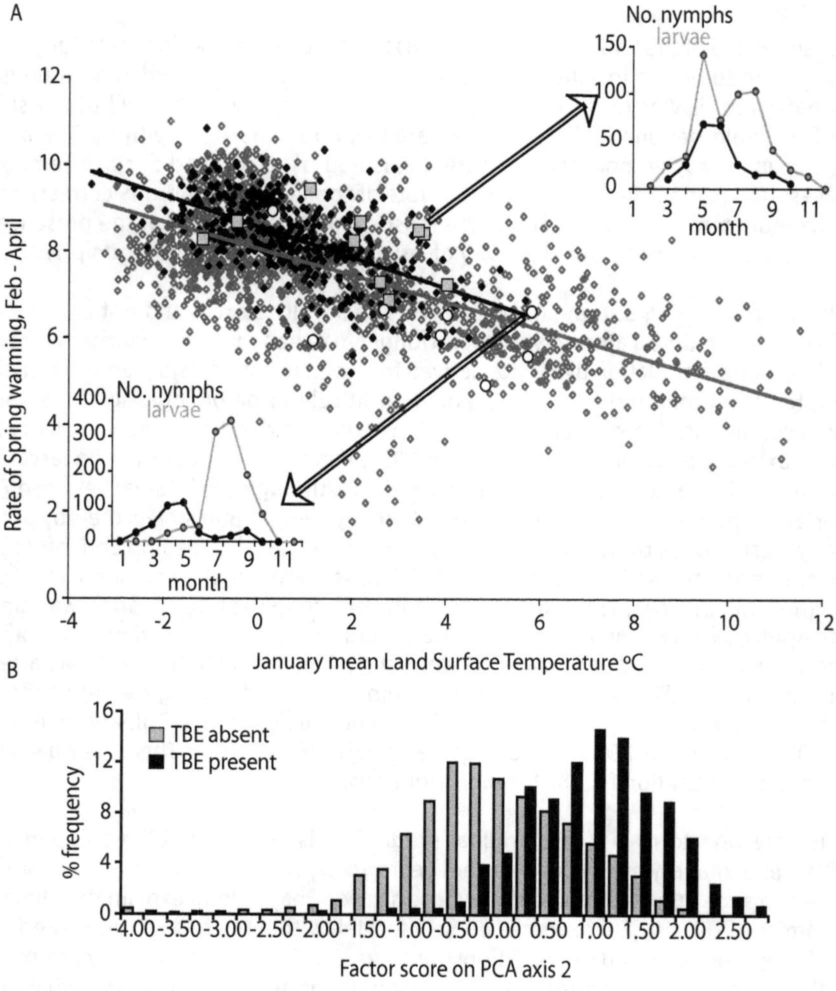

Figure 3. Satellite-derived temperature conditions across Europe and Baltic States, related to TBE presence and seasonal synchrony between larval and nymphal I. ricinus ticks. (A) Pixels where TBE is present (dark dots) or absent (grey dots), and sites where larval and nymphs show synchrony (large filled squares) or non-synchrony (large open circles), show significantly different rates of spring warming. (B) Results of principal components analysis. PCA axis 2 score: TBE-present, mean = 0.861, st. dev = 0.740, n = 418; TBE-absent, mean = -0.228, st. dev = 0.934, n = 1574; t = 25.23, P < 0.001.

Causes of upsurges in TBE incidence: artefact, environment or what?

Artefact due to changing public health activities?

In contrast to the reported reduced public health activities in Hungary (see above), general improvements in surveillance and diagnostic practices in many countries are commonly given as reasons to dismiss upsurges in TBE incidence in long-established foci as no more than artefacts. Such improved practices are likely to be applied more or less simultaneously across wide national regions, however, which does not match the marked spatio-temporal variation in TBE epidemiology, with increases, no change and even decreases in different parts of the same country. Although most countries in Eastern Europe suffered an upsurge in 1993, analysis at the county level in the three Baltic countries, for example, revealed increases starting in any year from 1990 to 1998 (Šumilo *et al.* 2007). There was also huge variation in the degree of increase from 2.5- to 706-fold, independent of previous levels and independent of the presence of *I. ricinus* alone or together with *I. persulcatus* (Figure 4). Exhaustive searches through paper and electronic archives, and face-to-face discussions with public health and medical personnel active in the Baltics both before and after the upsurge in TBE cases, have provided convincing evidence (Šumilo 2007) that any changes in public health activities could at most have exaggerated the upsurge a little, but cannot explain its full extent. Serological diagnosis was in place from the 1970s and laboratory confirmation of all notified cases was mandatory. This relied first on the haemaglutination inhibition method, later

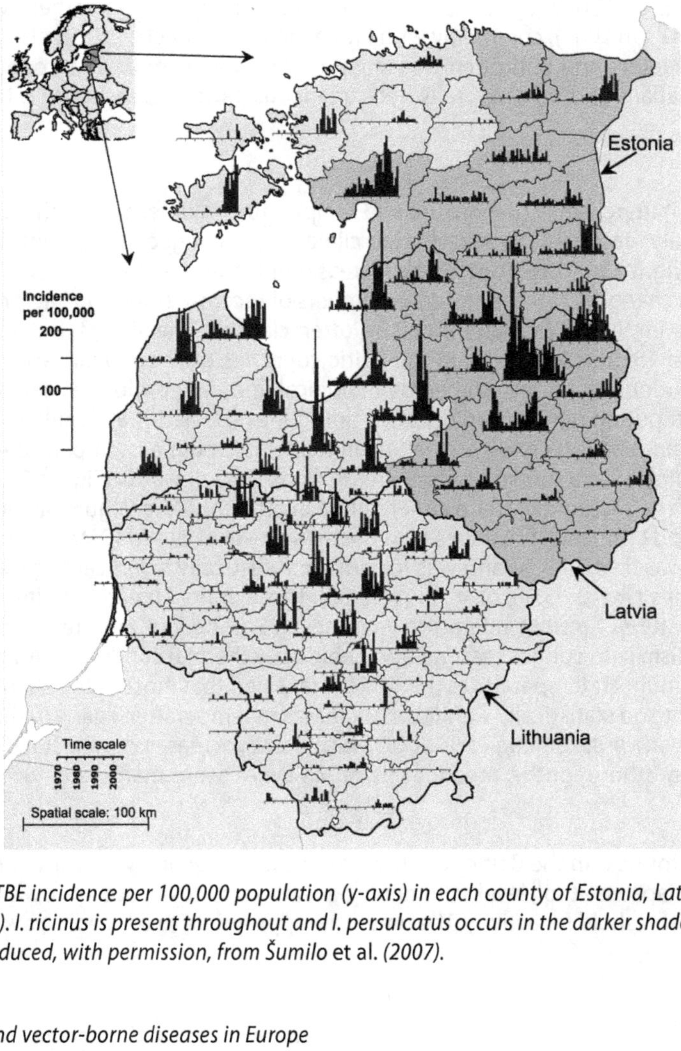

Figure 4. Annual TBE incidence per 100,000 population (y-axis) in each county of Estonia, Latvia and Lithuania, 1970-2004 (x-axis). I. ricinus is present throughout and I. persulcatus occurs in the darker shaded areas of Estonia and Latvia. Reproduced, with permission, from Šumilo et al. *(2007).*

superseded by the ELISA method which ran in parallel with the earlier method for a few years when introduced in 1993 in Estonia and Latvia, but not until 1994 in Lithuania. There is consistent denial, by both Baltic and Russian professionals, of any suppression of TBE case reports under Soviet rule. Furthermore, in Latvia national incidence has decreased again since 1999 (Figure 1). The conclusion is that the upsurge in TBE incidence was real and requires an explanation, not only in the Baltics but also elsewhere in Europe.

Environmental impacts on abundance of infected ticks

Despite the inherently dynamic nature of the Earth's living systems, environmental change is predominantly seen as due to human carelessness in this age of *mea culpa*. Nevertheless, in the absence of any major natural disruption in Europe, the rapidity of change, environmental and epidemiological, seen in Europe over the past few decades probably does indicate human agency. Climate is only one element of potentially important environmental change likely to have an impact on the incidence of all sorts of vector-borne disease, including TBE, but one that has attracted disproportionate emphasis (*inter alia* Epstein 2001, 2005, Lindgren and Gustafson 2001, WHO 2003, Menne and Ebi 2006). Landscape structure and land use is another significant abiotic factor (Patz and Olson 2006), commonly with knock-on biotic effects, principally concerning host availability for vectors and pathogen transmission. As soon as more than one causal factor is introduced, a spatially and/or temporally heterogeneous pattern of response is to be expected.

Climate change

Given the evident importance of moisture and temperature in determining the focal distribution of TBE, it is entirely reasonable to ask whether climate has changed at the right time, in the right place and in the right way to account for the increases in TBE incidence. The inevitable coincidence in time between changing climate and case numbers of recently emerging vector-borne diseases is not enough to imply the causality that is so often claimed. The abrupt, but spatially variable, discontinuities in TBE incidence within the Baltic countries offer an ideal case study (Šumilo *et al.* 2007). (Discontinuities in both the epidemiological and meteorological data can be formally identified by computing generalised distance time series (Davis 1986, Šumilo *et al.* 2007)). Over the past three and a half decades, annual mean daily maximum air temperatures (AMDMAT) in most parts of Europe have shown a step increase from 1989 onwards (Figure 5A for an example from Latvia), with no discernible trend either before or after this date (Figure 5B). At 20 sites within Estonia, Latvia and Lithuania, the increase in the mean level of the AMDMAT between 1970-1988 and 1989-2004 was 0.94-1.38 °C and was maximally statistically significant at this breakpoint in all cases (Student's t = 3.05-5.08, df = 33, $p<0.005$). These annual mean conditions highlight the discontinuous nature of past temperature increases in Europe, but are irrelevant to any type of free-living organism living under seasonally variable daily thermal conditions. Analysis of monthly mean daily maximum air temperatures (MMDMAT) revealed that most of the changes in AMDMAT arose from abrupt and statistically significant increases in temperatures early in the year (January-April) from 1989 onwards (Šumilo *et al.* 2007), with lesser increases during July-September from 1991 onwards. In other months, mean levels of MMDMAT were marginally lower after 1989 at many sites.

The critical time for ticks in the Baltic countries is not the winter and very early spring, when ticks are undergoing diapause at soil level, commonly beneath snow cover, but April, when temperatures start to cross the threshold for activity. The time scale appropriate to co-feeding by ticks of different

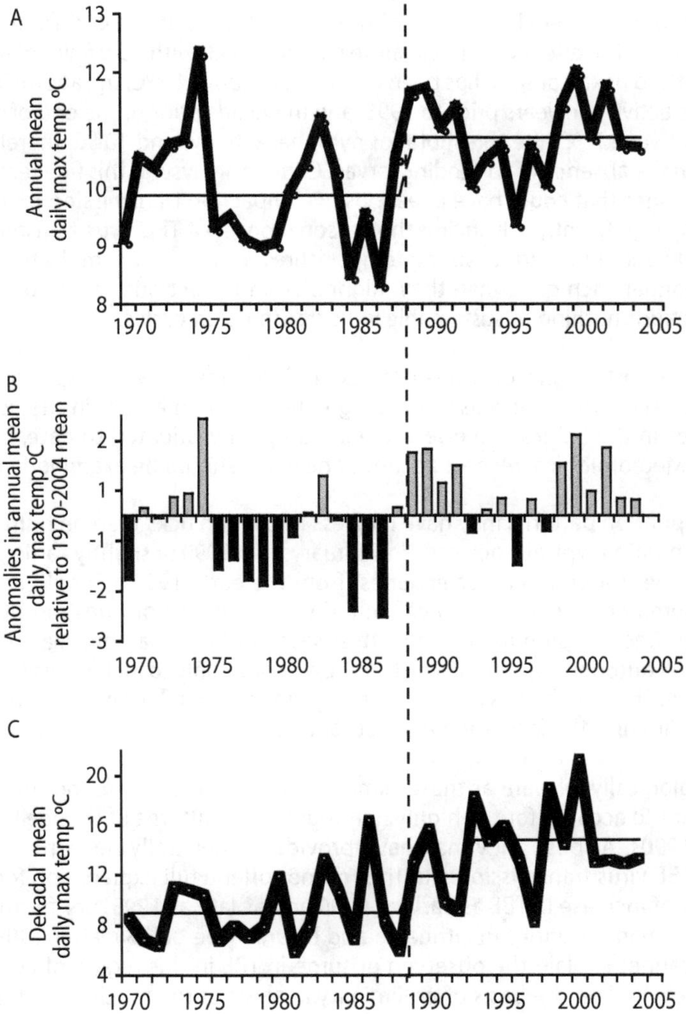

Figure 5. Annual mean daily maximum air temperature recorded at the meteorological station at Liepaja, Latvia. (A) Annual mean daily maximum air temperature, with mean temperatures 1970-1988 and 1989-2004 indicated by the horizontal dotted line. (B) Anomalies from the mean level 1970-2004. (C) Dekadal mean daily maximum air temperature for 21-30 April, with mean temperatures for 1970-1988 and 1989-1992 and 1993-2004 indicated by the horizontal dotted lines. The vertical dashed line indicates the timing of the marked increase from 1989 onwards (see text).

stages with different temperature sensitivities is dekadal (10-day) rather than monthly. At this scale, changes are less uniform. There was no step or trend in dekadal mean daily maximum air temperatures (DMDMAT) for the first dekad of April, but a significant step increase in 1989 for the last two dekads of April. Furthermore, exclusive to the third dekad of April, there was a much more significant step increase in 1993 than in 1989 (Figure 5C). Typically throughout the Baltics, this latter

DMDMAT increased on average by about 2 °C from 1989, but by about 4 °C from 1993, although slightly less in Estonia. The effect was that in all years since 1993 (with very few isolated exceptions) DMDMAT for the third dekad of April has been consistently above 10 °C, the approximate threshold for larval *I. ricinus* activity. In years prior to 1993, and in dekads prior to the end of April, DMDMAT commonly fell between 7 °C (the threshold for nymphal activity) and 10 °C, therefore permitting nymphs to feed in the absence of co-feeding larvae. Climate analysis at this fine temporal scale has thus revealed changes that could have a real positive impact on transmission potential (the bulk of the iceberg) by significantly advancing the seasonal onset of TBE virus transmission between co-feeding ticks. In addition, warmer spring temperatures would allow nymphs to be active earlier in the year and longer each day, when they might bite and infect humans, also active outdoors, even though this adds nothing to sustainable virus transmission cycles.

This change in spring temperatures has not resulted in a faster rate of spring warming from February to April (see above), because of even greater warming earlier in the year. This is true of a range of sites in the Baltics and elsewhere in Europe investigated to date, but may not be universally true. Meteorological records at sites of new foci should be examined in detail.

Changes in precipitation patterns may have had less impact on ticks than on humans: significant increases in precipitation typically occurred in February from 1990 or slightly earlier, and decreases in July or September (often after wetter Junes) from the early 1990s (Šumilo *et al.* 2007). Less rain in the late summer is unlikely to have limited either survival or questing by ticks through desiccation stress, because summer is the wettest season of the year, but may have allowed an increase in people's outdoor recreation, which is known commonly to include visits to tick-infested forests (see below). Snow cover was less deep during January-March from 1988 onwards, which, if anything, would diminish the insulating blanket for ticks.

Dramatic and biologically relevant as these climate changes are, they are too similar throughout the Baltic countries to account for the highly heterogeneous patterns of change in TBE incidence since the early 1990s. At best, they may have provided a generally more permissive climatic background for TBE virus transmission, but they do not offer a full explanation for the markedly different degrees of increase in TBE as early as 1990 and as late as 1998, nor for the decreases in Latvia, and also in many counties in Lithuania and Estonia (see below), after 1998. In summary, climate change cannot explain the observed upsurge in TBE in this region of Europe. The same conclusion was reached on the basis of preliminary analysis of the much cruder meteorological data taken from the interpolated climate surfaces at 0.5° longitude/latitude spatial resolution for 1960-2000 compiled by the UK University of East Anglia Climate Research Unit (CRU). There was no discernible difference in patterns of climate change between sites in the Czech Republic, Hungary and Poland where TBE incidence had or had not increased (Randolph 2004b).

Environmental changes other than climate

Instead, it is proposed that climate is just one of many different types of factors, many arising from the socio-economic transition associated with the end of Soviet rule, that have acted synergistically to increase the abundance of infected ticks (Šumilo *et al.* 2007). Relevant impacts on the environment include changes in land cover and abundance of livestock and wild hosts. The economic crisis in the Baltic countries during the 1990s was deeper than in most Central and Eastern European countries where reforms started earlier and the recession process had already stopped by 1992-1993 (Neiders and Eglīte 2005). In contrast to the predominantly private farming in some Soviet block countries such as Poland and Slovenia (Zīle and Kimelis 1996), most farms in the Baltics were owned, run or controlled by the state until collective farming was replaced by private ownership at the beginning of the 1990s. This was followed by the collapse of agricultural and livestock production, leading to widespread reversion from agricultural land and pastures to land covered in perennial weeds, shrubs and young trees, with probable positive effects on the abundance of rodents (TBE virus transmission hosts). This would have been enhanced by the very much lower usages of pesticides (including herbicides, fungicides and insecticides) once intensive farming ceased.

At the same time, populations of wildlife hosts, particularly those such as deer, elk and wild boar that feed adult ticks and therefore support tick populations, have fluctuated. The increase commonly recorded through the 1980s was abruptly, but temporarily, reversed in the early 1990s, several years after a sharp increase in abundance of wolves due possibly to hunters turning their attention to more edible targets. A rare long-term record of a tick population in Rigas county, Latvia, indicates increasing abundance of nymphs and adults from the mid 1990s, although nymphs (but interestingly not adults) had been at equally high levels in the 1970s. The relationship between density of questing ticks and hosts is complex: a reduction in host availability, as long as sufficient remains to support the tick population, may result in apparently greater numbers of host-seeking ticks as they remain on the vegetation for longer, thereby posing a greater risk to humans (Randolph and Steele 1985).

Or what? Socio-economic impacts on human exposure to infected ticks

National trends in socio-economic conditions

In addition to affecting the abundance of infected ticks, many features of the socio-economic transition also determine the degree of human exposure to infected ticks (Korenberg and Kovaleskii 1999, Randolph 2001, Bormane *et al.* 2003). In the Baltic countries, not only did large-scale agriculture collapse back to small-scale farming centred round the homestead, but industry also collapsed. Together, this resulted in a major diminution of the associated labour force, so that even by 1999 a large proportion of the population (approximately half of households in Lithuania, Latvia and Estonia) were without income from wages, despite an increase in service industries (Aasland and Tyldum 2000, Reutersward 2003, Eamets 2004). The unwaged includes both the job-seeking unemployed and the economically inactive such as students, pensioners, the disabled, single mothers on maternity leave, housewives, etc.

Many turned to subsistence farming as a survival strategy, which is likely to have brought them into closer contact with ticks in forests and rough land near to their homes. It is possible that the reduction in emission of industrial pollutants into the atmosphere was responsible for the observed increased solar radiation ('brightening') and temperature increases (Sharma *et al.* 2004, Wild *et al.* 2005) in the post-communist Baltic States (and elsewhere). Indeed, a reversal from decreasing to increasing atmospheric transmission at the end of the 1980s has been recorded at Tartu, Estonia and at other sites around the world (Wild *et al.* 2005).

Despite this increase in poverty (subjective and objective) (Lune 2001), material welfare improved in some parts of the population (i.e. income inequality increased), as indicated by the ownership of substantial items for long-term use, such as cars. This increased steadily from <150 per 100,000 in 1990 to 250-350 per 100,000 by 2002, with the lowest ownership seen in Latvia.

Not surprisingly, car ownership in Latvia was more likely in households with higher incomes, but also slightly more likely in rural than in urban households. Although private transport might increase potential access to tick- infested forests, this is only one factor amongst many determining the demographic, socio-economic and behavioural patterns of differential risk of exposure to tick bites.

Individual risk factors

Data from a survey conducted by the Marketing and Public Opinion Research Centre SKDS in August 2001 at the request of the Latvian Ministry of Welfare (SKDS, 2001) have been analysed to identify factors associated with differential individual risk of infection through entering tick-infested habitats and undertaking different activities there. Adjusted odds ratios (AOR) (calculated using multivariate binary logistic and ordered logit regressions) reveal the independent impact of a range of predictor variables while controlling for the most likely confounding factors (Šumilo 2007 and unpublished data). Those found to be significant include gender, age, nationality, education, employment, income, household size, size of settlement, place of residence and forest visit habits.

As expected from the natural habitat of both species of vector ticks, the SKDS survey confirmed that forest visitors in Latvia are 4-5 times more likely to encounter ticks than those who do not enter forests. Forest visits, however, are common activities. Amongst respondents to the survey, more than two-thirds of adults went to forests for leisure or work during 2000 or 2001, including town and even city dwellers, but more commonly rural dwellers. People more likely to make frequent visits (at least once a month) include males, those with lower educational achievements, and those on the lowest incomes.

The reason why so many people from all walks of life visit forests is that mushroom and berry harvesting has always been an important part of the culture of many Eastern European countries. This was given as the main purpose of forest visits in Latvia, either alone or combined with walking or recreation. Even those who went principally to work commonly combined this with collecting mushrooms or berries. Mushroom and berry collecting or working as the principal activities in forests were significantly associated with being unemployed, having lower incomes and visiting forests frequently. This suggests that these activities are likely to increase with falling socio-economic conditions.

Furthermore, these activities are associated with a much higher than average risk of having suffered a tick bite in the year prior to the SKDS survey. There is some evidence from the State Public Health Agency that the frequency of human visits to forests and associated exposure to ticks varies with the weather (Šumilo unpublished data), which is known to determine the abundance of these natural resources (Wallenius 1999, Bara 2004). Taken altogether, these data highlight the importance of varying human activities in tick habitats, dependent on both socio-economic factors and the weather, as important determinants of the risk of tick bite. At the IX International Jena Symposium on Tick-borne Diseases, 15-17 March 2007, Professor B. Kríz reported that an exceptional and prolonged mushroom season occurred in the Czech Republic in 2006, coinciding with unusually high numbers of cases of TBE during the autumn of that year.

Human habits can, of course, change quite suddenly for a variety of reasons, including in response to perceived risk of TBE (see below). A large proportion of the population questioned by the SKDS survey, but not the unemployed or forest workers, reported going to forests less frequently during 2000-01 than five years previously. Thus the majority of the population, albeit not the groups at highest risk, changed their behaviour in ways that could have reduced their exposure to infected ticks. This is consistent with, although by no means proof of the cause of, the decreased TBE incidence in Latvia from 1999.

Decreased incidence: reasons and lessons

Valuable lessons about the dynamics in TBE epidemiology may be learnt from exploring the reasons behind the decreases in incidence observed since 1999 in parts of the Baltic countries. One obvious way infection may have been controlled is through vaccination. There are at least four safe and highly effective vaccines against TBE available in Europe (produced by Baxter AG, Vienna, Austria and Chiron Behring GmbH & CoKG, Marburg, Germany) and Russia (produced by Virion, Tomsk and the Institute of Polyomyelitis and Viral Encephalitis, Moscow, Russia). The basic vaccination protocol comprises three initial doses at approximately 1-3 and 9-12 month intervals, followed by a booster every 3-5 years (Kunz 2003). Thus it takes two years to complete the primary course. For indirectly transmitted zoonoses, for which humans are dead-end hosts and contribute nothing to enzootic cycles of pathogen transmission, under homogenous conditions there is a direct one-to-one relationship between the percent vaccination in the human population and the percent reduction in incidence of infection. This is well illustrated in Austria, mentioned above, where c.85% reduction in annual TBE cases from c.300-700 in 1975-82 to 40-100 in 1998-2005 was achieved by improving the vaccination coverage of the at-risk population to nearly 90% (Kunz 2003). This is exceptional; vaccination in other European countries, including Switzerland, Sweden, and the Czech Republic, rarely exceeds c.10% (Beran 2004, Vene *et al.* 2007).

Vaccination in response to risk perception

Since the re-establishment of independence in the Baltic countries, vaccination is generally no longer covered by the state and must be paid for privately. Uptake has increased markedly over the past decade, most strongly correlated with TBE incidence two years previously in each of the Baltic countries (Šumilo unpublished data). This indicates that people seek vaccination in response to perceived risk, so that more start with their first primary dose the year following high annual numbers of TBE cases reported in the media over the past tick activity season.

The variable uptake of vaccination in each county in Latvia and Lithuania also indicates a response to perceived risk, as vaccination rates in 2002-05 were significantly positively correlated with previous TBE incidence (Šumilo unpublished data). Not surprisingly, many of the counties in Latvia with high uptake of vaccination across the population during 2002-05 were those whose children had been targeted by a vaccination campaign in 1998-2002 (Lucenko *et al.* 2004).

Nevertheless, there are reasons to think that the observed reduction in TBE in Latvia since 1999 was not due solely to vaccination. First, the degree of reduction (74% from 1993-98 to 2002-05) was far greater than the level of vaccination protection. This discrepancy is even more marked considering that the vaccination campaign in Latvia was targeted at children (Lucenko *et al.* 2004), making them disproportionately protected, especially in rural areas (Šumilo unpublished data), despite not being at any higher risk than adults for the country as a whole (Šumilo 2007). The result would have been to reduce the overall population impact of vaccination.

Secondly, vaccination was significantly less likely among people on lowest incomes (associated with lowest levels of education) (SKDS 2001, Šumilo unpublished data), even though these socio-economic groups were the most likely to visit forests frequently and to engage in the higher risk activity of mushroom or berry collecting (see above). Given that people who visit forests are more likely to be vaccinated than those who do not, indicating an awareness of ticks and the risk of TBE, this strongly suggests that people feel economic constraints, real or imagined, against being vaccinated. This factor could be addressed to achieve sustainable, sure protection against infection more effectively than by any other means (Kunz 2003).

Furthermore, when analysed at the county level, the degree of reduction in TBE was greatest where past incidence had been highest in both Latvia and Lithuania.

In conclusion, factors additional to, and of much greater magnitude than, vaccination must be responsible for the recent decreases in TBE in many parts of the Baltic countries, and appear to be responsive to TBE incidence. They are, therefore, most likely to be of conscious human agency rather than due to changes in the natural enzootic cycles of TBE virus, factors such as awareness and avoidance through changing behaviour. People have indeed reported the sorts of changes in their behaviour, reduced visits to forests (see above), which could reduce their exposure to infected ticks.

Conclusions

Although the second half of this chapter has focused on Baltics, where data collection and analysis are most advanced, the principles, even if not the specifics, could apply throughout Europe, despite the very different socio-economic conditions both historically and still today. For Eastern Europe at least, it is tempting to suggest that this tentative and as yet incomplete analysis offers some insight into the causes for the original increases in TBE incidence coincident with the end of Soviet rule, when political transition would have created socio-economic conditions likely to increase human exposure to ticks to a greater or lesser extent (Korenberg and Kovaleskii 1999, Randolph 2001, Bormane *et al.* 2003). At the same time, environmental changes also associated with political upheaval could have promoted increasingly robust enzootic cycles of TBE virus with which humans came into closer contact. The material presented in this chapter offers a conceptual framework for further detailed analysis of this complex epidemiological situation (Figure 6). Rather than seeking a single cause, we should focus on a nexus of indirectly linked changes in several different abiotic, biotic and non-biological factors, which, by chance, would have acted synergistically to cause an increased risk of tick-borne diseases (Šumilo *et al.* 2007). Variable contributions of each factor in different places at different times, due to normal regional variation in environmental and cultural conditions, alongside the more uniform changes in climate, could explain the spatio-temporal heterogeneity in upsurge in TBE across Europe, both Eastern and Western.

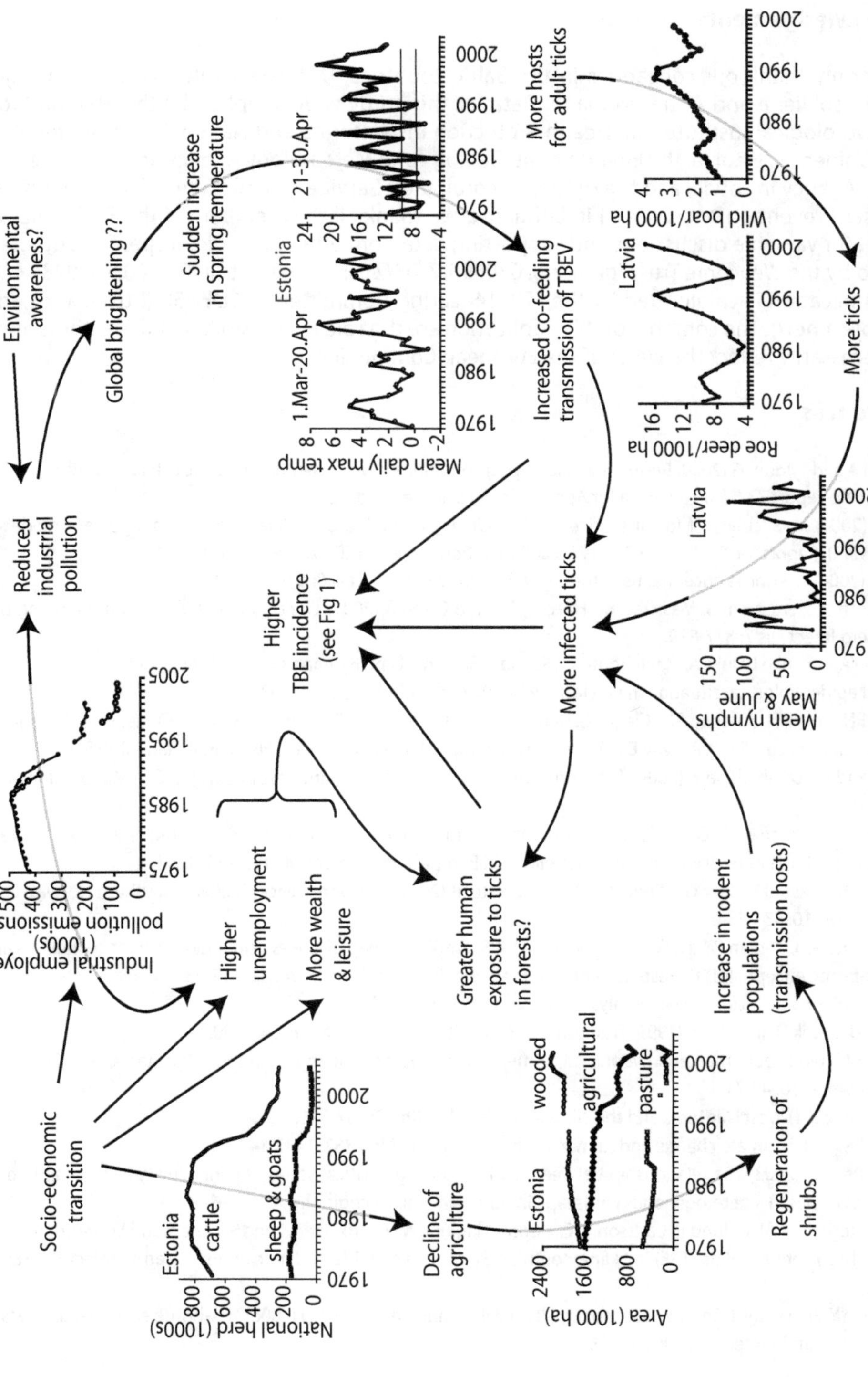

Figure 6. Hypothetical explanation for the epidemiology of TBE in the Baltic countries. Examples of data from Estonia, Latvia and Lithuania indicate some factors that may act independently but synergistically to cause the emergence of tick-borne diseases. Reproduced, with permission, from Šumilo et al. (2007).

Acknowledgements

We warmly thank our colleagues in the Baltic countries, Dr Veera Vasilenko, Irina Golovljova, Loreta Asokliene and Antra Bormane. Data for this study were supplied by the Meteorological and Hydrological Institute, the Health Protection Inspectorate and National Institute for Health Development in Estonia, the Environment, Geology and Meteorology Agency and the State Public Health Agency in Latvia, and the Hydrometeorological Service and the Centre for Communicable Diseases Prevention and Control in Lithuania. We thank all the members of the EDEN-TBD sub-project for valuable discussions and supporting data from other parts of Europe. The authors are funded by the Wellcome Trust (grant no. 070696/Z/03/Z) and EU grant GOCE-2003-010284 EDEN; this publication is catalogued by the EDEN Steering Committee as EDEN50 (http://www.eden-fp6project.net/). The contents of this publication are the sole responsibility of the authors and do not necessarily reflect the views of the European Commission.

References

Aasland A and Tyldum G (2000) Better or worse? Living conditions developments in Estonia, Latvia and Lithuania 1994-99. Fafo-report 334. Fafo Institute for Applied Social Science, Oslo. pp 32.

Bara G (2004) [Availability of forest services and products depending on different forest management models] (in Latvian). Report for the Latvian & UN FAO collaboration project (TCP/LAT/2901), Riga. pp 13.

Beran J (2004) Tickborne encephalitis in the Czech Republic. Euro surveill 8: 6-9.

Bjoersdorff A, Bergstrom S, Massung RF, Haemig PD and Olsen B (2001) *Ehrlichia*-infected ticks on migrating birds. Emerg Infect Dis 7: 877-879.

Bormane A, Lucenko I, Duks A, Mavtchoutko V, Ranka R, Salmina K and Baumanis V (2003) Vectors of tick-borne diseases and epidemiological situation in Latvia in 1993-2002. Int J Med Microbiol 293: 36-47.

Charrel RN, Attoui H, Butenko AM, Clegg JC, Deubel V, Frolova TV, Gould EA, Gritsun TS, Heinz FX, Labuda M, Lashkevich VA, Loktev V, Lundkvist A, Lvov DV, Mandl CW, Niedrig M, Papa A, Petrov VS, Plyusnin A, Randolph S, Süss J, Zlobin VI, and de Lamballerie X (2004) Tick-borne virus diseases of human interest in Europe. Clin Microbiol Infect 10: 1040-1055.

Comstedt P, Bergstrom S, Olsen B, Garpmo U, Marjavaara L, Mejlon H, Barbour AG and Bunikis J (2006) Migratory passerine birds as reservoirs of Lyme borreliosis in Europe. Emerg Infect Dis 12: 1087-1095.

Csango PA, Blakstad E, Kirtz GC, Pedersen JE and Czettel B (2004) Tick-borne encephalitis in southern Norway. Emerg Infect Dis 10: 533-534.

Daniel M, Kolar J, Zeman P, Pavelka K and Sadlo J (1998) Predictive map of *Ixodes ricinus* high-incidence habitats and a tick-borne encephalitis risk assessment using satellite data. Exp Appl Acarol 22: 417-433.

Davis JC (1986) Statistics and Data Analysis in Geology. Wiley, New York.

Dumpis U, Crook D and Oksi J (1999) Tick-borne encephalitis. Clin Infect Dis 28: 882-890.

Eamets R (2004) Labour market flows and adjustment to macroeconomic shocks in the Baltic States. Post-Communist Economies 16: 47-71.

Epstein PR (2001) West Nile Virus and the climate. J Urban Health 78: 367-371.

Epstein PR (2005) Climate change and human health. New Engl J Med 353: 1433-1436.

Estrada-Pena A (2003) The relationship between habitat topology, critical scales of connectivity and tick abundance *Ixodes ricinus* in a heterogeneous landscape in northern Spain. Ecography 26: 661-671.

Falt J, Lundgren A, Alsterlund R, Carlsson B, Eliasson I, Haglund M, Lundkvist A, Vene S, Vene S and Settergren B (2006) Tick-borne encephalitis (TBE) in Skåne, southern Sweden: a new TBE endemic region? Scand J Infect Dis 38: 800-804.

Hansmann Y, Pierre Gut J, Remy V, Martinot M, Allard Witz M and Christmann D (2006) Tick-borne encephalitis in eastern France. Scand J Infect Dis 38: 520-526.

Hay SI, Graham AJ and Rogers DJ, (eds.) (2006a) Global Mapping of Infectious Diseases. Academic Press, Oxford. pp 399.

Hay SI, Tatem AJ, Graham AJ, Goetz SJ and Rogers DJ (2006b) Global environmental data for mapping infectious disease distribution. Adv Parasitol 62: 37-77.

Hemmer CJ, Littman M, Loberman M, Lafrenz M, Bottcher T and Reisinger EC (2005) Tickborne meningo-encephalitis, first case after 19 years in northeastern Germany. Emerg Infect Dis 11: 633-634.

Jääskeläinen J, Tikkakoski T, Uzcategui NY, Alekseev AN, Vaheri A and Vapalahti O (2006) Siberian sub-type tick-borne encephalitis virus, Finland. Emerg Infect Dis 12: 1568-1571.

Kitron U (1998) Landscape ecology and epidemiology of vector-borne diseases: tools for spatial analysis. J Med Entomol 35: 435-445.

Korenberg EI and Kovaleskii YV (1999) Main features of tick-borne encephalitis eco-epidemiology in Russia. Zentrallbl Bakteriol 289: 525-539.

Kunz C (2003) TBE vaccination and the Austrian experience. Vaccine 21: S1/50-S51/55.

Labuda M, Jones LD, Williams T, Danielová V and Nuttall PA (1993a) Efficient transmission of tick-borne encephalitis virus between cofeeding ticks. J Med Entomol 30: 295-299.

Labuda M, Nuttall PA, Kozuch O, Eleckova E, Williams T, Zuffova E and Sabo A (1993b) Non-viraemic transmission of tick-borne encephalitis virus: a mechanism for arbovirus survival in nature. Experientia 49: 802-805.

Labuda M, Kozuch O, Zuffova E, Eleckova E, Hails RS and Nuttall PA (1997) Tick-borne encephalitis virus transmission between ticks co-feeding on specific immune natural rodent hosts. Virology 235: 138-143.

Lindgren E and Gustafson R (2001) Tick-borne encephalitis in Sweden and climate change. Lancet 358: 16-18.

Lucenko I, Jansone I, Velicko I and Pujate E (2004) Tickborne encephalitis in Latvia. Euro surveill 8: 10-12.

Lune E (2001) Household economic resources. Living conditions in Latvia. Central Statistical Bureau of Latvia, Riga. pp 99-114.

Macleod J (1935) *Ixodes ricinus* in relation to its physical environment. II. The factors governing survival and activity. Parasitology 27: 123-144.

Menne B and Ebi KL, (eds.) (2006) Climate Change and Adaptation Strategies for Human Health. Springer, Darmstadt. pp 449.

Neiders L and Eglīte S (2005) [New economics development policy and praxis in Latvia] (in Latvian). In: Škapars R, Šumilo E, (eds.). Problems of Restructuring Latvian Economics in the Conditions of New Economics. University of Latvia, Riga, pp 29-66.

Patz J and Olson SH (2006) Malaria risk and temperature: influences from global climate change and local land use practices. Proc Nat Acad Sci USA 103: 5635-5636.

Perret J-L, Guigoz E, Rais O and Gern L (2000) Influence of saturation deficit and temperature on *Ixodes ricinus* tick questing activity in a Lyme borrelosis-endemic area (Switzerland). Parasitol Res 86: 554-557.

Piesman J and Gern L (2004) Lyme borreliosis in Europe and North America. Parasitology 129: S191-S220.

Randolph SE and Steele GM (1985) An experimental evaluation of conventional control measures against the sheep tick *Ixodes ricinus* (L) (Acari: Ixodidae). II. The dynamics of the tick-host interaction. Bull Entomol Res 75: 501-518.

Randolph SE, Gern L and Nuttall PA (1996) Co-feeding ticks: epidemiological significance for tick-borne pathogen transmission. Parasitol Today 12: 472-479.

Randolph SE, Miklisová D, Lysy J, Rogers DJ and Labuda M (1999) Incidence from coincidence: patterns of tick infestations on rodents facilitate transmission of tick-borne encephalitis virus. Parasitology 118: 177-186.

Randolph SE and Storey K (1999) Impact of microclimate on immature tick-rodent interactions (Acari: Ixodidae): implications for parasite transmission. J Med Entomol 36: 741-748.

Randolph SE (2000) Ticks and tick-borne disease systems in space and from space. Adv Parasitol 47: 217-243.

Randolph SE, Green RM, Peacey MF and Rogers DJ (2000) Seasonal synchrony: the key to tick-borne encephalitis foci identified by satellite data. Parasitology 121: 15-23.

Randolph SE (2001) The shifting landscape of tick-borne zoonoses: tick-borne encephalitis and Lyme borreliosis in Europe. Phil Trans R Soc B 356: 1045-1056.

Randolph SE, Green RM, Hoodless AN and Peacey MF (2002) An empirical quantitative framework for the seasonal population dynamics of the tick *Ixodes ricinus*. Int J Parasitol 32: 979-989.

Randolph SE (2004a) Tick ecology: processes and patterns behind the epidemiological risk posed by ixodid ticks as vectors. Parasitology 129: S37-66.

Randolph SE (2004b) Evidence that climate change has caused 'emergence' of tick-borne diseases in Europe? Int J Med Microbiol 293, suppl 37: 5-15.

Reutersward A (2003) Labour markets and social policies in the Baltic Countries. Organisation for Economic Co-operation and Development, Paris. pp 177.

Sharma S, Lavoué D, Cachier H, Barrie LA and Gong SL (2004) Long-term trends of the black carbon concentrations in the Canadian Arctic. J Geophys Res - Atmos 109: D15203-15212.

Skarpass T, Sundoy A, Bruu AL, Vene S, Pedersen J, Eng PG and Csángó PA (2002) [Tick-borne encephalitis in Norway] (in Norwegian). Tidsskr Nor Laegeforen 122: 30-32.

SKDS (2001) [Investigation into the factors hindering immunisations of the population against diphtheria and tick-borne encephalitis] (in Latvian). National Public Health Agency SKDS, Latvia, Riga. pp 179.

Šumilo D (2007) PhD thesis, Oxford University.

Šumilo D, Asokliene L, Bormane A, Vasilenko V, Golovljova I and Randolph SE (2007) Climate change cannot explain the upsurge of tick-borne encephalitis in the Baltics. PLoS ONE 2: e500.

Vene S, Haglund M, Lundkvist A, Lindquist L and Forsgren M (2007) Study of the serological response after vaccination against tick-borne encephalitis in Sweden. Vaccine 25: 366-372.

Wallenius TH (1999) Yield variations of some common wild berries in Finland in 1956-1996. Ann Bot Fenn 36: 299-314.

Wild M, Gilgen H, Roesch A, Ohmura A, Long CN, Dutton EG, Forgan B, Kallis A, V. Russak V and Tsvetkov A (2005) From dimming to brightening: decadal changes in solar radiation at Earth's surface. Science 308: 847-850.

World Health Organization (2003) Climate Change and Human Health - Risks and Responses. WHO, Geneva. pp 297.

Zīle S and Kimelis A (1996) [Comparative analysis of restructurization and privatisation processes of agricultural sector in the Baltic States and Central and Eastern European countries during 1990-1995] (in Latvian). Development in Agricultural and Market Policies. Latvian State Institute of Agrarian Economics, Riga, pp 163-171.

12. Lyme disease in Europe: facts and no fiction

Fedor Gassner and Leo S. van Overbeek

Abstract

Lyme disease is an emerging disease in the Netherlands and elsewhere in Europe. *Ixodes ricinus* L. ticks infected with Lyme disease-causative agents have been found throughout Europe. Remarkable are the large differences in infection rates between different regions. It is supposed that the observed heterogeneity in infected tick numbers is characteristic of the way *Borrelia* species are acquired and transmitted by *I. ricinus* ticks. One factor directly affecting microbial communities in the tick gut, including populations of *Borrelia* species, is host blood. Smaller animals such as birds and rodents, often called 'reservoir hosts', play an especially important role in the transmission of *Borrelia* species between vector and hosts. Local environmental aspects may largely impact these groups of smaller animals and modulate their populations. Factors such as soil, vegetation, predation and human interference can have a large influence on local and temporal fluctuations of reservoir hosts and supposedly also on *Borrelia*-species-infected tick numbers. Local variations in these factors cause differences in infection rates in ticks and we propose that large fluctuations in infected tick numbers may occur at a small scale: so-called 'hot spots' of infected ticks. The concept of hot spots needs further attention, as it is important for the control of infected ticks in natural areas via local and integrative approaches.

Keywords: *Ixodes* tick species, *Borrelia* species, emerging diseases, Lyme borreliosis, infected tick distribution, biological control

Introduction

As mentioned in one of the most fundamental works on ticks 'The Biology of Ticks' (Sonenshine 1991), the first records of ticks as nuisances to humans date back hundreds of years BC, and were described by ancient Greek and Roman authors. Linnaeus first recorded ticks as species in his Systema Naturae of 1758. Ticks have been found encapsulated in fossil amber at least 50 million years old. Only since the early 1900s were ticks known to transmit diseases to cattle and humans. Although the symptoms of Lyme disease, also called Lyme borreliosis, were first described in Europe in 1883 (Matuschka *et al.* 1995), no link was made with ticks as vectors until the late 20th century, when it was revealed that the major vectors of Lyme borreliosis were species belonging to the genus *Ixodes*. The causative agent *Borrelia burgdorferi* was originally isolated in the USA in 1982 and a link was made between this pathogen and the Lyme disease symptom known as erythema migrans in the town of Old Lyme in 1975 (Burgdorfer *et al.* 1982). In Europe, *Borrelia* species have been detected in various *Ixodes* tick species that were obtained from museum specimens dating back to 1896 (Hubbard *et al.* 1998). Although historical evidence provides information on infected ticks up to about a century ago, genetic variation within *Borrelia* species isolates and *Ixodes scapularis* populations indicate that this relationship must have existed over a much longer period, at least since the last ice-age. Similar relationships are also expected between European *Ixodes* tick vectors, which are closely related to their Nearctic counterparts, and *Borrelia* species occurring in ticks, although it must be taken into account that European *Borrelia* species diversity is higher than found in the USA (Kurtenbach *et al.* 2006).

By the late 20th century, Lyme borreliosis had developed into the most prevalent tick-borne zoonosis in the northern hemisphere. The primary vectors of Lyme borreliosis are hard ticks belonging to the *Ixodes ricinus* species complex, which is widely distributed over the Palearctic as well as the Nearctic regions of the northern hemisphere. In North America, the westerly distributed black legged tick (*I. pacificus*) and the deer tick (*I. scapularis*) are the primary vectors of Lyme borreliosis (Piesman *et al.* 1999). In Eurasia, the sheep tick *I. ricinus* and the more easterly-distributed taiga tick *I. persulcatus* are considered to be primary vectors of Lyme borreliosis, although other tick species may also play a role in the enzootic cycle of the Lyme borreliosis agent, but have been rarely linked to clinical cases of Lyme disease (Gray 1998). *I. ricinus* also acts as a vector of other tick-borne diseases such as babesiosis, human granulocytic ehrlichiosis and tick borne encephalitis (Gray 2002), which are not as generally prevalent as Lyme borreliosis and limited to certain geographic regions (Piesman *et al.* 1999).

Lyme borreliosis, which can be acquired by the bite of an infected tick, is a disease with a wide array of symptoms, possibly causing skin, heart, joint and neurological disorders (Stanek and Strle 2003). An early symptom of infection is erythema migrans, a migrating, reddish skin rash that occurs in most, but not all infections after the bite of an infected tick. Different *Borrelia* species cause different symptoms in humans and not all *Borrelia* species have been proven to be pathogenic for humans. Occasionally, clinical symptoms have been observed in various domestic animals.

In this chapter, the emerging character of Lyme borreliosis will be discussed from the ecological perspective of *I. ricinus*. We will dicuss the aetiology of the Lyme-causative agents belonging to the *Borrelia* species, with an emphasis on the heterogeneous distribution of infected ticks in natural areas and possible methods of reducing incidences of Lyme borreliosis. Occasionally, other *Ixodes* species will be mentioned, because parallels in *Borrelia* species infections exist between Paleartic and Nearctic *Ixodes* species.

Ecology of *Ixodes ricinus*

The primary vector of Lyme borreliosis in Europe, *I. ricinus* (Figure 1) has a three-stage lifecycle: larva, nymph and adult, which is typical for all species of the hard tick family, *Ixodidae*.

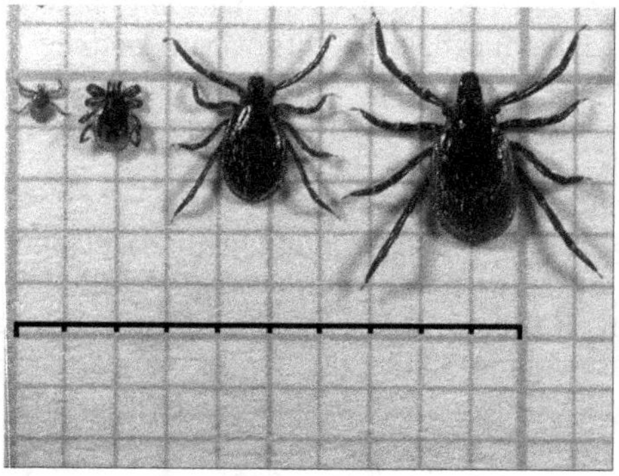

Figure 1. The three developmental stages of the common sheep tick Ixodes ricinus. *From left to right: larva; nymph; adult male; adult female. The scale bar indicates 1 cm.*

I. ricinus ticks depend strictly on a blood meal for moulting and entrance into the next developmental stage. The host is found by specific foraging behaviour called questing, during which the tick detects various cues from the host, such as CO_2, skin volatiles, movement and heat (Sonenshine 1991). In contrast to many other blood-feeding arthropods, *I. ricinus* ticks have a limited action radius and spatial dispersion is mainly dependent on transportation via the host. When questing, *I. ricinus* ticks climb into the vegetation and wait for suitable hosts. When a host passes, the tick grasps its fur and crawls to preferred feeding sites such as ears, neck and the skin between the legs. The *I. ricinus* attaches to the skin and uses specialised mouthparts (Figure 2) and salivary excretions to anchor itself to the skin, where it remains undetected until full engorgement (Sonenshine 1991). Engorgement usually lasts between 3 and 7 days.

Mating usually takes place within the vegetation (Sonenshine 2004) during questing or in some cases on the host. Females normally oviposit 3-27 days after feeding, and can lay 500 to 3000 eggs. Eggs are preferably deposited in leaf litter on the forest floor, and can remain there for months. Development of all tick life stages and eggs is optimal at temperatures between 8 and 11 °C. Once hatched, larvae prefer small rodents and birds for feeding, but they occasionally also feed on larger hosts (Gray 1998) including humans (Gray 2002). Occurrence of larvae is usually restricted to lower vegetation because of their host preference and susceptibility to desiccation, this in contrast to nymphs and adults, which reside higher in vegetation (Mejlon and Jaenson 1997). The occurrence of larvae close to the ground makes it difficult to collect them for research. The most commonly applied method is to collect ticks from the vegetation by dragging with a cotton blanket (Figure 3).

One must realise that the numbers of larvae collected will in general be underestimated in comparison with those of nymphs and adults. For example, in our studies at several locations within the Netherlands, no larvae were captured in a pine forest over a 5-month period, whereas many were found feeding on rodents in the same forest during that time. After their first blood meal, which takes generally 3-5 days, larvae will retreat into the forest floor and will develop into nymphs, which takes between 24 days to sometimes several months, depending on seasonal and climatic conditions.

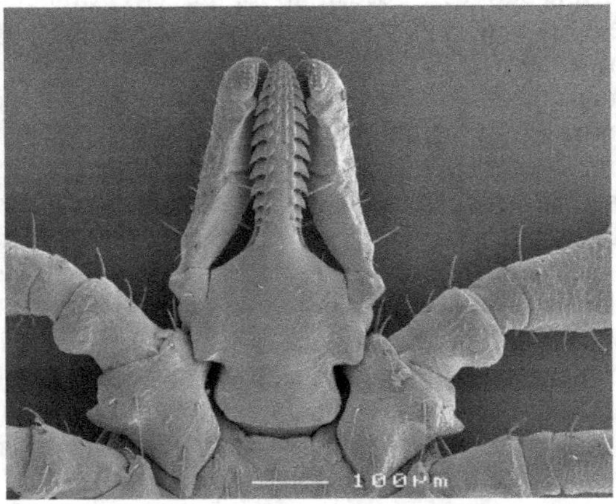

Figure 2. Scanning electron micrograph of nymphal Ixodes ricinus *hypostome (picture by A. van Aelst).*

Figure 3. Blanket dragging method used for collection of questing ticks in vegetation (picture by W. Takken).

The newly developed nymphs have a much wider host range than larvae, from small rodents and birds to reptiles and various ungulates. A wider host range in combination with relatively less susceptibility to desiccation allows nymphs to quest higher in the vegetation than larvae (Mejlon and Jaenson 1997). Once attached to a host, nymphs will feed during approximately 4-7 days, followed by a next moulting period in the forest floor, which can last up to 1 year before they appear as sexually mature adults. Adult ticks are generally found higher in the vegetation in comparison to both other stages and can even be found up to 1.5 meters above the forest floor, which is about the height of an ungulate (Mejlon and Jaenson 1997). Adult ticks have a different host range than larvae and nymphs, and prefer larger animals such as hedgehogs, hares and ungulates (Gray 1998), whereas they are rarely found on smaller rodents. These larger hosts are important to complete the tick life cycle and therefore these large animals are called 'reproduction' hosts. For all stages, questing periods are usually in spring, early summer and autumn depending on photoperiod, temperature and humidity. The duration of the life cycle is flexible because of temperature-dependent differences in development and the possibility to enter into an inactive stage, such as developmental diapause or quiescence during unfavourable climatic conditions or absence of hosts. The entire life cycle from egg to adult is usually between 2 and 4 years but may take up to 7 years, mainly depending on local climate and day length at different latitudes and altitudes. *I. ricinus* thrive in climates with cooler summers and milder winters, which explains why this species mainly occurs in the northern hemisphere (Gray 1998). A study on tick phenology in Western Europe revealed that all tick cohorts (i.e. individuals of the same developmental stage and age) always entered the next developmental stage during autumn (Randolph *et al.* 2002). Furthermore, this study revealed that all tick stages showed clear peaks in questing activity that were attributed to hatching or transition from inactive to active stages. In general, larvae were found questing from early May until the end of August, with a peak in the summer. Nymphs were generally found questing from mid February until the end of November, peaking between March and July, while adults were found questing throughout the year, without any clear peak in questing activity. These observations correspond with those found in the Netherlands, where larvae were most active from May until late November, nymphs from March until November and adults throughout the year with lowest activity during the winter (Smit *et al.* 2003).

Factors important for *I. ricinus* activity are temperature and relative air humidity. In general, ticks become active when the relative air humidity is above 80% and the temperature 4 °C or higher. Commonly, the saturation deficit is used as a parameter for tick activity. The saturation deficit combines air temperature with humidity and represents the 'drying power' to which ticks are exposed. Ticks are usually found to be active at saturation deficits of 4 mm Hg and higher (Perret *et al.* 2000). However, the susceptibility of ticks to desiccation not only depends on the saturation deficit, but also on the stored energy in tick bodies. It was found that ticks were more active at lower saturation deficits shortly after a blood meal (Randolph and Storey 1999). This can be explained by the fact that ticks can actively obtain water from the air and for this they need sufficient energy, which they can obtain from the nutrients in blood.

It is expected that *I. ricinus* ticks are more active during the night because the saturation deficit is optimal and many nocturnal host animals are active. It is plausible that ticks are at least as active at night as during the day (Mejlon and Jaenson 1997). However, most studies were done during daytime, and one may wonder whether tick numbers obtained are representative for the populations present in the areas studied. Daytime collection can lead to a serious underestimation of tick abundance (Randolph and Storey 1999). This was also confirmed by our observations, made in a forest near the village of Renkum, in The Netherlands, where higher tick numbers of all stages were found at night than during the day. However, most studies have been done in order to study the transmission of Lyme borreliosis-causative agents to humans and therefore daytime collection of ticks is more appropriate because most people visit natural areas during the day and not at night.

For the life cycle of *I. ricinus*, climatic conditions, vegetation cover, composition of the forest floor (soil and leaf litter layer), and the presence of suitable hosts are the most important parameters. It is clear that these parameters will fluctuate in time and space and thus it can not be expected that tick activity and abundance would be the same for all places at all times. This has important consequences for the transmission of tick-borne pathogens, because tick numbers fluctuate in time and space.

***Borrelia* species and transmission to hosts**

Lyme borreliosis-causative agents are the most important pathogens transmitted via *Ixodes* spp. to humans. A group of closely-related species belonging to the bacterial genus *Borrelia* is responsible for Lyme disease. This complex consists of three main clusters: Lyme borreliosis *Borrelia* species (consisting of approximately 12 species); new world Tick Borne Relapsing Fever *Borrelia* (TBRF) and old world TBRF *Borrelia*. The TBRF agents are widely distributed over the northern hemisphere and are mainly transmitted via body lice and ticks (Rebaudet and Parola 2006), and will not be further discussed in this chapter.

Globally, 33 species are known within the *Borrelia* species complex (http://www.bacterio.cict.fr/b/borrelia.html (Accessed 4 September 2007). At least seven of these are known to occur in Europe: *B. afzelii*, *B. burgdorferi* sensu stricto, *B. garinii*, *B. valaisiana*, *B. lusitaniae*, *B. spielmanii* and *B. bisettii*. *B. spielmanii* and *B. bisettii* have been recently described (Derdáková and Lencáková 2005). We cannot exclude that new *Borrelia* species will be found in the near future, for example species that have remained undetected to date, or more virulent strains of already known species that are newly formed through horizontal gene transfer (Kurtenbach *et al.* 2006). Within the *Borrelia* species complex, interspecies genetic variation is large, especially in *B. garinii*, which can be subdivided into several genospecies based on differential expression of the outer surface protein A (Osp-A). *Borrelia afzelii*, *B. burgdorferi* sensu stricto and *B. garinii* are in particular responsible for symptoms in humans (Stanek and Strle 2003), whereas the role of the other species in the development of Lyme borreliosis symptoms is more disputable. When dealing with disease transmission via *I. ricinus* ticks, one usually refers to the Lyme borreliosis-causative species of the *Borrelia* complex, whereas the other *Borrelia* species are often ignored. The term '*Borrelia* species' is preferred above '*B. burgdorferi* sensu lato', because it covers all species within the *Borrelia* species complex, including the ones that are supposed to be non-pathogenic and those from which pathogenic traits have not yet been elucidated.

Once inside *I. ricinus* ticks, the *Borrelia* bacteria establish in the midgut and can thus be transferred to the next developmental stage of the tick. In principle, tick infection with *Borrelia* can take place with each blood meal, which explains why the highest incidences of *Borrelia* infections are generally found in adults, followed by nymphs and lowest in larvae. Cells of *Borrelia* present in the blood meal can colonise the tick midgut, and from there cells migrate via the haemolymph to the salivary glands. Outer surface proteins (Osp) of *Borrelia* and TROSPA molecules expressed by the tick play an important role in the establishment of the bacteria and its migration from the midgut to the salivary glands. OspC was recently found to be indispensable in this process (Goettner *et al.* 2006, Fingerle *et al.* 2007). Once present in the salivary glands, other molecular mechanisms including factors expressed by the tick can affect the transmission of the *Borrelia* species as well as attachment of the tick to the host (Hovius *et al.* 2007). The parasite only begins to migrate to the salivary glands when the tick begins to feed; the time needed for migration of *Borrelia* from the midgut to the salivary glands in the tick differs per species: in *I. scapularis*, it was found to take place within 36-48 hours (De Silva and Fikrig 1995). However, for *B. afzelii* in *I. ricinus*, dissemination from the midgut to the salivary glands is much faster and can result in *Borrelia* transmission within 24 hours (Crippa *et al.* 2002). Therefore, it is generally advised to remove attached ticks within 24 hours after attachment.

From many different studies, it has become clear that there is a relationship between certain hosts and *Borrelia* species. For example, *B. afzelii*, *B. burgdorferi s.s.* and the *B garinii* Osp-A 4 genotypes are mainly associated with rodents, while *B. valaisiana* and other Osp-A types of *B. garinii* are more associated with different bird species (Kurtenbach *et al.* 2002). The suitability of particular hosts for the various *Borrelia* species may be explained by differences in host immune responses for different *Borrelia* species. In Europe, at least nine small and seven medium-sized mammals have been found able to transmit *Borrelia* species to *I. ricinus* ticks and are therefore considered reservoir hosts (Gern *et al.* 1998). Furthermore, mainly rodents (Humair *et al.* 1999) and birds (Hanincová *et al.* 2003) account for the natural circulation of the *Borrelia* species. Many reproduction hosts, including common hosts such as ungulates (Tälleklint and Jaenson 1994, Telford *et al.* 1988) and rabbits (Matuschka *et al.* 2000) are generally unsuitable as reservoirs. However, other reproduction hosts can serve as reservoirs, e.g. pheasants (Kurtenbach *et al.* 1998). These hosts are fed upon by all stages of the tick, and therefore could maintain a *Borrelia*-infected tick population in the absence of other reservoirs and reproduction hosts.

Another infection route for ticks can take place through co-feeding. Here, hosts that are poor reservoirs can acquire *Borrelia* species in the bloodstream near a feeding infected tick. Uninfected ticks feeding nearby could thereby pick up *Borrelia* species without the need of an systemic infection in the host (Ogden 1997). Such incidents can be expected to contribute significantly to the infection prevalence in tick populations, especially if co-feeding occurs in a combination of infected nymphs and uninfected larvae, which can sometimes feed in aggregations of over 50 larvae in one feeding area, such as the neck, the nose or the ears, together with one or more nymphs (F. Gassner, unpublished data).

Finally, a final process that may contribute to the prevalence *B. burgdorferi* s.l. in an area is transovarial transmission; it has not been established, however, that vertically-transmitted *Borrelia* species in larvae are infective to new hosts. This is certainly a phenomenon that needs further study, as 1.9% of larvae are infected Europe-wide (Rauter and Hartung 2005).

The role of various animal groups has major implications on the transmission of Lyme borreliosis-causative agents in nature. The group of smaller animals, such as birds and rodents, requires particular attention because of their presumed role in the transmission of Lyme borreliosis, ultimately leading to infections in humans.

The detection of *Borrelia* species in ticks and animal tissues

To investigate the behaviour of different *Borrelia* species in ticks and vertebrates it is important to distinguish the various Lyme disease-causing *Borrelia* species. Nowadays, a myriad of detection techniques are available to screen for the presence of different bacterial species in biological samples. However, *Borrelia* species need special attention for the following reasons: (1) *Borrelia* bacteria cannot always be recovered from biological samples by culturing, in spite of the fact that these cells are still often viable; (2) cells from different *Borrelia* species closely resemble each other and it is hard to differentiate between the species, especially those that are involved in Lyme borreliosis. Therefore, dark-field microscopy and cultivation-based detection methods are not always suitable for specific detection of *Borrelia* species in biological samples. Culture-independent detection techniques such as molecular (PCR-based) and serological techniques are therefore more appropriate. A commonly applied technique used for the detection of *Borrelia* species in tick samples is the so-called reverse line blot method (Rijpkema *et al.* 1995, Schouls *et al.* 1999). Reverse line blot detection is based on the extraction of total DNA from ticks, followed by PCR with *Borrelia* genus-specific primers, which incorporate biotin into the PCR product. The PCR products are hybridised to a membrane with *Borrelia* species-specific probes. This method has appeared to be applicable in many different studies involving *Borrelia* species infection in ticks, but it has some drawbacks. First, only a limited number of *Borrelia* species can be determined in a single sample and second, this method is qualitative and not quantitative. Because many different pathogens are supposed to be present in ticks, besides *Borrelia* species, it is often important to measure the relative abundance of all pathogens in the same sample. Therefore, detection in quantitative multiplex settings may be more appropriate. Although these techniques have not been applied in tick research yet, it is expected that they will soon become available in our laboratories and will provide deeper insight into interactions within microbial communities in ticks.

Microbial communities in *Ixodes ricinus*

Microbial communities in tick guts may influence the establishment and reproduction of invading pathogens, such as those belonging to the *Borrelia* species complex, in ticks. The microbial composition of the *I. ricinus* midgut may be influenced by various environmental factors (Figure 4).

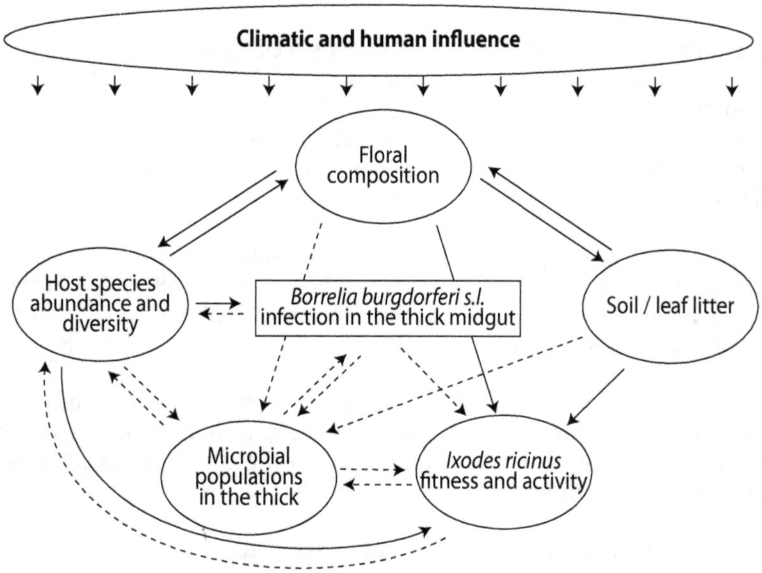

Figure 4. Natural factors affecting microbial composition and Borrelia *species establishment in* Ixodes ricinus *ticks. Solid lines indicate direct effects on microbial communities and dotted lines presumptive indirect effects.*

The most important factor influencing microbial communities in ticks is host blood. Blood composition and micro-organisms present in the blood will determine the microbial composition of tick gut communities. Because ticks are rather immobile, it is expected that local variation in animal hosts will strongly determine the microbial composition in ticks. Other environmental factors such as vegetation, soil type, predation and human activities are also important because these factors can locally influence the composition of small animal populations. The effect of local parameters on the microbial constitution of tick guts and their possible effect on pathogen invasion and establishment have so far rarely been explored; they could, however, explain more in detail why large fluctuations in *Borrelia* infection rates occur in different regions around the world.

An illustrative example of the indirect effect of vegetation on the occurrence of *Borrelia* species in ticks is shown in a study done on acorn mast. Acorn mast is a periodical phenomenon in which oak populations simultaneously produce larger than average amounts of acorns. A modelling study done in North America and based on 15 years of field observations indicated that a mast season can clearly affect reservoir host populations, resulting in higher than normal tick numbers and *Borrelia* infection rates (Ostfeld *et al.* 2006b). Because acorn mast is a common phenomenon in Europe as well, it is expected that it will have the same effect here. Not only does acorn mast sustain host populations, but also the presence of leaves, fruits and other seeds can cause large temporal fluctuations in tick densities through the alteration of host densities. Differences in *Borrelia* infection rates in ticks from areas with different plant communities has been found (Kampen 2004). Therefore, vegetation type must be considered as an important factor affecting *I. ricinus* tick populations and the *Borrelia* species inside these ticks.

Soil and leaf litter composition of forest floors may also be important factors affecting *Borrelia* infection rates in ticks. Soil type has a direct effect on plant composition and thus indirectly affects local animal communities. However, the leaf litter layer is an important environment for ticks as well, because ticks lay their eggs there and they also spend a considerable portion of their life span in this layer, e.g. for transstadial development. Size, pH and humidity of the leaf litter layer are therefore important factors directly affecting the success of tick reproduction and thus *Borrelia* infection rates in ticks.

As indicated in Figure 4, the role of many factors affecting *I. ricinus* - *Borrelia* species interactions are not yet fully understood. The most prominent one is the direct effect of the various micro-organisms present in ticks on tick survival. Although speculative, a limited number of studies have focused on this interaction and we must take into account that the presence of certain micro-organisms could be responsible for changes in tick reproduction and behaviour (Lefcort and Durden 1996). This has already been shown for several other pathogens in numerous arthropod vectors (Hurd 2003) and requires further attention in tick research.

Spatial and temporal variations in *Borrelia* infection rates in ticks

Local environmental factors have a strong influence on the survival and reproduction of ticks. However, an increase in tick numbers does not necessarily lead to higher infection rates. To understand how *Borrelia* species are able to accumulate in reservoir hosts and ticks, it is important to understand at which spatial and temporal scale the levels of infection rates vary in ticks.

Large spatial variations in *Borrelia* infection rates were found in different countries in Europe (Ferquel *et al.* 2006, Jouda 2004, Kampen 2004, Smit *et al.* 2003, Wielinga *et al.* 2006) and locally in North America (Van Buskirk and Ostfeld 1998). In a Europe-wide meta analysis study, large differences in infection rates were demonstrated (Rauter and Hartung 2005). Especially the difference between Western and Eastern European countries was striking: 14% tick infection rates in Eastern Europe and 24% tick infection rates in Western Europe. Infection rates in ticks ranged from low (1.5% in Slovenia) to very high (49% in Slovakia and Switzerland). Furthermore, only two out of 110 studies included in this meta analysis showed uninfected ticks, which emphasises that *Borrelia* infections in ticks are found almost all over Europe. Of all *Borrelia* species found, *B. afzelii* and *B. garinii* were the most common ones. However, these were not the most dominant species found in all studies. For example, a study in Germany revealed that *B. valaisiana* was more abundant than *B. garinii* and *B. afzelii* (Kampen 2004). It is questionable whether this meta study fully reflects the actual *Borrelia* infection rates throughout Europe, as most, if not all studies were performed in regions with a history of Lyme borreliosis. Within the Netherlands, differences in *Borrelia* infection rates in ticks were found at smaller spatial scale, infection rates varied per region, and were between 5 and 30% (de Boer *et al.* 1993, Rijpkema and Bruinink 1996, Rijpkema *et al.* 1994, 1996, Smit *et al.* 2003, Wielinga *et al.* 2006). Infected tick rates may vary locally, e.g. per forest. What is the minimal scale-level at which differences in *Borrelia* infections occur? One study revealed that tick densities and *Borrelia* infection rates fluctuated within a single woodland (Zeman 1999). Is it possible that incidences of high infection rates occur very locally, e.g. at a scale of a few square meters? These so-called 'hot spots' may explain why such variation exists between and within woodlands. Furthermore, it is important to understand how humans acquire Lyme borreliosis in natural areas and also to understand how the causative agents spread through natural areas and may disappear again. The question is why variations in tick infection rates at smaller spatial scales are usually not published in detail.

The existence of such a variation in numbers of infected ticks is fascinating and deserves further attention. From an ecological perspective, it may explain the role of reservoir hosts and the effect of ecological interactions of such populations on the spread of Lyme borreliosis. Practically, 'hot spots' may be important for the application of control measures aimed at reducing the spread and accumulation of pathogens belonging to the *Borrelia* species complex and possibly also of other pathogens present in *I. ricinus* ticks.

An extreme scenario: The Netherlands

A survey of erythema migrans and tick bite cases in The Netherlands showed a simultaneous, striking increase over a period of 10 years (Hofhuis *et al.* 2006). This study, performed country-wide, shows remarkable regional differences (Figure 5).

On average, the incidence of erythema migrans nearly tripled from 39 cases per 100,000 inhabitants in 1994 to 103 cases per 100,000 inhabitants in 2005. At the same time, tick-bite consultations at hospitals and general practitioners also increased, from 191 per 100,000 inhabitants in 1994 to 446 per 100,000 inhabitants 2005. The latter observation is most likely an underestimation, as many people do not seek medical help for a tick bite.

To date, there is no clear scientific explanation for this dramatic increase. There is a lack of nation-wide tick population and *B. burgdorferi* s.l. monitoring, which is common for most European countries. The available data of infected tick populations in The Netherlands do not show extreme differences with other European countries and vary between 5 and 30% with temporal as well as spatial differences (de Boer *et al.* 1993, Rijpkema and Bruinink 1996, Rijpkema *et al.* 1994, 1996, Smit *et al.* 2003, Wielinga *et al.* 2006). To date, no long-term study has been conducted, but it is expected that the infection rate of ticks may increase (W. Takken, unpublished data), with high local variations. Apart from an increase in populations of infected ticks, an increase in recreation, the creation of ecological corridors connecting nature areas, changes in host abundance and recent milder winters may have caused the increase in Lyme borreliosis. Perhaps the infection prevalence is locally boosted by the occurrence of transmission hotspots, as discussed earlier in this chapter.

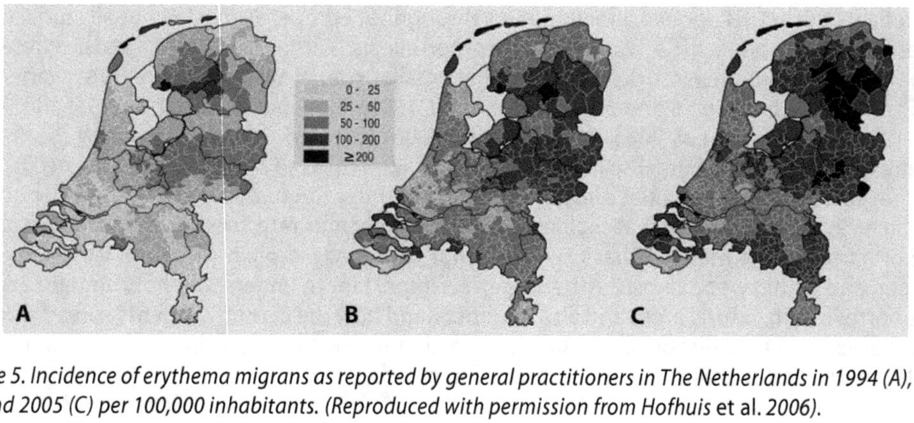

Figure 5. Incidence of erythema migrans as reported by general practitioners in The Netherlands in 1994 (A), 2001 (B) and 2005 (C) per 100,000 inhabitants. (Reproduced with permission from Hofhuis et al. *2006).*

The European distribution of *Borrelia burgdorferi* s.l. and *Ixodes ricinus* overlaps in residential areas

A possible explanation for the observed increase in *Borrelia* incidence in several European countries can be explained by the fact that infected *I. ricinus* ticks are found in highly populated areas. For example, an internet survey in The Netherlands in 2006 and 2007 revealed that many people acquired a tick bite in their own gardens (W. Takken, personal communication). Europe-wide, infected *I. ricinus* are found in recreational areas near or in cities (Table 1).

This phenomenon brings people more into contact with infected ticks; but, how has this developed? For example, did we build our cities in areas where infected ticks were already present, or did they advance into these areas after the expansion of the city boundaries? If city parks and private gardens have been newly constructed, the ticks present would have had to be brought in by various hosts. These urban hosts could be, for example rats, mice, squirrels and various birds such as thrushes and blackbirds (Humair *et al.* 1998, Humair and Gern 1998, Kurtenbach *et al.* 1998, Matuschka *et al.* 1997).

The question remains whether we are expanding our living and recreation areas into tick habitats, or whether the ticks and their pathogens are advancing into human territory. A combination of both may also be true, but it shows that monitoring of tick and *B. burgdorferi* s.l. populations in residential areas is of great importance for monitoring and predicting Lyme borreliosis.

Table 1. Lyme borreliosis vector Ixodes ricinus in European urban areas.

Country	Urban area type	Reference
Finland	Urban recreational areas	Junttila (1999)
Germany	City outskirts, recreational sites and private gardens	Maetzel *et al.* (2005)
Poland	Recreational sites near city	Michalik *et al.* (2003)
The Netherlands	Urban park	Wielinga *et al.* (2006)
United Kingdom	Urban park	Guy and Farquhar (1991)
	Recreational site	Robertson (2000)

Lyme borreliosis and climate change

Over the past few decades, especially in the past few years, climate change has become a very popular topic in the media, in science and for the general public. Evidently, the climate is changing, with various effects worldwide, and threatening human health through direct effects of thermal stress and extreme weather events. Many other indirect effects can be noted, such as the effect on agriculture, air pollution and changes in various infectious (vector-borne) diseases (McMichael *et al.* 2006). In Europe, climate change has been predicted to cause increased drought in the southern parts, and warmer winters with wetter summers in more northern countries. To date, the effects of climate change are most prominently seen in increased winter temperatures and the increased length of the growing season (Lindgren and Jaenson 2006).

As described earlier in this chapter, local climatic conditions are important for the survival and behaviour of *I. ricinus* in a direct as well as indirect way (Figure 4). Consequently, the effects of climate change can be expected to have amplified effects on populations of infected ticks as climate affects several life stages of the tick at the same time. These effects can be positive as well as negative, for example considering habitat suitability for the vector. A recent modelling study that correlated habitat suitability showed that climatic factors such as temperature and summer rainfall can have different effects depending on the region; some areas were predicted to face stable or increasing *I. ricinus* densities and other areas showed opposite trends (Estrada-Pena and Venzal 2006).

Climatic conditions depend on the latitude and altitude of an area and thereby determine the boundaries of Lyme borreliosis-endemic regions. With a changing climate, these boundaries are expected to shift to other altitudes and latitudes. For *I. ricinus,* an increase in distribution and abundance has been linked to climate change in various studies (Daniel *et al.* 2003, Hubálek 2005, Lindgren *et al.* 2000).

Because an increase in ticks is positively correlated to the incidence of Lyme borreliosis (Hubálek *et al.* 2003, Stafford *et al.* 1998), climate change that increases tick distribution and density is expected to increase the incidence of Lyme borreliosis (Bennet *et al.* 2006). Additionally, the behaviour of the general public should also be taken into account, as warmer summers generally increase the exposure of people to ticks, if an increase in tick numbers coincides with an increase in recreational activity (Joss *et al.* 2007, Mavin *et al.* 2005). Recently, a report by WHO has also stressed the role of climatic change in Europe (Lindgren and Jaenson 2006), but concluded that the emergence of Lyme borreliosis cannot exclusively be attributed to climatic factors. Based on climate changes alone, the distribution of *I. scapularis* and thereby the incidence of Lyme borreliosis in North America has been predicted to increase substantially in the coming decades (Brownstein 2005). Further research is needed in Europe to make reliable predictions.

Control of ticks in nature

Regarding the emerging character of Lyme borreliosis and other tick-borne diseases such as tick borne encephalitis (see Chapter 11), tick control is still a relatively neglected subject in Europe. Surprisingly little scientific attention is paid to direct control of ticks in Europe compared to the USA and Africa, where ticks transmit livestock diseases which can cause severe economic damage regionally (Jongejan and Uilenberg 2004).

In Europe, little information is available on the annual economic damage caused by tick borne diseases, taking into account, e.g. damage to livestock, health-care costs for humans as well as domestic animals, and economic loss due to a decrease in income from recreation in known endemic areas. An accurate assessment, together with the alarming signals described earlier in this chapter, could lower the threshold for European governments to take action.

Tick control does not necessarily need to be aimed at large-scale applications of chemical pesticides such as dichloro-diphenyl-trichloroethane (DDT) or various pyrethroid pesticides. Nowadays, Integrated Pest Management (IPM) is widely applied in agriculture and may be adopted for application in nature as well. Several strategies can be combined to control ticks; e.g. biological agents in combination with forestry management practices. Biological control of ticks can be done with entomopathogenic fungi, which have been shown to be effective against *I. scapularis* (Benjamin *et al.* 2002, Kaaya and Hassan 2000). Application of these fungi has not been tested before on a large scale, but may offer opportunities in an IPM-like approach.

Other tick control strategies are: the use of insecticide-impregnated nesting materials with which the host may make contact, four posters[12] and bait boxes with a chemical or biological control agent. These methods are already applied and offer opportunities to control ticks in various developmental stages (Lane 2001, Ostfeld *et al.* 2006a). The control of potential reservoirs for *Borrelia* species and vegetation management are alternative measures to control ticks, although these measures are more invasive (Buskirk 1995, Wilson *et al.* 1988). Other methods, such as trapping ticks with pheromones or other chemical attractants, would be most effective on the host, for example by preventing ticks from mating on a reproduction host (Sonenshine 2004). The effect of other types of traps may be limited due to the relatively short-range motility of ticks from the place where they moulted compared to, for example, free-flying mosquitoes. An IPM-like approach may be the best option to control ticks in nature. However, most methods are in an experimental stage and not yet ready for larger-scale applications in Europe.

Apart from control focussing on ticks and tick hosts, reducing the risk of Lyme disease can also be focussed on education and information of the general public, including professional groups such as forest workers. Indeed, the general public has heard of Lyme disease, but very few people know exactly what the risks are. For example, two common misconceptions are that ticks can jump like a flea and that they drop from trees. Knowing that ticks will only cling if one touches it in its questing position, and that ticks rarely drop from trees and generally do not climb up higher than approximately 1.5 meters (Mejlon and Jaenson 1997) can alter peoples' behaviour and awareness.

One effective way to inform the general public is through the media, through which information about times of high risk can be given, together with methods for personal protection and basic information on Lyme borreliosis (Lindgren and Jaenson 2006). For example, tick removal needs clear instructions as it can be a precarious action, due to the strongly barbed tick hypostome (Figure 2) and inadequate removal of the tick may cause faster transmission of the disease agent.

A potentially interesting measure could be a short tick warning item that could be embedded in national weather forecasts, in newspapers, on radio, television and the internet during high-risk periods. Information on personal protection could also be focussed on risk groups such as forest workers and (youth) groups. Possibly, once local hotspots are identified, local warnings in newspapers and at the entrances of nature areas could also be effective.

Concluding remarks

Lyme disease is emerging and difficult to control. The causative agents and the primary vectors of Lyme borreliosis are present in many countries in Europe. Occasional high incidences of *Borrelia* infections in ticks reveal the heterogeneous nature of these infections in *I. ricinus* ticks. It is

[12] Four posters are bait stations with four pesticide-impregnated bristles which rub the ears, antlers and neck of deer that are attracted to the food offered. Four posters are commercially available in the USA.

important to understand the local environmental factors affecting the accumulation of *Borrelia* species in ticks and their natural hosts in order to develop tick control measures in nature. There is a clear need for long-term studies that describe year-round fluctuations in ticks and their pathogens. Additionally, spatial heterogeneity needs to be further assessed in order to assess the risk of Lyme borreliosis. Microbial communities associated with ticks may play an important role in the modulation of *Borrelia* populations in the vector. The role of various microbial species in the establishment and reproduction of *Borrelia* species in *I. ricinus* ticks is so far unexplored and may reveal new opportunities to combat Lyme borreliosis.

References

Benjamin MA, Zhioua E and Ostfeld RS (2002) Laboratory and field evaluation of the entomopathogenic fungus *Metarhizium anisopliae* (Deuteromycetes) for controlling questing adult *Ixodes scapularis* (Acari: Ixodidae). J Med Entomol 39: 723-728.

Bennet L, Halling A and Berglund J (2006) Increased incidence of Lyme borreliosis in southern Sweden following mild winters and during warm, humid summers. Eur J Clin Microbiol Infect Dis 25: 426-432.

Brownstein JS (2005) Effect of Climate Change on Lyme Disease Risk in North America. Ecohealth 2: 38.

Burgdorfer WW, Barbour AAG, Hayes SSF, Benach JJL, Grunwaldt EE and Davis JJP (1982) Lyme disease-a tick-borne spirochetosis? Science 216: 1317-1319.

Buskirk JV (1995) Controlling lyme disease by modifying the density and species composition of tick hosts. Ecol Appl 5: 1133.

Crippa M, Rais O and Gern L (2002) Investigations on the mode and dynamics of transmission and infectivity of Borrelia burgdorferi sensu stricto and Borrelia afzelii in Ixodes ricinus ticks. Vector-Borne Zoon Dis 2: 3-9.

Daniel M, Danielová V, Kríz B, Jirsa A and Nozicka JJ (2003) Shift of the tick *Ixodes ricinus* and tick-borne encephalitis to higher altitudes in central Europe. Eur J Clin Microbiol Infect Dis 22: 327-328.

De Boer R, Hovius KE, Nohlmans MK and Gray JS (1993) The woodmouse (*Apodemus sylvaticus*) as a reservoir of tick-transmitted spirochetes (*Borrelia burgdorferi*) in The Netherlands. Zentralbl Bakteriol 279: 404-416.

De Silva AM and Fikrig E (1995) Growth and migration of *Borrelia burgdorferi* in Ixodes ticks during blood feeding. Am J Trop Med Hyg 53: 397-404.

Derdáková M and Lencáková D (2005) Association of genetic variability within the Borrelia burgdorferi sensu lato with the ecology, epidemiology of Lyme borreliosis in Europe. Ann Agric Environ Med 12: 165-172.

Estrada-Pena A and Venzal JM (2006) Changes in habitat suitability for the tick *Ixodes ricinus* (Acari: Ixodidae) in Europe (1900-1999). EcoHealth 3: 154-162.

Ferquel E, Garnier M, Marie J, Bernède-Bauduin C, Baranton G, Pérez-Eid C and Postic D (2006) Prevalence of *Borrelia burgdorferi* Sensu Lato and Anaplasmataceae members in *Ixodes ricinus* ricks in Alsace, a focus of Lyme borreliosis endemicity in France. Appl Environ Microbiol 72: 3074-3078.

Fingerle V, Goettner G, Gern L, Wilske B and Schulte-Spechtel U (2007) Complementation of a *Borrelia afzelii* OspC mutant highlights the crucial role of OspC for dissemination of Borrelia afzelii in *Ixodes ricinus*. Int J Med Microbiol 297: 97-107.

Gern L, Estrada-Peña A, Frandsen F, Gray JS, Jaenson TG, Jongejan F, Kahl O, Korenberg E, Mehl R and Nuttall PA (1998) European reservoir hosts of Borrelia burgdorferi sensu lato. Zentralbl Bakteriol 287: 196-204.

Goettner G, Gern L, Schulte-Spechtel U, Wilske B and Fingerle V (2006) OspC of *B. afzelii* is crucial for dissemination in the vector as shown by transformation and complementation of a European OspC-deficient *B. afzelii* strain. Int J Med Microbiol 296S1: 122-124.

Gray JS (1998) The ecology of ticks transmitting Lyme borreliosis. Exp Appl Acarol 22: 249-258.

Gray JS (2002) Biology of Ixodes species ticks in relation to tick-borne zoonoses. Wiener Klin Wochenschr 114: 473-478.

Guy EC and Farquhar RG (1991) *Borrelia burgdorferi* in urban parks. Lancet 338: 253.

Hanincová K, Taragelova V, Koci J, Schafer SM, Hails R, Ullmann AJ, Piesman J, Labuda M and Kurtenbach K (2003) Association of *Borrelia garinii* and *B. valaisiana* with songbirds in Slovakia. Appl Environ Microbiol 69: 2825-2830.

Hofhuis A, van der Giessen JW, Borgsteede FH, Wielinga PR, Notermans DW and van Pelt W (2006) Lyme borreliosis in the Netherlands: strong increase in GP consultations and hospital admissions in past 10 years. Euro surveill 11: E060622 060622.

Hovius JWR, van Dam AP and Fikrig E (2007) Tick-host-pathogen interactions in Lyme borreliosis. Trends Parasitol 23: 434-438.

Hubálek Z (2005) North Atlantic weather oscillation and human infectious diseases in the Czech Republic, 1951–2003. Eur J Epidemiol 20: 263-270.

Hubálek Z, Halouzka J and Juricová Z (2003) Longitudinal surveillance of the tick Ixodes ricinus for borreliae. Med Vet Entomol 17: 46-51.

Hubbard MJ, Baker AS and Cann KJ (1998) Distribution of *Borrelia burgdorferi s.l.* spirochaete DNA in British ticks (Argasidae and Ixodidae) since the 19th century, assessed by PCR. Med Vet Entomol 12: 89-97.

Humair PF, Postic D, Wallich R and Gern L (1998) An avian reservoir (*Turdus merula*) of the Lyme borreliosis spirochetes. Zentralbl Bakteriol 287: 521-538.

Humair PF, Rais O and Gern L (1999) Transmission of *Borrelia afzelii* from *Apodemus* mice and *Clethrionomys* voles to *Ixodes ricinus* ticks: differential transmission pattern and overwintering maintenance. Parasitol 118: 33-42.

Humair PF and Gern L (1998) Relationship between *Borrelia burgdorferi sensu lato* species, red squirrels (*Sciurus vulgaris*) and *Ixodes ricinus* in enzootic areas in Switzerland. Acta Trop 69: 213-227.

Hurd H (2003) Manipulation of medically important insect vectors by their parasites. Ann Rev Entomol 48: 141-161.

Jongejan F and Uilenberg G (2004) The global importance of ticks. Parasitol 129: S3-14.

Joss A, Mavin S and Ho-Yen D (2007) Increased incidence of Lyme borreliosis following mild winters and during warm, humid summers. Eur J Clin Microbiol Infect Dis 26: 79-79.

Jouda F (2004) *Ixodes ricinus* density, and distribution and prevalence of *Borrelia burgdorferi sensu lato* infection along an altitudinal gradient. J Med Entomol 41: 162-169.

Junttila J (1999) Prevalence of *Borrelia burgdorferi* in *Ixodes ricinus* ticks in urban recreational areas of Helsinki. J Clin Microbiol 37: 1361-1365.

Kaaya GP and Hassan S (2000) Entomogenous fungi as promising biopesticides for tick control. Exp Appl Acarol 24: 913–926.

Kampen HD (2004) Substantial rise in the prevalence of Lyme borreliosis spirochetes in a region of western Germany over a 10-year period. Appl Environ Microbiol 70: 1576.

Kurtenbach K, De Michelis S, Etti S, Schafer SM, Sewell H-S, Brade V and Kraiczy P (2002) Host association of *Borrelia burgdorferi sensu lato* - the key role of host complement. Trends Microbiol 10: 74-79.

Kurtenbach K, Hanincová K, Tsao JI, Margos G, Fish D and Ogden NH (2006) Fundamental processes in the evolutionary ecology of Lyme borreliosis. Nature Rev Microbiol 4: 660-669.

Kurtenbach K, Peacey M, Rijpkema SGT, Hoodless AN, Nuttall PA and Randolph SE (1998) Differential transmission of the genospecies of *Borrelia burgdorferi sensu lato* by game birds and small rodents in England. Appl Environ Microbiol 64: 1169-1174.

Lane RS (2001) Research on animal-borne parasites and pathogens helps prevent human disease. Calif Agricul 55: 13.

Lefcort H and Durden LA (1996) The effect of infection with Lyme disease spirochetes (*Borrelia burgdorferi*) on the phototaxis, activity, and questing height of the tick vector *Ixodes scapularis*. Parasitol 113 97-103.

Lindgren E, Tälleklint L and Polfeldt T (2000) Impact of climatic change on the northern latitude limit and population density of the disease-transmitting European tick Ixodes ricinus. Environ Health Perspect 108: 119-123.

Lindgren EJ and Jaenson TG (2006) Lyme borreliosis in Europe: Influences of climate and climate change, epidemiology, ecology and adaptation measures. World Health Organization, Copenhagen.

Maetzel D, Maier WA and Kampen H (2005) *Borrelia burgdorferi* infection prevalences in questing *Ixodes ricinus* ticks (Acari: Ixodidae) in urban and suburban Bonn, western Germany. Parasitol Res 95: 5-12.

Matuschka FR, Ohlenbusch A, Eiffert H, Richter D and Spielman A (1995) Antiquity of the Lyme-disease spirochaete in Europe. Lancet 346: 1367.

Matuschka FR, Endepols S, Richter D and Spielman A (1997) Competence of urban rats as reservoir hosts for Lyme disease spirochetes. J Med Entomol 34: 489-493.

Matuschka FR, Schinkel TW, Klug B, Spielman A and Richter D (2000) Relative incompetence of european rabbits for Lyme disease spirochaetes. Parasitol 121: 297-302.

Mavin S, Joss AWL and Ho-Yen DO (2005) Incidence of Scottish Lyme disease appears to be related to the effects of weather on tick survival and human behaviour. J R Coll Phys Edinb 35: 112-115.

McMichael AJ, Woodruff RE and Hales S (2006) Climate change and human health: present and future risks. Lancet 367: 859-869.

Mejlon HA and Jaenson TGT (1997) Questing behaviour of *Ixodes ricinus* ticks (Acari: Ixodidae). Exp Appl Acarol 21: 747-754.

Michalik J, Hofman T, Buczek A, Skoracki M and Sikora B (2003) *Borrelia burgdorferi s.l.* in *Ixodes ricinus* (Acari: Ixodidae) ticks collected from vegetation and small rodents in recreational areas of the city of poznan. J Med Entomol 40: 690-697.

Ogden NH (1997) Natural Lyme disease cycles maintained via sheep by co-feeding ticks. Parasitol 115: 591.

Ostfeld RS, Price A, Hornbostel VL, Benjamin MA and Keesing F (2006a) Controlling ticks and tick-borne zoonoses with biological and chemical agents. BioScience 56: 383-394.

Ostfeld RS, Canham CD, Oggenfuss K, Winchcombe RJ and Keesing F (2006b) Climate, deer, rodents, and acorns as determinants of variation in Lyme-disease risk. PLoS Biol 4: e145.

Perret JL, Guigoz E, Rais O and Gern L (2000) Influence of saturation deficit and temperature on Ixodes ricinus tick questing activity in a Lyme borreliosis-endemic area (Switzerland). Parasitol Res 86: 554-557.

Piesman J, Clark KL, Dolan MC, Happ CM & Burkot TR (1999) Geographic survey of vector ticks (*Ixodes scapularis* and *Ixodes pacificus*) for infection with the Lyme disease spirochete, *Borrelia burgdorferi*. J Vector Ecol 24: 91-98.

Randolph SE, Green RM, Hoodless AN and Peacey MF (2002) An empirical quantitative framework for the seasonal population dynamics of the tick *Ixodes ricinus*. International Journal for Parasitol 32: 979-989.

Randolph SE and Storey K (1999) Impact of microclimate on immature tick-rodent host interactions (Acari: Ixodidae): implications for parasite transmission. J Med Entomol 36: 741-748.

Rauter C and Hartung T (2005) Prevalence of *Borrelia burgdorferi sensu lato* genospecies in *Ixodes ricinus* ticks in Europe: a metaanalysis. Appl Environ Microbiol 71: 7203-7216.

Rebaudet S and Parola P (2006) Epidemiology of relapsing fever borreliosis in Europe. FEMS Immunol Med Microbiol 48: 11-15.

Rijpkema SG and Bruinink HH (1996) Detection of *Borrelia burgdorferi sensu lato* by PCR in questing *Ixodes ricinus* larvae from the Dutch North Sea island of Ameland. Exp Appl Acarol 20: 381-385.

Rijpkema SG, Molkenboer MJ, Schouls LM, Jongejan F and Schellekens JF (1995) Simultaneous detection and genotyping of three genomic groups of *Borrelia burgdorferi sensu lato* in Dutch *Ixodes ricinus* ticks by characterization of the amplified intergenic spacer region between 5S and 23S rRNA genes. J Clin Microbiol 33: 3091-3095.

Rijpkema SG, Nieuwenhuijs J, Franssen FFJ and Jongejan F (1994) Infection rates of *Borrelia burgdorferi* in different instars of *Ixodes ricinus* ticks from the Dutch North Sea Island of Ameland. Exp Appl Acarol 18: 531-542.

Rijpkema SGT, Herbes RG, Verbeek-de Kruif and Schellekens J (1996) Detection of four species of *Borrelia burgdorferi sensu lato* in *Ixodes ricinus* ticks collected from roe deer (*Caprolus capreolus*) in The Netherlands. Epidemiol Infect 117: 563-566.

Robertson JN (2000) Tick bite and Lyme borreliosis risk at a recreational site in England. Eur J Epidemiol 16: 647.

Schouls LM, Van De Pol I, Rijpkema SGT and Schot CS (1999) Detection and identification of *Ehrlichia*, *Borrelia burgdorferi sensu lato*, and *Bartonella* species in Dutch *Ixodes ricinus* ticks. J Clin Microbiol 37: 2215-2222.

Smit R, Dijk F, Kruidbos F, Schouls LM, Pol Ivd, Docters van Leeuwen B, Helinski M, Holtkamp R and Takken W (2003) Populatiedynamiek en fenologie van teken in Nederland. Infectieziekten Bull 14: 167-170. [in Dutch]

Sonenshine DE (1991) Biology of ticks. Oxford University Press, New York.

Sonenshine DE (2004) Pheromones and other semiochemicals of ticks and their use in tick control. Parasitol 129: S405-425.

Stafford KC III, Cartter ML, Magnarelli LA, Ertel S-H and Mshar PA (1998) Temporal correlations between tick abundance and prevalence of ticks infected with *Borrelia burgdorferi* and increasing incidence of Lyme disease. J Clin Microbiol 36: 1240-1244.

Stanek G and Strle F (2003) Lyme borreliosis. Lancet 362: 1639-1647.

Tälleklint L and Jaenson TG (1994) Transmission of *Borrelia burgdorferi s.l.* from mammal reservoirs to the primary vector of Lyme borreliosis, *Ixodes ricinus* (Acari: Ixodidae), in Sweden. J Med Entomol 31: 880-886.

Telford SR III, Mather TN, Moore SI, Wilson ML and Spielman A (1988) Incompetence of deer as reservoirs of the Lyme disease Spirochete. Am J Trop Med Hyg 39: 105-109.

Van Buskirk JAM and Ostfeld RS (1998) Habitat heterogeneity, dispersal, and local risk of exposure to Lyme disease. Ecol Appl 8: 365-378.

Wielinga PR, Gaasenbeek C, Fonville M, de Boer A, de Vries A, Dimmers W, Akkerhuis Op Jagers G, Schouls LM, Borgsteede F and van der Giessen JWB (2006) Longitudinal analysis of tick densities and *Borrelia*, *Anaplasma*, and *Ehrlichia* infections of *Ixodes ricinus* ticks in different habitat areas in The Netherlands. Appl Environ Microbiol 72: 7594-7601.

Wilson ML, Telford SR, Piesman J and Spielman A (1988) Reduced abundance of immature *Ixodes dammini* (Acari: Ixodidae) following elimination of deer. J Med Entomol 25: 224-228.

Zeman PE (1999) Mosaic pattern of *Borrelia* infection in a continuous population of the tick *Ixodes ricinus* (Acari: Ixodidae). Exp Appl Acarol 23: 327-335.

Emerging arthropod pests

13. Psoroptic mange: rising prevalence in UK sheep flocks and prospects for its control

Richard Wall

Abstract

Psoroptic mange in sheep, known as scab, is caused by infestation by the astigmatid mite *Psoroptes ovis* (Hering). The disease has been widely prevalent and economically important in Europe for many hundreds of years. In the UK, intensive control measures over the first half of the twentieth century, involving compulsory treatment of infested flocks with organochlorine insecticides and movement controls, resulted in apparent eradication of the disease by 1952. However it was inadvertently reintroduced with imported stock in 1972. Twenty years of unsuccessful attempts to re-eradicate the disease then followed. Eventually, concern over the effects of the widespread use of acaricides on human health and the environment, and the failure of compulsory treatment programmes to control the disease, led in 1992 to deregulation of sheep scab and removal of compulsory national treatment programmes. Following this change, scab has very quickly become increasingly common and problematic. Nationally, there were estimated to be only 40 and 120 outbreaks per year in 1989 and 1992, but prevalence was estimated to have risen to 3,000 in 1997 and then 7,000 outbreaks by 2004. This has led to calls to re-impose eradication measures. Here, the biology of *Psoroptes* mites and the spread of the disease following reintroduction are described; the formidable obstacles to the success any proposed new eradication programme in the UK are then discussed.

Keywords: Acari, disease, eradication, insecticide, pest management, prevalence, mites, *Psoroptes*, scab, veterinary parasitology

Introduction

Psoroptes mites (Acari: Psoroptidae) are non-burrowing, obligate ectoparasites of mammals. They infest a wide variety of hosts, including sheep, cattle, goats, buffalo, horses, mules, camelids and rabbits (Bates 1999). The pathology of infestation varies between hosts and between populations of mites, but essentially occurs in two forms: a chronic otoacariasis or a more acute body-mange (Bates 1999). The latter, in particular, is a disease of considerable importance causing economic loss and welfare problems in livestock, particularly sheep, in many areas of the world (O'Brien 1999, Van den Broek and Huntley 2003).

Clinical disease is thought to be the result of a hypersensitivity reaction by the host to antigenic mite faecal material (Sinclair and Kirkwood 1983, Mathieson and Lehane 1996), although skin abrasion by the chelicerae may also be contributory (Mapstone *et al.* 2002). The presence of actively feeding mites leads to inflammation and the exudation of lymph. The mites then feed superficially on the lipid emulsion of lymph, skin cells, secretions and bacteria on the surface of the host skin (Sinclair and Filian 1989). The lymph dries to form yellowish crusts and scabs which lift away from the skin as the wool grows. This presence of the mites results in severe irritation leading to restlessness, biting and scratching of infested areas and wool loss (Kirkwood 1980, Sargison *et al.* 1995).

About 40-50 days after initial infestation the mite population reaches its peak, by which time the infestation may cover as much as three quarters of the host's body. Transmission may occur directly via infested sheep. The probability of direct transmission from sheep to sheep peaks at about the time when the population of adult mites reaches its peak abundance (Beriatua *et al.* 1999, 2001). Transmission may also occur indirectly from the environment; mites may remain alive and reinfective in the environment for up to 18-20 days, depending on ambient conditions of temperature and humidity (O'Brien *et al.* 1994a, Smith *et al.* 1999).

In some sheep, infestation may lead rapidly to death; in others, however, the immune responses of the sheep affect mite fecundity and survival causing the mite population to decline. Mites, which were dispersed over the body, may only remain in sites such as the orbital fossae or external auditory canal (Downing 1936a,b).The infested sheep appears to recover and wool regrows. The mite population may die out or sheep may carry small populations of mites for up to two years, during which time they are asymptomatic (Babcock and Black 1933). This has important implications for the spread of sheep scab, since what appear to be clean, disease-free sheep, may transmit infection to naïve hosts and flocks.

Sheep scab is currently prevalent throughout Europe, South America, Asia and South Africa (Colebrook and Wall 2004). However, it was successfully eradicated in 1884 from Australia, in 1924 from Canada, in 1927 from Scandinavia and, it is claimed, from the United States in 1973 (Seddon 1964, Graham and Hourigan 1977). The Australian eradication was achieved through an intense control programme in which all infested sheep and their contacts were destroyed and all carcasses, fencing and infested pastures burned (Seddon 1964).

The infestation of cattle is less common than that of sheep, although when present it may cause a very severe dermatitis, characterised by extensive lesions and more often by secondary bacterial infestation than is commonly seen in sheep. There appears to be a marked difference in cattle breed susceptibility, with the infestation of cattle in Belgium (predominantly Belgian White and Blue breed) by *P. ovis* being particularly common; it is estimated that in Belgium 400,000 of 3.2 million cattle are treated for infestation each year (Losson *et al.* 1999).

Psoroptes ovis and scab epidemiology

Psoroptes ovis has five life-cycle stages: egg, larva, two nymphal stages, the protonymph and tritonymph, and adult (Figure 1). Species of astigmatid mite often have an unusual deutonymph, modified for dispersal. These dispersal stages do not have functional mouthparts and usually have an incomplete gut. In most taxa the formation of the deutonymph is facultative and is dependent on the occurrence of appropriate environmental conditions. When it is not formed, or when it is entirely suppressed, as in the parasitic species of astigmatid mite, the protonymph moults directly into the tritonymph skipping the deutonymphal instar entirely, as is the case in *P. ovis*.

In the adult male there are jointed pretarsi and pulvilli on the first three pairs of legs (Figure 1). The fourth pair of legs are much shorter than others. The male possesses a very characteristic pair of adanal suckers and paired posterior lobes. The adult female has two pairs of metapodosomal setae on the ventral surface and a characteristic inverted U-shaped vulva in the propodosomal region.

Both male nymphal stages lack dorsoposterior tubercles (Figure 1). However, male tritonymphs have two pairs of cuticular pits on the central metapodosoma while the protonymph carries a

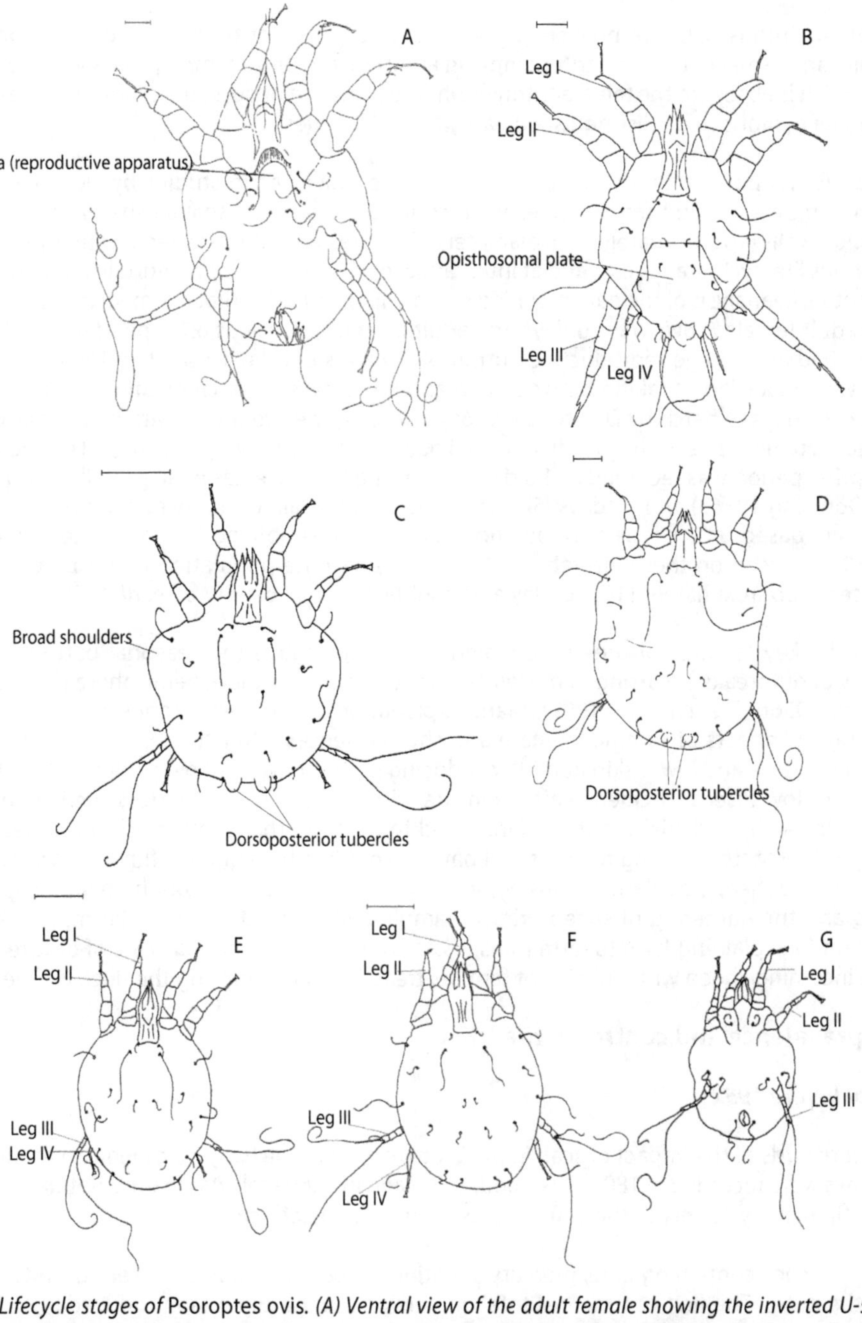

Figure 1. Lifecycle stages of Psoroptes ovis. *(A) Ventral view of the adult female showing the inverted U-shaped vulva. (B) Dorsal view of an adult male showing the opisthosomal plate and short IVth leg. (C) Dorsal view of a female protonymph showing the dorsoposterior tubercles, broad shoulders and tapered waist. (D) Dorsal view of a female tritonymph showing the dorsoposterior tubercles and oval shape. (E) and (F) Dorsal view of a male protonymph and tritonymph respectively. (G) Ventral view of a larva showing only three pairs of legs. Bar = 100µm (modified from Sanders* et al. *2000).*

single pair. Both female nymphal stages possess dorsoposterior tubercles, but the protonymph is significantly smaller than the tritonymph. In addition, the protonymph possesses a pulvillus on leg IV which is absent in the female tritonymph. The presence of dorsoposterior tubercles enables the female nymphs to be distinguished easily from the males.

Data on *P. ovis* population dynamics *in vivo* are rare. However, in studies by Downing (1936a), known numbers of adult female mites were confined to areas of shaved sheep skin, in circular aluminium cells (approximately 3 cm diameter). The mites were then inspected and the number of the various life-cycle stages counted at intervals, over which the rates of mortality and oviposition were noted. Reanalysis of the data provided by Downing, indicates that the median life expectancy for an adult female *P. ovis* was 16 days, the adult mortality rate was 6.2% per day and that adult females had an average reproductive output of 2.9 eggs per day (Wall *et al.* 1999). The eggs of *P. ovis* were recorded as taking between two and three days to hatch by Shilston (1915) and on average 2.7 days to hatch by Downing (1936a). Similarly, the average duration of the larval stage was shown to be 2.2 days, the protonymph 2.3 days and the tritonymph 2.2 days. The average pre-oviposition period was recorded as 1.3 days. The entire life-cycle has been described as taking 11 days (Downing 1936a) and 16 days (Shilston 1915). An overall mortality rate of 0.91 per day was calculated, based on analysis of population age structures obtained from infested sheep (Wall *et al.* 1999). Based on these life-table data, a *Psoroptes* mite population might be expected to increases at approximately 11% per day and double every 6.3 days (Wall *et al.* 1999).

One of the key features of the epidemiology of sheep scab is the seasonal pattern, with the majority of outbreaks occurring in the winter months in the northern hemisphere (Shilston 1915, Downing 1936b, French *et al.* 1999). Many explanations have been proposed for this pattern, including the effects of summer climate and shorter fleeces (after shearing), leading to higher skin temperature and lower skin humidity, reducing mite survival and oviposition rates. However, patterns of low disease incidence also coincide with management practices such as the use of insecticides with acaricidal properties, intended to control other summer ectoparasites such as blowfly. Other factors driving the seasonal pattern could be the annual influx of new susceptible lambs into the sheep population, housing, sheep sales, the loss of ewe condition during gestation, mating and the gathering of sheep prior to lambing. Movement in particular may be critically important in explaining farm to farm transmission; one study found that 73% of outbreaks were due to incoming sheep with only 3% of farmers treating sheep entering the flock (O'Brien 1999).

Scab prevalence and control in the UK

Control before 1952

The first records of the impact of scab in the UK date back to the early ninetheenth century: 2,573 outbreaks were recorded in 1807 (Kirkwood, 1986) and in 1869 scab was made a notifiable disease. At that time every county in the country was thought to be effected.

Initially, washes containing lime, mercury, nicotine, turpentine or arsenic were used to attempt to kill the mites. The first commercial total immersion plunge-dip was introduced in 1843 when William Cooper marketed a wettable powder containing arsenic and sulphur (Kirkwood 1986). Other formulations, including coal tar, creosote, crysallic acid, rotenone and lime have also been used, but many of these also have adverse effects on sheep health (O'Brien 1999). In the early part of the last century in the USA, Babcock and Black (1933) recommended both a lime-sulphur based dip and a nicotine sulphate-based dip for killing mites. When compulsory dipping was introduced in 1906 for all infested sheep in the UK, lime, sulphur, arsenical or phenolic compounds were used. However, the formulations used did not persist in the fleece for long enough and did not kill the eggs, so two dips 10 to 14 days apart, were required (Kirkwood 1985).

The major advance in the insecticidal control of sheep scab was made in the 1940s when the organochlorines, particularly Lindane (gamma-hexachlorocyclohexane, a pure form of the active isomer of benzene-hexachloride), were shown to be effective in dips, giving up to three months of residual activity from a single treatment. This dramatically shifted the balance against the mites. Eventually, through (a) a combination of efforts by farmers, veterinarians and the police for many years, (b) the use of organochlorine insecticides, (c) a centralised policy of statutory total immersion dipping, and (d) the imposition of statutory movement controls, the disease was believed to be eradicated in 1952. Britain then enjoyed 20 years, apparently without any scab, until 1973 when an outbreak was confirmed in Lancashire.

Scab control: 1972-1992

Mites were inadvertently reintroduced in 1972, probably on sheep imported from Ireland, but the infection was initially misdiagnosed as mycotic dermatitis (Loxam 1974). The disease rapidly spread, with 27 outbreaks discovered in January 1973, followed by a further 21 during the winter of 1973-74 (Loxam 1974, Tarry 1974). As a result, scab was once again made a notifiable disease and compulsory insecticidal treatment of all in contact sheep was reintroduced. Infested flocks were placed under movement restrictions and all flocks in the local area of an outbreak had to be dipped (Loxam, 1974). However, regional compulsory dipping was unable to prevent rapid expansion of outbreaks as the disease became widespread throughout the UK, particularly the south-west and the north-west of England. This spread continued until introduction of a compulsory national autumn dip in approved acaricides in 1976 which initiated a continuous 16-year effort to eradicate the disease (Figure 2). A four year period of decline in number of reported outbreaks followed the introduction of the national treatment programme, although the south-west still reported large numbers of outbreaks (French *et al.* 1999).

In 1972, gamma-HCH was the only proven treatment (Tarry 1974). However, growing concern over the environmental effects of the use of organochlorine insecticides led to calls for their withdrawal. Nevertheless, it was not until late 1985 that organochlorines were finally fully withdrawn from use in sheep dips (Kirkwood 1985, O'Brien 1999). By this time they had been replaced by the less persistent organophosphates, diazinon and propetamphos (Kirkwood and Quick 1981, 1982). If used correctly diazinon will give 6-8 weeks protection. Subsequently, the pyrethroids flumethrin (1987), and later high-cis cypermethrin, were also registered for use.

In 1982, the national autumn dip was replaced by a national summer dip, but this led to an immediate increase in outbreaks, with nearly 160 reported in 1983 (Figure 2). The summer dip only strategy, was followed therefore by the introduction of twice yearly, summer and autumn dips, in 1984. There was then a rapid decline in the number of cases in most counties, reaching as low as 16 in 1988. It is clear from this pattern that major changes in treatment policy appeared to coincide with the change in prevalence of the disease (Figure 2). However, by the late 1980s, it appeared that less than 0.1% of flocks at risk in the UK were infested and that, for various reasons, the ongoing insecticidal treatment programme did not appear to be able to finally eliminate the last residual outbreaks. In addition, since scab was endemic in all countries of Europe the further

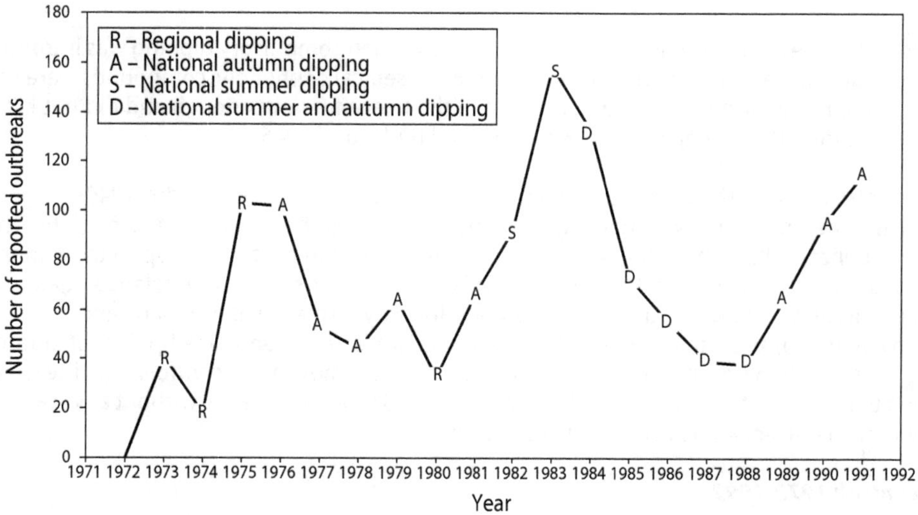

Figure 2. The number of reported outbreaks recorded by the State Veterinary Service and the Veterinary Laboratory Agency for the Ministry of Agriculture, Fisheries and Food (MAFF), between 1973 and 1992, showing the different dipping protocols used at the time (redrawn from French et al. *1999).*

import of potentially infested animals was likely to be impossible to prevent altogether. Concern was also growing about toxicological effects of organophosphate sheep dips on human operators and the environmental effects of pyrethroid dipping. Political thinking also appeared to favour a move away from government-imposed control, towards industry-led responsibility for disease and animal health problems. As a result, the compulsory national summer dip was phased out in 1988 and the final compulsory autumn dip removed in 1992 when accurate record collection also ceased.

In 1989, the year after removal of the summer dip, the number of outbreaks rose to 40, with a greater involvement of counties in the east of England (French *et al.* 1999). In 1991, when the final compulsory dip was abolished, there were approximately 120 outbreaks in that year. Overall, a total of 1480 outbreaks had been reported over the 20 years between 1973 and 1992 with most counties in England, Wales and Scotland affected by sheep scab and the largest number of outbreaks reported in the south-west, particularly Devon and Cornwall and the north-west, particularly Lancashire (French *et al.* 1999).

Scab control: 1992 to the present

After deregulation in 1992, sheep scab was treated by farmers on a case-by-case basis. As a result, accurate figures for the incidence of the disease are hard to obtain. However, in 1997 up to 3,000 outbreaks were estimated to have occurred nationally (Lewis 1997) and between October 1998 and September 1999, estimates suggested that there may have been as many as 5,000 outbreaks of sheep scab (Corke and Broom 2000).

To obtain more accurate prevalence estimate, a stratified retrospective questionnaire of 3,500 sheep farmers was used to obtain information relating to the period between March 2003 and

February 2004 (Bisdorff *et al.* 2006). The results of this questionnaire survey showed that overall, 9% (± 1.72) of farmers reported at least one outbreak of scab between March 2003 and February 2004. Given the number of flocks in the UK, this prevalence equates to approximately 6,750 outbreaks in that year. There was a significant regional difference in the prevalence (Figure 3).

The proportion of farms with at least one outbreak of scab was highest in Wales, followed by Scotland and Northern England, with 17%, 14% and 11% of flocks affected, respectively. The percentage of farms affected in England was substantially lower and similar in all regions, at between 3% and 6%. In addition, scab outbreaks were significantly more common between October and March than during the summer months, confirming its status as a predominantly winter disease. While the cause of the regional variation was not determined in the questionnaire study, the geographic distribution is suggestive of an association between prevalence and upland farms employing common grazing.

The pyrethroid dips were removed as approved products for scab control in the UK in 2006, largely as a result of environmental concerns; pyrethroid dips are highly toxic in the aquatic environment and have been responsible for a number of inadvertent pollution incidents in sheep-producing areas. Propetamphos was withdrawn for economic reasons and following temporary withdrawal to allow the development of safer packaging, diazinon remains the only available insecticide for use in dips in the UK, although pressure for its removal remains because of concerns over human exposure. Currently, only the injectable macrocyclic lactones are available as therapeutic alternatives to dips, and they have seen a growing application as ectoparasiticides. Ivermectin is effective against sheep scab following two subcutaneous injections of 200 μg/kg, seven days apart (Soll *et al.* 1992). Similarly, one injection of moxidectin at 200 μg/kg is effective (O'Brien *et al.* 1994b, 1996), as is doramectin at 300 μg/kg injected intramuscularly (Bates *et al.* 1995). To achieve satisfactory results, however, the use of these compounds requires considerable care. This is particularly the case with ivermectin where only limited persistent activity has been reported. Hence each sheep must be carefully injected with the correct dose for its weight, marked after treatment, and then moved to clean housing. With ivermectin this process must then be repeated.

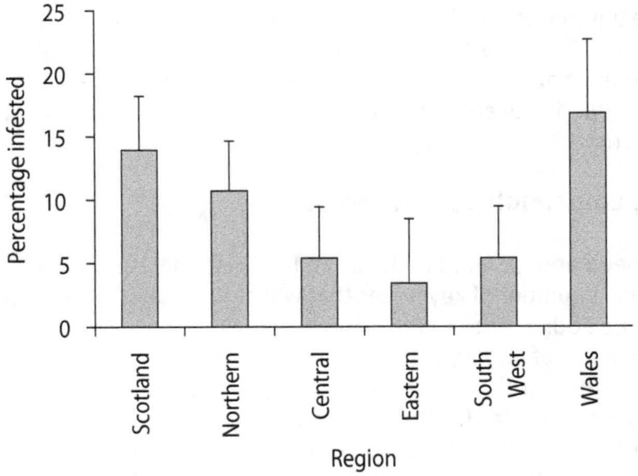

Figure 3. The percentage (± 95% confidence interval) of sheep farms that reported at least one case of psoroptic mange in 2003/2004, in six regions of the UK (redrawn from Bisdorff et al. *2006).*

Scab mites may remain fully infective off host for periods of 2-3 weeks, so reinfection of treated sheep from contaminated housing is a persistent threat; a single untreated sheep may reinfest the entire flock. In contrast, moxidectin has been found to have prophylactic action for at least 28 days following treatment (O'Brien *et al.* 1996); no clear reports of the persistence of doramectin exist but it is believed to at least equal that of moxidectin.

There are concerns that resistant strains of *P. ovis* may develop; the first case of organophosphate resistance was identified in 1996 and resistance is likely to become increasingly common (Bates 2004). Simultaneous resistance to formulations of organophosphates, synthetic pyrethroids and ivermectin has been reported in Ireland (O'Brien 1999). Synge *et al.* (1995) reported resistance to the pyrethroid flumethrin, and Clark *et al.* (1996) found resistance to the organophosphate, propetamphos.

The future for sheep scab

The history of sheep scab prevalence and its control in the UK illustrates the development of a widespread and growing economic and animal welfare problem, resulting from the spread of mite infestation in sheep flocks and failure to impose adequate consistent control, exacerbated by pressure to discontinue the use of treatment with neurotoxic insecticides.

Concern over the growing prevalence of this disease is leading to renewed calls for a national statutory scab eradication programme in the UK and, in the first instance, this has resulted in the development of a number of user and industry-led initiatives. The most ambitious, mediated by the National Sheep Association (NSA) and SCOPS (Sustainable Control of Parasites in Sheep), aims to progress through three phases. Phase 1 was launched in August 2006 and it is being promoted through an awareness-raising scheme. In its initial step, the NSA has sent a fact sheet and control advice to all UK sheep farmers. It is proposed that a second phase (probably in autumn of 2007) will involve a continuation of the awareness raising campaign plus targeted eradication campaigns to focus on 'hotspots' as they have become known. The phase 2 programme will also start to involve a more concerted effort to raise peer pressure to encourage farmers to report other flocks in the area which have the disease. Localised initiatives to clear 'hotspots' will be encouraged and facilitated through veterinary practises and other service providers. In phase 3 (possibly Autumn 2008) pressure on remaining 'hotspots' will be intensified. But, at this stage, a more legislation-based approach may also be required in extreme cases. While, as yet there has been no move towards government support to develop a compulsory scheme, this is being advocated by the various co-ordinating bodies.

Key issues in the epidemiology and spread

Given the evident need and growing enthusiasm for a national campaign in the UK designed to eradicate sheep scab, a number of key issues that will have a major impact on its likely success or failure, will need to be addressed:

1. What is the prevalence of undiagnosed, sub-clinical infection in the national flock? In addition to the acute body-form of the disease, sheep may also harbour mites in sites such as the ear, where they may remain as a chronic, often sub-clinical infection for long periods. Standard dipping procedures do not always result in dip reaching these sites, particularly deep in the ear (Bates 1996a). The prevalence of parasitic otitis associated with *Psoroptes* mite infestation was considered in a survey which showed that 28% of the 64 British farmers contacted reported

clinical signs of the disease (ear lesions); within-flock prevalence was between 1 and 60% (mean 15.5%) (Morgan 1992). However, no evidence of body mange was reported in these flocks. In a more comprehensive study, 2,952 sheep heads from England, Wales and Scotland were examined over a 24 month period (Bates 1996b). There was no evidence of clinical sheep scab on the heads or any traumatic damage, but 1.4% (42) were infested with psoroptic ear mites. Prevalence appeared to increase yearly, with 0.3, 1.9 and 3.9% of examined sheep infested in 1989, 1990 and 1991 respectively. Between January 1989 and May 1991, seven flock-level cases of sub-clinical psoroptic otoacariasis were reported by the UK state veterinary service. Examination of these flocks for sheep scab, headshaking, aural lesions and other evidence of infestation obtained via skin scrapings and ear swabs, revealed that the ears of 3.1% of the 2,676 sheep were infested with *Psoroptes* spp. Prevalence within flocks ranged between 1.3 and 23.9% (Bates 1996b).

These results indicate that ear infestation with *Psoroptes* mites, which is often sub-clinical, may be far more prevalent than is usually anticipated. Clearly, if these infections are overlooked in any national eradication campaign which targets only therapeutic treatment of clinically diagnosed cases, the potential for reinfection will remain high. Hence it is likely that all sheep and all flocks would need to be inspected and/or treated; but this is likely to be an expensive and time-consuming requirement.

2. A further major question facing the eradication attempt is the degree to which transmission is possible from feral sheep and non-ovine hosts.

Over the last 170 years, the classification of *Psoroptes* mites has had a long and complex history. The initial species descriptions of what we now know as *Psoroptes* mites were given by E. Hering in 1838, originally under the genus name *Sarcoptes*. Subsequently, many early authors believed there to be only a single species; host-derived populations were referred to simply as *P. communis* with varietal names such as *ovis* or *bovis*, depending on the host species from which they were obtained (Stockman and Berry 1913, Shilston 1915). Hirst (1922) described only two species: the general *P. communis* and *P. natalensis* from specimens found on cattle in South Africa. On the other hand, later authors gave full species status to a wide range of host-derived *Psoroptes* and the number of putative species proliferated (see Bates 1999, for a review).

Sweatman (1958) attempted to rationalise the taxonomy by arguing that there were in fact five good species within this genus, based on the host infested, the infestation site and differences in length of the outer opisthosomal setae of adult male mites. Sweatman's species were: *Psoroptes cuniculi* and *P. cervinus* from the ears of rabbits and bighorn sheep respectively, and *P. equi*, *P. ovis* and *P. natalensis*, which were described as primarily body mites of horses, sheep/cattle and buffalo, respectively. Sweatman's classification achieved widespread acceptance in the veterinary literature, despite which some parasitologists still feel the need to designate new host-derived species for slight morphological variants such as *Psoroptes auchinae* in alpacas (Foreyt *et al.* 1992) or *Psoroptes pienaari* in African Buffalo (Fain 1970).

Recently, however, studies that have undertaken extensive comparisons of morphology and molecular characterisation using sequence data from the mitochondrial ITS-2 region and microsatellite markers, have found little or no consistent host-related variation between the mite population samples. The results of these studies strongly suggest that there is no case for considering the *Psoroptes* mites from different hosts as separate species (Zahler *et al.* 1998, Bates 1999, Pegler *et al.* 2005) and the name *P. ovis* (Hering 1838) should carry taxonomic priority for the synonomised genus (Wall and Kolbe 2006). It is suggested that *Psoroptes* mites should now be considered to represent a single species, which can infest a range of hosts,

but which show local phenotypically-adapted populations on different hosts, perhaps due to differing conditions of temperature or immune challenge on that host (Bates 1999).
Studies of genetic relatedness of mite populations on different hosts also provide some evidence that cross-species transmission may occur (Pegler at al. 2005). Hence, a critical question is: to what extent is cross-infection possible in the field in the UK? While direct host-to-host transmission may be unlikely between different host species, possible indirect transmission via the environment clearly has important practical epidemiological implications for control of this disease. If other hosts, such as feral sheep, goats or llamas, can act as reservoirs and these are not included in a control programme, even if cross-transmission is a relatively rare event, then expensive and time-consuming eradication attempts are likely to fail. At present there is simply no good data on this risk.

3. The prevalence data collected by Bisdorff *et al.* (2006), highlighted the association between high prevalence and upland areas with common grazing. Co-ordinating the treatment of sheep on such areas will be difficult to achieve.
   If treatment is not synchronised, the potential for reinfection from the environment remains high. This is particularly problematic if the therapeutic treatment of choice has a residual period of activity below the off host survival period, i.e. below three weeks, such as with the macroyclic lactones, since all sheep would need to be kept isolated and off the areas of common grazing for this period. Similarly, populations of feral sheep and goats in these areas would also need to be treated or eliminated. In practice, this is likely to be difficult to achieve. It has been suggested that a lack of full co-operation by farmers contributed to the failure to eradicate the disease between 1972 and 1992 (O'Brien 1999).
4. Finally, if control is voluntary and regionally focused, the transport and movement of sub-clinical but infective sheep will become a particularly important issue, since movement of these sheep will undermine regional control activities.
   The ability to trace the likely source of infections will therefore become of increased value. The ability to trace sources of infection also has important implications for the ability to designate 'hotspots' where control activities should be focused. It is unclear who will control and co-ordinate movement restrictions; and with no government-imposed legislative enforcement, it is difficult to envisage how this will be achieved.

Conclusions

Despite growing need, the obstacles to success for any national eradication programme, based on traditional control technologies, remain formidable. Problem issues include: the growth in insecticide resistance and the slow rate of development of new actives, the national scale over which insecticide would need to be applied (40 million sheep), the need to impose quarantine, mixing and movement restrictions, the widespread presence of potential reservoir species, coupled with environmental and health concerns associated with the continued extensive use of some of the older neurotoxic insecticides. Good epidemiological data remains unavailable for many of these issues, such as cross-species transmission, so their practical significance cannot be fully evaluated. This catalogue of difficulties suggests that any hopes of eradication based on traditional diagnose-and-treat approaches may be unrealistic. It may perhaps be more appropriate to consider more sophisticated approaches to scab control, with less focus on the elusive goal of national eradication. A more effective use of time and resources might be achieved by the development of proactive local or regional programmes which aim to allow scab problems to be managed at acceptable levels, while conserving the compounds that remain available. The development of such local management programmes in Scotland has been described by Sargison *et al.* (2006). The development of integrated approaches, in which cascades of management tactics are deployed, with insecticides available as one component to be used in requisite circumstances, may also be appropriate. High on the list of management approaches should be good flock biosecurity and the isolation, inspection and, if necessary, treatment of all bought-in sheep. For upland flocks, collaborative, community-based efforts are likely to be essential. However, for any such management programmes to be developed, considerably better information is required relating to the epidemiology of scab, in addition to the greater availability of effective alternative control tools.

Acknowledgements

I am extremely grateful to current and former students whose work and ideas have contributed to this paper, particularly Alexandra Brooks, Amy Sanders and Betty Bisdorff.

References

Babcock OG and Black WL (1933) The common sheep scab mite and its control. Texas Agric Exp St Bull 479: 1-34.

Bates P, Groves, BA, Courtney, SA and Coles, GC (1995) Control of sheep scab (*Psoroptes ovis*) on artificially infested sheep with a single injection of doramectin. Vet Rec 137: 491-492.

Bates PG (1996a) Epidemiology of subclinical ovine psoroptic otoacariasis in Great Britain. Vet Rec 138: 388-393.

Bates PG (1996b) Ovine psoroptic otoacariasis: an abattoir survey. Vet Rec 139: 235-236.

Bates, PG (1999) Inter- and intra-specific variation within the genus *Psoroptes* (Acari: Psoroptidae). Vet Parasitol 83: 201-217.

Bates, PG (2004) Therapies for ectoparasiticism in sheep. In Practice 26: 538-547.

Berriatua E, French, NP, Wall R, Smith KE and Morgan KL (1999) Within-flock transmission of sheep scab in naive sheep housed with single infested sheep. Vet Parasitol 83: 277-289.

Berriatua E, French NP, Broster CE, Morgan KL and Wall R (2001) Effect of infestation with *Psoroptes ovis* on the nocturnal rubbing and lying behaviour of housed sheep. Appl Anim Behav Sci 71: 43-55.

Bisdorff B, Milnes A and Wall R (2006) Prevalence and regional distribution of scab, lice and blowfly strike in sheep in Great Britain. Vet Rec 158: 749-752.

Clark AM, Stephen FB, Cawley GD, Bellworthy SJ and Groves BA (1996) Resistance of the sheep scab mite *Psoroptes ovis* to propetamphos. Vet Rec 139: 451.

Colebrook E and Wall R (2004) Ectoparasites of livestock in Europe and the Mediterranean region. Vet Parasitol 120: 251-274.

Corke MJ and Broom DM (2000) Sheep Scab: a questionnaire survey of sheep farmers. Proc Sheep Vet Soc 24: 115-121.

Downing W (1936a) The life history of *Psoroptes communis* var *ovis* with particular reference to latent or suppressed scab. J Comp Path and Therapeutics 49: 183-209.

Downing W (1936b) The life-history of the scab mite in relation to control measures. Vet Rec 48: 1065-1066.

Fain A (1970) Sur une nouvelle espèce de genre *Psoroptes* produisant la gale chez le buffle africain (Acarina: Sarcoptiformes) Rev Zool Bot Afr 81: 95-100.

Foreyt WJ, Rickard LG and Boyce W (1992) *Psoroptes* sp. in two llamas (*Lama glama*) in Washington. J Parasitol 78: 153-155.

French NP, Berriatua E, Wall R, Smith K and Morgan KL (1999) Sheep scab outbreaks in Great Britain between 1973 and 1992: spatial and temporal patterns. Vet Parasitol 83: 187-200.

Graham OH, Hourigan JL (1977) Eradication programs for arthropod parasites of livestock. J Med Entomol 13: 639-641.

Hering E (1838) Die Kräzmilben der Thiere und einige verwandte Arten, nach eigenen Untersuchungen beschrieben. *Nova Acta Physico-Medica Acadamiae Caesareae Leopoldino-Carolinae Naturae Curiosorum.* Tomi 18, Pars 2, Vratislaviae und Bonn.

Hirst S (1922) Mites injurious to domestic animals. In: British Museum (Natural History) Economic Series No. 13. London.

Kirkwood AC (1980) Effect of *Psoroptes ovis* on the weight of sheep. Vet Rec 107: 469-470.

Kirkwood AC (1985) Some observations on the biology and control of the sheep scab mite *Psoroptes ovis* (Hering) in Britain. Vet Parasitol 18: 269-279.

Kirkwood AC (1986) History, biology and control of sheep scab. Parasitol Today 2: 302-307.

Kirkwood AC and Quick MP (1981) Diazinon for the control of sheep scab. Vet Rec 108: 279-280.

Kirkwood AC and Quick MP (1982) Propetamphos for the control of sheep scab. Vet Rec 111: 367.

Lewis C (1997) Sheep Scab- an update on the history, control and present situation. Proc Sheep Vet Soc 21: 99-102.

Losson BJ, Lonneux J-F and Lekimme M (1999) The pathology of *Psoroptes ovis* infestation in cattle with a special emphasis on breed difference. Vet Parasitol 83: 219-229.

Loxam JG (1974) Sheep scab epidemic: January 1973. State Vet J 29: 1-10.

Mapstone SC, Beasley A and Wall R (2002) Structure and function of the gnathosoma of the mange mite, *Psoroptes ovis*. Med Vet Entomol 16: 378-385.

Mathieson BRF and Lehane MJ (1996) Isolation of the Gram-negative bacterium, *Serratia marcescens*, from the sheep scab mite, *Psoroptes ovis*. Vet Rec 138: 210-211.

Morgan KL (1992) Parasitic otitis in sheep associated with *Psoroptes* infestation: a clinical and epidemiological study. Vet Rec 130, 530-532.

O'Brien DJ (1999) Treatment of psoroptic mange with reference to epidemiology and history. Vet Parasitol 83: 177-185.

O'Brien DJ, Gray JS, O'Reilly PF (1994a) Survival and retention of infectivity of the mite *Psoroptes ovis* off the host. Vet Res Com 18: 27-36.

O'Brien DJ, Gray JS, O'Reilly PF (1994b) The use of moxidectin 1% injectable for the control of psoroptic mange in sheep. Vet Parasitol 52: 91-96.

O'Brien DJ, Parker LD, Menton C, Keavney C, McCollum E and O'Laoide S (1996) Treatment and control of psoroptic mange (sheep scab) with moxidectin. Vet Rec 139: 437-439.

Pegler KR, Evans L, Stevens JR and Wall R (2005) Morphological and molecular comparison of host-derived populations of parasitic *Psoroptes* mites. Med Vet Entomol 19: 392-403.

Sanders A, Froggatt P, Wall R and Smith KE (2000) Life-cycle stage morphology of I mange mites. Med Vet Entomol 14: 131-141.

Sargison ND, Scott PR, Penny CD and Pirie RS (1995) Effect of an outbreak of sheep scab (*Psoroptes ovis* infestation) during mid-pregnancy on ewe body condition and lamb birthweight. Vet Rec 136: 287-289.

Sargison N, Taylor D and Dun K (2006) Regional control of sheep scab in UK flocks. In Practice 28: 62-69.

Seddon HR (1964) Eradication of sheep scab from New South Wales. Aust Vet J 40: 418-421.

Shilston AW (1915) Sheep scab. Observatons on the life-history of *Psoroptes communis* var *ovis*, and some points connected with the epizootiology of the disease in South Africa. The Third and Fourth Reports of the Director of Veterinary Research. Pretoria. Department of Agriculture, Union of South Africa. Pp. 70-98.

Sinclair AN and Filian SJ (1989) Lipid ingestion from sheep epidermis by *Psoroptes ovis* (Acari: Psoroptidae). Vet Parasitol 31: 149-164.

Sinclair AN and Kirkwood AC (1983) Feeding behaviour of *Psoroptes ovis*. Vet Rec 112: 65.

Smith KE, Wall R, Berriatua E and French NP (1999) The effects of temperature and humidity on the off-host survival of *Psoroptes ovis* and *Psoroptes cuniculi*. Vet Parasitol 83: 265-275.

Soll MD, Carmichael IH, Swan SE and Abrey A (1992) Treatment and control of sheep scab (*Psoroptes ovis*) with ivermectin under field conditioning in South Africa. Vet Rec 130: 572-574.

Stockman S and Berry AH (1913) The *Psoroptes communis ovis*: Some observations on ova and ovipositing. J Comp Path Therap 26: 45-50.

Sweatman GK (1958) On the life history and validity of the species in *Psoroptes*, a genus of mange mites. Can J Zool 36: 905-929.

Synge BA, Bates PG, Clark AM and Stephen FB (1995) Apparent resistance of *P. ovis* to flumethrin. Vet Rec 137: 51.

Tarry DW (1974) Sheep scab: Its diagnosis and biology. Vet Rec 95: 530-532.

Van den Broek AHM, Huntley JF, Halliwell REW, Machell J, Taylor M, Miller HRP (2003) Cutaneous hypersensitivity reactions to *Psoroptes ovis* and Der p 1 in sheep previously infested with *P. ovis* - the sheep scab mite. Vet Immunol Immunopathol 91, 105-117.

Wall R, Smith KE, Berriatua E and French NP (1999) Simulation analysis of the population dynamics of the mite, *Psoroptes ovis*, infesting sheep. Vet Parasitol 83: 253-264.

Wall R and Kolbe K (2006) Taxonomic priority in *Psoroptes* mange mites: *P. ovis* or *P. equi*? Exp Appl Acarol 39: 159-162.

Zahler M, Essig A, Gothe R and Rinder H (1998) Genetic evidence suggests that *Psoroptes* isolates of different phenotypes, hosts and geographic origins are conspecific. Int J Parasitol 28: 1713-1719.

14. Waiting for the tiger: establishment and spread of the *Aedes albopictus* mosquito in Europe

Ernst-Jan Scholte and Francis Schaffner

Abstract

The Asian tiger mosquito *Aedes albopictus* Skuse is so called because of its black and white patches which adorn its legs and body. It is well known throughout the World, apart from the biting nuisance, especially because of its vectorial role in Chikungunya and dengue outbreaks, and its ability to invade new territories. In this chapter the focus is on its recent distribution and spread in Europe. Furthermore, introduction routes are discussed, as well as recommendations regarding options for control.

Keywords: *Aedes albopictus*, Culicidae, exotic species, distribution, control

Global distribution

Historically, *Aedes (Stegomyia) albopictus* (Skuse) (=*Stegomyia albopicta sensu* Reinert *et al.* 2004) (Diptera: Culicidae) (Figure 1) occurs as a tree breeding 'forest-mosquito' throughout the Oriental Region from the tropics and subtropical regions of Southeast Asia, the Pacific and most Indian Ocean Islands, north through China and Japan and west to Madagascar. In Korea it is found to about the latitude of Seoul and in Japan as far north as Sendai (Hawley 1988). During the 20th century it has expanded into at least 28 other countries around the world (Benedict *et al.* 2007). Establishments could succeed because of frequent introductions and ability for the species to adapt to local conditions owing to induction of diapause at the egg stage in temperate zones and ability to breed in numerous types of artificial containers (tyres, diverse containers, catch basins, etc.) and its propensity to feed on a diverse range of hosts.

Probably the first areas to be invaded were the Hawaiian Islands (late 19th century) and islands of the southern Pacific (e.g. Guam) as well as of the Indian Ocean (e.g. La Réunion). The large expansion, however, started in the last few decades of the 20th century. In North America, the first substantial population was found in 1985 in Texas, USA (CDC 1986, Sprenger and Wuithiranyagool 1986), although on a few occasions a handful of larvae and pupae were discovered in used tyres

Figure 1. Aedes albopictus *female (CDC/ J. Gathany).*

returning from South East Asian ports as early as 1946, and again in 1972 (Madon *et al.* 2002). The species is currently found in 866 counties in 26 states in the continental USA, mostly along the eastern and southern states. The northernmost established infestation in the USA is Chicago, Illinois, although an infestation was found in Minnesota in 1997 (CDC 2005). The invasion pattern was similar in Central and South America, where the species was first introduced into São Paolo, Brazil, in 1986 (Forattini 1986), where it has now spread over large parts of the Southeast of Brazil, covering at least 14 states (Santos and La 2003). After being established in both South and North America, many countries of Latin America and the Caribbean started surveillance programmes. After Brazil, *Ae. albopictus* populations were reported from the northern States of Mexico in 1993 (Ibanez-Bernal 1994), from where it spread along the coastal areas to the south (Casas-Martinez and Torres-Estrada 2003). In 1995 populations were reported from Guatemala (Ogata and Lopez Samayoa 1996), Honduras, and El Salvador (Benedict *et al.* 2007), followed in 1997 by Bolivia (Benedict *et al.* 2007), and in 1998 from the Amazon area of Colombia (Vélez *et al.* 1998), Paraguay (Benedict *et al.* 2007), from the Missiones Province in Argentina near Brazil (Rossi *et al.* 1999). In 2002 the species was found in Panama (Benedict *et al.* 2007), and in 2003 in Nicaragua (Lugo Edel *et al.* 2005). From the Caribbean, reports arrived from established populations in the Dominican Republic (Pena 1993), followed by Barbados (Rodhain 1996), the Cayman Islands in 1997 (A. Harris, personal communication, Lounibos *et al.* 2003), and Cuba (Broche and Borja 1999). In Africa, the species has been reported from car-tyre import inspections into South Africa (Cornel and Hunt 1991), and it became established in Cameroon (Simard *et al.* 2005), Equatorial Guinea (Toto *et al.* 2003), Nigeria (Savage *et al.* 1992), and Gabon (Krueger and Hagen 2007). The lack of surveys probably masks other infested areas. In New Zealand the species has been intercepted on the North and the South island (Laird *et al.* 1994), but it has not become established. Also in Australia the species has been intercepted several times at sea ports, but so far is established only on several of the Torres islands north of Queensland (Ritchie *et al.* 2005). In Europe, the species was reported for the first time in 1979 from Albania, and its presence was subsequently reported from eleven other countries as well, which will be described in detail below.

Pathways of introduction

Aedes albopictus has the dubious honour to be ranked as the 4th species on the top 100 list of World's Worst Invasive Alien Species from the Global Invasive Species Database (ISSG, which is part of the Global Invasive Species Programme (GISP)) (ISSG website). It is the highest ranking insect species on the list, and shows just how capable this mosquito species is to invade new areas. Long distance spread of this mosquito is possible due to the combination of the cold-resistance of its eggs that can survive long periods without water, and the extensive commercial shipment of used tyres, a preferred breeding site of *Ae. albopictus* (Hawley 1988, Grist 1993, Reiter 1998, Figure 2A).

High impact-introductions such as the ones into Texas (USA) in 1985, Sao Paŏlo, Brazil, in 1986, and Padua, Italy, in 1991, were tracked down to the introduction pathway by importation of egg-infested used tyres. The concerned international tyre trade deals with tyres for trucks, heavy vehicles, and aircraft or of unusual size (i.e. metro, military vehicles) and are resold either directly or after being recapped. National transport of used tyres is also considered a pathway for local distribution in northern Italy (Knudsen *et al.* 1996). Occasionally, long-distance spread is linked to the importation of ornamental plants from *Ae. albopictus*-endemic areas. Examples of this pathway were the introductions into southern California in 2001 (Madon *et al.* 2002), and into the Netherlands in 2005 (Scholte *et al.* 2007b). In both cases, the species was imported from Southeast

Figure 2. Main Aedes albopictus *inroduction routes: (A) Used tyres. (B),(C) Lucky Bamboo (*Dracaena *spp.).*

Asia in maritime container shipments with 'Lucky Bamboo' (*Dracaena* spp.), packaged in standing water or on gel (Figure 2B, C). In several cases (Bosnia-Herzegovina, Croatia, France (Corsica), Greece), large maritime ferryboats with trucks and cars coming from *Ae. albopictus*-infested areas are considered as likely pathways, although it remains uncertain whether these mosquitoes were 'trapped' in the cars and trucks coming from infested areas, which is more likely, or if mosquitoes accidentally got on board of the ferryboat at an infested sea-port (e.g. Venice, Italy). There are strong indications that transportation of *Ae. albopictus* by car and/or trucks occasionally happens: from the Ticino Canton of southern Switzerland and from the Alpes-Maritimes department in France, isolated populations of *Ae. albopictus* have been reported along roads coming from neighbouring *Ae. albopictus*-infested areas in northern Italy (Flacio *et al.* 2006, F. Schaffner, personal observation). Local distribution, or short-distance dispersal, is probably based on passive transport of adult mosquitoes in cars and trucks, or by local movement of infested artificial oviposition sites such as used tyres or flowerpots brought from garden centres to gardens.

Role as (potential) disease vector

Aedes albopictus is a competent vector for about 24 arboviruses as well as for *Dirofilaria* parasites, which are all listed and described in detail elsewhere (Shroyer 1986, Mitchell 1991, 1995, Moore and Mitchell 1997, Gratz 2004, Eritja *et al.* 2005), but in the field it has proven so far an efficient vector only for Chikungunya and dengue viruses. Although about 7 of these arboviruses are present in Europe (Eritja *et al.* 2005; Medlock *et al.* 2007), no cases of human arboviral infections, transmitted by *Ae. albopictus* were reported from Europe until August 2007, when an outbreak of Chikungunya, vectored by this species, occurred in Italy (see Chapter 10).

Distribution in Europe per country, chronologically.

Albania

The first observation of *Ae. albopictus* in Europe was reported from Albania, in August 1979, when the Kërkimor Institute of Hygiene and Epidemiology received complaints that large numbers of black-and-white mosquitoes were biting voraciously in broad daylight at a truck car park in Laç, a small town on the northern coastal plain (Adhami and Murati 1987). In the months after this initial discovery, surveys revealed that the mosquito was already abundant in six towns from north-eastern Albania as well as at a dump site for discarded tyres, 2 km from the nearest human habitation. This, and the fact that employees at the first cited autopark were emphatic that the nuisance had been with them for at least four years, suggested that the species may have been introduced at least as early as 1975 (Adhami and Reiter 1998). As no used tyres were ever imported into Albania, it is presumed that the species arrived in some other item or items of cargo. The survey indicates also that the only infested seaport was Durrës, the principal point of entry for goods from China, the principal trading partner until 1980. In October 1979, the institute of Hygiene and Epidemiology recommended a series of measures that should limit the production of *Ae. albopictus*. The most effective measure has been to prevent accumulation of rainwater in used tyres by storage measures. This approach has resulted in a marked reduction of the *Ae. albopictus* population in car parks and their surroundings. Unfortunately, no updated information on surveillance and control are available.

Italy

In Italy, the first finding of *Ae. albopictus* was in September 1990 at the city of Genoa (Region of Liguria, north-western Italy), when a few adult specimens were found at a kindergarten (Sabatini *et al.* 1990). One year later, in August 1991, following nuisance reports from daytime biting mosquitoes in the southern part of the city of Padua (Region of Veneto, North-eastern Italy), the first breeding population was discovered (Dalla Pozza and Majori 1992). From a thorough investigation on the introduction pathway of this population, Dalla Pozza *et al.* (1994) stated that at least one source from which the mosquitoes originated, was a shipment of used tyres imported from Atlanta (Georgia), USA. Reiter (1998) specified that between 1988 and 1995 Italy had imported, on multiple occasions, a total of several thousands of used tyres from countries where *Ae. albopictus* occurs (44,687 from the USA, 48,032 from Japan, and 1,550 from Taiwan). As for the *Ae. albopictus* found in the southern USA, the *Ae. albopictus* in Italy are closely linked to populations from Japan, as shown by a genetic study by Urbanelli *et al.* (2000). It is believed that from the second focus (Padua) the species has spread to other areas in Italy, exclusively through the local and inter-regional transport of used tyres (Knudsen *et al.* 1996). By 1993 the mosquito had spread to other areas in (mostly) the northern regions, including the provinces of Treviso, Venice, Vicenza and Brescia. Starting in 1994, a relatively large increase in foci was observed (Romi and Majori 1998, Dalla Pozza *et al.* 1994), still mostly confined to the northern regions. By the end of 1995, the species had spread to scattered foci in 10 regions including 20 provinces (Romi 1995, Romi 2001). Especially some rural cities in North-eastern Italy (Veneto region) were heavily infested (Knudsen *et al.* 1996). In 1997, more than 20 provinces had reported infestations of the mosquitoes, including Rome (DiLuca *et al.* 2001, Romi 2001, Figure 3). By 2000, *Ae. albopictus*-infested foci were known from 30 provinces (Romi *et al.* 1999, Romi 2001). Apart from the regions north of the Apennines, the species had also been found in northern Tuscany and the western coastal provinces from Pisa southwards to the province of Napoli, with the exception of Livorno

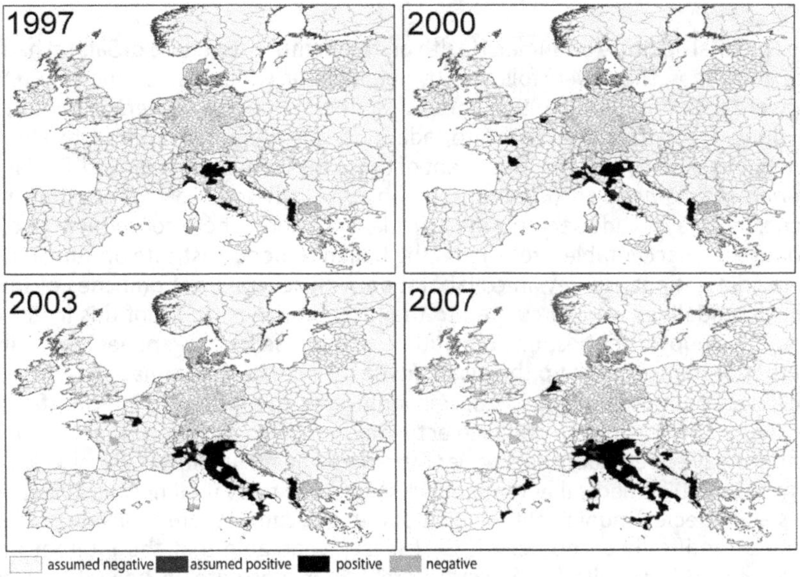

Figure 3. Presence of Aedes albopictus *in Europe per province for the years 1997-2007. Data to complete this figure were kindly made available by Roberto Romi (Italy), Roger Eritja and David Roiz (Spain), Eleonora Flacio (Switzerland), Charles Jeannin (France), Anna Klobučar (Croatia), Zoran Lukac (Bosnia and Herzegovina), Igor Pajovic and Dusan Petrić (Serbia and Montenegro), Bjoern Pluskota (Germany), Anna Samanidou-Voyadjoglou (Greece). The map was made by Patrizia Scarpulla. The 2007 outbreak of Chikungunya virus in Italy is indicated with an arrow in the 2007 box.*

and Viterbo (Anonymous 2003, Romi 2001). However, more than 90% of the infested areas were still concentrated in the north-east of the country (Romi 2001). In those areas, nuisance of *Ae. albopictus* is often considered by residents as more than 'inconvenient'. One of the results from a study at Rovigo (region of Emilia Romagna) in 2004, was that more than 54% of the interviewed people declared to spend less time outdoors due to the nuisance of biting *Ae. albopictus* (Carrieri *et al.* 2006). In tourist areas the economy is suffering from the presence of this aggressive mosquito, e.g. in the old city of Venice, where people are 'chased away' from terraces due to extensive daytime mosquito biting (R. Bellini, personal communication). Until 2000, the species had not been reported from the central, mostly mountainous areas, the eastern coastal areas, and the southern regions of Italy. Mathematical models based on climatic data predict that *Ae. albopictus* has a difficult time surviving in hot and dry areas such as the southern and central mountainous areas of Italy (Romi *et al.* 1999). In 2003, however, the mosquito had been reported from several central mountainous provinces such as Perugia and Terni, many eastern areas, covering almost the entire Adriatic coastal provinces southwards until Foggia (region of Puglia), and several southern areas in the provinces of Taranto, Cosenza, and Salerno (Romi *et al.* 2003). It is true, however, that population densities of this mosquito species in the southern areas of Italy are, with the exception for Rome, much lower than those in the north-eastern areas of Italy, which receive more rainfall. Apart from mainland Italy, the species is present also on the islands of Sicily (provinces of Palermo, Messina and Ragusa), Sardinia (cities of Olbia and Sassari), and Elba.

In 1991, the National Institute of Human Health of Italy (Istituto superiore di Sanità) had produced surveillance guidelines. These were followed by guidelines in which also protocols and techniques for control were included (Romi 1996). These describe the species, where it breeds, and what should be done to control the mosquito, adapting an integrated approach: elimination of potential breeding sites, covering or treatment of those that cannot be removed. For larval control *Bacillus thuringiensis israeliensis* (*Bti*) and temephos are advised. Control of adult mosquitoes using insecticidal sprays is advised only in particular situations where consequences of excessive mosquito bites are unacceptable. From 1994, the Italian National Institute of Human Health (ISS) has coordinated a national survey, in collaboration with several local branches of this institute (USLs). The ISS guidelines were circulated to the local branches (USLs) of this institute all over the country. It was up to the responsible authority of these local branches of the institute to decide whether or not to follow up the given advises. Since then, information is more and more widely available. Countless documents on recognition of the mosquito species, fact sheets on its biology, its spread in Italy, surveillance activities, and 'what can you do for prevention and control?' options have been published, either by national health institutes (Romi 1996, Romi *et al.* 2007, Anonymous 2007), medical entomology research institutes (Bellini *et al.* 2005), and almost all provinces and affected municipalities. Most of these documents are available on the internet. Although not scientifically acceptable, it is at least remarkable that the internet programme 'Google' produces 197,000 hits for 'zanzara tigre' ('tiger mosquito' in Italian), against around 410,000 hits for '*albopictus*', suggesting that almost half of all that is written on this mosquito species worldwide on the internet, is found in Italian documents. Apart from information on the internet, Italy has also produced a large amount of widely distributed information on almost anything one can imagine: flyers, bill-boards, television spots, articles in magazines, newspapers, information evenings, lessons on primary schools, etc. The province of Reggio Emilia, consisting of 32 municipalities, has even launched a 'tiger mosquito day' (21 April). Countless consumer products are available; T-shirts, stickers, posters, coffee-mugs, plates, screensavers, just about anything one can think of. Although to some this may seem as overkill, it has caused the vast majority of inhabitants of heavily infested areas to be aware of the problem, children and elderly included, which is important if inhabitants are to be included in control programmes.

Before any control action is undertaken, the presence of *Ae. albopictus* is verified, either by collecting actively flying adult mosquitoes (only in case of heavy infestation and complaints of citizens), or by collecting eggs using oviposition traps. The latter is the standard method for monitoring in Italy (Fay and Eliason 1966, Bellini *et al.* 1995, 1996, 2005, Romi 1996). Based on the outcome of these studies, control strategies are planned, often aided by GIS maps in which infestation levels are depicted (see also the paragraph on 'control options and recommendations').

France

In France, the main French mosquito control agencies, gathered among the 'Adege', warned in 1998 the French ministry of health against the risk of introduction of *Ae. albopictus* and the related sanitary risk. As a result, a committee in charge of surveillance and control has been established and a first field surveillance was implemented in the fall of 1999, allowing the discovery of established populations on two used tyres storages centres (Schaffner and Karch 2000, Schaffner *et al.* 2001). These related sites, belonging to the same company, were located in the département (dép.) Orne, région (rég.) Normandie and dép. Vienne, rég. Poitou-Charentes, whereas used tyres storages located in Lyon and Marseille, as well as cities close to Italy on the Côte d'Azur, did not show any introductions of *Ae. albopictus* during these years (M. Babinot, personal communication, Fauran *et al.* 1998). Introduction was clearly related to the tyre trade originating either in the USA or in Japan. Thereafter, a permanent surveillance and control programme was implemented, based on vectorial risk assessment and management related to exotic mosquito species and funded by the French Ministry in charge of Health. All pathways for the introduction were surveyed (by larval search and adult trapping) and control measures (by larviciding and adulticiding) were immediately implemented after discovery of specimens. Thus, three more used tyres storage centres experienced an introduction of *Ae. albopictus*: in dép. Val-d'Oise, rég. Île-de-France, in 2002, dép. Seine-et-Marne, rég. Île-de-France, in 2004 and dép. Calvados, rég. Normandie, in 2006. In addition, six other exotic mosquito species (almost all Nearctic species) were collected at these sites, some of them intercepted in tyres recently imported from USA. After these observations, control measures successfully eradicated the exotic species from these sites, as no specimen was found afterwards, except in occasional cases of new introductions. Preventive insecticide treatments are still regularly applied at these sites. From 2000 to 2006, 142 of 185 tyre importing companies, which handle 99.9% of the import volume, were investigated. Despite a complaint of biting-nuisance in a Lucky Bamboo company in 2002, no evidence of introduction of mosquitoes through the pathway of plant trade was found. In contrast, road traffic has been confirmed a major risk for the spreading of *Ae. albopictus* in Europe. A first introduction occurred in Corsica in 2002 in a holiday village of the Oriental coast (dép. Haute-Corse). The survey set up at the village and its surrounding had remained negative until 2005, suggesting that the population had not become established, probably due to a lack of flooded breeding sites during summer. Thereafter, major possible introductions routes into France were surveyed from 2002 to 2005 with a network of ovitraps: ports in Corsica, car parks on the motorways coming from Catalonia, Spain and Liguria, Italy, and the first towns bordering Italy on the Côte d'Azur. As a result, introduction of *Ae. albopictus* was observed in September 2004 in a botanic garden in the city of Menton, dép. Alpes-Maritimes. Control measures were applied immediately and the ovitrap network was intensified. No other specimen was observed until July 2005. Then, despite new treatments, other mosquito populations were rapidly detected in other towns of the Côte d'Azur, from Menton to Nice.

The Chikungunya epidemic in the French overseas department La Réunion, Indian Ocean, in 2005-2006 (>266,000 human cases), led to an increased concern of the sanitary authorities. A risk assessment and management programme for CHICK and DEN (dengue) viruses was implemented for metropolitan France in 2006. It includes (1) the mandatory notification of imported CHIK and DEN human cases, (2) the extension of the entomological surveillance using ovitraps, (3) the application of local control measures against *Ae. albopictus* in the neighbourhood of imported CHIK and DEN cases, and (4) the initiation of research programmes on local mosquito vector competence and *Ae. albopictus* vector capacity. Entomological surveillance was extended and intensified on the whole French Mediterranean coast, the Rhone Valley, Corsica, Paris, and to international ports and airports. As a result, an established population was detected in Bastia, dép. Haute-Corse, which rapidly spread to neighbouring towns, despite the application of control measures by local public agencies. At the end of 2006, two areas were colonised by *Ae. albopictus*, in dép. Alpes-Maritimes, covering 23 municipalities and 266 km^2, and in dép. Haute-Corse, covering 31 municipalities and 475 km^2. The surveillance and control programme is pursued at national level for the sanitary concern, funded by the Ministry in charge of Health, as well as at local level for the nuisance induced by this new pest species, funded by local authorities (départements). The aim of these programmes will be to diminish the vectorial transmission risk and the nuisance, and to slow down the spread of the species.

Belgium

The only record of *Ae. albopictus* in Belgium was made in 2000 in a village in Oost-Vlaanderen province, close to Antwerp (Schaffner *et al.* 2004). A few specimens were collected in the autumn in the used tyre stock of a recycling company that imports tyres from the USA and Japan. The species had reproduced on site, and the local environmental conditions were identified as favourable to its establishment. Despite this and the absence of any control measure initiated by local authorities, the species has not been observed again in the autumn of 2002, suggesting that climate conditions were unfavourable, at least for its proliferation (F. Schaffner personal observation). A tree-year mosquito survey is being undertaken covering the entire country in 2007-09 (MoDiRisk programme). This programme is expected to provide information on the presence of exotic and invasive species (W. van Bortel, personal communication).

Montenegro

A first specimen of *Ae. albopictus* was registered in August 2001 in the suburbs of Podgorica, capital city of Montenegro (formerly Yugoslavia till February 2003, then Serbia and Montenegro till June 2006). Several used tyre shops were inspected and more than 30,000 tyres checked, but only one specimen was found (Petrić *et al.* 2001). The batch of tyres came from Germany and the species was found in the municipality of Skutari, only 10 km away from Podgorica in Albania. Other shops that import tyres from Albania where checked but did not show any introduction of the species. Until September 2002 the species spread to other suburbs of Podgorica and central parts of the city (Petrić *et al.* 2006). From 2003 to 2006, several new foci were recorded (Petrić *et al.* 2006): in 2003 along the coast in the cities of Bar, Sutomore and Ulcinj; in 2004 in Peninsula Lustica (Radovici, Rose), in the gorge of Boka Kotorska bay close to the Croatian border, with further spreading in Podgorica and Ulcinj; in 2005 in Budva (central coast) and Andrijevica (mountain region in the north), with confirmed findings and a distribution in previously infested places; in 2006 the species extended its range in the coastal region (Prcanj, Herceg Novi) and remained present in the previous foci. This situation results probably from several introductions by road traffic from Albania (Podgorica – 2001) or by ferry traffic coming from Bari, Italy (Bar – 2003). There is also evidence of dissemination by road traffic along the coast as well as inland. *Aedes albopictus* is now established and spreading throughout the country. The surveys were made by scientists from Novi Sad, Serbia. To date, no information or control measures have been implemented in Montenegro.

Switzerland

In southern Switzerland (canton Ticino) an *Ae. albopictus* surveillance and control programme is implemented since 2000, which exists of surveying potential introduction sites using ovitraps and applying control measures in case of observed mosquito introductions. The monitoring is concentrated on the main traffic axes coming from Italy. Ovitraps are positioned at traffic stops along the highways such as gas stations, restaurants, boarder crossings, as well as ports, airports, camping sites and tyre storage centres. The first record of *Ae. albopictus* occurred in 2003 (Flacio *et al.* 2004) and since this date, each year of monitoring has revealed new foci (Flacio *et al.* 2006). Immediate control measures are implemented, based on larviciding with *Bacillus thuringiensis israeliensis* (*Bti*) and adulticiding with permethrin, accompanied also by community participation in order to prevent breeding opportunities on private properties. Up to now, this scheme seems to be successful for the control of *Ae. albopictus* in Ticino as identified introduction foci became negative, some of them remaining so for several years, and as there are no complaints from the population about nuisance caused by this mosquito. This strategy should prevent the establishment of pockets of *Ae. albopictus* along the boarder region with Italy and reduce the risk of a spread of the tiger mosquito to areas north of the Alps including southern Germany. Unfortunately, the federal government did not yet support the extension of the surveillance programme to the south-north traffic axes north of the Alps.

Greece

The first reports of confirmed *Ae. albopictus* in Greece were from the island of Corfu and the nearby city Igoumenitsa on the mainland in 2003 (Samanidou-Voyadjoglou *et al.* 2005, Patsoula *et al.* 2006), although inhabitants of the area allege that the species had been present for 'a few years' before that. A surveillance programme, launched at Corfu in 2006, verified establishment of the species at the island. The current status regarding presence of the species at the city of Igoumenitsa is unknown. It is believed that there have been several introductions, most probably by boat from Himara (Albania) or Italy (A. Samanidou-Voyadjoglou, personal communication): each summer large ferryboat cruise-ships cross the Adriatic Sea from *Ae. albopictus*-infested cities such as Venice, Bari, and Ancona. As many of these ships also enter Athens, a surveillance programme is planned for the area around the port of Athens (A. Samanidou-Voyadjoglou, personal communication). After several years in which no control programmes against *Ae. albopictus* could be carried out, experts at the Greek National School of Public Health in Athens are currently discussing control plans with the local Public Health Services in Corfu.

Spain

The first observation of *Ae. albopictus* in Spain occurred in 2004 in Sant Cugat del Valles, close to Barcelona, Catalonia (Aranda *et al.* 2006). The species was already established and probably present since at least 2002 considering citizen's complaints. Two years later, it is present in more than 20 municipalities, including a jump to Altafulla near Tarragona in the south of Catalonia, demonstrating establishment and spread. In 2005, a second population was reported in the south-east of Spain (Orihuela, Alicante), located more than 500 km from the first one (R. Eritja, personal communication). Relationship between mosquitoes from the two other foci cannot be confirmed and the possibility of a multiple introduction cannot be discarded. The tyre trade was surveyed in 2003-2004 throughout Spain and no presence of *Ae. albopictus* was reported. Introductions may not be related to tyre or Lucky Bamboo trade but more likely to road traffic, as a major Mediterranean motorway is going round the two infested areas.

In the Baix Llobregat area, Catalonia, where a Mosquito Control Service was already present, control measures have been established immediately. A complete integrated control programme has been implemented, including media management, some biocide applications but mostly focusing on public awareness because there are very few feasible pesticide applications. Information campaigns conducted door-to-door have been set up using teams of trained technicians, and marketing operations designed by public communication specialists, including media advertisments. Educational actions in primary schools are being prepared as well. These operations are considered to be very positive but efficiency of such actions can only be evaluated over a long term. Unfortunately in this situation other measures are not possible as the vast majority of breeding sites are in closed private properties and the main problem here is the lack of cooperation by homeowners. In the tourist area of Orihuela (Alicante), insecticide treatments were carried out during 2005-06 (R. Eritja, personal communication). Despite all these actions and the current status of *Ae. albopictus* in these areas, there is lack of interest by most policy makers and therefore scarce funding in Spain.

The case of Spain is interesting in that, even with a surveillance system in place on used tyres storages, *Ae. albopictus* colonised the country, maybe introduced by vehicles from Italy. It proves that an efficient system of surveillance should combine monitoring of used tyres and Lucky Bamboo trades, and monitoring by oviposition traps along highways.

Croatia

The first record of *Ae. albopictus* in Croatia was made in Zagreb, in the autumn of 2004. Larvae were collected in a discarded ceramic toilet bowl in a forested area in the southwest suburbs, during a regular control of mosquito breeding sites (Klobučar *et al.* 2006b). In 2005, other larvae were found 4 km away from the first site, in used tyres stored by a company importing tyres from north-eastern Italy. At this time the mosquito was also recorded at several locations along the Adriatic coast, from Istria in the north to Dubrovnik in the south. The survey made on Istria peninsula reported eight colonised locations among the 43 that were checked, in different types of artificial breeding sites including tyres. In Dalmatia, several locations were found in Zadar, Split and Dubrovnik. As many locations were close to harbours, especially in Dalmatia, it is suggested that introductions have occurred by boat traffic (ferries) directly from Italy, but it is not excluded that the road traffic via Slovenia allowed the mosquito to reach Istria and Zagreb, including through the trade in tyres. The low number of specimens collected at all these sites suggested recent introductions. Climatic conditions are likely to allow the establishment of the species along the Adriatic coast, but observations confirmed also the possibility for the species to survive in northern Croatia. Public health and university institutes are strengthening control measures as well as educational programmes (Klobučar *et al.* 2006a).

The Netherlands

In July 2005, a routine inspection of the Dutch National Plant Protection Service (NPPO) at a horticultural company in the municipality of Haarlemmermeer (Netherlands) was carried out. The inspector reported the presence of large numbers of an unknown mosquito species on the premises of the company where the plants were kept. Several mosquitoes were caught and tentatively identified as *Ae. albopictus*. Shortly after these findings, *Ae. albopictus* was found at two other horticulture companies as well, both in large glasshouses. People working at these companies had experienced mosquito nuisance by 'black and white mosquitoes'. Follow-up investigation revealed that these insects were introduced through the importation of *Dracaena sanderiana* ('Lucky Bamboo') from southern China (Scholte *et al.* 2007a,b). Introduction of the species through this introduction route had also occurred in California in 2001 (Linthicum *et al.* 2003). Following reports of the findings of *Ae. albopictus* in the Netherlands to the National Institute of Public Health and the Environment, a joint risk assessment was started in the summer of 2006 to assess the spread of the species, potential establishment, and potential risks to human health. All 18 Lucky Bamboo importing companies were included in the survey. During the 2006/2007 survey, more than 700 *Ae. albopictus* adults were collected at 15 companies (Scholte *et al.*, unpublished results) in three provinces (Zuid Holland, Noord Holland and Utrecht), mostly in the area around Aalsmeer. Although during the survey in 2005 several adults had been collected outdoors, always close to the premises of the Lucky Bamboo-importing companies (Scholte *et al.* 2007b), it is yet unknown whether the species can become established in the Netherlands. Since outdoor presence of the species is only sporadic, and in close proximity of the Lucky Bamboo importing companies, no control programmes for outdoor environments have been carried out. In the view of the Dutch government it is the Lucky Bamboo importing companies' responsibility to make sure that these mosquitoes do not enter the Netherlands.

Slovenia

Several reports on the presence of *Ae. albopictus* from sites on the Adriatic coast of North-western Istria (cities of Izola and Piran), have been mentioned in a local newspaper in the summer of 2005 ('Delo' newspaper, 07.09.2005, in Petric *et al.* 2006). No more precise information is available yet but it is likely that *Ae. albopictus* may have been introduced in this area by road traffic originating from north-eastern Italy.

Bosnia and Herzegovina

As for Slovenia, scarce information is available for Bosnia and Herzegovina. Only one record of *Ae. albopictus* is known from the city of Banja Luka (Adriatic coast) in late autumn 2005 (Z. Lukac, personal communication *in* Petric *et al.* 2006).

Other countries: Hungary, Germany, UK, Serbia, etc.

In 2001, the presence of *Ae. albopictus* was suspected in Hungary. Unfortunately, this information could not be confirmed and no precise data is available to date.

In Germany a surveillance programme based on ovitraps is implemented in Baden-Württemberg along the motorway from Basel to Heidelberg, as well as surveillance of trade activities (ornamental flowers, tyres). At the time of writing, no *Ae. albopictus* has been reported. In parallel, studies on the bionomics of *Ae. albopictus* are undertaken at the University of Heidelberg. These programmes are funded and implemented by the German mosquito control agency (KABS) in collaboration with the University of Heidelberg (N. Becker and B. Pluskota, personal communication).

When *Ae. albopictus* was found in Montenegro, a survey was also implemented in Vojvodine province, Serbia. Tyre importers were alerted by broadcasting stations and are regularly checked; no introduction has been observed to date (D. Petrić, personal communication).

In countries where no active surveillance is conducted, passive surveillance schemes or routine mosquito surveillance may result in detection of *Ae. albopictus*, but probably only when well established and already responsible for causing nuisance. In Portugal (C. Sousa, personal communication), scientists have proposed to national funding agencies to implement a surveillance programme, but no funding has yet become available. In the United Kingdom (J. Medlock, personal communication), a passive surveillance scheme for nuisance biting mosquitoes is in place, which involves environmental health officers involved in nuisance mosquito biting issues to submit samples for identification; so far no *Ae. albopictus* has been found.

Options and recommendations for mosquito control

All these experiences confirm the relevance of the entomological surveillance based on ovitrap networks and the effectiveness of control programmes applied immediately after the detection of the species in small-scale demarcated areas. By contrast, infestation of large areas by *Ae. albopictus* seems to be irreversible, at least in southern Europe. The recent CHIK epidemic which occurred on islands in the Indian Ocean revealed a higher vectorial capacity of *Ae. albopictus* than previously assumed, at least in a tropical climate, and despite its lack of host specificity. Therefore, in Europe, further studies on vectorial capacity of local *Ae. albopictus* populations as well as conducting surveillance and control programmes are highly desirable.

Although occasionally immature stages of *Ae. albopictus* are intercepted during tyre import inspections (South Africa [Cornel and Hunt 1991], New Zealand [Laird *et al.* 1994], USA [Hawley 1988], France [eggs, F. Schaffner, personal communication], Australia [Ritchie *et al.* 2005]), many countries have experienced how easily *Ae. albopictus* can enter their territories undetected and become established before the authorities are aware of its existence. Since most introduction pathways are through the international trade of used tyres, a ban on the import of used tyres from *Ae. albopictus*-infested areas seems the best option to reduce risks of introduction. Once imported, tyres could be stored under shelters or under tarpaulins in order to reduce the possibility for the introduced eggs to hatch. Considering the diversity of tyre trade routes, tyres originating from countries where *Ae. albopictus* is established should be identified and marked in order to allow their surveillance even if they transit through several companies and countries. In order to reduce the risk of *Ae. albopictus* introduction through import of Lucky Bamboo plants, stringent mosquito-control treatments should be implemented at the export location, and major import locations should be checked through continuous surveillance (Scholte *et al.* 2007a).

Once an area is infested with *Ae. albopictus* it is not easy to eradicate the species. The list of successful eradications (Jardina 1990, F. Schaffner, personal observation, Cristo *et al.* 2006) is much shorter than the list of unsuccessful eradication projects. In Europe, eradication was possible on a few occasions only, of relatively small, local, isolated populations: in some locations in France, Switzerland and Italy: in Cagliary (Sardinia) and Bientina (Pisa) the species was eradicated between 1996 and 1997, and in a few municipalities in the regions of Veneto and Piemonte (Romi and Majori 1998; Cristo *et al.* 2006). In two cases, Belgium in 2000 and Corsica in 2002, the species did not become established without any intervention, suggesting that unfavourable environmental conditions have prevented its establishment and proliferation. Once an infested area is surrounded by neighbouring populations and not all breeding sites can be reached for control measures, eradication becomes almost impossible. In those cases, the best strategy is to keep population levels of *Ae. albopictus* as low as possible, and try to keep further spreading under control or at least to reduce the speed of dispersal. This can only be achieved by an effective and integrated mosquito control programme. Such a control programme must be based on carefully collected and analysed entomological surveillance data, with particular emphasis on ecological factors that determine where, how, and when to initiate vector control (Service 1993, Bellini *et al.* 1996, Romi and Majori 1998, Romi *et al.* 1999, Carrieri *et al.* 2006). The standard for such entomological surveys is by using weekly or bi-weekly-checked oviposition traps in both public and private areas (Figure 4). Data analysis and decision making of where to carry out control measures is greatly facilitated when GIS software programmes are used (Albieri *et al.* 2003).

Elimination, covering, and/or treatment of breeding sites is generally accepted as the most effective control methodology (Romi 1996, Bellini *et al.* 2005). Breeding sites of *Ae. albopictus* include a range of small outdoor water bodies such as manholes, flower pots, buckets, discarded tyres, bathtubs, tarpaulins, drainpipes, gutters, basins, sinks, tree holes, catch basins, etc. (Figure 5). Large natural waterbodies such as canals, marshes, lakes and river borders are not used by this

*Figure 4. Oviposition trap as used in France: (A) Black bucket (1 litre) with wood infusion water and insecticide; (B) The labelled bucket with a piece of polystyrene (5*5*2 cm) as oviposition support; (C),(D) The labelled oviposition support which is regularly collected and checked for eggs counting.*

Figure 5. Main Aedes albopictus *breeding sites: (A) Used tyres; (B) Catch basins; (C),(D) Flour pots and dishes in gardens and cemeteries.*

mosquito species for breeding. From the listed oviposition sites, especially used tyres are favourite sites of *Ae. albopictus*. Not only the introduction of the species into the country, but also the subsequent spread within Italy has been attributed largely to transport of egg-infested tyres. The effectiveness of efforts to stop the spread of the species in Italy has been greatly hampered by the lack of regulations concerning storage (e.g. covering) and transportation of used tyres (Urbanelli *et al.* 2001). Larval control can be achieved by either elimination, covering, or treatment of breeding sites. Treatment can roughly be divided into biological control and chemical control. Biological control is usually based on products containing *Bti* as active substance. In Italy households of heavily infested areas are often given free *Bti* tablets to use in their gardens or on balconies. Occasionally, when the water bodies are relatively large, small fish such as *Gambusia* can be used (Donati *et al.* 2006). However, almost all organised control programmes use chemical control as their choice of treatment of public areas. Temephos has been mostly used until 2006, especially for treatment of manholes, but European legislation ('Biocide' directive 98/8/CE) does not allow further utilisation of this insecticide except when national authorities have obtained a prolongation for essential use. It is fundamental that treatment of breeding sites includes both public and private areas, as was shown by Carrieri *et al.* (2006): larvicidal control of manholes in public areas alone was significantly less effective in suppressing the existing population as compared to areas where manholes were treated in both public and private areas, with a difference of approximately 60% in population density.

For a successful mosquito control programme in a heavily infested area, participation of local citizens is essential. To achieve this, massive public awareness campaigns were launched in many affected areas in Italy. Citizens were informed about the presence, nuisance, potential health threats from *Ae. albopictus*, on what is being done or can be done to stop the mosquito, and 'what can you as a citizen do to control and prevent further spread of this mosquito'. Information came from various sources, such as national and local authorities (municipalities, provinces, and regions), the national and local branches of the Institute of Human Health, Universities, and mosquito control agencies. Different approaches to reach the public were used, e.g. through widely distributed flyers, information evenings in community buildings and club houses, billboards, and information to children in primary schools. An important and frequently used tool to reach, inform, and involve citizens with the problems associated with *Ae. albopictus* is the media. In Italy, articles in local, regional and national newspapers, magazines, television, radio, and internet are continuously used to keep people informed and involved. In most of the heavily affected areas the authorities have asked, and in some cases even obliged, the public to participate in control methods such as elimination of potential breeding sites, and the use of larvicidal products such as the biological control agent *Bti*. Tablets of *Bti* are often provided for free by municipalities to all households. The importance of getting citizens voluntarily and actively involved in mosquito control was shown in a recent study in a village near Bologna (region of Emilia Romagna), where a team of a mosquito control technician and a mosquito control specialist, employed by a regional mosquito control agency, went 'door to door' in a village to carry out mosquito control and involve the citizens actively with their activities. These activities existed of (1) direct chemical treatment of manholes at the private property area (using temephos), (2) providing *Bti* products to treat their manholes during the whole summer season, elimination of all micro-larval habitats in gardens/balconies, and (3) sensitising inhabitants about preventive measures. It was shown that this approach reduced the *Ae. albopictus* population by at least 50% compared to the 'standard strategy' in that village, namely biocidal treatment of manholes only (Carrieri *et al.* 2006). Results from the same study on the use of adulticidal treatment using a ULV treatment of Permex 22E, suggest that this treatment had no effect on the population density. This is just one of many examples where scientific studies are designed to optimise surveillance strategies and control options or open up new, alternative control strategies. Among other scientific studies on *Ae. albopictus* in Italy that are directly linked to optimise control are studies on the sterile insect technique (SIT), use of different types of ovitraps, the use of metallic copper to inhibit larval development, the use of *Gambusia* fish and indigenous copepods for larval control, improving delivery techniques of larvicides, determining adulticide thresholds and the efficacy of insecticides, and studying spatial pattern analysis to optimise monitoring activities (Bellini *et al.* 1996, 1998, 2002, Bellini and Veronesi 2006, Donati *et al.* 2006, Romi *et al.* 2003, Toma *et al.* 2003).

Further spread?

The fact alone that *Ae. albopictus* is nominated as the highest listed insect species on the top 100 list of the world's most invasive species is sufficient reason to be cautious about predictions of the dispersal and distribution of this mosquito species. Several mathematical models have used climatic data from areas where the mosquito is endemic and where it is not, in order to predict where it might occur in future. Most models use factors such as the photoperiod that trigger females to produce diapausing eggs, coldest and warmest month isotherms, annual mean temperatures, minimum and maximum precipitation, and humidity (Focks *et al.* 1994, Kobayashi *et al.* 2002, Medlock *et al.* 2006, Benedict *et al.* 2007). All these factors seem to affect the potential of the species to establish itself in a new region. As a rule of thumb regarding Europe, it can be said that very dry areas such as the southernmost parts of Italy are unfavourable, and in relatively cold climates such as mountainous areas and the North of Europe the larvae will not succeed in developing fast enough for a 'viable population'. The question of course is where the distribution limits will be. Among the first models is one by Mitchell (1995), which predicted that large areas of Mediterranean countries are at risk. Regarding potential spread to northern areas in Europe, the distribution of *Ae. albopictus* in the north of Asia occasionally reaches the -5 °C isotherm. Even assuming a conservative 0 °C isotherm, this means that the species might become established in northern Europe as far as the southern coast of Sweden and Norway (Mitchell 1995). Kobayashi *et al.* (2002) write that the northern limits of *Ae. albopictus* in Japan are between latitudes 38-40°N. Using Geographical Information System (GIS) software, they proposed that *Ae. albopictus* could survive in global regions where the annual mean temperature is higher than 11 °C and the mean temperature of the coldest month is not lower than -2 °C. Furthermore, the period with temperatures above 11 °C should successively continue for more than 186 days per year. This means that the species would have its northernmost distribution limit somewhere from southern England southeastwards along the southern border of Belgium, south of Luxembourg, similar to the areas depicted in Figure 6. The northernmost record of outdoor collected adults in Europe is in the Netherlands at approximately 52.15-18° N (Scholte *et al.* 2007b), which is considerably further north than previous northernmost records in Japan (Kobayashi *et al.* 2002) and in the USA (Moore and Mitchell 1997), although it should be noted here that it is not sure if these mosquitoes had become established in the Netherlands or had escaped from nearby greenhouses: The model of Benedict *et al.* (2007), based on a genetic algorithm predicts that the species could establish in relatively cool climates such as in Ireland, the South of England, the Netherlands, Belgium, and, in warmer areas: large parts of the Balkan countries, Greece and the western part of Turkey. The model of Medlock *et al.* (2006) predicts that abiotic risk factors would permit establishment of the species throughout large parts of lowland of the United Kingdom, with at least four to five months of adult activity. In contrast, Spain shows a very ragged suitable area for *Ae. albopictus* establishment (Eritja *et al.* 2005), mostly because of lack of rainfall (< 500 mm) and rain distribution (<60 rainy days). Further studies are ongoing for Europe as well as for country scales. Using the same criteria that were used for the model published by Medlock *et al.*

(unpublished data) for the UK, a potential distribution map of *Ae. albopictus* in Europe was provided by the same authors (Figure 6).

Practical obstacles and perspectives

In many cases the finding of an *Ae. albopictus* population in a 'new' area or country does not immediately result in the setting up of a surveillance system or control activities. Funds for surveillance activities and/or control are often lacking or insufficient, either because policy makers are not aware of the potential consequences of an *Ae. albopictus* invasion and give priority to other problems, or funds may simply not be available. Some countries do not have organised mosquito control, and specific knowledge about surveillance and control may not be immediately available; personnel needs to be trained and equipment needs to be brought in. Other obstacles include legislation/regulations regarding spraying activities/product licences, and the willingness of the public to engage in control programmes.

Although Europe as a cohort of individual countries may be getting more and more linked by regulatory rules issued by the European Commission, there are still differences between the countries regarding the control of *Ae. albopictus*: some countries in which the species is absent already have active well-organised surveillance programmes, whereas other countries, which experience a sudden infestation, are struggling to face the problem at hand. It may be an option for 'Brussels' to provide standard surveillance and control guidelines and to persuade all countries to be prepared for outbreaks.

Similarly, the recent Chikungunya outbreaks in tropical areas were reason for the ECDC to give recommendations for reducing the risk of transmission on Europe (Depoortere *et al.* 2006), in

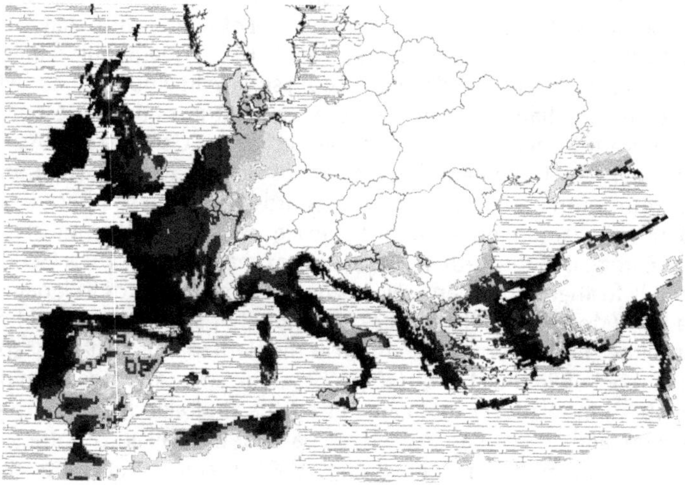

Figure 6. Areas for possible establishment of Aedes albopictus *in Europe based on 5 climate scenarios. The image shows likelihood for establishment. Scenario 1 ( ) = 450mm annual rainfall, -1 °C January isotherm; Scenario 2 ( ) = 500mm, 0 °C; Scenario 3 ( ) = 600mm, 1 °C; Scenario 4 ( ) = 700mm, 2 °C; Scenario 5 ( ) = 800mm, 3 °C. Map made by Jolyon Medlock on a 10 minute resolution and based on the same model criteria as outlined for UK (Medlock* et al. *2006).*

particular (1) to develop further studies and documentation of vector competence and capacity of *Ae. albopictus* in areas in Europe where this vector is known to be present, (2) to identify areas at risk of vector establishment and implement or strengthen regularly monitor in these areas, and (3) to consider measures to prevent the introduction of *Ae. albopictus* through the used tyre trade and plants transported in water (e.g. *Dracaena* spp.). Considering the recent outbreak of Chikungunya in Italy, such recommendations are to be taken very seriously indeed.

Acknowledgements

The authors wish to thank the following people for providing up-to-date information: Norbert Becker, Romeo Bellini, Roger Eritja, Eleonora Flacio, Charles Jeannin, Anna Klobučar, Zoran Lukac, Giancarlo Majori, Jolyon Medlock, Igor Pajovic, Dusan Petrić, Bjoern Pluskota, David Roiz, Roberto Romi, Anna Samanidou-Voyadjoglou, Patrizia Scarpulla, Carla Sousa and Rodolfo Veronesi.

References

Adhami J and Murati N (1987). Prani e mushkonjës *Aedes albopictus* në Shqipëri. [Presence of the mosquito *Aedes albopictus* in Albania.] Rev Mjekësore 1: 13-16. (In Albanian)

Adhami J and Reiter P (1998) Introduction and establishment of *Aedes* (*Stegomyia*) *albopictus* Skuse (Diptera: Culicidae) in Albania. J Am Mosq Control Assoc 14: 340-343.

Albieri A, Carrieri M and Bellini R (2003) Monitoring method and spatial analysis of *Aedes albopictus* in an urban area. Annual Meeting of the Society of Vector Ecology 2003, Sept. 3-6, 2003, Bellinzona, Switzerland.

Anonymous (2003) Zanzare a Firenze. http://www.comune.firenze.it/servizi_pubblici/ ambiente/zanzaratigre.htm (Accessed 29 August 2007)

Anonymous (2007) Zanzara tigre. Italian National Centre of Epidemiology, Surveillance, and Health Promotion; Epicentro. http://www.epicentro.iss.it/problemi/zanzara/zanzara.asp (Accessed 29 August 2007).

Aranda C, Eritja R and Roiz D (2006) First record and establishment of *Aedes (Stegomyia) albopictus* in Spain. Med Vet Entomol 20: 150-152.

Bellini R, Medici A, Carrieri M, Calvitti M, Cirio U, Maini S (2002) Applicazione della tecnica del maschio sterile nella lotta contro *Aedes albopictus* in Italia: ottimizzazione delle fasi di allevamento massale e sessaggio. Atti XIX Congresso nazionale italiano di Entomologia Catania 2002: 993-998.

Bellini R, Carrieri M, Burgio G and Bacchi M (1996) Efficacy of different ovitraps and binomial sampling in *Aedes albopictus* surveillance activity. J Am Mosq Control Assoc 12: 632-636.

Bellini R, Carrieri M and Benedetti MA (1995) Primo anno di attivitá del programma di lotta a *Aedes albopictus* a Desenzano del Garda, Sirmione e Lonato. Disinfestazione 13: 55-61.

Bellini R, Carrieri M, Bacchi M, Fonti P and Celli G (1998) Possible utilization of metallic copper to inhibit *Aedes albopictus* (Skuse) larval development. J Am Mosq Control Assoc 14: 451-456.

Bellini R and Veronesi R (2006) Utilizzo del rame metallico come larvicida per il controllo delle zanzare. Ig Alim-Disinfest Ig Amb 23: 45-46.

Bellini R, Veronesi R, Venturelli C and Angelini P (2005) Linee guida per la sorveglianza e la lotta alla zanzara tigre (*Aedes albopictus*). Report of the Servizio Sanitario Regionale Emilia-Romagna, 24 pp.

Benedict MQ, Levine RS, Hawley WA and Lounibos LP (2007) Spread of the tiger: global risk of invasion by the mosquito *Aedes albopictus*. Vector Borne Zoon Dis 7: 76-85.

Broche RG and Borja EM (1999) *Aedes albopictus* in Cuba. J Am Mosq Control Assoc 15: 569-570.

Casas-Martínez M and Torres-Estrada JL (2003) First evidence of *Aedes albopictus* (Skuse) in Southern Chiapas, Mexico. Emerg Infect Dis 9: 606-607.

Carrieri M, Colonna R, Gentile G (2006) Lotta alla zanzara tigre: strategie a confronto. Ig Alim-Disinfest Ig Amb 23: 45-50.

Centers for Disease Control (1986) *Aedes albopictus* introduction--Texas. Morbidity and Mortality Weekly Report 35:141-142.

Centers for Disease Control (2005) Information on *Aedes albopictus*. Arboviral Encephalitides. http://www.cdc.gov./ncidod/dvbid/arbor/albopic_new.htm (Accessed 29 August 2007).

Cornel A and Hunt RH (1991) *Aedes albopictus* in Africa? First records of live specimens in imported tires in Cape Town. J Am Mosq Control Assoc 7: 107-108.

Cristo B, Loru L, Sassu A, Pantaleoni RA (2006) The Asian tiger mosquito again in Sardinia. Bull Insectol 59: 161-162.

Dalla Pozza GL and Majori G (1992) First record of *Aedes albopictus* establishment in Italy. J Am Mosq Control Assoc 8: 318-320.

Dalla Pozza GL, Romi R and Severini C (1994) Source and spread of *Aedes albopictus* in the Veneto region of Italy. J Am Mosq Control Assoc 10: 589-592.

Depoortere E and Coulombier D and ECDC Chikungunya risk assessment group (2006) Chikungunya risk assessment for Europe: recommendations for action. Euro Surveill 11: E060511.2. www.eurosurveillance.org/ew/2006/060511.asp#2 (Accessed 29 August 2007)

DiLuca M, Toma L, Severini F, D'Ancona F and Romi R (2001) *Aedes albopictus* a Roma: monitoraggio nel triennio 1998-2000. Ann Ist Super Sanitá 37: 249-254.

Donati L, Veronesi R, Colonna R and Bellini R (2006) Gambusia nella lotta alle zanzare nei bidoni degli orti. Ig Alim-Disinfest Ig Amb 23: 39-44.

Eritja R, Escosa R, Lucientes J, Marquès E, Molina R and Ruiz S (2005) Worldwide invasion of vector mosquitoes: present European distribution and challenges for Spain. Biol Invasions 7: 87-97.

Fauran P, Marty P, Izri MA and Le Fichoux Y (1998) Results of a survey to detect the mosquito *Aedes albopictus* in the French Riviera. Parassitologia 40: 321-323.

Fay RW, Eliason DA (1966) A preferred oviposition site as a surveillance method for *Aedes aegypti*. Mosq News 26: 531-535.

Flacio E, Patocchi N, Peduzzi R and Lüthy P (2006) Strategies to stop the establishment of *Aedes albopictus* in Switzerland. Abstracts of the SOVE 15th European Meeting, Serres, Greece: 57.

Flacio E, Lüthy P, Patocchi N, Guidotti F, Tonolla M and Peduzzi R (2004) Primo ritrovamento di *Aedes albopictus* in Svizzera. Boll Soc Ticinese Sci Nat 92: 141-142.

Focks DA, Linda SB, Craig GB, Hawley WA and Pumpuni CB (1994) *Aedes albopictus* (Diptera: Culicidae): A statistical model of the role of temperature, photoperiod, and geography in the induction of egg diapauze. J Med Entomol 31: 278-286.

Forattini OP (1986) Identificacao de *Aedes* (*Stegomyia*) *albopictus* (Skuse) no Brasil. Rev Saude Púb São Paulo 20: 244-245.

Gratz NG (2004) Critical review of the vector status of *Aedes albopictus*. Med Vet Entomol 18: 215-227.

Grist NR (1993) *Aedes albopictus*: the tire-traveling tiger. J Infect 27: 1-4.

Hawley AH (1988) The biology of *Aedes albopictus*. J Am Mosq Control Assoc 4: 2-39.

Ibanez-Bernal S, Martinez-Campos C (1994) *Aedes albopictus* in Mexico. J Am Mosq Control Assoc 10: 231-232.

ISSG. http://www.invasivespecies.net/database/species/search.asp?st=100ss (Accessed 29 August 2007).

Jardina BJ (1990) The eradication of *Aedes albopictus* from Indianapolis, Indiana. J Am Mosq Control Assoc 6: 310-311.

Klobučar A, Merdić E, Benić N, Baklaić Ž and Krćmar S (2006b). First record of *Aedes albopictus* in Croatia. J Am Mosq Control Assoc 22: 147-148.

Klobučar A, Benić N, Johnn V, Čulig J, Merdić E, Boca I, Landeka N and Žitko T (2006a) *Stegomyia albopicta*, Asian tiger mosquito in Croatia. Poster at the 15th European Society for Vector Ecology Meeting, 10-14 April 2006, Serres, Greece.

Knudsen AB, Romi R and Majori G (1996) Occurrence and spread in Italy of *Aedes albopictus*, with implications for its introduction into other parts of Europe. J Am Mosq Control Assoc 12: 177-183.

Kobayashi M, Nihei N and Kurihara T (2002) Analysis of Northern Distribution of *Aedes albopictus* (Diptera: Culicidae) in Japan by Geographical Information System. J Med Entomol 39: 4-11.

Krueger A and Hagen RM (2007) First record of *Aedes albopictus* in Gabon, Central Africa, during the EUFOR RD CONGO Mission. Trop Med Int Health 12: 1105-1107.

Laird M, Calder L, Thornton RC, Syme R, Holder PW and Mogi M (1994) Japanese *Aedes albopictus* among four mosquito species reaching New Zealand in used tires. J Am Mosq Control Assoc 10: 14-23.

Linthicum KJ, Kramer VL, Madon MB and Fuijoka K (2003) Introduction and potential establishment of *Aedes albopictus* in California in 2001. J Am Mosq Control Assoc 19: 301-308.

Lounibos LP, Escher RL, Lourenco-de-Oliveira R (2003) Asymmetric evolution of photoperiodic diapause in temperate and tropical invasive populations of *Aedes albopictus* (Diptera: Culicidae). Ann Entomol Soc Am 96: 512-518.

Lugo Edel C, Moreno G, Zacharia MA, Lopez MM, Lopez JD, Delgado MA, Valle SI, Espinosa PM, Salgado MJ, Perez R, Hammond SN and Harris E (2005) Identification of *Aedes albopictus* in urban Nicaragua. J Am Mosq Control Assoc 21: 325-327.

Madon MB, Mulla MS, Shaw MW, Kluh S and Hazelrigg JE (2002) *Aedes albopictus* (Skuse) introduction into sourthern California and potential for its establishment. J Vector Ecol: 149-154.

Medlock JM, Avenell D, Barrass I, Leach S (2006). Analysis of the potential for survival and seasonal activity of *Aedes albopictus* (Diptera: Culicidae) in the United Kingdom. J Vector Ecol 31: 292-304.

Medlock JM, Snow KR and Leach S (2007) Possible ecology and epidemiology of medically important mosquito-borne arboviruses in Great Britain. Epidemiol Infect 135: 466-482.

Mitchell CJ (1991) Vector competence of North and South American strains of *Aedes albopictus* for certain arboviruses. J Am Mosq Control Assoc 7: 446-451.

Mitchell CJ (1995) Geographic spread of *Aedes albopictus* and potential for involvement in arbovirus cycles in the Mediterranean Basin. J Vector Ecol 20: 44-58.

Moore CG and Mitchell CJ (1997) *Aedes albopictus* in the United States: ten-year presence and Public Health implications. Emerg Infect Dis 3: 329-334.

Ogata K and Lopez Samayoa A (1996) Discovery of *Aedes albopictus* in Guatemala. J Am Mosq Control Assoc 12: 503-506.

Patsoula E, Samanidou-Voyadjoglou A, Spanakos G, Kremastinou J, Nasioulas G and Vakalis N (2006) Molecular and Morphological Characterization of *Aedes albopictus* in Northwestern Greece and Differentiation from *Aedes cretinus* and *Aedes aegypti*. J Med Entomol 43: 40-54.

Pena CJ (1993) First report of *Aedes* (*Stegomyia*) *albopictus* (Skuse) from the Dominican Republic. Vector Ecol Newsl 24: 68.

Petrić D, Pajovic I, Ignjatović Ćupina A and Zgomba M (2001) *Aedes albopictus* (Skuse, 1894) nova vrsta komaraca (Diptera: Culicidae) u entomofauni Jugoslavije. Abstract Volume, Symposia of Serbian Entomologists, Entomological Society of Serbia, Goc, 26-29 September 2001, p. 29.

Petrić D, Zgomba M, Ignjatović Ćupina A, Pajovic I, Merdić E, Boca I, Landeka N, (2006) Invasion of the *Stegomyia albopicta* to a part of Europe. Presentation at the 15th European Society for Vector Ecology Meeting, 10-14 April 2006, Serres, Greece.

Reinert JF, Harbach RE and Kitching IJ (2004). Phylogeny and classification of Aedini (Diptera: Culicidae), based on morphological characters of all life stages. Zool J Linn Soc 142: 289-368.

Reiter P (1998) *Aedes albopictus* and the world trade in used tires, 1988-1995: the shape of things to come. J Am Mosq Control Assoc 14: 83-94.

Ritchie SA, Moore P, Carruthers M, Williams C, Montgomery B, Foley P, Ahboo S, Van den Hurk AF, Lindsay MD, Cooper B, Beebe N, Russell RC (2005) Discovery of a widespread infestation of *Aedes albopictus* in the Torres Strait, Australia. J Am Mosq Control Assoc 22: 358-365.

Rodhain F (1996) Problems posed by the spread of *Aedes albopictus*. Bull Soc Pathol Exot 89: 137-140.

Romi R and Majori G (1998) Commercio di copertoni usati e importazione di zanzare: un aggiornamento della distribuzione de *Aedes albopictus* e *Aedes atropalpus* in Italia. Notiz Ist Super Sanitá 11: 1-5.

Romi R (1995) History and updating of the spread of *Aedes albopictus* in Italy. Parassitologia 37:99-103.

Romi R (1996) Linee guida per la sorveglianza ed il controllo di *Aedes albopictus* in Italia. Rapporto ISTISAN, 96/4: 51 pp.

Romi R (2001) *Aedes albopictus* in Italia: un problema sanitario sottovalutato. Ann. Ist Super Sanità 37: 241-247.

Romi R, Toma L, Severini F and di Luca M (2003) Susceptibility of *Aedes albopictus* Italian populations to temephos and to other insecticides. J Am Mosq Control Assoc 19: 419-423.

Romi R, di Luca M and Majori G (1999) Current status of *Aedes albopictus* and *Ae. atropalpus* in Italy. J Am Mosq Control Assoc 15: 425-427.

Romi R, DiLuca M, Severini F and Toma L (2007) Linee guida per la sorveglianza. Istituto Superiore della Sanita, Roma. http://www.dsp-auslbo.it/pdf/profil/zanzara_tigre.pdf (Accessed 29 August 2007)

Rossi GC, Pascual NT, Krsticevic FJ (1999) First record of *Aedes albopictus* (Skuse) from Argentina. J Am Mosq Control Assoc 15: 422.

Sabatini A, Raineri V, Trovato G and Coluzzi M (1990) *Aedes albopictus* in Italia e possibile diffusione della specie nell'area Mediterranea. Parassitologia 32: 301-304.

Samanidou-Voyadjoglou A, Patsoula E, Spanakos G and Vakalis NC (2005) Confirmation of *Aedes albopictus* (Skuse) (Diptera: Culicidae) in Greece. Eur Mosq Bull 19: 10-12.

Santos R and La C (2003) Actualização da distribuição de *Aedes albopictus* no Brasil (1997-2002). Rev Saúde Públ 37, Oct. 2003.

Savage HM, Ezike VI, Nwankwo ACN, Spiegel R and Miller BR (1992) First record of breeding populations of *Aedes albopictus* in continental Africa: implications for arboviral transmission. J Am Mosq Control Assoc 8: 101-103.

Schaffner F, Boulétreau B, Guillet B, Guilloteau J and Karch S (2001) *Aedes albopictus* (Skuse, 1894) established in metropolitan France. Eur Mosq Bull 9: 1-3.

Schaffner F and Karch S (2000) Première observation d'*Aedes albopictus* (Skuse, 1894) en France métropolitaine. Compt Rend Acad Sci (Paris), Sciences de la vie / Life Sciences 323: 373-375.

Schaffner F, Van Bortel W and Coosemans M (2004) First record of *Aedes (Stegomyia) albopictus* in Belgium. J Am Mosq Control Assoc 20: 201-203.

Scholte EJ, Dijkstra E, Ruijs H, Jacobs F, Takken W, Hofhuis A, Reusken C, Koopmans M and De Boer A (2007a) The Asian tiger mosquito (*Aedes albopictus*) in the Netherlands: should we worry? Proc Neth Entomol Soc 18: 131-136.

Scholte EJ, Jacobs F, Linton YM, Dijkstra E, Fransen J and Takken W (2007b) First record of *Aedes* (*Stegomyia*) *albopictus* in the Netherlands. Eur Mosq Bull 22: 5-9.

Service MW (1993) Mosquito Ecology: Field Sampling Methods. Elsevier Science Publ., Essex, 988 pp.

Shroyer DA (1986) *Aedes albopictus* and arboviruses: a concise review of the literature. J Am Mosq Control Assoc 2: 424-428.

Simard F, Nchoutpouen E, Toto JC and Fontenille D (2005) Geographical distribution and breeding site preference of *Aedes albopictus* and *Aedes aegypti* (Diptera: Culicidae) in Cameroon, Central Africa. J Med Entomol 42: 726-731.

Sprenger D, Wuithiranyagool T (1986) The discovery and distribution of *Aedes albopictus* in Harris County, Texas. J Am Mosq Control Assoc 2: 217-219.

Toma L, Severini F, Bella A and Romi R (2003) A semifield evaluation of Vectobac DT (ABG-6499), a new formulation of *Bacillus thuringiensis israeliensis* for control of *Aedes albopictus*. J Am Mosq Control Assoc 19: 424-429.

Torres-Martínez M, Torres-Estrada JL (2003) First evidence of *Aedes albopictus* (Skuse) in Southern Chiapas, Mexico. Emerg Infect Dis 9: 606-607.

Toto JC, Abaga S, Carnevale P and Simard F (2003) First record of the oriental mosquito *Aedes albopictus* on the West African island of Bioko, Equatorial Guinnea. Med Vet Entomol 17: 343-346.

Urbanelli S, Bellini R, Carrieri M, Sallicandro P and Celli G (2000) Population structure of *Aedes albopictus* (Skuse): the mosquito which is colonizing Mediterranean countries. Heredity 84: 331-337.

Vélez ID, Quiñones ML, Suárez M, Olano V, Murcia LM, Correa E, Arévalo C, Pérez L, Brochero H and Morales A (1998) Presencia de *Aedes albopictus* en Leticia Amazonas, Colombia. Biomedica 18: 192–198.

15. Bed bugs: research and resurgence

Clive Boase

Abstract

Bed bugs have long been associated with human habitation. They are not known to transmit disease, but their bites cause considerable nuisance, and may trigger allergic reactions in some people. Infestations in the hospitality industry have significant economic impact for those managing such premises. Bed bugs became much less common in developed countries during the period c. 1950–1990, but commencing in the late 1990's, there was evidence of a substantial resurgence of the temperate bed bug (*Cimex lectularius*) in parts of Europe, the Americas, and Australia, coupled with increasing difficulty in controlling the infestations. The recent discovery of widespread resistance to some insecticides commonly used to control bed bugs is the most likely explanation for the increase, although other factors may also play a part. A variety of technologies have been developed to help control the resurgence, and Codes of Practice are being developed to standardise and improve control measures.

Keywords: bed bugs, resurgence, control, insecticide, resistance

Introduction

Bed bugs (Order: Hemiptera) are a small group of nocturnal, blood-feeding ectoparasites, that have been associated with humans and human dwellings since pre-history. However a decline in their incidence occurred in the mid 20th century and in developed countries at least, the period from the late 1960s to the early 1990s was characterised by the relative scarcity of bed bugs. Unexpectedly, in the late 1990's something changed, and it became apparent that in parts of Europe, the Americas and Australia, a synchronised resurgence of bed bugs was underway.

Recent surveys indicate that the rate of increase is substantial. A 2006 survey in London, revealed an annual 25% increase in the number of bed bug reports from 2000-2006. In Australia there appears to have been a 45 fold increase in bed bug treatments over the seven years since 1999. The upsurge has raised a broad range of questions spanning insect physiology, behaviour and ecology, public communication, and management of residential premises.

When the resurgence first became apparent, the process of understanding and dealing with the upsurge was based largely on information generated decades previously, when bed bugs were last a problem (Usinger 1966). However the urban pest research community has now swung into action, and new information addressing the challenges facing those responsible for dealing with the current outbreak, is starting to emerge. In this overview, new information is cited wherever possible. Some is from peer-reviewed sources, some is from trade publications, and much is personal communication. However it is hoped that during 2007 and 2008, much of the anecdotal or informally acquired information included here will appear at conferences and in journals, and will establish a more comprehensive evidence-based approach to addressing and resolving the current bed bug problem.

This article will concentrate mainly on the temperate bed bug *Cimex lectularius*, but mention will be made of *C. hemipterus* and other species, where appropriate.

Background bed bug biology

Most true bugs (Order: Hemiptera) feed on plant fluids, but a few groups of bugs have lost the phytophagous habit, and feed instead on invertebrates or vertebrates. The two main vertebrate-feeding Families are the Reduviidae, which are responsible for the transmission of Chagas disease in Central and South America, and the Cimicidae. There are around 90 recognised species of cimicids, with two species called 'bed bugs'; the common or temperate bed bug *Cimex lectularius*, and in the tropics the very similar *C. hemipterus*. In addition to these two species, there are other cimicid bugs associated with other hosts such as birds or bats, but which will also bite people when the opportunity occurs.

Adult bed bugs are wingless, typically chestnut brown in colour, but will appear darker when freshly fed. The adult bugs are around 5-6 mm in length, while the first instar nymphs measure between 1 and 2 mm. All stages are dorso-ventrally flattened, although when freshly fed they are distended.

Bed bug harbourage and distribution

Bed bugs are nocturnal, spending most of their time concealed in cracks, crevices and other harbourages, close to where the host sleeps. Bugs typically occur as loose groups of individuals, with the aggregation being maintained by chemical substances (Levinson and Bar Ilan 1971). Well-used harbourage areas may be recognised by black faecal spots on surfaces around the harbourage. In bedrooms, typical harbourage areas include the seams, buttons and handles of mattresses, joints in bed-frames, behind headboards, behind skirting boards, in furniture, inside electrical fittings, behind pictures, in pleats in curtains, under fitted carpets, under floorboards, and in wall voids. Bed bugs have very occasionally been found harbouring on the person, typically in cases of neglect and deprivation (Burgess 1990).

Bed bug infestations may be found wherever people sleep, including domestic properties, hotels, hostels, hospitals, ships, etc. (Table 1). Data collected in the UK by Cornwell (1974) and in Canada by Hwang *et al.* (2005), on the distribution of infestation across different property types, shows broad similarities.

Table 1. Distribution of bed bug infestation across different property types.

Property type	Distribution of infestations	
	Cornwell 1974 (UK)	Hwang *et al.* 2005 (Canada)
Residential (Homes and apartments)	61%	71%
Accommodation (Hotels, hostels, rooming houses)	23.2%	24.7%
Education (e.g. schools)	1%	2.7%
Healthcare (e.g. hospitals)	2.2%	0%
Transport (e.g. trains)	2.3%	0%
Other	10.1%	0.4%

The view that 'bug infestation is usually associated with low hygienic standards' (Busvine 1980), may possibly have been true in the mid 20th century, but at the present time bed bugs occur in a broad cross section of property types (Doggett 2005).

Data from nine similarly sized towns in the north-west of England, each with populations of from c. 200–450 thousand people, were collected in 1971 and 1991 at a time when bed bug incidence was at a relatively low level (G. Dunn - unpublished data). In Table 2, the towns are arbitrarily labelled as A – I for anonymity. The data show that bed bugs were very unevenly distributed, with a variation in infestation rates of up to 67 fold between towns.

This uneven distribution of bed bug infestations has also been seen in other situations, such as in hotels. Data collected in 2006 from several hundred UK hotels, shows that although about 10% of the hotels were treated for bed bug infestation in that year, 55% of the call-outs were to 1% of the hotels. (C. Boase – unpublished data). The reason for this much skewed distribution is not clear, but suggests that problems may be related more to difficulties in ridding infested premises of infestation, than to a regular flux of incoming bugs.

Table 2. Number of bedbug infestations reported per 100 thousand inhabitants, across 9 UK towns (G. Dunn – unpublished data).

Town:	A	B	C	D	E	F	G	H	I
1971	39	42	41	56	17	28	60	221	194
1991	0.7	2.8	9.0	11.4	12.1	14.6	15.3	37.1	46.9

Bed bug feeding

Both male and female bed bugs feed on vertebrate blood. Hosts appear to be detected by a number of cues including heat, carbon dioxide, and a number of host-produced compounds (Reinhardt and Siva-Jothy 2007). When not feeding, the proboscis is carried beneath the thorax in a rearward facing position. When preparing to feed the proboscis is swung forward, and the insect then forces the mandibles and maxillae through the skin, and the maxillary bundle then probes within the tissue until it encounters a suitable capillary (Dickerson and Lavoipierre 1959; Figure 1). Feeding time for an adult bug is around 10–15 minutes, but shorter for small nymphs. The volume of blood ingested by an adult bug is around 10 μl (Titschack 1930).

Although bed bugs are most often reported as biting humans, they also occur as occasional ectoparasites of poultry (Axtell and Arends 1990). Bed bugs are also occasionally found as parasites of other animals, such as cats (Clark *et al.* 2002). In the laboratory, bed bugs feed readily on rodents (Johnson 1941), although it is not known if bed bug infestations in houses are maintained by rat or mouse infestations.

At a temperature of c. 20 °C, the bed bug will feed about once per week. However if deprived of a blood-meal, the bed bug is capable of surviving for extended periods. Bacot (1914) placed fed bed bugs in an outbuilding without access to a blood-meal, and found survivors 18 months later.

Figure 1. Feeding bed bugs (Courtesy R. Naylor).

Bed bug reproduction and development

Mating in bed bugs is an unusual process. Upon mounting the female, the male's intromittant organ pierces the female's abdominal wall, and the sperm enter the mesospermalege from where they migrate into the paragenital tract. The female possesses a genital opening, but this is only used for oviposition, never during copulation. The female mates after every blood meal (Reinhardt and Siva-Jothy 2007).

The number of eggs laid will vary according to temperature being around 1–2 per week at 17 °C, and 4–10 per week at 23 °C (Johnson, 1941). The eggs are attached to the surface of the harbourage and surrounding area, where the empty cases will remain long after hatching. Having been laid, the egg goes through a period of incubation, the length of which is temperature dependent, before hatching (Table 3). Many hotels in the UK maintain bedrooms at around 21-23 °C. Temperatures in domestic property are more variable. In the UK, 24 °C is believed to be the upper comfortable limit for bedroom temperature, 18 °C is seen as the ideal, while there are several million households with winter bedroom temperatures below the ideal (Hartley 2007). Temperature should be ascertained in broad terms before establishing bed bug treatment guidelines in particular regions.

The eggs hatch into small nymphs, which go through four further instars before moulting to the adult. Each nymphal stage requires a single blood feed before moulting to the next stage. Although warmer environments clearly favour development, some bed bugs can nonetheless survive 1 hour exposure to temperatures as low as -17 °C (Kemper 1936).

Table 3. Relationship between bed bug development and temperature (from Johnson 1941).

	13 °C	15 °C	18 °C	23 °C	25 °C	28 °C
Duration of egg, from laying to hatch (days)	49	34	20.2	9.2	7.1	5.5
Total duration of nymphal stage (days)	No development below 13 °C	119	66	29	24	14

In the UK, the number of infestations reported varies with the season. Cornwell (1974) reported a seasonal cycle in the numbers of infestations across the UK, with a peak from August to October. Some other urban invertebrate pests in the UK show similar cycles, and the cycle is believed to be a simple reflection of more rapid multiplication in the warmer months.

There is little published data to indicate the size of infestations that occur in practice. Work in Zulu huts in South Africa indicated an average of 91.5 nymphal and adult *C. lectularius* per hut (Newberry 1990). Anecdotal evidence from the UK indicates that in extreme cases bed bug numbers may reach tens of thousands of bugs in one home (R. Naylor, personal communication). At the other end of the spectrum, complaints may be received from premises where intensive inspection reveals only 2 or 3 bugs.

Bed bug dispersion behaviour

Data from other cimicid species suggests that long-distance dispersion in bugs is not entirely an accidental or passive process. Research on the swallow bug (*Oeciacus vicarious*) shows that dispersing bugs are not a random selection of those in the harbourage, but comprise mostly adult female bugs with substantial fat bodies (Reinhardt and Siva-Jothy 2007). This suggests that even so-called 'passive' bed bug dispersion may actually be an active process, driving colonisation of new sites. However it is not yet clear whether *C. lectularius* adopts the same strategy.

Problems resulting from bed bug infestation

Nuisance biting

The degree of reaction to the bed bug bite varies greatly from person to person. Some people have either not developed a sensitivity to the proteins in bed bug saliva, or have lost the sensitivity altogether, and so may be completely unaware of bites. Kemper (1936) found that around 20% of the people he examined showed no reaction to bed bug bites. The proportion of people who currently do not react to bites is not known, although it is not uncommon to find people who show little or no dermal reaction to bites. The aetiology of the lesions resulting from bed bug bites is difficult to diagnose, and it is not uncommon to find cases where the host's medical practitioner has diagnosed a cause other than insect bites (Doggett 2005).

In most cases however, the bed bug's bite, although normally painless whilst actually feeding, is followed by the appearance of a localised itchy swelling. The interval between the bite and the appearance of the reaction may vary, from perhaps a few minutes to several hours (Sansom *et al.* 1992). The bed bug prefers to feed on exposed skin, so in temperate regions at least, bites tend to occur predominantly on the face, neck, shoulders and arms, often in irregular rows. In some cases the allergic response may develop into a more generalised widespread skin reaction (Fletcher *et al.* 2002), and in extreme cases anaphylaxis may occur (Parsons 1955).

Apart from the direct effects of bites, bed bug infestation can cause considerable distress to those living in infested premises. This distress may be manifested by loss of sleep, general anxiety, and by discarding bedding, furniture and other items suspected to be harbouring infestation (Klein 2001).

Disease transmission

Given their blood-feeding behaviour, the potential for the transmission of human disease by bed bugs has been investigated closely. For example, laboratory work has shown that hepatitis B surface antigen (but not hepatitis C) has been shown to persist in both *C. lectularius* and *C. hemipterus* for up to six weeks after an infected blood meal, and to be excreted in the faeces, from where infection of a susceptible person via contamination of skin lesions is theoretically possible (Silverman *et al.* 2001). However, in West Africa, children living in villages subject to intensive bed bug control did not show a decline in hepatitis B conversion rates, relative to those in untreated villages (Vall Mayans *et al.* 1994). This indicates that in practice, bed bugs are not a significant vector of hepatitis among children. Early work with HIV and bed bugs indicated that the virus could survive on bed bug mouthparts for up to an hour after feeding (Lyons *et al.* 1986), so creating the possibility that infection could be passed between an infected and an uninfected person using the same bed. However subsequent work (Webb *et al.* 1989) strongly indicated that HIV is not transmitted by biting insects such as bed bugs. Unlike most other blood-feeding arthropods (such as mosquitoes, fleas, lice and ticks), bed bugs do not therefore appear to be involved in the transmission of pathogens.

Bed bug dispersion within and between buildings

Bed bugs appear capable of movement over modest distances, and it is not unusual for bed bugs to be harbouring at a distance of 3–4 m from the bed (C. Boase – personal observation). In addition to this regular host-seeking movement, different but consistent patterns of bed bug dispersal have sometimes become apparent in buildings in which the author has worked:

In residential premises which are constructed to a high standard, and where there are no connections between rooms, then any bed bug infestations that do occur typically remain within individual rooms.

By contrast, in buildings where there are lateral connections between rooms, e.g. where plumbing or electrical services run from room to adjoining room, then it is common for bed bug infestation to spread over time into adjoining rooms. Pest control records will often indicate a chronology of infestation appearing to spread from room to room along one side of a corridor (Figure 2).

In buildings such as hotels, it is common practice for the linen to be removed from each room and dumped on the floor in the corridor prior to collection onto a trolley. In such cases the infestation also appears to spread to rooms across the corridor (C. Boase, personal observation).

Regarding movement of bed bugs over longer distances, there are numerous reports of bed bugs being found on a very wide range of personal effects including luggage, toys, books, clothing, furniture, beds, electrical items, etc. It is likely that the movement of such infested items will deposit bed bugs at the destination, and so result in the spread of infestation (Simon 2004). Doggett *et al.* (2004) cites records collected by the Australian Quarantine and Interception Service from 1990 to 2003, of bed bugs found on incoming goods. The records show that the majority of bugs found were arriving by air, and were associated with personal baggage. In addition to the transport of infested items, infested vehicles may also infest goods in transit.

Wk	Room numbers (rooms adjoining) 225	227	229	231	233	235
1				X		
2				X		
3			X	X	X	
4			X	X	X	
5			X	X	X	
6			X	X	X	
7			X	X	X	
8			X	X	X	
9	X	X	X	X	X	
10	X	X	X	X	X	
11	X	X	X	X	X	
12	X	X	X	X	X	
13	X	X	X	X	X	
14	X	X	X	X	X	X
15	X	X	X	X	X	X

Figure 2. Dispersion of bed bugs across adjoining rooms in a hotel in which plumbing ran through a void between rooms. X = bed bug infested room. C. Boase (unpublished data).

The time taken for a bed bug-free building to acquire an infestation will depend on many factors, including the overall level of infestation in the community, the number of occupants, and the movement patterns of occupants and their belongings. In South Africa, working with both *C. lectularius* and *C. hemipterus*, Newberry (1991b) estimated that huts in KwaZulu were re-infested at a rate of 11% per year when the nearest infested properties were distant, but rose to 35% per year when nearby huts were infested. Examination of data sets on bed bug infestations in previously uninfested UK hotels suggests low infestation rates, of the order of 5% per year (C. Boase, personal observation).

Fear of acquiring infestation

There have been concerns raised by those working in infested premises, that they might inadvertently acquire bed bugs in the course of their work, and take them home. At times such concerns have created problems for those organising maintenance or healthcare work in buildings known to be infested. An informal survey of c. 60 pest control technicians in the UK, each with up to 15 years involvement in pest control, indicated that only one was able to recall ever having inadvertently taken bed bugs home (C. Boase, personal communication). It would appear that those working in infested premises during the daytime are at very low risk of acquiring infestation.

Financial impact of infestation

Although the immediate impact of bed bug infestation will be the discomfort of bites, for anyone responsible for residential property, the longer term impact is likely to be financial. Costs resulting from a bed bug infestation arise from several different sources:

a. The cost of using professional pest control services to eradicate bed bugs from any residential property, can be considerable. In residential blocks, the pest control organisation may recommend that adjoining rooms, both horizontally and vertically, should also be treated. One infested room may therefore require six or more rooms to be treated and paid for.
b. Pest control organisations may recommend that the mattress and bed is replaced in infested rooms.
c. Pest control organisations will typically recommend that all clothing in the infested rooms is laundered, or dry-cleaned if conventional hot washing is not possible.
d. In commercial properties, preparing rooms for treatment and re-instating rooms after treatment, can incur significant costs.
e. Temporary closure of rooms undergoing treatment will mean loss of income to an organisation providing accommodation on a commercial basis.
f. Finding accommodation for residents or guests elsewhere during the treatment process will result in additional costs.
g. For a hotel, waiving charges for guests complaining of bites will generate additional costs.
h. Negative publicity resulting from bed bug infestation in commercial accommodation premises, is likely to impact negatively on business.

The total costs will vary enormously, but for an infestation in a private house being treated by a professional organisation, the total cost of dealing with an infested bedroom may exceed several hundred Euros. For a single large hotel with an established infestation, total annual costs have been known to exceed one million Euros/year.

Liability arising from bed bug infestation

For the hospitality industry, the issue of liability with respect to bed bug infestations is a major concern. There have been numbers of claims against hotel organisations in the USA, for damages and costs resulting from bed bug bites received. The financial sums involved tend to be larger than in other countries, because of the ability of US courts to award punitive damages. For example, in a case brought against Accor Econ. Lodging Inc, a court in Illinois awarded each of two trade show attendees a sum of $5000 for compensatory damages, plus a sum of $186,000 as punitive damages. In January 2007, an American couple stayed in a hotel in London and claim to have been bitten by bed bugs (Anonymous 2007a). They are bringing a multi-million dollar claim against the headquarters of the hotel group in the USA, where damages awarded may be higher. This case emphasises that cross-border litigation could place many organisations at much greater liability than previously anticipated.

Successful claims against hotels have typically been characterised by the hotel failing to respond adequately to the risk of bed bugs within its premises.

Changes in bed bug status

Evidence for a change in status

Data on bed bug prevalence in the first half of the 20th century is limited, but all points to bed bugs being very common at that time. For example, it was reported in London, UK that 'in many areas all the houses are to a greater or lesser degree infested with bed bugs' (Ministry of Health 1933). However, in more developed countries there then appears to have been a gradual reduction in bed bug problems over the mid-20th century. Busvine (1957, 1964) presents data from the UK and Denmark showing a steady decline over this period. The factors responsible for this are not known with certainty, but may include improvements in the housing stock (Medical Research Council 1942), legislation, and synthetic insecticides. In the UK, section 83 of the 1936 Public Health Act gave local authorities powers and responsibilities regarding 'verminous premises'. The introduction of powerful synthetic insecticides such as dichloro-diphenyl-trichloroethane (DDT) is often cited as an important factor, and undoubtedly played a role. However, the data presented by Busvine (1964) in the UK shows that bed bug incidence fell by c. 80% over the 6 years before DDT was introduced, indicating that other measures played a crucial role. Cornwell (1966) shows that for the pest control sections of local authorities in the UK, bed bugs were by then the 7th most important insect, after houseflies, cockroaches, silverfish etc. By the 1980s, bed bugs were considered relatively uncommon in many developed countries, although as Table 2 shows, significant populations did nonetheless persist in a few urban centres. However in less developed countries, the available information indicates that tropical bed bug infestation remained at relatively high levels, with research in Hyderabad, India (King *et al.* 1989) indicating that 65% of homes were infested, studies in Tanzania (Temu *et al.* 1999) indicated that 56% of homes were infested, while in the Gambia, Lindsay *et al.* (1989) showed that 37% of children's beds were infested.

What appears to be the first recognition of the current bed bug resurgence, was an article about bed bugs in the UK, by Birchard (1998), published in a medical journal. This report was quickly followed by two more publications relating to the UK; Paul and Bates (2000) and Boase (2001). The latter publication suggested that the increase may have commenced in the mid 1990's. In the USA, the first report of an increase was published in 1999 (Pinto 1999). In Australia, Coghlan (2002) also reported an increase, and Doggett *et al.* (2003) presented data showing that the increase had been underway from the mid 1990's. Although these authors tend to agree that the increase first became apparent in the last few years of the 20th century, it is possible that an increase may have been underway but undetected for some years previously.

In many European countries, such as France, Ireland, Czech Republic, and Hungary, there are anecdotal reports of increased problems, but little concrete evidence. In Germany, Liebold *et al.* (2003) state that 'Bed bugs seem to have become more common due to international trade and travelling'. By contrast, Leverkus *et al.* (2006) also working in Germany, states that 'bites from the common bed bug (*Cimex lectularius*) are nowadays rather uncommon. Nevertheless infestations are sometimes observed…due to international travel and migration'. In Italy, Masetti and Bruschi (2007) reported on cases of bed bug infestation.

Estimates of the extent of increase vary considerably. One detailed study (Richards, cited in Anonymous 2007b) involved examination of the records of requests for bed bug treatment received by the pest control sections of eight London local authorities, over the period 2000–2005. The data show an annual increase of 24.7%. In Australia, data from a survey of pest control companies indicated a c. 45 fold increase over 7 years (Doggett and Russell 2007).

Bed bug infestations as a meta-population

In addition to viewing the bed bug upsurge as an increasing number of individual infestations, the individual infestations within a given city, area or country may be collectively viewed as a metapopulation (Stejskal 2002). The status of the metapopulation results from the overall balance of a number of local processes including dispersion, establishment, detection and control activities, acting on each individual infestation. Considering bed bug infestations as one metapopulation, helps highlight key aspects of the control programme. It emphasises for example, the importance of early detection of infestations, and of achieving complete elimination of each identified infestation, in order to reduce the time available for each infestation to disperse and give rise to new populations.

Changes in the response of bed bugs to control measures

Coinciding with the upsurge in the numbers of bed bug infestations has been an increased difficulty in eradicating infestations. Doggett (2005) describes an infestation that required 8 treatments to achieve eradication. Feedback from pest control technicians in the UK indicate that call-back rates from single bed bug treatments vary considerably, but will often be around 50%, and can be as high as 90% (C. Boase, personal observation). This issue has not received the same publicity as the overall increase, but is likely to be very important in understanding the resurgence.

Explanations for the resurgence

As the scale and extent of the current bed bug resurgence became apparent, a wide range of possible explanations for the increase were forthcoming, both from those involved in pest control, and the media (Boase 2001, Doggett 2005), shown in Table 4.

Table 4 outlines the debate concerning the various explanations for the bed bug resurgence. Establishing and understanding the cause of the upsurge is important, as it then allows the introduction of appropriate and specific changes that may improve bed bug control, and so turn the upsurge into a decline. It is possible that there may be a combination of factors that have coincidentally changed and caused the upsurge. However the simplest explanation for (a) the increased difficulty in controlling bed bugs, and (b) the increase in the number of infestations, would appear to be the development of resistance to commonly used insecticides. This resistance results in a greater proportion of infestations surviving treatment, which then allows them to continue to disperse and give rise to additional infestations.

Addressing the resurgence

Introduction

Managing the increase in bed bugs typically requires action in a variety of different areas. In broad terms, individual infestations may be addressed by either the resident, or a professional pest control organisation. Although both these options have the same aim of eradicating the bed bugs, the overall strategy, the tools available, and the likelihood of success, are likely to be different. Although no data on the effectiveness of amateur bed bug control activities are available, the perception is that in most cases professional inputs are much more likely to be successful.

The tools available to manage bed bug infestation may be divided into the non-chemical, and the chemical options. The following sections outline the current issues regarding the various components of bed bug management.

Table 4. 'Explanations' for the bed bug resurgence.

Issue	'Explanation'	Comment
Social and human issues	Increased sales of second hand items.	Although bed bugs can certainly be dispersed on infested second-hand items, there is no evidence that sales of such items have increased substantially over the last 15 years. In any case, this does not address the increased occurrence of bed bugs in modern hotels.
	Decline in public awareness of bed bugs.	There are no data on this, although it appears plausible. From an epidemiological perspective, a community that is slow to recognise the presence of infestation would facilitate an increase through allowing more time for entrenchment and dispersion.
	Increased overcrowding in residential properties.	Undoubtedly congested domestic premises make bed bug control difficult. However bed bug problems have also increased substantially in hotels, where congestion is seldom an issue.
	Increased human migration, as a result of political changes (EU expansion) and conflict (Balkans, Middle East). Increased human movement for business and leisure.	Undoubtedly bed bugs are dispersed in the course of the travel, and the number of trips taken grows steadily year on year. Although metapopulation dynamics might allow an upsurge to be driven by increased human movement rates alone, the importation hypothesis typically presupposes that there are large reservoirs of bed bug in some countries, which are being exported to those countries experiencing an increase. To date there is no evidence for large reservoirs of *C. lectularius* overseas, from which those found in recipient countries are being steadily imported. Indeed Table 2 shows that there were anyway long-term reservoirs of bedbugs in the UK which could have initiated the resurgence. Similarly, if export of bed bugs from tropical countries was driving the increase, then we would expect to see a large increase in *C. hemipterus* in recipient countries. Although tropical bed bugs have been found in southern Australia (Doggett *et al.* 2003), and in the UK, there is no suggestion that they are responsible for the overall increase.
Environ-mental issues	More central heating and double glazing. Global warming.	If indoor temperature increases have been sufficient to accelerate bed bug reproduction, then it would be reasonable to expect that other indoor invertebrate pests might also show similar increases. However, there is no indication of such an increase.
Pest control issues	Inadequate training of pest control technicians.	It is true that pest control practices have had to change in response to the resurgence, which has required some re-training. However, there is no evidence that an initial decline in competence actually caused the increase.
	Changes in the pattern of insecticide use.	Starting in the late 1980s, there has been a switch away from use of broad spectrum residual sprays for ant and cockroach control (that may also have collaterally killed some bed bugs), to species-specific bait products. The timing of this change fits, but it is questionable whether sprays applied primarily to kitchens and bathrooms will have had sufficient impact on bed bugs concealed on beds and mattresses.
	Effective insecticides withdrawn from use.	The withdrawal of DDT has sometimes been cited as the reason for the bed bug resurgence. However in most countries bed bugs were widely resistant to DDT before it was withdrawn.
	Insecticide resistance.	Bed bug resistance to DDT was first recognised in 1958. However, the recent discovery that there is now widespread and sometimes high levels of resistance to at least two classes of commonly used conventional insecticides, would explain both the current difficulty in controlling bed bugs, and the increasing numbers of infestations.

Surveying and monitoring

As with the management of any other pest or vector, thorough pre- and post-treatment surveying is essential, to define the initial problem, and to assess the impact of the intervention. Residents' feedback alone cannot be relied on to provide an objective view, as many people are either unaware of infestation, or are reluctant to report it.

With many pests within buildings, specific monitoring devices (e.g. adhesive traps) are available that may be left on site and which will detect the pest, if present. For bed bugs, no effective devices are yet available, although development work is underway (Boase 2001). As a result, bed bug inspections rely heavily on detailed manual inspections. Doggett (2007) outlines the procedures to be followed. As an alternative, sniffer dogs are being tried for bed bug detection in the USA. In accommodation such as hotels and hostels, it is crucial that inspections are carried out not only in the complainant's room, but in nearby rooms also. Such proactive inspection will often reveal that the actual number of infested rooms is several-fold greater than that reported by residents (C. Boase, personal observation).

Denial of harbourage

Denying the pest access to potential harbourage areas is a classical component of indoor pest control. With respect to bed bug control, caulking materials are commonly used to seal the crevices around headboards or bedroom furniture. In practice, such proofing is seldom applied and maintained comprehensively, and harbourage may still exist. In most situations, easy-to-move or free-standing bedroom items allow for more convenient movement, inspection and treatment, as compared with partially proofed items (C. Boase, personal observation).

Protective mattress covers will prevent bed bugs from gaining access to the beading, buttons and handles of mattresses (Potter *et al.* 2007), but can still allow infestation to develop in folds on the cover itself (C. Boase, personal observation).

Processing of potentially infested clothing and textiles

Clothing, linen and removable furnishings within the infested room may have bed bugs harbouring or entrained within it. Most pest control organisations recommend that such materials are laundered, although until recently such advice was not based on specific evidence. Potter *et al.* (2007) presented some data on the effectiveness of laundry procedures against bed bugs. Further work (Table 5) has examined the efficacy of various laundry procedures in detail. In these tests, small cloth bags containing egg, nymph and adult bed bugs were fixed within clothing items, which were then combined with a variety of other clothing items to form a standard load, which was then processed in various ways. The actual temperature profiles within the washing machine, tumble drier and freezer were recorded using a data-logger.

Room treatment using extreme physical conditions

As an alternative, or as a complement to insecticide use, a variety of bed bug control systems based on physical means, have been introduced. In general these systems do not offer any lasting protection of the treated room. However, if generally effective, these systems are likely to be

Table 5. Effectiveness of processing of clothing in killing bed bugs (Naylor and Boase, cited in Anonymous 2007b).

Technique	Setting	% Mortality
Soak in bath	Cold water, 2 hr	Eggs: 0% Active stages: 16-23%
	Cold water, 24 hr	Eggs: 0% Active stages: 100%
Wet wash in domestic washing machine	30 °C	Eggs: 0% Active stages: 80–96%
	40 °C	Eggs: 24% Active stages: 100%
	60 °C	All stages: 100%
Domestic tumble drier. (Dry clothes placed straight in drier – no preceding wash.)	'Hot', 10 mins	Eggs: 75% Active stages: 75-84%
	'Hot', 30 mins (Reached 43 °C)	All stages: 100%
Dry cleaning at commercial dry cleaners.	Perchloroethylene. No pressing.	All stages: 100%
Freezing in domestic deep freeze.	-17 °C for 2 hr. (However a 2.5 kg load of clothes in a domestic freezer requires at least 8 hr to reach -17°C)	All stages: 100%

effective against insecticide-resistant bed bugs. At present very limited data are available on the operational use of these systems.

In Australia, use of solar energy to kill bed bugs on mattresses, by enclosing the infested mattress in a black plastic bag and leaving it in full sun, has been widely recommended. However, critical evaluation of the technique has shown that temperatures reached in the bag are not high enough to kill all bugs (Doggett *et al.* 2006).

Hand-held systems using steam to treat mattresses and other sensitive items within the bedroom are widely used in the USA and Australia. The limited available information suggests that they are effective (Meek 2003), although somewhat laborious. The practical challenge is that any bed bugs remaining hidden within areas that are inaccessible to the steam, still need to be eliminated, probably using conventional insecticides.

Some organisations use a large insulated bag in which the mattress, bed frame, loose furniture, curtain etc are all placed. The bag is connected to a hot air blower, which raises the temperature in the bag to a lethal level. A different methodology is likely to be required to eliminate bed bugs from the rest of the room.

Use of hot-air blowers that heat the whole room, rather than just some of the items within it, are also being evaluated. If shown to be effective, it is possible that this whole-room technique could largely obviate the need for insecticide use. However the success of such systems depends on being able to achieve a uniformly high temperature within the treated room, and ensuring that insects are not able to retreat into protected voids and so escape the elevated temperatures (Potter *et al.* 2007).

At the other end of the temperature range, a commercial system using carbon dioxide is available. The device applies a narrow swath of low temperature solid carbon dioxide to surfaces and harbourages, and is claimed to kill all stages of the bug (Dawe 2006).

Insecticide activity

Although there are a variety of non-chemical bed bug control options available, in most cases, insecticides form at least part of the overall bed bug control work at any one site. A wide variety of insecticidal substances is in principle available for control of insect pests in buildings (Boase 2006). However as a result of commercial practices and regulatory constraints, only a subset is actually available for bed bug control in any one country. An even smaller subset will be approved for mattress treatment. No comparative evaluation of the activity against bed bugs of the full range of insecticides is available. Probably the most complete is that presented by Fletcher and Axtell (1993) in which several organophosphates, carbamates and pyrethroids, were evaluated against a strain of bed bugs collected from a poultry farm in the USA. The study did not indicate whether the strain was considered resistant or susceptible (see Table 6).

Results showed that at the LC_{90} level, the activity of the various insecticides classes ranked approximately as follows: organophosphates>carbamates>pyrethroids. This general efficacy ranking of the conventional residual insecticides against bed bugs has been reported, although not so clearly, in other publications (Newberry 1991a).

In addition to the conventional insecticides, various insect growth regulators (IGRs) are also commonly used, depending on the regulatory position, for bed bug control. Efficacy of the juvenile hormone analogue (JHA) pyriproxyfen has been briefly reported (Boase 2001), and work on other JHAs is underway. The other class of IGRs, the chitin synthesis inhibitors, is in Europe represented by flufenoxuron, which is regularly used in combination with pyrethroids for bed bug control. Compared to conventional neurotoxic insecticides, the IGRs are typically slow acting, although in practice they are often used in combination with a conventional fast-acting insecticide. Although there are very little data currently available on the efficacy of these compounds against field

Table 6: Efficacy of insecticides against C. lectularius *(Fletcher and Axtell 1993).*

Insecticide class	Insecticide	LC_{90} (mg ai/m^2)
Organophosphate	Dichlorvos	0.72
	Pirimiphos-methyl	3.76
	Malathion	30.89
	Tetrachlorvinphos	59.6
Carbamate	Bendiocarb	12.09
	Carbaryl	45.9
Pyrethroid	Permethrin	25.4
	Lambdacyhalothrin	45.1
	Fenvalerate	5269

strains of bed bugs, they are interesting as they have a mode of action very different from that of conventional fast-acting insecticides, and so may remain active against field strains.

In the USA, the novel compound chlorfenapyr has recently been approved for bed bug control. It also has a different mode of action from conventional insecticides, and so may be active against field strains, although it is relatively slow in action against susceptible bed bugs (Moore and Miller 2006).

Finally, desiccant dusts, based on finely divided silica, are used in some countries, including the UK, for bed bug control. They are believed to work by damaging the wax layer on the insect's cuticle, causing uncontrolled loss of water, desiccation and death (Ebeling 1971). Recent laboratory work on UK field strains of bed bug resistant to conventional insecticides, shows that the bugs remain susceptible to desiccant dusts (R. Naylor, personal communication).

The various active substances outlined above are typically applied as residual sprays to surfaces in the infested area, although dry dusts do have a role for treatment of voids.

Reduced insecticide susceptibility

Most insects that have been the target of regular insecticide use have, over time, evolved resistance to some of the active substances used against them. As far back as 1980, 42 species of Hemiptera were recognised as being resistant to one or more insecticides (Georghiou and Lagunes-Tejeda 1991).

The first account of strains of insecticide resistant *C. lectularius* was in 1958, when resistance to DDT was reported (Busvine 1958). Cyclodiene resistance was reported in 1963 (Guneidy and Busvine 1963), organophosphate resistance in 1968 (Feroz 1968), carbamate resistance in 2006 (Boase *et al.* 2006), and pyrethroid resistance in 2006 (Boase *et al.* 2006, Romero *et al.* 2007). In *C. hemipterus,* field resistance to organochlorines was reported in 1991 (Newberry 1991a), resistance to pyrethoids was reported in 2003 (Myamba *et al.* 2003), and resistance to organophosphates and carbamates was reported in 2007 (Karunaratne *et al.* 2006).

Myamba's work in Tanzania neatly showed the relationship between treatment and resistance. Pyrethroid-treated bednets were distributed and used in selected villages, and at first provided good control of bed bugs. However after a few years villagers reported that the bed bugs had returned. Tests showed that in the villages using treated bednets the bed bugs had become resistant, while in villages in which treated bednets had not been used, the bed bugs remained susceptible.

The recent data on the extent of resistance suggests that it is very widespread. In the UK, ten field strains of bed bugs were collected during 2006 from within the London area, and tested for susceptibility to carbamate and pyrethroid insecticides using a discriminating dose technique. All 10 strains showed little or no mortality on the respective LC_{95} discriminating doses (R. Naylor, personal communication). In the USA, Romero *et al.* (2007) reported results on 8 strains of bed bug collected from five states across the USA. Results showed that all strains except one were highly resistant to deltamethrin and lambdacyhalothrin. Resistance factors of >1000x were reported for these USA strains.

Prior resistance to organochlorines may have predisposed the bugs to pyrethroid resistance through a common resistance mechanism (Farnham 1977). Carbamates have a different site of action from pyrethroids, but do share a common site of action with the organophosphates. It is possible therefore that the apparently swift appearance of carbamate and pyrethroid resistance in bed bugs may have been facilitated by prior resistance to organochlorine and organophosphate insecticides. It is not clear whether this resistance was selected through the use of insecticides to control bed bugs in homes, or in the poultry industry, nor is it clear whether the resistance arose at one point and dispersed internationally, or arose almost simultaneously in several parts of the world.

Results of these susceptibility tests raise concerns over the ability of pest control organisations to control bed bugs using insecticides. In the UK, this concern is supported by feedback from operational bed bug control, which shows that many organisations that initially apply a single residual insecticide treatment will experience call-back rates of around 50% (C. Boase, personal observation).

Suggested strategies (Rust 1996) for managing insecticide resistance include management by:

a. Moderation. This approach involves use of techniques that apply a low selection pressure, and which will conserve insects with susceptible genes. However as Rust states: 'In many urban situations…a strategy of moderation will be unacceptable because there is an extremely low pest threshold'. This is likely to be especially true for bed bugs.
b. Saturation. This approach assumes that resistance genes are still rare and thus exist almost entirely as heterozygotes, that a high dose will kill these heterozygotes, and thus the resistance gene will have very little selective advantage.
c. Alternation. The alternation approach assumes that resistance genes to each of the insecticides are different and that each has sub-normal fitness when the relevant insecticide is not in use. Thus, while insecticide A is in use natural selection should reduce the frequency of resistance to B and vice versa.

In the UK, feedback from operational bed bug control adds weight to the case for managing bed bugs by both saturation and alternation. Pest control organisations that routinely carry out a sequence of two or three treatments, using a series of different active ingredients, experience call-back rates that are much lower than those using one treatment as their standard. In one large hotel in the UK with severe and long-established bed bug problems with which the author was involved, a pragmatic saturation/alternation strategy was adopted. The previous largely pyrethroid-based control programme was changed to a carefully planned sequence of three treatments, involving insecticides from different chemical classes. All treatments were applied within the statutory label conditions. This change has resulted in a dramatic reduction in bed bug incidence. In addition to the genetic basis for this strategy, the apparent success of a sequence of treatments may also stem from the prolongation of the treatment period, so ensuring that young nymphs that eventually emerge from eggs are exposed to a fresh active insecticide deposit.

Gaining public co-operation and compliance

Preparation of premises prior to any treatment is essential in order to achieve full efficacy. The type of preparation required will depend very much on the proposed treatment and the layout of the room, but typically this will include removal of bed linen, lifting the mattress from the bed base, separating the headboard from the bed, emptying drawers and wardrobes, possibly lifting the edges of fitted carpets, loosening the cover plates to electrical sockets, and removing pictures from the walls. In addition, it is normally recommended that clothing and linen be laundered.

Preparation of hotel premises is normally relatively simple, but for domestic property, the challenges are much greater. To encourage residents to carry out these works prior to the arrival of the pest control technician, a number of approaches are used by those carrying out pest control:

a. A verbal explanation to the resident over the telephone, at the time of booking the pest control visit.
b. An explanatory leaflet posted to the resident.
c. A preliminary visit by a representative of the pest control organisation, to explain what preparations need to be done.

To reinforce the importance of this preparation by the resident, some public sector pest control organisations will use financial leverage to gain compliance. For example they may require payment of an advance fee, which will be forfeited if preparation has not been carried out. Many of those carrying out bed bug treatments in congested urban residential properties will say that persuading residents to carry out thorough pre-treatment preparation is one of the most challenging aspects of bed bug control. There may be much to learn from other well-researched campaigns that use psychological models to change public behaviour, e.g. in health education.

Bed bug control strategies and Codes of Practice

The recent increase in bed bug problems has focussed critical attention on the techniques used for bed bug control, and attempts are being made to identify and agree best practice.

Close examination of the practice of bed bug control has revealed that there is a considerable diversity of thinking and understanding among pest control practitioners, and as a result, a diversity of techniques being used. Identifying best practice has been somewhat hampered by the lack of operational research that might enable effective practices to be selected on the basis of clear evidence. As a result, there is a tendency for the Codes and Regulations to attempt to address and standardise every aspect of bed bug management, rather than focussing on those specific aspects which may be predominantly responsible for the upsurge. It is hoped that as the reasons underlying the increase become clearer, then the Codes of Practice will become more targeted. To date none of these approaches explicitly addresses the issue of insecticide resistance, which is likely to be one of the main factors driving the resurgence.

The Australian Code of Practice for Bed bug Control (Doggett 2007) was the first national code to appear. This Code first appeared in July 2006, and an updated version is scheduled to appear in summer 2007. The Code was assembled by a panel consisting of entomologists, representatives of the pest control servicing industry, and of insecticide manufacturers. Compliance with the document is not mandatory in Australia. This document runs to over 60 pages, and stresses the need for:

a. Training of pest control staff and those responsible for managing accommodation.
b. Preventative procedures such as use of metal bed frames.
c. Detailed pre-treatment inspection procedures, both in infested and adjoining rooms.
d. Use of both chemical and non-chemical control procedures.
e. At least one follow-up visit and treatment after 7–10 days.

In September 2006, in the USA, the San Francisco Department of Public Health produced Regulations on how to control bed bug infestations (Ojo 2006). These Regulations set out succinctly the responsibilities of residents, those responsible for managing residential properties, and pest control operatives. Key steps include the need for:

a. Bed bug awareness training for those managing residential premises,
b. A rapid response to bed bug complaints,
c. Inspection of premises and units adjoining the known infested property,
d. Proper pre-treatment preparation of premises by the resident
e. A sequence of 3 treatments at 2-weekly intervals,
f. A 45 day guarantee on all treated premises

The measure of the effectiveness of such Codes and Regulations will be the extent to which they impact on the rate of appearance of new bed bug infestations. Good data collection, and time, will clarify their impact.

Conclusion

The recent international bed bug upsurge demonstrates the fragility of our relationship with urban pests and vectors. Problems that were considered to have been 'solved' decades ago, and which were considered to be unlikely to return due to improvements in the built environment, have been shown to be quite capable of resurgence, within a period as short as three to five years. However, the urban pest control community found that the information which it needed to make informed decisions and recommendations was last updated almost half a century ago. The intensity of the problem and the shortage of recent information have spurred new research, to understand the causes of the problem, and to design specific solutions. It is anticipated that much of what is currently anecdote and conjecture will shortly become peer-reviewed science, and will enable evidence-based processes to (once again) bring about the demise of this unpleasant insect. Fortunately, what has prevented the resurgence from becoming a major public health crisis is that unlike most other blood-feeding insects, the bed bug apparently does not transmit human disease.

References

Anonymous (2007a) Exclusive London hotel sued over bed bug attack. The Times. January 16, 2007.

Anonymous (2007b) The goblin in our midst. Professional Pest Controller, Summer 2007, 10-11.

Axtell RC and Arends JJ (1990) Ecology and management of arthropod pests of poultry. Ann Rev Entomol 35: 101-126.

Bacot AW (1914) The influence of temperature, submersion and burial on the survival of eggs and larvae of *Cimex lectularius*. Bull Entomol Res 5: 111-117.

Birchard K (1998) Bed bugs biting in Britain: only rarely used pesticides are effective. Med Post 34: 55.

Boase CJ (2001) Bed bugs – back from the brink. Pestic Outlook Aug: 159-162.

Boase CJ (2006) The number of insecticides is decreasing – fact or factoid? Int Pest Contr 48: 34-36.

Boase CJ, Small G and Naylor R (2006) Interim report on insecticide susceptibility status of UK bed bugs. Prof Pest Contr, Summer 2006: 6-7.

Burgess N (1990) Public Health Pests. Chapman and Hall, London.

Busvine JR (1957) Recent progress in the eradication of bedbugs. Sanitarian 65: 365-369.

Busvine JR (1958) Insecticide resistance in bed bugs. Bull World Health Organ 19: 1041-1052.

Busvine JR (1964) Medical Entomology in Britain. Ann Appl Biol 53: 190-199.

Busvine JR (1980) Insects and Hygiene. Chapman and Hall, London.

Clark S, Gilleard JS and McGoldrick J (2002) Human bed bug infestation of a domestic cat. Vet Rec 151: 336.

Coghlan A (2002) Bed bugs bite back. New Scientist 176: 10.

Cornwell PB (1966) The incidence of pests dealt with by Local Authorities. Environ Health 78: 267-272, 282.

Cornwell PB (1974) The incidence of fleas and bed bugs in Britain. Int Pest Contr 16: 17-20.

Dawe H (2006) Laboratory bioassay to assess the efficacy of the Cryonite system, against bed bug adult, nymph and egg stages.http://www.cryonite.net (accessed 22 August 2007).

Dickerson G and Lavoipierre MMJ (1959) Studies on the methods of feeding of blood sucking arthropods. II. The method of feeding adopted by the bed bug (*Cimex lectularius*) when obtaining a blood meal from the mammalian host. Ann Trop Med Parasitol 53: 347-357.

Doggett SL, Geary MJ, Crowe WJ, Wilson P and Russell RC (2003) Has the tropical bed bug, *Cimex hemipterus* (Hemiptera: Cimicidae), invaded Australia? Environ Health 3: 80-82.

Doggett SL, Geary MJ and Russell RC (2004) The resurgence of bedbugs in Australia, with notes on their biology and control. Environ Health 4: 30-38.

Doggett SL (2005) Bed bug ecology and control. In: Pests of Disease and Unease. Westmead Hospital, New South Wales, April 2005. pp 7.1-7.69.

Dogget SL, Geary MJ and Russell RC (2006) Encasing mattresses in black plastic will not provide thermal control of bed bugs *Cimex* spp. (Hemiptera: Cimicidae). J Econ Entomol 99: 2132-2135.

Doggett SL and Russell RC (2007) Bed bugs: Recent trends and developments. Australian Environmental Pest Managers Association Annual Conference. Synopsis of Papers. Coffs Harbour.

Doggett SL (2007) Code of Practice for the control of bed bug infestations in Australia. 2nd Edition. http://www.bed bug.org.au (Accessed 22 August 2007).

Ebeling W (1971) Sorptive dusts for pest control. Ann Rev Entomol 16: 123-158.

Farnham AW (1977) Genetics of resistance of houseflies (*Musca domestica* L.) to pyrethroids. I. Knockdown resistance. Pestic Sci 8: 631-636.

Feroz M (1968) Toxicological and genetical studies of organophosphorous resistance in *Cimex lectularius* L. Bull Entomol Res 59: 377-382.

Fletcher CL, Arden-Jones MR and Hay RJ (2002) Widespread bullous eruption due to multiple bed bug bites. Clin Exp Dermatol 21: 74-75.

Fletcher MG and Axtell RC (1993) Susceptibility of the bed bug *Cimex lectularius* to selected insecticides and various treated surfaces. Med Vet Entomol 7: 69-72.

Georghiou GP and Lagunes-Tejeda A (1991) The occurrence of resistance to pesticides in arthropods. Food and Agriculture Organization of the United Nations, Rome.

Guneidy AM and Busvine JR (1963) The inheritance of dieldrin resistance in *Cimex lectularius* L and *Pediculus h. humanus* L. Bull Entomol Res 55: 509-517.

Hartley A (2007) West Midlands Health Issues – Fuel Poverty. http://www.wmpho.org.uk/information/fuelpovertyshort.pdf (Accessed 22 August 2007).

Hwang SW, Svoboda TJ, De Jong IJ, Kabasele KJ and Gogosis E (2005) Bed bug infestations in an urban environment. Emerg Infect Dis 11: 533-538.

Johnson CG (1941) The ecology of the bed bug, *Cimex lectularius*. J Hyg 41: 345–461.

Karunaratne SHPP, Damayanthi BT, Fareena MHJ, Imbuldeniya V and Hemingway J 2007. Insecticide resistance in the tropical bed bug, *Cimex hemipterus*. Pestic Biochem Physiol 88: 102-107.

Kemper H (1936) Die Bettwanze und ihre Bekampfung. Kleintiere & Pelztiere 12: 1-107.

King F, Dick I and Evans P (1989) Parasitol Today 5: 100-102.

Klein N (2001) Don't let the bed bugs bite. The Times, 24 April 2001.

Liebold K, Schliemann-Willers S and Wollina U (2003) Disseminated bullous eruption caused by *Cimex lectularius*. J Eur Acad Dermatol Venerol 17: 461-463.

Leverkus M, Jochim RC, Schad S, Brocker EB, Andersen JF, Valenzuela JG and Trautmann A (2006) Bullous allergic hypersensitivity to bed bug bites mediated by IgE against salivary nitrophorin. J Invest Dermatol 126: 91-96.

Levinson HZ and Bar Ilan AR (1971) Assembling and alerting scents produced by the bed bug *Cimex lectularius* L. Experientia 27: 102-103.

Lindsay SW, Snow RW, Armstrong JRM and Greenwood BM (1989) Permethrin impregnated bednets reduce nuisance arthropods in Gambian houses. Med Vet Entomol 3: 377-383.

Lyons SF, Jupp PG and Schoub BD (1986) Survival of HIV in the common bed bug. Lancet 2(8497): 45.

Masetti M and Bruschi F (2007) Bed bug infestation recorded in central Italy. Parasitol Int 56: 81-83.

Medical Research Council (1942) Bed bugs and slums. Report on the Committee on bed bug infestation. HMSO, London.

Meek F (2003) Bed bugs bite back. Pest Contr Techn 31: 43-52.

Ministry of Health Reports No. 72. (1933) Report of Committee on the Eradication of Bed Bugs. HMSO, London.

Moore DJ and Miller DM (2006) Laboratory evaluations of insecticide product efficacy for control of *Cimex lectularius*. J Econ Entomol 99: 2080-2086.

Myamba J, Maxwell CA and Curtis CF (2002) Pyrethroid resistance in tropical bed bugs (*Cimex hemipterus*) associated with use of treated bed nets. Med Vet Entomol 16: 448-451.

Newberry K (1990) The tropical bed bug near the southernmost edge of it's range. Trans R Soc Trop Med Hyg 84: 745-747.

Newberry K (1991a) Field trials of bendiocarb, deltamethrin and fenitrothion to control DDT resistant bed bugs in KwaZulu, South Africa. Int Pest Contr May-June, 64-68.

Newberry K (1991b) Bed bug re-infestation rates in rural Africa. Med Vet Entomol 5: 503-505.

Ojo J (2006) Director's Regulations. How to control bed bug infestation. San Francisco Department of Public Health, California.

Parsons DJ (1955) Bed bug bite anaphylaxis misinterpreted as coronary occlusion. Ohio State Med J 51: 669.

Paul J and Bates J (2000) Is infestation with the common bedbug increasing? BMJ 320: 1141.

Pinto L (1999) Bed bugs…they're back. Pest Contr 67: 10-12.

Potter M, Romero A, Haynes KF and Hardebeck E (2007) Killing them softly. Pest Contr Techn, January, 24-32.

Reinhardt K and Siva-Jothy MT (2007) Biology of the Bed Bugs (Cimicidae). Ann Rev Entomol 52: 351-374.

Romero A, Potter MF, Potter DA and Haynes KF (2007) Insecticide resistance in the bed bug – a factor in the pest's sudden resurgence? J Med Entomol 44: 175-178.

Rust M (1996) Managing insecticide resistance in urban pests. Proceedings of the 2nd International Conference on Urban Pests. K.B. Wildey; editor. pp 11-15.

Sansom JE, Reynolds NJ and Peachey RD (1992) Delayed reaction to bed bug bites. Arch Dermatol 128: 272-273.

Silverman AL, Qu LH and Blow J (2001) Assessment of hepatitis B virus DNA and hepatitis C virus DNA in the common bed bug (*Cimex lectularius*) and the kissing bug (*Rhodnius prolixus*). Am J Gastroenterol 96: 2194-2198.

Simon EY (2004) Re-emergence of bed bugs creates pest control issues. Hotel Motel Manag 219: 42-44.

Stejskal V (2002). Metapopulation concept and the persistence of urban pests in buildings. Proceedings of the 4th International Conference on Urban Pests (Jones S, Zhai J and Robinson W (Eds)), pp75-85.

Temu EA, Minjas JN, Shiff CJ and Majala A (1999) Bed bug control by permethrin-impregnated bed nets in Tanzania. Med Vet Entomol 13: 457–459.

Titschack E (1930). Untersuchungen über das Wachstum, den Narhringsverbrauch und die Eierzeugung. 3. *Cimex lectularius* L. Z Morphol Oekol Tiere 17: 471-551.

Usinger RL (1966) Monograph of Cimicidae. Entomological Society of America, College Park, Maryland, USA. 585 pp.

Vall Mayans M, Hall AJ, Inskip HM, Lindsay SW, Chotard J, Mendy M and Whittle HC (1994) Do bed bugs transmit Hepatitis B? Lancet 343(8900): 761-763.

Webb PA, Happ CM, Maupin GO, Johnson BJB, Ou CY and Monath TP (1989) Potential for insect transmission of HIV: experimental exposure of *Cimex hemipterus* and *Toxorhynchites amboinensis* to human immunodeficiency virus. J Infect Dis 160: 170-177.

16. Houseflies, annoying and dangerous

Renate C. Smallegange and Cornelis J. den Otter

Abstract

Musca domestica (Diptera: Muscidae), the common housefly, is found in close conjunction with humans and their livestock around the world. The flies cause annoyance to people and are vectors of several diseases. Commonly-used control methods (cultural, chemical, biological, traps and baits) do not result in reducing fly populations to acceptable levels. Research on integrated control approaches is required. Models based on expected climate changes predict a substantial increase in fly populations over the next decades, which may result in a considerable expansion in the incidence of fly-borne diseases, thereby endangering health and well-being of people globally. This strengthens the need for intensive studies on the behaviour, ecology, and vectorial capacity of houseflies.

Keywords: *Musca domestica*, common housefly, biology, behaviour, ecology, pest, control

Introduction

Musca domestica L. (Diptera: Muscidae), the common housefly, is one of the most widespread fly species in the world. The insect belongs to a group of domestic flies often called 'filth flies'. *M. domestica* is an eusynanthropic, endophilic species, i.e. it lives closely with humans and is able to complete its entire life-cycle within residences of humans and their domestic animals. The fly can be found in human dwellings, in dairies, poultry houses, horse stables, food processing factories, landfills, and in other domesticated areas and buildings (Hewitt 1910, 1912, West 1951, Hansens 1963, Lillie and Goddard 1987, Axtell and Arends 1990, Kettle 1995, Howard 2001, Steenberg and Jespersen 2002, Lole 2005).

Biology

M. domestica is a non-biting fly that undergoes a complete metamorphosis in its development from egg to adult. Its developmental time depends on food availability and temperature, and may take as little as 10 days. The reproductive potential of the flies is very high. A female may lay four to six batches of eggs consisting of 75 to 150 eggs each at three- to four-day intervals. The pearly white, ca. 1.2 mm long eggs are deposited in clumps in cracks and crevices of a moist medium to protect them from desiccation. Manure and food waste are known to be the principal breeding media for houseflies. The eggs hatch within 24 hours after oviposition. The whitish, legless, saprophagous larvae (maggots) develop through three larval stages within approximately a week. Fully-grown larvae migrate to drier conditions and bury themselves into the substrate where they pupate. After approximately 5 days, the adults, about 7 mm long, emerge from the reddish brown or almost black puparia. Females are usually bigger than males. The sexes can easily be distinguished by the dorsal space between the eyes which is wider in females. Adult houseflies may live 15 to 30 days (Hewitt 1910, 1912, West 1951, Kettle 1995, Anonymous 2000, Steenberg and Jespersen 2002). Males may already mate on the day of their emergence. Mating readiness of females which, contrary to males, are monogamous, is highest when they are three days old (Saccà 1964). Oviposition takes place a few days after copulation. Overwintering occurs in the larval or pupal stage (Hewitt 1910, 1912, West 1951, Kettle 1995).

During warm weather the life cycle, from egg to egg, takes two to three weeks. Because of this high rate of development and the large numbers of eggs produced by a female, large populations can rapidly build up. In temperate regions of the world ten to twelve generations a year may occur. In colder regions breeding is restricted to the warmer months, resulting in four to six generations a year (Hewitt 1910, 1912, West 1951, Barnard and Geden 1993, Kettle 1995). Organic waste materials and the relatively high temperatures in livestock farms promote rapid development and continuous presence of flies (Howard and Wall 1996b).

The activity of adult houseflies is affected by several physical variables such as temperature, humidity, light intensity, air currents, barometric pressure, and electrostatic fields. The flies are diurnal and are more active when temperature is high and humidity low, but they tend to become sluggish when both temperature and humidity are high. They remain active at lower temperatures down to about 7 °C, regardless of humidity; their activity optimum lies around 33 °C. Near 45 °C symptoms of heat paralysis become evident (West 1951, Kettle 1995). Although insects are usually tolerant of variations in barometric pressure which occur in their home range, their activities may be markedly affected by abrupt barometric changes. Flight activity of houseflies has been seen to increase instantaneously when a storm approaches and there is a rapid fall in atmospheric pressure (West 1951). Electric fields also affect the behaviour of insects and have been shown to deter houseflies and change their circadian rhythm of locomotion (Perumpral *et al.* 1978, Engelmann *et al.* 1996, McGonigle and Jackson 2002).

The flies are positively anemotactic, i.e. they tend to fly upwind possibly induced by airborne odours which direct them towards suitable food or oviposition substrates. Flight velocity is about 2 m/s (West 1951).

Houseflies as a pest

Houseflies are usually found within 500 m from a breeding source but are able to disperse over distances up to 30 kilometres. They are a nuisance to people living near breeding sites, resulting in poor community relations and legal conflicts; they irritate people and animals by flying around or landing on them or on their food and leave regurgitation and faecal spots on surfaces (Anonymous 2000, Hinkle 2002, Steenberg and Jespersen 2002, Winpisinger *et al.* 2005). High population densities of houseflies in poultry farms may cause nuisance to such an extent that chickens reduce their egg production. Furthermore, the faeces of houseflies decreases the aesthetic appearance and value of the eggs (Howard and Wall 1996b). Economical losses caused by *M. domestica* in poultry houses are reported to exceed 60 million US dollars per year in the United States (Anonymous 1976). These losses do not at all counterbalance the fact that the flies may serve as a protein source for chickens (Ocio *et al.* 1979, El Boushy 1991).

In addition to causing annoyance, houseflies may be vectors of several diseases. Their movements between human and animal food, organic waste, garbage, faeces, manure, and other sources of filth on which they may feed and breed make them ideal transmitters of human and animal pathogens. About a hundred different pathogen species have been found in and on houseflies. The flies may transmit pathogens in different ways. The surface of their body, particularly the legs and proboscis, may be contaminated; and because houseflies suck food after it has been liquefied in regurgitated saliva, pathogens may be deposited onto food with the vomit drops. Thirdly, pathogens may pass through the gut of the flies and be deposited with their faeces. Washes from the surface of houseflies yielded total bacterial counts from 2.5 to 29.5 million per fly and in the digestive system the presence of 84 thousand to 2 million bacteria per fly was demonstrated (Ostrolenk and Welch 1942).

Pathogens that may be transmitted by houseflies are, for example, viruses causing diarrhoea, cholera bacteria, *Salmonella* spp. and *Escherichia coli* bacteria causing enteric infections, haemolytic streptococci, and agents of typhoid, diphtheria, tuberculosis, leprosy and yaws. In addition, they may carry cysts of Protozoa, including those causing amoebic dysentery, and the eggs of nematodes. A relation between the necrotic enteritis causing bacteria *Clostridium perfringens* isolated from houseflies and an increase in chicken mortality has recently been suggested by Dhillon *et al.* (2004). Flies are also suspected to transmit the bacteria *Campylobacter jejuni* to chickens (Hald *et al.* 2004). Every year millions of people in Europe are seized by severe gastro-enteritis caused by this bacterium. Finally, houseflies may be vectors and intermediate hosts of certain cestodes of poultry and nematodes of horses (Hewitt 1910, 1912, Ostrolenk and Welch 1942, West 1951, Saccà 1964, Kettle 1995, Grübel *et al.* 1997, Tan *et al.* 1997, Kurahashi *et al.* 1998, Li and Stutzenberger 2000, Sasaki *et al.* 2000, Fotedar 2001, Howard 2001, Steenberg and Jespersen 2002, Graczyk *et al.* 2005, Kinde *et al.* 2005, Barro *et al.* 2006). Fly larvae have also been found to carry pathogenic bacteria, both externally and internally (Banjo *et al.* 2005).

Control of houseflies

Several techniques are used to control houseflies. Screening of windows and doors with gauze screens is an effective method of preventing houseflies from entering buildings. By establishing air currents in doorways houseflies can also be kept outside (Carlson *et al.* 2006). Proper management of locations where manure or garbage is present may prevent fly infestation. Outdoors and indoors baits and traps, and chemical and biological methods are used to reduce fly populations to acceptable levels.

Cultural control

Removal, sanitation and ventilation of possible breeding sites are probably the most effective control methods. Moist manure is highly attractive to female houseflies (Smallegange 2003) and provides a good breeding and feeding site for larvae (Achiano and Giliomee 2005). Hence, regular removal of manure in poultry houses, diaries, pig-sties and horse stables is advisable. Dry manure does not only prevent oviposition and larval development but also favours the development of housefly predators and parasitoids. Ventilation systems that control temperature may be used to create an airflow over the manure surface to facilitate drying. In addition this may diminish oviposition in the manure because houseflies avoid relatively strong air currents (Geden *et al.* 1999).

Garbage containers should have tight-fitting lids and should be cleaned regularly. Manure and spilled feed have to be removed at least twice a week and waste disposal sites covered with a layer of soil or other inorganic material every week (Kettle 1995, Anonymous 2000, Hinkle 2002, Steenberg and Jespersen 2002).

Traps and baits

A simple way to catch flies indoors is the use of sticky papers suspended from the ceiling or a lamp. However, apart from their unaesthetic appearance, they are unhygienic and not suitable for controlling fly pests. Large sticky traps may be effective for application at farms, but their use is often limited because of rapid accumulation of dust on the sticky material (Kaufman *et al.* 2001, 2005b). Despite their disadvantages, sticky traps are commonly used for indoor monitoring purposes (see Chapter 19; Geden *et al.* 1999, Kaufman *et al.* 2001, 2005b, Cornelius 2002, Jacobs *et al.* 2003, Lole 2005).

Light sources attract flies and other insects. Sources emitting light in the ultraviolet (UV: emittance peaks below 400 nm) attract much higher numbers of houseflies than light sources emitting wave lengths longer than 400 nm; within the UV region no preferences are found (Cameron 1938, Deay and Taylor 1962, Thimijan *et al.* 1973, Roberts *et al.* 1992, Smallegange 2003, Smallegange and Den Otter, unpublished data). Therefore, most commercially available light traps consist of UV lamps in front of which high-voltage electrocutor grids are mounted. However, outdoors and in illuminated areas the effectiveness of these traps is very low. In addition, we found that flies younger than 3 days of age are hardly or not at all attracted to the light. Even in a dark chamber only a maximum of 60% of the older flies is attracted to an UV lamp, which does not seem to be sufficient to reduce fly populations to acceptable levels (Smallegange 2003, Smallegange and Den Otter, unpublished data).

Since UV-light traps also attract other insect species, they often kill beneficial insects (Nabli *et al.* 1999, Mohan 2002). An additional disadvantage of electrocuting light traps is that bacteria and viruses present on the flies' bodies are spread into the air upon electrocution of the flies (Urban and Broce 2000). Moreover, enteric bacteria may survive to up to five weeks after the flies have been killed (Cooke *et al.* 2003). This implies that the traps should be designed in such a way that all parts of the electrocuted flies stay in or on the trap and cannot disperse from it into the surroundings. In addition, the traps should be emptied on a regular basis.

Oviposition substrates and food sources, especially putrefying and fermenting substances emanating amines, aldehydes, ketones and alcohols, and dairy products and sugar-containing substances have been shown to attract houseflies (Awati and Swaminath 1920, Brown *et al.* 1961, Künast and Günzrodt 1981, Skoda and Thomas 1993 in Cossé and Baker 1996, Smallegange 2003, Smallegange and Den Otter, unpublished data). However, commercially available baits show variable and often contradictory results (Browne 1990). This may result from the fact that in areas where houseflies are usually found, the flies are confronted with several olfactory stimuli, which may influence the effectiveness of odour traps negatively. Moreover, the attractiveness of odour sources to the flies depends on the physiological state of the flies. Although, in contrast to UV traps, odour-baited traps are equally attractive to mature as well as young flies, male and female flies and food-deprived and well-fed flies may respond differently to odours (Smallegange 2003, Smallegange and Den Otter, unpublished data).

(Z)-9-tricosene has been identified as a component of the female sex pheromone of *M. domestica* (Carlson *et al.* 1971). This compound, also known as muscalure, has been found to be attractive not only to male, but also to female houseflies in laboratory and field trials (Carlson *et al.* 1971, Van Oosten and Persoons 1981, 1982, 1983, Chapman *et al.* 1998, 1999, Noorman 2001, Hanley *et al.* 2004). Muscalure is often used to lure houseflies to electrocuting light traps or to toxic targets, although it was shown to be only a weak and short-range attractant (Morgan *et al.* 1974, Fletcher and Bellas 1988, Smallegange 2003, Van Deventer and Griepink 2006). Furthermore, it was shown that a combination of UV light and attractive odours decreases the luring capacity of the odours considerably (Smallegange 2003, Smallegange and Den Otter, unpublished data).

Finally, electrostatic charge built up on traps may deter houseflies. It has been shown that in a choice chamber houseflies prefer an uncharged cage over a highly charged one (Perumpral *et al.* 1978, McGonigle and Jackson 2002).

Chemical control

Application of insecticides may initially be effective, but muscids, such as houseflies, readily develop resistance to persistent insecticides either because enzymes enable the flies to break down the insecticides or because behavioural adaptations enable the flies to avoid these substances (Scott and Georghiou 1985, Meyer *et al.* 1987, Kettle 1995, Pospischil *et al.* 1996, Keiding 1999, Anonymous 2000, Scott *et al.* 2000, Akiner and Çağlar 2006). Because only a limited number of genetic factors are involved in the development of resistance which are not strictly bound to specific molecules, cross resistance to novel insecticides exists (Plapp 1986). In addition, these substances are toxic to non-target, beneficial animals.

A higher specificity was aimed with the development of insect growth regulators (IGRs). IGRs show some selectivity (Miyamota *et al.* 1993) and can be divided into three categories: juvenile hormones which suppress metamorphosis, chitin synthesis inhibitors which disrupt moulting and pupation, and cyromazine which interferes with moulting by disrupting the sclerotisation process (Graf 1993, Howard and Wall 1996a). However, widespread resistance against IGRs has developed (Silhacek *et al.* 1976, Pap and Farkas 1994).

Not only the increase of tolerance and resistance of flies to insecticides but also the increasing costs of the use of insecticides and their toxicity to other organisms make them less desirable for fly control. Moreover, it appears to be difficult and expensive to develop new insecticides (Pickens and Miller 1987, Scott *et al.* 2000, Wall 2002).

Biological control

Many natural enemies may kill housefly eggs, larvae or adults: entomopathogenic fungi (e.g. *Entomophthora muscae*, *Beauveria bassiana*), bacteria (e.g. *Brevibacillus laterosporus*) and nematodes; parasitoids (e.g. *Muscidifurax raptorellus*, *Spalangia cameroni*); and predatory mites (e.g. *Macrocheles muscaedomesticae*), flies (e.g. *Hydrotaea aenescens*), beetles (histerid and staphylinid species), spiders and birds. Mites and beetles pray on eggs and first instar larvae, predatory fly larvae eat housefly larvae. Various parasitoids oviposit on fly pupae which serve as food for the parasitoid larvae. Bacteria and fungi may also reduce fecundity or longevity of *M. domestica* (Geden *et al.* 1993, Glofcheskie and Surgeoner 1993, Møller 1993, Kettle 1995, Farkas and Pap 1996, Geden 1999, Hogsette and Jacobs 1999, Skovgård and Jespersen 1999, Anonymous 2000, Hinkle 2002, Hogsette *et al.* 2002, Skovgård 2004, 2006, Skovgård and Nachman 2004, Kaufman *et al.* 2005a, Lecuona *et al.* 2005, Achiano and Giliomee 2006a, 2006b, Geden and Hogsette 2006, Rossi and Godoy 2006, Ruiu *et al.* 2006).

Several authors examined these biological agents for control purposes. It appeared that release of the fungus *E. muscae* does not significantly reduce housefly populations (Geden *et al.* 1993), possibly because female flies are hardly attracted to infected conspecifics (Møller 1993) and the flies readily recover from the infection at higher temperatures (Hajek and St Leger 1994). *B. bassiana* treatments seem to be more promising since they were found to be able to decrease fly numbers to a significantly greater extent than the insecticide pyrethrin (Kaufman *et al.* 2005a). Muscovy ducks may also significantly affect housefly populations (Glofcheskie and Surgeoner 1993). The usefulness of the dump fly, *H. aenescens*, appears questionable since it does not seem to be able to keep housefly populations low when no additional control methods are used. In addition, its establishment and development depend on manure type and moisture, and it is more susceptible to insecticides than *M. domestica* (Farkas and Pap 1996, Hogsette and Jacobs 1999, Hogsette *et al.* 2002). Likewise the efficiency of predatory mites and beetles, and parasitoids greatly depends on moisture, composition and location of the manure. Finally, the effectiveness of the pupal parasitoid *S. cameroni* in stables appears to have unpredictable and variable effects on housefly populations (Skovgård 2004, 2006, Skovgård and Nachman 2004, Geden and Hogsette 2006).

Since in several cases significant reductions of housefly populations with natural enemies have been achieved, biological control appears to be an efficient alternative for or addition to chemical control. It may be expected that combinations of natural enemies will be more successful when circumstances are heterogeneous (Geden 1999, Skovgård and Jespersen 1999, Achiano and Giliomee 2006b, Skovgård 2006). However, it is evident that more research is needed to be able to design effective control programmes adjusted to local demands.

Integrated control

Integrated control programmes consider all available control tactics and evaluate the interaction among them. They use cultural practices, traps, natural enemies, messenger chemicals, not individually exclusive but in the best combination imaginable. They may also use pesticides, but only after natural control factors indicate a need. The programmes are dynamic, and may vary from field to field, from year to year, because no two pest situations are the same. Hence, to achieve successful housefly reductions in locations with fly problems, it is advisable to combine (some of) the above mentioned strategies into an integrated programme. Which methods have to be used depends on the local situation, density and age composition of the fly population, the degree of annoyance, economic costs, and the availability of biological control agents. As for the latter, very moist manure, thorough removal of manure and short fly reproduction cycles hamper establishment of fly predators and parasitoids.

Computer simulation models and expert systems have been developed for pest population management, including housefly pests in livestock and poultry facilities. The programme Fly Management Simulator incorporates chemical control measures, manure manipulation, and housefly predators and parasitoids (Axtell 1999, see also http://cipm.ncsu.edu/ent/vetent/expert.html, accessed 29 August 2007).

To keep track of housefly population growth and to know when special measures are needed, several monitoring tools are available such as scudder grids, baited jug traps, spot cards, and various types of sticky traps (Pickens and Miller 1987, Axtell and Arends 1990, Axtell 1999, Anonymous 2000, Geden and Hogsette 2001, Anonymous 2003, Jacobs *et al.* 2003, Williams 2003, Geden 2005, 2006, Kaufman *et al.* 2005b, Lole 2005). These devices are described in more detail in Chapter 19.

Conclusion

Adult houseflies are not only annoying for people and animals but also carry many pathogens which may cause a reduction in animal production resulting in considerable financial losses.

Moreover, flies are involved in the transmission of human diseases. In spite of this, basic knowledge of fly behaviour, ecology and vectorial capacity is still lacking.

Worldwide much money is spent on filth fly control. However, more scientific studies should focus on this topic. Until now it is hardly understood why biological control of houseflies sometimes fails, and sometimes is successful. In-depth studies on putative biological control agents have never been undertaken. The effects of sanitation and manure manipulation have hardly been investigated; these methods are time-consuming and not always as effective as expected. So far, the impact of traps and baits on fly populations is not satisfactory. To increase their effectiveness, the effects of trap design and siting, and of various visual and olfactory attractants and repellents, should be investigated in more detail. The use of chemical insecticides is restricted and limited due to the development of resistance and the negative effects on beneficial arthropods. Research on integrated control is required. Geden and Hogsette (2001) published a list of recommended research topics for integrated control applications. In addition, a shift in mind-set of people dealing with housefly control may be necessary to get these programmes accepted. Also a better insight in intervention thresholds is required (Wall 2007).

In 1993 the European Union decreed (Directive 93/43/EEC) that member states should adopt the main principles of the so-called Hazard Analysis of Critical Control Points (HACCP). HACCP is a total quality management system for food safety which, if implemented fully, would result in a minimum level of food contamination (Howard 1999). This requires that insect pests have to be considerably reduced in number or even totally eliminated, because only one single insect may transmit pathogens. With the present control methods (local) elimination cannot be achieved.

It is now evident that global temperature is rising (IPCC 2007). Models based on expected climate changes predict a substantial increase in *M. domestica* populations in the next decades, which may result in a considerable expansion in the incidence of fly-borne diseases (Goulson *et al.* 2005). This will endanger health and life of people all over the world, which strengthens the need of intensive studies on houseflies, which are and will not only be annoying but also dangerous.

References

Achiano KA and Giliomee JH (2005) Diptera breeding in poultry manure and the abiotic factors affecting their numbers. Afr Entomol 13: 239-248.

Achiano KA and Giliomee JH (2006a) Feeding behaviour of the potential predators of the house flies, *Musca domestica* L. and *Fannia canicularis* (L.) (Diptera: Muscidae). Afr Entomol 14: 69-75.

Achiano KA and Giliomee JH (2006b). House fly predators in poultry manure and environmental factors affecting them. Afr Entomol 14: 349-355.

Akiner MM and Çağlar SS (2006) The status and seasonal changes of organophosphate and pyrethroid resistance in Turkish populations of the house fly, *Musca domestica* L. (Diptera: Muscidae). J Vector Ecol 31: 58-64.

Anonymous (1976) Control of insects affecting livestock. USDA Agricultural Research Service Natural Research Program no. 2048.

Anonymous (2000) Pest Management Recommendations for Poultry. A Cornell and Penn State Cooperative Extension Publication, Pennsylvania State University. 16 pp.

Anonymous (2003) The Spot Card Technique for sampling *Musca domestica*. Report of Santa Clara County Vector Control District. 18 pp.

Awati PR and Swaminath CS (1920) Bionomics of houseflies. III. A preliminary note on attraction of houseflies to certain fermenting and putrefying substances. Indian J Med Res 7: 560-567.

Axtell RC (1999) Poultry Integrated Pest Management: Status and Future. Integr Pest Manag Rev 4: 53-73.

Axtell RC and Arends JJ (1990) Ecology and management of arthropod pests of poultry. Ann Rev Entomol 35: 101-126.

Banjo AD, Lawal OA and Adeduji OO (2005) Bacteria and fungi isolated from housefly (*Musca domestica* L.) larvae. Afr J Biotechnol 4: 780-784.

Barnard DR and Geden CJ (1993) Influence of larval density and temperature in poultry manure on development of the house fly (Diptera: Muscidae). Environ Entomol 22: 971-977.

Barro N, Aly S, Tidiane OCA and Sababénédjo TA (2006) Carriage of bacteria by proboscises, legs, and feces of two species of flies in street food vending sites in Ouagadougou, Burkina Faso. J Food Prot 69: 2007-2010.

Brown AWA, West AS and Lockley AS (1961) Chemical attractants for the adult house fly. J Econ Entomol 54: 670-674.

Browne LE (1990) The use of pheromones and other attractants in house fly control. In: L.R. Ridgway, R.M. Silverstein and M.N. Inscoe (eds.), Behavior-modifying chemicals for insect management. Marcel Dekker, Inc., New York, pp 531-537.

Cameron JWM (1938) The reactions of the housefly, *Musca domestica* Linn., to light of different wave-lengths. Can J Res 16: 307-342.

Carlson DA, Hogsette JA, Kline DL, Geden CJ and Vandermeer RK (2006) Prevention of mosquitoes (Diptera: Culicidae) and house flies (Diptera: Muscidae) from entering simulated aircraft with commercial air curtain units. J Econ Entomol 99: 182-193.

Carlson DA, Mayer MS, Silhacek DL, James JD, Beroza M and Bierl BA (1971) Sex attractant pheromone of the house fly: isolation, identification, and synthesis. Science 174: 76-78.

Chapman JW, Knapp JJ and Goulson D (1999) Visual responses of *Musca domestica* to pheromone impregnated targets in poultry units. Med Vet Entomol 13: 132-138.

Chapman JW, Knapp JJ, Howse PE and Goulson D (1998) An evaluation of (Z)-9-tricosene and food odours for attracting houseflies, *Musca domestica*, to baited targets in deep-pit poultry units. Entomol Exp Appl 89: 183-192.

Cooke EA, O'Neill G and Anderson M (2003) The survival of ingested *Serratia marcescens* in houseflies (*Musca domestica* L.) after electrocution with electric fly killers. Curr Microbiol 46: 151-153.

Cornelius ML (2002) Sticky traps. In: D. Pimentel (ed.), Encyclopedia of Pest Management. Marcel Dekker, Inc., New York, pp 795-797.

Cossé AA and Baker TC (1996) House flies and pig manure volatiles: Wind tunnel behavioral studies and electrophysiological evaluations. J Agric Entomol 13: 301-317.

Deay HO and Taylor JG (1962) Response of the house fly, *Musca domestica* L., to electric lamps. Proc Indiana Acad Sci 72: 161-166.

Dhillon AS, Roy P, Lauerman L, Schaberg D, Weber S, Bandli D and Wier F (2004) High mortality in egg layers as a result of necrotic enteritis. Avian Dis 48: 675-680.

El Boushy AR (1991) House-fly pupae as poultry manure converters for animal feed: A review. Bioresource Technol 38: 45-49.

Engelmann W, Hellrung W and Johnsson A (1996) Circadian locomotor activity of Musca flies: Recording method and effects of 10 Hz square-wave electric fields. Bioelectromagn 17: 100-110.

Farkas R and Papp L (1996) Susceptibility of *Hydrotaea aenescens* (Wiedemann) (Diptera: Muscidae) to selected insecticides and its importance in integrated fly management. Parasitol Res 82: 170-173.

Fletcher BS and Bellas TE (1988) Pheromones in Diptera. In: E.D. Morgan and N.B. Mandava (eds.). CRC Handbook of natural pesticides. IV. Pheromones. Part B. CRC Press Inc., Florida, pp 1-57.

Fotedar R (2001) Vector potential of houseflies (*Musca domestica*) in the transmission of Vibrio cholerae in India. Acta Trop 78: 31-34.

Geden CJ (1999) Host location by house fly (Diptera: Muscidae) parasitoids in poultry manure at different moisture levels and host densities. Environ Entomol 28: 755-760.

Geden CJ (2005) Methods for monitoring outdoor populations of house flies, *Musca domestica* L. (Diptera: Muscidae). J Vector Ecol 30: 244-250.

Geden CJ (2006) Visual targets for capture and management of house flies, *Musca domestica* L. J Vector Ecol 31: 152-157.

Geden CJ and Hogsette JA (eds.) (2001) Research and Extension Needs for Integrated Pest Management for Arthropods of Veterinary Importance. Proceedings of a Workshop in Lincoln, Nebraska, April 12-14, 1994. Last Updated October 2001.

Geden CJ and Hogsette JA (2006) Suppression of house flies (Diptera: Muscidae) in Florida poultry houses by sustained releases of *Muscidifurax raptorellus* and *Spalangia cameroni* (Hymenoptera: Pteromalidae). Environ Entomol 35: 75-82.

Geden CJ, Hogsette JA and Jacobs RD (1999) Effect of airflow on house fly (Diptera: Muscidae) distribution in poultry houses. J Econ Entomol 92: 416-420.

Geden CJ, Steinkraus DC and Rutz DA (1993) Evaluation of two methods for release of *Entomophtora muscae* (Entomophtorales: Entomophtoraceae) to infect house flies (Diptera: Muscidae) on dairy farms. Environ Entomol 20: 1201-1208.

Glofcheskie BD and Surgeoner GA (1993) Efficacy of Muskovy ducks as an adjunct for house fly (Diptera: Muscidae) control in swine and dairy operations. J Econ Entomol 86: 1686-1692.

Goulson D, Derwent LC, Hanley ME, Dunn DW and Abolins SR (2005) Predicting calyptrate fly populations from the weather, and probable consequences of climate change. J Appl Ecol 42: 795-804.

Graczyk TK, Knight R and Tamang L (2005) Mechanical transmission of human protozoan parasites by insects. Clin Microbiol Rev 18: 128-132.

Graf J-F (1993) The role of insect growth regulators in arthropod control. Parasitol Today 9: 471-474.

Grübel P, Hoffman JS, Chong FK, Burstein NA, Mepani C and Cave DR (1997) Vector potential of houseflies (*Musca domestica*) for *Helicobacter pylori*. J Clin Microbiol 35: 1300-1303.

Hajek AE and St Leger RJ (1994) Interactions between fungal pathogens and insect hosts. Ann Rev Entomol 39: 293-322.

Hald B, Skovgård H, Duong Bang D, Pedersen K, Dybdahl J, Jespersen JB and Madsen M (2004) Flies and Campylobacter infection of broiler flocks. Emerg Infect Dis 10: 1490-1492.

Hanley ME, Dunn DW, Abolins SR and Goulson D (2004) Evaluation of (Z)-9-tricosene baited targets for control of the housefly (*Musca domestica*) in outdoor situations. J Appl Entomol 128: 478-482.

Hansens EJ (1963) Fly populations in dairy barns. J Econ Entomol 56: 842-844.

Hewitt CG (1910) The housefly, *Musca domestica* Linnaeus. A study of its structure, development, bionomics and economy. Publications of the University of Manchester. Biological series- no.1. 195 pp.

Hewitt CG (1912) House-flies and how they spread disease. The Cambridge Manuals of Science and Literature. Cambridge University Press. 122 pp.

Hinkle NC (2002) Poultry pest management (arthropods). In: D. Pimentel (ed.), Encyclopedia of Pest Management. Marcel Dekker, Inc., New York, pp 657-660.

Hogsette JA, Farkas R and Coler RR (2002) Development of *Hydrotaea aenescens* (Diptera: Muscidae) in manure of unweaned dairy calves and lactating cows. J Econ Entomol 95: 527-530.

Hogsette JA and Jacobs RD (1999) Failure of *Hydrotaea aenescens*, a larval predator of the housefly, *Musca domestica*, to establish in wet poultry manure on a commercial farm in Florida, USA. Med Vet Entomol 13: 349-354.

Howard J (2001) Nuisance flies around a landfill: Patterns of abundance and distribution. Waste Manag Res 19: 308-313.

Howard J and Wall R (1996a) Autosterilization of the housefly, *Musca domestica*, using the chitin synthesis inhibitor triflumuron on sugar-baited targets. Med Vet Entomol 10: 97-100.

Howard J and Wall R (1996b) Control of the house fly, *Musca domestica*, in poultry units: Current techniques and future prospects. Agricul Zool Rev **7**: 247-265.

Howard M (1999) HACCP and food hygiene law in England: the implications for pest control strategies. In: Wm. H. Robinson, F. Rettich & G.W. Rambo (eds.), Proceedings of the 3rd International Conference on Urban Pests, Prague, pp 311-315.

IPCC (2007) Summary for Policymakers. In: Climate Change 2007: The Physical Science Basis. Contribution of Working Group 1 to the Fourth Assessment Report of the Intergovernmental Panel on Climate Change (Solomon S, Qin D, Manning M, Chen Z, Marquis M, Averyt KB, Tignor M and Miller HL (eds.). Cambridge University Press, Cambridge, United Kingdom and New York, NY, USA.

Jacobs RD, Hogsette JA and Miller RW (2003) Using sticky cards to monitor fly populations in poultry houses. Series of Animal Science Department, Florida Cooperative Extension Service, Institute of Food and Agricultural Sciences, University of Florida. http://edis.ifas.ufl.edu (Accessed 27 August 2007).

Kaufman PE, Reasor C, Rutz DA, Ketzis JK and Arends JJ (2005a) Evaluation of *Beauveria bassiana* applications against adult house fly, *Musca domestica*, in commercial caged-layer poultry facilities in New York state. Biol Contr 33: 360-367.

Kaufman PE, Rutz DA and Frisch S (2001) Sticky traps for large scale house fly (Diptera: Muscidae) trapping in New York poultry facilities. J Agricul Urban Entomol 18: 43-49.

Kaufman PE, Rutz DA and Frisch S (2005b) Large sticky traps for capturing house flies and stable flies in dairy calf greenhouse facilities. J Dairy Sci 88: 176-181.

Keiding J (1999) Review of the global status and recent development of insecticide resistance in field populations of the housefly, *Musca domestica* (Diptera: Muscidae). Bull Entomol Res 89: S7-S67.

Kettle DS (1995) Medical and Veterinary Entomology. Second edition. CAB International. University Press, Cambridge. 725 pp.

Kinde H, Castellan DM, Kerr D, Campbell J, Breitmeyer R and Ardans A (2005) Longitudinal monitoring of two commercial layer flocks and their environments for *Salmonella enterica* serovar enteritidis and other salmonellae. Avian Dis 49: 189-194.

Künast C and Günzrodt C (1981) Vergleichende Laboruntersuchungen über Lockstoffe und Köder bei der Stubenfliege (*Musca domestica* L.). Anzeiger Schädlingskunde, Pflanzensch, Umweltsch 54: 131-135.

Kurahashi H, Hayashi T, Moribayashi A, Kobayashi M and Agui N (1998) The house-fly - a mechanical vector for verotoxin-producing E.coli O157: H7 associated with some outbreaks of food poisoning in Japan. In: Proceedings 4th International Congress of Dipterology, Oxford, 116 pp.

Lecuona RE, Turica M, Tarocco F and Crespo DC (2005) Microbial control of *Musca domestica* (Diptera: Muscidae) with selected strains of *Beauveria bassiana*. J Med Entomol 42: 332-336.

Li S and Stutzenberger FJ (2000) The housefly (*Musca domestica*) as a possible vector for *Helicobacter pylori* at agricultural sites. Int J Environ Health Res 10: 141-152.

Lillie TH and Goddard J (1987) Operational testing of electrocutor traps for fly control in dining facilities. J Econ Entomol 80: 826-829.

Lole MJ (2005) Nuisance flies and landfill activities: an investigation at a West Midlands landfill site. Waste Manag Res 23: 420-428.

McGonigle DF and Jackson CW (2002) Effect of surface material on electrostatic charging of houseflies (*Musca domestica* L). Pest Manag Sci 58: 374-380.

Meyer JA, Georghiou GP and Hawley MK (1987) House fly (Diptera: Muscidae) resistance to permethrin on southern California dairies. J Econ Entomol 80: 636-640.

Mohan S (2002) Light traps. In: D. Pimentel (ed.), Encyclopedia of Pest Management. Marcel Dekker, Inc., New York, pp 455-459.

Møller AP (1993) A fungus infecting domestic flies manipulates sexual behaviour of its host. Behav Ecol Sociobiol 33: 403-407.

Morgan PB, Gilbert IH and Fye RL (1974) Evaluation of (Z)-9-tricosene for attractancy for *Musca domestica* in the field. Florida Entomol 57: 136-140.

Miyamota J, Hirano M, Takimoto Y and Hatakoshi M (1993) Insect growth regulators for pest control, with emphasis on juvenile hormone analogs. Present status and future prospects. In: S.O. Duke, J.J. Menn & J.R. Plimmer (eds.), Pest Control with Enhanced Environmental Safety, ACS Symposium Series 524, San Francisco, pp 144-168.

Nabli H, Bailey WC and Necibi S (1999) Beneficial insect attraction to light traps with different wavelengths. Biol Contr 16: 185-188.

Noorman N (2001) Pheromones of the housefly. A chemical and behavioural study. PhD Thesis University of Groningen, The Netherlands. 127 pp.

Ocio E, Vinaras R and Rey JM (1979) House fly larvae meal grown on municipal organic waste as a source of protein in poultry diets. Anim Feed Sci Technol 4: 227-231.

Ostrolenk M and Welch H (1942) The common house fly (*Musca domestica*) as a source of pollution in food establishments. Food Res 7: 192-200.

Pap L and Farkas R (1994) Monitoring of resistance of insecticides in house fly (*Musca domestica*) populations in Hungary. Pestic Sci 40: 245-258.

Perumpral JV, Earp UF and Stanley JM (1978) Effects of electrostatic-field on locational preference of house-flies (Diptera: Muscidae) and flight activities of cabbage loopers (Lepidoptera: Noctuidae). Environ Entomol 7: 482-486.

Pickens LG and Miller RW (1987) Techniques for trapping flies on dairy farms. J Agricul Entomol 4: 305-313.

Plapp FW Jr. (1986) Genetics and biochemistry of insecticide resistance in arthropods: Prospects for the future. In: Pesticide Resistance: Strategies and Tactics for Management. National Academy Press, Washington, pp 74-86.

Pospischil R, Szomm K, Londershausen M, Schröder I, Turberg A and Fuchs R (1996) Multiple resistance in the larger house fly *Musca domestica* in Germany. Pestic Sci 48: 333-341.

Roberts AE, Syms PR and Goodman LJ (1992) Intensity and spectral emission as factors affecting the efficacy of an insect electrocutor trap towards the house-fly. Entomol Exp Appl 64: 259-268.

Rossi MN and Godoy WAC (2006) Prey choice by *Nesticodes rumpes* (Araneae, Theridiidae) on *Musca domestica* (Diptera, Muscidae) and *Dermestes ater* (Coleoptera, Dermestidae). J Arachnol 34: 186-193.

Ruiu L, Delrio G, Ellar DJ, Floris I, Paglietti B, Rubino S and Satta A (2006) Lethal and sublethal effects of *Brevibacillus laterosporus* on the housefly (*Musca domestica*). Entomol Exp Appl 118: 137-144.

Saccà G (1964) Comparative bionomics in the genus Musca. Ann Rev Entomol 9: 341-358.

Sasaki T, Kobayashi M and Agui N (2000) Epidemiological potential of excretion and regurgitation by *Musca domestica* (Diptera: Muscidae) in the dissemination of Escherichia coli O157: H7 to food. J Med Entomol 37: 945-949.

Scott JG, Alefantis TG, Kaufman PE and Rutz DA (2000) Insecticide resistance in house flies from caged-layer poultry facilities. Pest Manag Sci 56: 147-153.

Scott JG and Georghiou GP (1985) Rapid development of high-level permethrin resistance in a field-collected strain of the housefly (Diptera: Muscidae) under laboratory selection. J Econ Entomol 78: 316-319.

Silhacek DL, Oberlander H and Zettler JL (1976) Susceptibility of malathion-resistant strains of *Plodia interpunctella* to juvenile hormone treatments. J Stored Prod Res 20: 201-203.

Skovgård H (2004) Sustained releases of the pupal parasitoid *Spalangia cameroni* (Hymenoptera: Pteromalidae) for control of house flies, *Musca domestica* and stable flies *Stomoxys calcitrans* (Diptera: Muscidae) on dairy farms in Denmark. Biol Contr 30: 288-297.

Skovgård H (2006) Search efficiency of *Spalangia cameroni* and *Muscidifurax raptor* on *Musca domestica* pupae in dairy cattle farms in Denmark. BioContr 51: 49-64.

Skovgård H and Jespersen JB (1999) Activity and relative abundance of hymenopterous parasitoids that attack puparia of *Musca domestica* and *Stomoxys calcitrans* (Diptera: Muscidae) on confined pig and cattle farms in Denmark. Bull Entomol Res 89: 263-269.

Skovgård H and Nachman G (2004) Biological control of house flies *Musca domestica* and stable flies *Stomoxys calcitrans* (Diptera: Muscidae) by means of inundative releases of *Spalangia cameroni* (Hymenoptera: Pteromalidae). Bull Entomol Res 94: 555-567.

Smallegange RC (2003) Attractiveness of different light wavelengths, flicker frequencies and odours to the housefly (*Musca domestica* L.). PhD Thesis University of Groningen, The Netherlands. 140 pp.

Steenberg T and Jespersen JB (2002) Control of house flies and 'filth' flies. In: D. Pimentel (ed.), Encyclopedia of Pest Management. Marcel Dekker, Inc., New York, pp 144-147.

Tan SW, Yap KL and Lee HL (1997) Mechanical transport of rotavirus by the legs and wings of *Musca domestica* (Diptera: Muscidae). J Med Entomol 34: 527-531.

Thimijan RW, Pickens LG and Morgan NO (1973) Responses of the house fly, stable fly and face fly to electromagnetic radiant energy. J Econ Entomol 66: 1269-1270.

Urban JE and Broce A (2000) Killing of flies in electrocuting insect traps releases bacteria and viruses. Curr Microbiol 41: 267-270.

Van Deventer P and Griepink FC (2006) Kort literatuuronderzoek stand van zaken Lure & Kill. Nota 407. Plant Research International B.V., Wageningen. 28 pp.

Van Oosten AM and Persoons CJ (1981) Onderzoek naar de toepasbaarheid van feromonen bij de bestrijding van de huisvlieg (*Musca domestica*) onder praktijkomstandigheden. TNO-rapport. 20 pp.

Van Oosten AM and Persoons CJ (1982) Onderzoek naar de toepasbaarheid van feromonen bij de bestrijding van de huisvlieg (*Musca domestica*) onder praktijkomstandigheden. TNO-rapport. 10 pp.

Van Oosten AM and Persoons CJ (1983) Onderzoek naar de toepasbaarheid van feromonen bij de bestrijding van de huisvlieg (*Musca domestica*) onder praktijkomstandigheden. TNO-rapport. 19 pp.

Wall R (2002) Livestock pest management (insects). In: D. Pimentel (ed.), Encyclopedia of Pest Management. Marcel Dekker, Inc., New York, pp 460-462.

Wall R (2007) Ectoparasites: Future challenges in a changing world. Vet Parasitol, in press.

West LS (1951) The housefly. Its natural history, medical importance, and control. Comstock Publishing Company, Ithaca, New York. 584 pp.

Williams RE (2003) Livestock & Poultry. Control of poultry pests. Purdue University Cooperative Extension Service. 8 pp. http://www.ces.purdue.edu/extmedia (Accessed 25 August 2007).

Winpisinger KA, Ferketich AK, Berry RL and Moeschberger ML (2005) Spread of *Musca domestica* (Diptera: Muscidae), from two caged layer facilities to neighboring residences in rural Ohio. J Med Entomol 42: 732-738.

17. Novel strategies in head lice control

Hilde Lapeere, Jean-Marie Naeyaert†,[13] and Robert H. Vander Stichele

Abstract

Head lice are persistent ectoparasites of humans, affecting children as well as adults on all continents. Even though the literature on the subject is extensive, there is still controversy on many crucial elements such as diagnostic procedures and treatment options. In this paper, an overview of different diagnostic methods is given. Furthermore, topical and systemic treatments, mechanical treatments with combs and devices, and alternative treatments such as suffocating agents and essential oils are discussed. One of the elements influencing the treatment of choice is resistance to topical chemical treatments, which has been observed in various European countries. The removal of nits, the eggshells of head lice, is often done for cosmetic purposes. An objective method to measure forces needed to remove nits is described. Finally we discuss the community framework of co-ordinated detection and treatment in which treatments of pediculosis need to be implemented.

Keywords: *Pediculus humanus capitis*, pediculicide, resistance, wet combing, lice infestations, co-ordinated approach

Introduction

Pediculosis capitis is a highly prevalent infestation in children between three and twelve years of age, with sporadic local epidemics. Adolescents and adults can also be infested or serve as reservoirs. Findings about the prevalence of head lice differ from country to country, depending on the diagnostic criteria and measurement methods used. In industrialised countries, active infestation rates of around 10% are frequently reported in school children (Counahan *et al.* 2004, Mumcuoglu *et al.* 1990, Vander Stichele *et al.* 2002, Willems *et al.* 2005). Epidemics with even higher infestation rates are not exceptional (Downs *et al.* 2000b, Speare and Buettner 1999). Infestations often cause physical discomfort and sometimes clinical lesions, but the medical importance derives mainly from the psychological, financial and social impact of the infestation. Head lice epidemics cause social distress at school and in the family, are associated with feelings of shame and disgust in infested individuals, and may trigger stigmatisation in communities (Chunge *et al.* 1991, Koch *et al.* 2001). The direct costs for pediculicides and louse combs associated with pediculosis are impressive. Estimates for the costs of pediculicides in the USA (with a total population of almost 300 million) are between $240 million (Hansen and O'Haver 2004) and $367 million. In Belgium (total population just over 10 million) over €4 million was spent on pediculicides in 2002 (IMS Health Care 2003, West 2004). Pediculosis has a large financial and social impact on the child, its family and the community, necessitating a specific and well-designed public health approach.

Finding the appropriate treatment for pediculosis is a real challenge. The success rate of the treatment depends not only on its intrinsic efficacy but also on the patient's preference, which might be influenced by financial issues, time investment, motivation, concern about toxicity, application skills needed, and consideration of environmental consequences. Topical treatments with pediculicides used to be the cornerstone of the treatment of pediculosis. Other topical

[13] This chapter is dedicated to the late Prof. Naeyaert, who passed away in Ghent, Belgium, on 26 July 2007.

treatments such as natural products, suffocating agents and household remedies have gained popularity. Various mechanical methods such as dry combing, wet combing, and bug busting have been advocated (the latter within the framework of a public health approach). The search for new treatments continues with testing of systemic pharmacological treatments and the development of devices such as a comb attached to a vacuum cleaner or an appliance producing hot air.

Diagnosis of head lice and determination of the outcome of treatement (cure rate)

Diagnostic techniques are a crucial element in epidemiological research and clinical trials (Burgess 2005). An accurate diagnosis is also the cornerstone of the management of infectious diseases. Undetected cases of pediculosis can act as a source of (re)infestation and hamper the effective management, even if detected cases are treated correctly (Figueroa 2000). False positive diagnoses on the other hand will lead to unnecessary treatment. Furthermore, if a patient is evaluated post treatment and a false positive diagnosis is made, one could wrongly conclude that the treatment has failed. Therefore false positive diagnoses can also be a potential reason for 'treatment failure' (Takano-Lee *et al.* 2004). The diagnosis of pediculosis is based on two key elements: the diagnostic criteria and the diagnostic method.The only correct definition of an active infestation is the presence of a living, moving louse and/or viable eggs. The presence of egg shells (either empty or containing a dead egg) is a sign of a past infestation and thus the latter obviously should not be used to diagnose active pediculosis (Figueroa 2000, Koch *et al.* 2001).

Several methods have been described to diagnose an infestation. The oldest, traditional method is visual inspection (VI). The hair is parted with the fingers or applicator sticks and the scalp is inspected (Burgess 2004, Mumcuoglu *et al.* 2001). Another method is dry combing (DC). First the hair is untangled using an ordinary comb. Then the hair is combed section per section using a fine toothed comb (Mumcuoglu *et al.* 2001). An ordinary shampoo or a pediculicide is sometimes used to facilitate the detection of head lice. In 'shampooing and straining' the hair is washed with an ordinary shampoo and the rinse water is collected and strained through a towel (Meinking *et al.* 2002). Hair can be shampooed with a pediculicide before combing it with a fine toothed comb (Speare and Buettner 1999). Finally, wet combing (WC) is a detection method in which water, conditioner, and a fine-toothed comb are used. First the hair is thoroughly wetted and ample conditioner is applied to the hair. The scalp is systematically combed lock by lock using the fine-toothed comb while gently touching the scalp. After each stroke the comb is wiped on a tissue and checked for lice. The hair is rinsed when the entire scalp surface is combed. Then the procedure is repeated but this time the hair is combed from the nape of the neck towards the forehead. This method was derived from Bug Busting wet combing in which the hair is shampoed before applying the conditioner (Figueroa 2000, Fry *et al.* 2002). The methods described above are used in epidemiological research and therapeutic trials (Catala *et al.* 2005, Heukelbach *et al.* 2005, Hill *et al.* 2005, Sim *et al.* 2003, Speare and Buettner, 1999, Vander Stichele *et al.* 2002). However, there are only two published studies comparing the diagnostic performance of three, frequently used, detection methods. DC was found to be four times more effective than VI (Mumcuoglu *et al.* 2001). In a comparison of VI with WC, VI resulted in 10% false negatives and 30% false positives (De Maeseneer *et al.* 2000). Recently we compared dry combing to wet combing in a sample of 608 children between two and fourteen years of age attending four primary schools and kindergartens in Ghent, Belgium. All children were consecutively screened on the same day with visual inspection, dry combing and wet combing by three different, blinded screening teams. The McNemar test was used to compare related samples and the threshold for statistical significance was chosen at $\alpha=0.05$. An active infestation with head lice was found in 7.6%, 15.0% and 18.8%

of the children with respectively VI, DC and WC. The sensitivity of wet combing was significantly higher than that of dry combing and of visual inspection. The sensitivity of dry combing was almost two times higher than the sensitivity of visual inspection.

It seems that wet combing has a higher diagnostic performance than dry combing and visual inspection but is more expensive and human resources are needed. Wet combing should be advocated as method of choice to diagnose pediculosis and to assess cure after a topical treatment in individuals. This method should also be preferred for epidemiologic purposes. It should be noted that this diagnostic procedure disrupts the head louse population. Furthermore, the conditioner used might also have insecticidal properties. Therefore dry combing is better suited to assess the cure rate of topical treatments within clinical trials.

Topical chemical treatments

Since the 1940s, most research on the treatment of pediculosis has revolved around natural or synthetic insecticides (Burgess 2004). The oldest insecticides are the organochlorines of which dichloro-diphenyl-trichloroethane (DDT) is the most notorious member. DDT has been used extensively for the treatment of pediculosis but also in food production. It was banned in 1973 in the USA because of human safety reasons and environmental issues (Ware and Whitacre 2004) and, for the same reasons, not long after that in many other countries. Lindane is another insecticide belonging to the organochlorines. A 1% shampoo formulation was first marketed more than 30 years ago. It has become less popular and even forbidden in a number of countries because of environmental concerns, reports about resistance, and severe side effects on the central nervous system. Carbaryl is an insecticide which is available as a headlouse product only on prescription in the UK. In the 19th century flowers of *Chrysanthemum cinerariaefolium* were used for their insecticidal properties. These flowers contain the natural insecticidal product pyrethrin. Soon after the discovery of the chemical structure of pyrethrin, a first synthetic pyrethroid, allethrin was produced. It was not long before other more active pyrethroids were developed, such as d-phenothrin, bioallethrin and permethrin (Taplin and Meinking 1990). Pediculicides based on pyrethroids often also contain the synergist piperonyl butoxide. It inhibits mixed-function oxidases and reduces metabolic resistance to pyrethroids (Picollo *et al.* 2000). Malathion 0.5% lotion was developed more than 30 years ago. It is an organophosphate, which irreversibly inhibits cholinesterase (Meinking 1999). These pediculicides have been used massively for the treatment of head lice. However, resistance to these products is emerging, limiting their usefulness in head lice control. Other agents for topical use in the treatment of head lice have been suggested. Imidacloprid and fipronil are effective against several insects and their activity against head lice has been demonstrated in bioassays. The pediculicidal actvitity of compounds such as indoxacarb and chlorfenapyr has not yet been tested but they have a good insecticidal activity against other species and a favourable safety profile (Hill 2006b). These products might be useful in head lice control and merit further research.

When evaluating the efficacy of these treatments the best available evidence should be consulted. A major problem in this field of research is the lack of randomised controlled trials of good quality (Lapeere *et al.* 2003). Furthermore, the two systematic reviews that have been published on the topic have conflicting conclusions. The first by Vander Stichele *et al.* concluded that only for permethrin sufficient evidence of efficacy was available and that more evidence was needed for treatments such as malathion and carbaryl. Pyrethroids were assessed as not sufficiently efficacious (Vander Stichele *et al.* 1995). In subsequent comments, the criteria to assess the quality of the clinical trials were questioned (Dixon *et al.* 1996). A second systematic Cochrane review by Dodd concluded that effectiveness was proven for permethrin, malathion and synergised pyrethrins (Dodd 2001). However, this review included only three clinical trials on pediculicides. Several methodological concerns were raised regarding the selection process in this review and the trial quality assessment process, biased by a high number of personal communications. The review also ignored two trials referenced in the previous systematic review (Lapeere *et al.* 2003).

The safety profile of the different treatments is mainly based on observations of cases or case series. Natural pyrethrins can elicit local side effects such as skin irritation, rash, redness and swelling. This natural plant extract should not be used in patients allergic to chrysanthemums, ragweed and other related plants because of a possible cross reaction. Permethrin, on the other hand, does not cause allergic reactions in people with plant allergies. In a postmarketing surveillance study of permethrin 1% cream rinse, 2.2 adverse events per 1000 treatments were reported. All recorded adverse events (such as oedema, erythema, hypoesthesia and pruritus) were mild (Burkhart and Burkhart 2006b). Malathion has a potential of toxicity, when ingested, but is well tolerated as a topical treatment, with only minor adverse events (scalp irritation and conjunctivitis). However, alcoholic lotions containing malathion are potentially flammable and may cause accidents. Mutagenic and carcinogenic effects have been found *in vitro* and in animal studies (Cox, 2003) but these effects have not been reported in humans. Severe adverse events have been associated with the use of lindane. These include neurological effects such as dizziness, seizures, headache and paresthesia and even some deaths have been reported (Burkhart and Burkhart, 2006a). One case-control study found a statistically significant association between acute leukaemia and the use of home insecticides, garden insecticides and insecticidal treatment for pediculosis (Menegaux *et al.* 2006). However, these results may be biased by a 'recall bias' and were not yet confirmed by further research.

Insecticide resistance

The cure rate of chemical treatments depends on local resistance patterns. DDT was formerly used for head louse control but widespread use of the product led to world-wide resistance. Similarly, lindane has become ineffective in many parts of the world due to extensive use. Clinical trials with permethrin showed a cure rate of 99% back in 1986 when the product was first introduced. Less than 10 years later *in vitro* studies from Israel and the Czech Republic demonstrated that lice collected from local school children were significantly less sensitive to permethrin compared to a fully susceptible reference strain (Downs 2004). Since then, varying levels of permethrin and/or pyrethrin resistance have been reported. in France, Argentina, the UK, Australia, the USA and Denmark (Chosidow *et al.* 1994, Downs *et al.* 1999b, Hunter and Barker 2003, Kristensen *et al.* 2006, Vassena *et al.* 2003, Yoon *et al.* 2003). Cure rates with malathion have remained high (close to 100%) until the early 1990's but recently resistance to malathion has been reported in the UK, Australia and Denmark (Downs *et al.* 1999b, Hunter and Barker 2003, Kristensen *et al.* 2006).

Resistance is usually monitored with bioassays. There is a large heterogeneity in the way susceptibility assays are performed, making it difficult to compare resistance rates between different countries. In 1994, resistance to permethrin was documented in head lice in Israel, only 2.5 years after the introduction of permethrin to the Israeli pediculicide market (Mumcuoglu *et al.* 1995). In Argentina, in 1998, a sample of lice collected from children in different schools in Buenos Aires showed resistance to permethrin, however with considerable variation in the resistance ratios according to the school where the lice originated from. There was cross-resistance to deltamethrin and sumithrin, two other pyrethroids, but not to carbaryl (Picollo *et al.* 1998). Permethrin-resistant lice have also been detected in the USA in 2003. This sample was still susceptible to malathion, a product that is only available on prescription and is available on the US market since 1999 (Yoon *et al.* 2003). In-vitro studies on head lice collected in Bath and Bristol in 1999 showed double resistance to both malathion and permethrin. Head lice from these samples remained fully sensitive to carabaryl (Downs *et al.* 1999b). In a later study in 2002, head lice from four regions in the UK were found to be resistant to permethrin and malathion. Head lice collected in Leeds were also resistant to carbaryl (Downs *et al.* 2002). Susceptibility tests by Hunter *et al.* demonstrated substantial variability in susceptibility to malathion, pyrethrum an permethrin between different schools in Australia in 2003 (Hunter and Barker 2003). Permethrin and malathion resistance has been demonstrated in samples of head lice in Denmark in 2006 (Kristensen *et al.* 2006). Resistance to permethrin is widespread, and resistance to malathion, combined resistance and resistance to carbaryl is starting to emerge. The pattern of resistance varies from region to region and even from school to school. It seems plausible that the resistance pattern in a region is influenced or shaped by the specific pediculicides used in that specific region (Hunter and Barker 2003, Lee *et al.* 2000, Picollo *et al.* 1998).

In general, there are two major resistance mechanisms: target-site modification and enzyme-based detoxification. In target-site modification, the pediculicide no longer binds to the target site. In enzyme-based detoxification the pediculicide is degraded before it exerts its effect. Genetic adaptation with upgrading of detoxifying enzymes such as monooxygenases, esterase and glutathione S-transferase (GST) is the major mechanism for enzyme-based resistance to organophosphates (such as malathion), carbamates (such as carbaryl) and pyrethroids (such as natural pyrethrum, phenothrin and permethrin). Reduced neuronal sensitivity or 'knockdown-resistance' is a form of target-site modification and is a common resistance mechanism for both DDT and permethrin resistance. This mechanism has been documented in several species, it is unaffected by synergists which inhibit insect esterases and monooxygenases. It is caused by reduction in the sensitivity of the insect nervous system to pyrethroids. Most of the research on resistance mechanisms in head lice has focussed on pyrethroid resistance. Through molecular cloning and sequencing, two point mutations (T929I and L932F) have been located in the IIS5 transmembrane segment in permethrin-resistant lice collected in Florida and Bristol (Lee *et al.* 2000). In a later study, permethrin resistance was again confirmed and a strong correlation was found between the presence of T929I and L932F mutations and permethrin resistance (Yoon *et al.* 2003). In Denmark, 95% of the tested permethrin resistant lice had both T929I and L932F mutation (Kristensen *et al.* 2006). These two mutations and the more recently detected D11E and M850T mutations are potential genetic factors responsible for kdr type resistance (Tomita *et al.* 2003). The mechanisms conferring malathion resistance have only recently been investigated. Possible mechanisms are reduced target-site sensitivity but also enhancement of metabolism by cytochrome p450 monooxygenase, GST and phosphotriesterases. It is hypothesised that an elevated level of malathion cholinesterase activity with possibly a unique isoform that has an elevated hydrolysis activity is the major cause for malathion resistance. Altered target site is not likely to be involved in malathion resistance (Gao *et al.* 2006).

There is an urgent need for standardised protocols for sampling, collecting and analysis of specimens in geographical regions, in order to make comparable estimates of the prevalence of the resistance rates to various commonly used topical treatments. Hopefully, improved genetic routine assays will facilitate continuous and comparable monitoring of resistance in small health areas, linked to continuous monitoring of pesticide and pediculicide use in these regions.

Mechanical methods with combing

Mechanical treatments are based on the removal of adult lice and nymphs using a head louse comb. Only the Bug Busting method has been tested in clinical trials (Bingham *et al.* 2000, Downs *et al.* 1999a, Hill *et al.* 2005, Olsen 2005, Resnik 2005, Roberts *et al.* 2000). In the Bug Busting method, wet, conditioned hair is combed using a fine toothed comb. This combing procedure needs to be repeated four times, leaving three days in between. Until now Bug Busting has been studied as a treatment option in four clinical trials, of which the first was a pilot trial and the second, an open trial on efficacy (Bingham *et al.* 2000, Plastow *et al.* 2001). A third, pragmatic trial found that the cure rate with Bug Busting was only 38% which was significantly lower than the cure rate of 78% in the group treated with malathion (Roberts *et al.* 2000). Finally, in a recent effectiveness trial with an improved version of the Bug Buster Kit, a significantly greater proportion was cured with the Bug Buster Kit than with over the counter pediculicides (57% versus 13%) (Hill *et al.* 2005). This study was criticised because the way patients were recruited might have resulted in an inclusion bias. Furthermore it is unclear whether allocation of treatment was adequately concealed. Other comments are that the pediculicides were inadequately dosed and that the conditioner used in Bug Busting might kill lice, leading to bias in favour of Bug Busting (Chosidow 2006, Hill 2006a).

Mechanical methods with electrical devices

Electric head louse combs and a system connected to a vacuum cleaner ('Lice snatcher') have been developed but were not tested in clincal trials. Based on the observation that lice are killed in hot environments, the effect of hot air on lice and eggs was studied. Six different hot air treatment methods (of which one was a custom-built machine called the LouseBuster) were studied in 169 school children with head lice. The LouseBuster was the most successful method resulting in 80% mortality in hatched lice and almost complete mortality of eggs. The results of this small study seem promising but warrant further research (Goates *et al.* 2006).

Natural products

Over the last decade there has been an increasing interest for the insecticidal properties of essential oils and natural compounds (Priestley *et al.* 2006). Essential oils such as tea tree oil, *Eucalyptus globulus* leaf oil, *Eugenia caryophyllata* oil, *Lippia multiflora* oil, have shown some pediculicidal and ovicidal activity *in vitro* (Amr and Nusier 2000, Oladimeji *et al.* 2000, Yang *et al.* 2003, 2004). Two small clinical studies attributed pediculicidal activities to extracts of *Annona squamosa* seed and henna (El-Basheir and Fouad 2002, Tiangda *et al.* 2000). A natural remedy containing coconut and ylang-ylang oil claims the same effectiveness as a classic pediculicide in an open clinical trial (Mumcuoglu *et al.* 2002).

Essential oils and natural compounds are extracted from plant material and usually contain many different chemical constituents. There is little control on the quality and composition of these oils. Their composition may vary according to the manufacturer or batch. These products are also at risk of contamination, for example with pesticides used during the cultivation of the plants (Barnes, 2003). Little is known about the mechanisms responsible for the pediculicidal effect of essential oils and other plant extracts (Downs *et al.* 2000a). Tea tree oil was recently shown to inhibit acetylcholinesterase (Mills *et al.* 2004), which is also the mode of action of malathion. More information about the mechanism of action and safety profile of these products is needed before they can be assessed in clinical trials (Downs *et al.* 2000a). Until then, these products should not be used in the treatment of pediculosis.

Suffocating agents

Another approach to the treatment of head lice is based on the hypothesis that lice can be suffocated with local agents (Mumcuoglu 1999, Schachner 1997). One study reported astonishing results with Cetaphil® cleanser (a gentle soap to clean the face) but raised ethical concerns and was heavily biased (Pearlman 2004). Dimethicone is a silicone which is used in many cosmetics (creams, conditioners) for its lubricating properties. A 4% dimethicone lotion is being promoted in the UK as a treatment for pediculosis. It does not have any neurotoxic properties but according to Burgess *et al.* it irreversibly immobilises lice. This product was compared to phenothrin lotion in a single blinded randomised controlled equivalence trial. In the group treated with dimethicone lotion a cure rate 70% was found, which was comparable to the cure rate of phenothrin (Burgess *et al.* 2005). This treatment might be valuable if the results are confirmed in future trials.

Systemic treatments

Several systemic treatments have been suggested for the treatment of head lice. Two small clinical trials of limited quality report on the use of the antibiotic combination of sulfamethoxazole and trimetoprim (Hipolito *et al.* 2001, Pollack 2001, Sim *et al.* 2003). Rare but severe side effects such as aplastic anemia and toxic epidermal necrolysis have been reported after the use of this compound. This antibiotic should be reserved for the treatment of serious bacterial infections (Meinking 1999). Anthelmintic compounds, such as thiabendazole, albendazole and levamisole have been suggested as effective treatments, based on uncontrolled open studies or dubious clinical trials (Akisu *et al.* 2006, Namazi 2001, 2003). Side effects such as cutaneous reactions, hematologic abnormalities and gastro-intestinal problems have been reported (Bagheri *et al.* 2004, Gupta and Gupta 2005). These agents should be used for severe parasitic infestations and not for pediculosis as safer, local treatments are available. Recently ivermectin has been suggested as a treatment option for pediculosis (Burkhart 2004). However, this drug is not recommended for children younger than five years of age or less than 15 kg in weight (Dourmishev *et al.* 2005). Furthermore, the safety and efficacy of ivermectin in the treatment of pediculosis has not been established (Ko and Elston 2004). For these reasons, ivermectin should not be used in the treatment of pediculosis.

Removal of nits

The presence of egg shells (with or without viable eggs) poses several problems. The first problem is that chemical treatments for pediculosis should have optimal pediculicidal but also ovicidal properties in order to effectively control pediculosis. But the latter is often their downfall. Most formulations are not sufficiently ovicidal and must rely on regular repetition of application to kill the newly hatched nymphs (Downs 2004, Meinking 1999). Even if a patient is successfully treated, the empty eggshells can remain for a long time in the hair because they are firmly stuck to the hair and difficult to remove. In these circumstances, the presence of obvious nits is not only a cosmetic problem, but also a cause of mistake in diagnosis of an active infestation (false positive), resulting in false positive diagnoses and unnecessary treatment (Williams *et al.* 2001). The egg laying process has been extensively studied in body lice. Egg laying by head lice shows great resemblance to the process in body lice. The egg and its membranes develop in the ovarioles of the female louse. They are separated from the oviduct by a plug of tissue. First, the louse threads a hair between the gonopods. Then an adhesive substance is secreted by the glue glands. The substance fills the oviduct and is spread around the human hair, forming a tube. Then the plug of tissue breaks and the egg is released into the oviduct and pushed into the adhesive substance. The substance sets and forms a solid tube or cylinder around the hair which extends into an eggshell (which is also known as a 'nit') surrounding and protecting the egg (Carter 1990).

Several authors have tried to unravel the composition of the egg shell. Histochemical examination of the adhesive substance using the glue glands dissected from living female lice revealed that the adhesive mainly consist of proteins (Carter 1990). This finding was confirmed by Burkhart *et al.* who performed pyrolysis-gas chromatography and mass spectrometry on the solid nit cylinders (Burkhart *et al.* 1999a, 1999b). This study revealed a high relative abundance of components derived from phenylalanine, tyrosine, tryptophan and glutamic acid. It is possible that the histochemical constitution of the eggshell and the cylinder around the hair shaft is composed of material which is so similar to the constituents of the hair itself, that any chemical aggression against the eggshell will also damage the hair. The same authors investigated whether products such as vinegar, acetone, vodka, bleach, all purpose degreaser and petrolatum facilitated the removal of nits *in vitro*. Only bleach, vodka and vinegar assisted mildly but were not therapeutically efficacious (Burkhart *et al.* 1998). Mumcuoglu reported that acid shampoos, 5% acetic acid, conditioners and vegetable oils are effective in detaching eggs from the hair (Mumcuoglu 1999). However, none of these authors describe how they measured the effect of the studied products. In a small un-blinded clinical study beneficial results were demonstrated with a nit removal system containing a formic acid 8% cream rinse and metal comb. The control site was only combed with a plastic comb (DeFelice *et al.* 1989). However, difference in the type of combs used could account for a better result in the treated site (Elston 1999). Thorough *in vitro* and *in vivo* studies are lacking to support the use of products to facilitate nit removal. Nevertheless, home remedies such as vinegar and the formic acid cream rinse are cited over and over again in traditional literature reviews (Jones and English 2003, Ko and Elston 2004, Mumcuoglu 1999).

Research on this topic has been hampered by a lack of precise measurement of the forces needed to remove a nit, until recently a reliable method was developed. In this method a force transducer is used to which a small metal cylinder was connected. A hair with a nit was put with its scalp end through the metal cylinder and knotted to a bent needle connected to a surgical thread. The thread was attached to the axis of a slow-spinning electrical motor. The measurement procedure started when the motor was activated and began pulling the hair through the cylinder. At the point where the nit was halted by the metal cylinder, force started to build up, until the nit cylinder lost grip on the hair and the hair began sliding through the immobilised nit. The device was calibrated and tests showed that the method had an acceptable level of precision. The force from the force transducer was recorded with a pen recorder, producing graphs of developed force as a function of time. The following parameters were determined: (1) the first peak in the plot (Fpeak), (2) the average force (Fav) and (3) the maximum force (Fmax) (Lapeere *et al.* 2005). The device described above is easy to handle and cheap to construct. Furthermore, it proved to be a feasible and reliable method to measure the force parameters that determine the difficulty or ease with which nits can be removed from human hairs. A limitation is that for technical reasons only nits attached 17 mm from the proximal end of the hair can be used. The procedure was further optimised to improve data processing. The method was used to make a descriptive study of the forces needed to remove nits, and their correlations with various characteristics of the eggshell and the hair. It was also used to evaluate the efficacy of several products to facilitate the removal of nits from human hairs.

In a first explorative study a total of 104 hairs from 18 infested children were cut as close to the scalp as possible. Only single nits attached 17 mm from the proximal end of the hair were used for technical reasons. The distal end of the hair was cut 20 mm behind the nit. The diameter of the hair, the distance of the nit from the scalp end of the hair and the length of the nit cylinder were also determined. A Spearman's rho correlation coefficient was used to explore correlations and the Kruskal-Wallis test was used to compare medians of multiple groups. The mean ± SEM of Fpeak, Fav and Fmax was respectively 66 ± 3.7, 51 ± 3.1 and 130 ± 5.3 mN (n=93). There were no significant between-patient differences in the measured forces. A positive correlation was found between all three force parameters and the length of the nit cylinder, a negative correlation was found between Fmax and the distance of the nit from the scalp end side of the hair. There was no correlation with the diameter of the hair (Lapeere *et al.* 2005). In a second study, we examined the effect of different interventions to reduce the forces needed to remove nits. Hairs were obtained from 6 children between 7 and 11 years (average age 9). We tested commercial nit removal products, available in Belgium or The Netherlands, short and long application of ordinary conditioner, a 4% formic acid gel, and a 2% and 3% HPMC gel, all against pure deionised water as control. Data were analysed using SPSS software (version 12). The Kruskal-Wallis test was used to compare medians of multiple groups and the Mann-Whitney test was used to compare medians of two groups. The threshold for statistical significance was at $P<0.05$. We observed that ordinary conditioner facilitated the nit removal. A short application of conditioner was less efficacious than the continuous application of the conditioner. It has been demonstrated that a conditioner acts as a lubricant and decreases friction. Presumably the hair is surrounded by a thicker layer if conditioner is left in place during the force measurements (La Torre and Bhushan 2006). We also observed no difference between ordinary conditioner and commercial nit removal products. Furthermore, the addition of formic acid to deionised water did not reduce the forces needed to remove the nit compared to pure deionised water. On the contrary, forces were higher in formic acid solutions with pH 4, 5 and 6. This might be caused by acid damage to the hair, resulting in a higher friction coefficient, hampering the removal of nits.

This research suggests that the continuous application of an ordinary conditioner significantly enhances nit removal. Furthermore, it is a user-friendly and non-toxic product, already commercially available. It is not known if this *in vitro* research is representative for the *in vivo* setting. For this reason, clinical trials are also needed to assess the effectiveness of the tested products.

Strategies

Several policies have been proposed for the management of pediculosis.

No-nit policy: In this policy regular screenings (at least 3 times a year) are advised to facilitate early detection. The no-nit policy means that a child is expelled from school as soon as signs of head lice infestation are found. Children are not allowed back to school until all egg shells or nits are removed because they are regarded as a source of potential reinfestation and may cause confusion in the diagnosis. The school nurse is in a key position and is responsible for the spread of information, detection, follow up of treatment and prevention (Donnelly *et al.* 1991). This type of control strategy has received much criticism because the responsibility for pediculosis rests entirely on the shoulders of the school nurse (Mumcuoglu *et al.* 2006). It has not been demonstrated that regular screenings actually reduce prevalence (Frankowski 2004, Underwood 1990). Data are also lacking to substantiate the benefits of a no-nit policy (Dolianitis and Sinclair 2002, Mumcuoglu *et al.* 2006). Furthermore, exclusion from school (which might be prolonged in some cases) disrupts the educational process (Schoessler 2004).

Co-ordinated approach: The opposite of the no-nit policy is a co-ordinated approach involving the whole community. A well-planned public education campaign directed to professionals in first instance and consecutively to the general public is a prerequisite. Emphasis is put on co-ordinated detection (and treatment of all cases with an active infestation) by all parents on a pre-determined day. The effect of a co-ordinated approach on head lice prevalence was first illustrated in Teesside in the early 1970's where a public education programme reduced the level of head lice infestations by half (Donaldson 1979). More examples of a 'whole-school approach' or community approach have been published (Brainerd 1998, Downs 2004, Schoessler 2004, Underwood, 1990). An elimination programme consisting of education, assessment of baseline prevalence and a collective 'elimination day' conducted on the Isle of Man led to a spectacular reduction of prevalence and incidence of head lice infestations (Vermaak 1996). Co-ordinated detection (and treatment) has been applied in different settings in the UK, Belgium and Denmark (Ibarra *et al.* 2006). The management of pediculosis is a concern of the whole community. Children, parents, school nurses, teachers, school administrators, pharmacists and physicians should be involved (Downs 2004). Everyone should join forces to create an open, non-stigmatising atmosphere in order to keep pediculosis manageable (Brainerd 1998).

Effective and safe treatments should be implemented within the community approach. There is a longstanding tradition of topical chemical treatments for pediculosis. Their usefulness is being threatened by resistance (see above), which has been documented in several countries. Reliable, fast and cheap detection methods are needed for the repeated epidemiological assessment of local resistance patterns, based on bio-assays and if possible, on genetic screening for mutations in head lice. Resistance to pediculicides can no longer be ignored and will probably become more and more important. One way to overcome insecticide resistance is to develop new pediculicides. However, history has illustrated that these products have a relatively short life span. Mechanical methods such as Bug Busting, are not influenced by resistance and are therefore a valid treatment option, especially in regions with high resistance. The effectiveness of this method compared to pediculicides is not yet definitely established and deserves research with well constructed clinical trials.

Conclusion

Harnessing the prevalence of head lice below acceptable levels will not be achieved by relying on the efficacy of one treatment option, but requires a multi-faceted public health approach, based on sound epidemiologic methods.

References

Akisu C, Delibas SB and Aksoy U (2006) Albendazole: single or combination therapy with permethrin against pediculosis capitis. Pediatr Dermatol 23: 179-182.

Amr ZS and Nusier MN (2000) Pediculosis capitis in northern Jordan. Int J Dermatol 39: 919-921.

Bagheri H, Simiand E, Montastruc JL and Magnaval JF (2004) Adverse drug reactions to anthelmintics. Ann Pharmacother 38: 383-388.

Barnes J (2003) Quality, efficacy and safety of complementary medicines: fashions, facts and the future. Part I. Regulation and quality. Br J Clin Pharmacol 55: 226-233.

Bingham P, Kirk S, Hill N and Figueroa J (2000) The methodology and operation of a pilot randomized control trial of the effectiveness of the Bug Busting method against a single application insecticide product for head louse treatment. Public Health 114: 265-268.

Brainerd E (1998) From eradication to resistance: five continuing concerns about pediculosis. J Sch Health 68: 146-150.

Burgess I (2005) Head lice. Clin Evid: 2044-2049.

Burgess IF (2004) Human lice and their control. Annu Rev Entomol 49: 457-481.

Burgess IF, Brown CM & Lee PN (2005) Treatment of head louse infestation with 4% dimeticone lotion: randomised controlled equivalence trial. BMJ 330: 1423.

Burkhart CG (2004) Relationship of treatment-resistant head lice to the safety and efficacy of pediculicides. Mayo Clin Proc 79: 661-666.

Burkhart CG and Burkhart CN (2006a) Asphyxiation of lice with topical agents, not a reality...yet. J Am Acad Dermatol 54: 721-722.

Burkhart CG and Burkhart CN (2006b) Safety and efficacy of pediculicides for head lice. Expert Opin Drug Saf 5: 169-179.

Burkhart CN, Arbogast J, Smythe P and Burkhart CG (1999a) Histochemical analysis of the nit of *Pediculus humanus capitis* (Anoplura: Pediculidae). J Med Entomol 36: 530-532.

Burkhart CN, Burkhart CG, Pchalek I and Arbogast J (1998) The adherent cylindrical nit structure and its chemical denaturation in vitro: an assessment with therapeutic implications for head lice. Arch Pediatr Adolesc Med 152: 711-712.

Burkhart CN, Stankiewicz BA, Pchalek I, Kruge MA and Burkhart CG (1999b) Molecular composition of the louse sheath. J Parasitol 85: 559-561.

Carter D (1990) Insect Egg Glue. An investigation of the nature and secretion of insect egg glues, with special reference to the human louse, *Pediculus humanus* and the cabbage white butterfly, *Pieris brassicae*., Vol. Doctor of Philosophy: Department of Applied Biology (ed. Cambridge University, Cambridge, p. 217.

Catala S, Junco L and Vaporaky R (2005) Pediculus capitis infestation according to sex and social factors in Argentina. Rev Saude Publica 39: 438-443.

Chosidow O (2006) Bug Buster for Head Lice: Is It Effective? Arch Dermatol 142: 1635-1637.

Chosidow O, Chastang C, Brue C, Bouvet E, Izri M, Monteny N, Bastuji-Garin S, Rousset JJ and Revuz J (1994) Controlled study of malathion and d-phenothrin lotions for *Pediculus humanus var capitis*-infested schoolchildren. Lancet 344: 1724-1727.

Chunge RN, Scott FE, Underwood JE and Zavarella KJ (1991) A review of the epidemiology, public health importance, treatment and control of head lice. Can J Public Health 82: 196-200.

Counahan M, Andrews R, Büttner P, Byrnes G and Speare R (2004) Head lice prevalence in primary schools in Victoria, Australia. J Paediatr Child Health 40: 616-619.

Cox C (2003) Insecticide factsheet. Malathion. Journal of Pesticide Reform 23: 10-15.

De Maeseneer J, Blokland I, Willems S, Vander Stichele R and Meersschaut F (2000) Wet combing versus traditional scalp inspection to detect head lice in schoolchildren: observational study. BMJ 321: 1187-1188.

DeFelice J, Rumsfield J, Bernstein JE and Roshal JY (1989) Clinical evaluation of an after-pediculicide nit removal system. Int J Dermatol 28: 468-470.

Dixon R, Glenny AM, Eastwood A, Sharp DL, Fullerton D, Petticrew M and Sheldon T (1996) Topical treatments for head lice. Review's weaknesses may undermine its conclusions. BMJ 312: 123; author reply 123.

Dodd CS (2001) Interventions for treating headlice. Cochrane Database Syst Rev: CD001165.

Dolianitis C and Sinclair R (2002) Optimal treatment of head lice: is a no-nit policy justified? Clin Dermatol 20: 94-96.

Donaldson R (1979) Head Lice: Parasites and Western Men (ed. by R Donaldson) MTP Press Limited, Lancaster, pp. 57-77.

Donnelly E, Lipkin J, Clore ER and Altschuler DZ (1991) Pediculosis prevention and control strategies of community health and school nurses: a descriptive study. J Community Health Nurs 8: 85-95.

Dourmishev AL, Dourmishev LA and Schwartz RA (2005) Ivermectin: pharmacology and application in dermatology. Int J Dermatol 44: 981-988.

Downs AM (2004) Managing head lice in an era of increasing resistance to insecticides. Am J Clin Dermatol 5: 169-177.

Downs AM, Stafford KA and Coles GC (1999a) Head lice: prevalence in schoolchildren and insecticide resistance. Parasitol Today 15: 1-4.

Downs AM, Stafford KA and Coles GC (2000a) Monoterpenoids and tetralin as pediculocides. Acta Derm Venereol 80: 69-70.

Downs AM, Stafford KA, Harvey I and Coles GC (1999b) Evidence for double resistance to permethrin and malathion in head lice. Br J Dermatol 141: 508-511.

Downs AM, Stafford KA, Hunt LP, Ravenscroft JC and Coles GC (2002) Widespread insecticide resistance in head lice to the over-the-counter pediculocides in England, and the emergence of carbaryl resistance. Br J Dermatol 146: 88-93.

Downs AM, Stafford KA, Stewart GH and Coles GC (2000b) Factors that may be influencing the prevalence of head lice in British school children. Pediatr Dermatol 17: 72-74.

El-Basheir ZM and Fouad MA (2002) A preliminary pilot survey on head lice, pediculosis in Sharkia Governorate and treatment of lice with natural plant extracts. J Egypt Soc Parasitol 32: 725-736.

Elston DM (1999) What's eating you? *Pediculus humanus* (head louse and body louse). Cutis 63: 259-264.

Figueroa JI (2000) Head lice: is there a solution? Curr Opin Infect Dis 13: 135-139.

Frankowski BL (2004) American Academy of Pediatrics guidelines for the prevention and treatment of head lice infestation. Am J Manag Care 10: S269-272.

Fry F, Ibarra J, Smith J and Wickenden C (2002) Wet combing to eradicate head lice. J R Soc Med 95: 630-631.

Gao J-R, Yoon KS, Frisbie RK, Coles GC and Clark JM (2006) Esterase-mediated malathion resistance in the human head louse, Pediculus capitis (Anoplura: Pediculidae). Pestic Biochem Physiol 85: 28-37.

Goates BM, Atkin JS, Wilding KG, Birch KG, Cottam MR, Bush SE and Clayton DH (2006) An effective nonchemical treatment for head lice: a lot of hot air. Pediatrics 118: 1962-1970.

Gupta R and Gupta S (2005) Drug rash due to levamisole. Indian J Dermatol Venereol Leprol 71: 428-429.

Hansen RC and O'Haver J (2004) Economic considerations associated with Pediculus humanus capitis infestation. Clin Pediatr (Phila) 43: 523-527.

Heukelbach J, Wilcke T, Winter B and Feldmeier H (2005) Epidemiology and morbidity of scabies and pediculosis capitis in resource-poor communities in Brazil. Br J Dermatol 153: 150-156.

Hill N (2006a) Clinical evaluation of bug busting: response to suggestions of bias. Archives of dermatology 142: 1651-1653.

Hill N (2006b) Control of head lice: past, present and future. Expert review of anti-infective therapy 4: 887-894.

Hill N, Moor G, Cameron MM, Butlin A, Preston S, Williamson MS and Bass C (2005) Single blind, randomised, comparative study of the Bug Buster kit and over the counter pediculicide treatments against head lice in the United Kingdom. BMJ 331: 384-387.

Hipolito RB, Mallorca FG, Zuniga-Macaraig ZO, Apolinario PC and Wheeler-Sherman J (2001) Head lice infestation: single drug versus combination therapy with one percent permethrin and trimethoprim/sulfamethoxazole. Pediatrics 107: E30.

Hunter JA and Barker SC (2003) Susceptibility of head lice (*Pediculus humanus capitis*) to pediculicides in Australia. Parasitol Res 90: 476-478.

Ibarra J, Fry F, Wickenden C, Olsen A, Vander Stichele RH, Lapeere H, Jenner M, Franks A and Smith J (2007) Overcoming health inequalities by using the Bug Busting 'whole-school approach' to eradicate head lice. J Clin Nurs 16: 1955-1965.

IMS Health Care (2003.) Sales figures of pediculicides in Belgium. Personal communication.

Jones KN and English JC, 3rd (2003) Review of common therapeutic options in the United States for the treatment of pediculosis capitis. Clin Infect Dis 36: 1355-1361.

Ko CJ and Elston DM (2004) Pediculosis. J Am Acad Dermatol 50: 1-12; quiz 13-14.

Koch T, Brown M, Selim P and Isam C (2001) Towards the eradication of head lice: literature review and research agenda. J Clin Nurs 10: 364-371.

Kristensen M, Knorr M, Rasmussen AM and Jespersen JB (2006) Survey of permethrin and malathion resistance in human head lice populations from Denmark. J Med Entomol 43: 533-538.

La Torre C and Bhushan B (2006) Investigation of scale effects and directionality dependence on friction and adhesion of human hair using AFM and macroscale friction test apparatus. Ultramicroscopy 106: 720-734.

Lapeere H, Brochez L, Haeghen YV, Mabilde C, Stichele RV, Leybaert L and Naeyaert JM (2005) Method to measure force required to remove *Pediculus humanus capitis* (Phthiraptera: Pediculidae) eggs from human hair. J Med Entomol 42: 89-93.

Lapeere H, Vander Stichele RH and Naeyaert JM (2003) Evidence in the treatment of head lice: drowning in a swamp of reviews. Clin Infect Dis 37: 1580-1582.

Lee SH, Yoon KS, Williamson MS, Goodson SJT-L, M, Edman JD, Devonshire AL and Clark JM (2000) Molecular Analysis of kdr-like Resistance in Permethrin-Resistant Strains of Head Lice, *Pediculus capitis*. Pestic Biochem Physiol 66: 130-143.

Meinking TL (1999) Infestations. Curr Probl Dermatol: 77-118.

Meinking TL, Clineschmidt CM, Chen C, Kolber MA, Tipping RW, Furtek CI, Villar ME and Guzzo CA (2002) An observer-blinded study of 1% permethrin creme rinse with and without adjunctive combing in patients with head lice. J Pediatr 141: 665-670.

Menegaux F, Baruchel A, Bertrand Y, Lescoeur B, Leverger G, Nelken B, Sommelet D, Hemon D and Clavel J (2006) Household exposure to pesticides and risk of childhood acute leukaemia. Occup Environ Med 63: 131-134.

Mills C, Cleary BJ, Gilmer JF and Walsh JJ (2004) Inhibition of acetylcholinesterase by Tea Tree oil. J Pharm Pharmacol 56: 375-379.

Mumcuoglu KY (1999) Prevention and treatment of head lice in children. Paediatr Drugs 1: 211-218.

Mumcuoglu KY, Friger M, Ioffe-Uspensky I, Ben-Ishai F and Miller J (2001) Louse comb versus direct visual examination for the diagnosis of head louse infestations. Pediatr Dermatol 18: 9-12.

Mumcuoglu KY, Hemingway J, Miller J, Ioffe-Uspensky I, Klaus S, Ben-Ishai F and Galun R (1995) Permethrin resistance in the head louse *Pediculus capitis* from Israel. Med Vet Entomol 9: 427-432, 447.

Mumcuoglu KY, Meinking TA, Burkhart CN and Burkhart CG (2006) Head louse infestations: the 'no nit' policy and its consequences. Int J Dermatol 45: 891-896.

Mumcuoglu KY, Miller J, Gofin R, Adler B, Ben-Ishai F, Almog R, Kafka D and Klaus S (1990) Epidemiological studies on head lice infestation in Israel. I. Parasitological examination of children. Int J Dermatol 29: 502-506.

Mumcuoglu KY, Miller J, Zamir C, Zentner G, Helbin V and Ingber A (2002) The in vivo pediculicidal efficacy of a natural remedy. Isr Med Assoc J 4: 790-793.

Namazi MR (2001) Levamisole: a safe and economical weapon against pediculosis. Int J Dermatol 40: 292-294.

Namazi MR (2003) Treatment of pediculosis capitis with thiabendazole: a pilot study. Int J Dermatol 42: 973-976.

Oladimeji FA, Orafidiya OO, Ogunniyi TA and Adewunmi TA (2000) Pediculocidal and scabicidal properties of Lippia multiflora essential oil. J Ethnopharmacol 72: 305-311.

Olsen A (2005) Head Lice: standards in clinical trial design: BMJ (editorial).

Pearlman DL (2004) A simple treatment for head lice: dry-on, suffocation-based pediculicide. Pediatrics 114: e275-279.

Picollo MI, Vassena CV, Casadio AA, Massimo J and Zerba EN (1998) Laboratory studies of susceptibility and resistance to insecticides in *Pediculus capitis* (Anoplura; Pediculidae). J Med Entomol 35: 814-817.

Picollo MI, Vassena CV, Mougabure Cueto GA, Vernetti M and Zerba EN (2000) Resistance to insecticides and effect of synergists on permethrin toxicity in *Pediculus capitis* (Anoplura: Pediculidae) from Buenos Aires. J Med Entomol 37: 721-725.

Plastow L, Luthra M, Powell R, Wright J, Russell D and Marshall MN (2001) Head lice infestation: bug busting vs. traditional treatment. J Clin Nurs 10: 775-783.

Pollack RJ (2001) Head lice infestation: single drug versus combination therapy. Pediatrics 108: 1393.

Priestley CM, Burgess IF and Williamson EM (2006) Lethality of essential oil constituents towards the human louse, Pediculus humanus, and its eggs. Fitoterapia 77: 303-309.

Resnik KS (2005) A non-chemical therapeutic modality for head lice. J Am Acad Dermatol 52: 374.

Roberts RJ, Casey D, Morgan DA and Petrovic M (2000) Comparison of wet combing with malathion for treatment of head lice in the UK: a pragmatic randomised controlled trial. Lancet 356: 540-544.

Schachner LA (1997) Treatment resistant head lice: alternative therapeutic approaches. Pediatr Dermatol 14: 409-410.

Schoessler SZ (2004) Treating and managing head lice: the school nurse perspective. Am J Manag Care 10: S273-276.

Sim S, Lee IY, Lee KJ, Seo JH, Im KI, Shin MH and Yong TS (2003) A survey on head lice infestation in Korea (2001) and the therapeutic efficacy of oral trimethoprim/sulfamethoxazole adding to lindane shampoo. Korean J Parasitol 41: 57-61.

Speare R and Buettner PG (1999) Head lice in pupils of a primary school in Australia and implications for control. Int J Dermatol 38: 285-290.

Takano-Lee M, Edman JD, Mullens BA and Clark JM (2004) Home remedies to control head lice: assessment of home remedies to control the human head louse, *Pediculus humanus capitis* (Anoplura: Pediculidae). J Pediatr Nurs 19: 393-398.

Taplin D and Meinking TL (1990) Pyrethrins and pyrethroids in dermatology. Arch Dermatol 126: 213-221.

Tiangda CH, Gritsanapan W, Sookvanichsilp N and Limchalearn A (2000) Anti-headlice activity of a preparation of *Annona squamosa* seed extract. Southeast Asian J Trop Med Public Health 31 Suppl 1: 174-177.

Tomita T, Yaguchi N, Mihara M, Takahashi M, Agui N and Kasai S (2003) Molecular analysis of a para sodium channel gene from pyrethroid-resistant head lice, *Pediculus humanus capitis* (Anoplura: Pediculidae). J Med Entomol 40: 468-474.

Underwood EJ (1990) Effectiveness of screening for headlice. Can J Public Health 81: 240.

Vander Stichele RH, Dezeure EM and Bogaert MG (1995) Systematic review of clinical efficacy of topical treatments for head lice. BMJ 311: 604-608.

Vander Stichele RH, Gyssels L, Bracke C, Meersschaut F, Blokland I, Wittouck E, Willems S and De Maeseneer J (2002) Wet combing for head lice: feasibility in mass screening, treatment preference and outcome. J R Soc Med 95: 348-352.

Vassena CV, Mougabure Cueto G, Gonzalez Audino P, Alzogaray RA, Zerba EN and Picollo MI (2003) Prevalence and levels of permethrin resistance in *Pediculus humanus capitis* De Geer (Anoplura: Pediculidae) from Buenos Aires, Argentina. J Med Entomol 40: 447-450.

Vermaak Z (1996) Model for the control of *Pediculus humanus capitis*. Public Health 110: 283-288.

Ware G and Whitacre D (2004) An Introduction to insecticides: Radcliffe's IPM World Textbook (ed. by BE Radcliffe & WD Hutchison) University of Minnesota, Saint-Paul.

West DP (2004) Head lice treatment costs and the impact on managed care. Am J Manag Care 10: S277-282.

Willems S, Lapeere H, Haedens N, Pasteels I, Naeyaert JM and De Maeseneer J (2005) The importance of socio-economic status and individual characteristics on the prevalence of head lice in schoolchildren. Eur J Dermatol 15: 387-392.

Williams LK, Reichert A, MacKenzie WR, Hightower AW and Blake PA (2001) Lice, nits, and school policy. Pediatrics 107: 1011-1015.

Yang YC, Choi HY, Choi WS, Clark JM and Ahn YJ (2004) Ovicidal and adulticidal activity of Eucalyptus globulus leaf oil terpenoids against *Pediculus humanus capitis* (Anoplura: Pediculidae). J Agric Food Chem 52: 2507-2511.

Yang YC, Lee SH, Lee WJ, Choi DH and Ahn YJ (2003) Ovicidal and adulticidal effects of Eugenia caryophyllata bud and leaf oil compounds on *Pediculus capitis*. J Agric Food Chem 51: 4884-4888.

Yoon KS, Gao JR, Lee SH, Clark JM, Brown L and Taplin D (2003) Permethrin-resistant human head lice, Pediculus capitis, and their treatment. Arch Dermatol 139: 994-1000.

18. Novel methods for the identification and control of *Culicoides* midges as vectors of emerging diseases

A. Jennifer Mordue (Luntz), John F. Dallas, Damien V. Nolan and James G. Logan

Abstract

Two factors requiring research in the emerging scenario of exotic diseases entering Europe relate to (1) the identification of *Culicoides* vectors and (2) methods for their control. Classical morphological identification of members of the *Culicoides obsoletus* and *Culicoides pulicaris* families, in which vector competence has been shown, is not possible in all cases and new methods of identification are required. DNA-based approaches as an efficient means of pursuing the systematics and identification of animal species are described. The development of multiplex rapid throughput DNA-barcoding assays together with classical taxonomy will provide survey tools for mapping midge distribution across Europe. Linking such tools with vector competence studies in individual specimens will provide direct evidence of competence within species. In addition, the use of molecular techniques in the field will identify larval habitats at farms that may be linked to control. Control of midges as vectors of disease has not been examined at any level within the EU. Classical control methods by insecticide use and changes in husbandry practise are now being looked at again. However new approaches are required to create an integrated pest management strategy. Research into the use of semiochemicals is revealing their potential. The use of pheromones and host kairomones in traps together with exploitation of the phenomenon of 'differential attractiveness' between animals in the same herd will provide new push-pull approaches for midge vector monitoring and control.

Keywords: midge vectors, DNA taxonomy, semiochemicals

Introduction

Vector-borne animal diseases are endemic in regions worldwide and several such as bluetongue are now encroaching northwards from Europe's southern and eastern borders with consequent threats to animal production. There are multiple reasons for this shift. Economic and cultural drivers are causing trade in animals and animal products to be conducted on larger and larger scales. Importing animals and animal products clearly increases the risk of importing novel animal pathogens, including vector-borne pathogens. Vectors of animal pathogens can make incursions into Europe, via passive transport with imported animals, carriage on changed wind patterns, or even range expansion, providing an entry route for pathogens as well as the means of their subsequent transmission. In addition, some animal pathogen vectors are closely related to widespread, native species in Europe, and these can possess unrecognised vector competence that becomes effective following exposure to novel vector-borne pathogens entering Europe.

Perhaps the most significant impact is due to climate change, which is expected to have an increasingly pervasive influence on vector-pathogen systems. This is because many system components, e.g. animal production sources and the range expansions, population demographies and competence levels of vectors, are influenced by factors in the physical environment, and these are expected to undergo systematic trends under different climate change scenarios.

The bluetongue epidemic of cattle and sheep in Belgium, Luxemburg, France, Germany and the Netherlands in winter 2006-spring 2007 provides a graphic, recent example of a vector-borne animal disease emerging in Europe (see Chapter 6). The epidemic involved a novel pathogen (a bluetongue virus (BTV) serotype known previously from sub-Saharan Africa, South America and the Indian subcontinent) and a novel vector or vectors (*Culicoides* midges native to and widespread throughout Europe). The goal of the present chapter is to use the bluetongue-*Culicoides* system to illustrate promising new methodologies for detecting, monitoring and controlling vector species that pose risks of emerging animal diseases in Europe. There are several areas where new knowledge is needed on vectors: species identification, competence assessment, distribution and habitats of key species, and monitoring and vector control. This chapter will address current knowledge on two of these topics; species identification and novel methods of control.

Midge biology

In order to fully understand the epidemiology of diseases such as bluetongue (BT) vectored by *Culicoides* midges now entering Europe it is essential to have detailed knowledge of the biology of potential *Culicoides* vectors. The family Ceratopogonidae within the Diptera contains about 5,500 species comprising 125 genera. One of the four blood-sucking genera is *Culicoides* which has within it more than 1,400 species of biting midge. Midges inhabit all land regions with the exception of New Zealand, Patagonia, the Hawaiian Isles and Antarctica. They thrive equally well in the tropics and temperate regions. All species share a common pattern to their life cycle that comprises the egg, four larval stages, a brief pupal stage and a short-lived adult.

Larvae

The larval stages of *Culicoides* may last as little as four to five days or in temperate climes last throughout winter as a diapausing fourth instar. Larval and pupal stages require moisture, often as free water, but this is found in a surprisingly wide range of habitats including streams, pools, bogs, beaches, areas around leaking irrigation pipes, tree holes, rotting fruit, vegetation and animal dung.

For a group with virtually worldwide distribution it is not surprising that the diversity of larval habitats is large involving aquatic and semi-aquatic habitats. As for many biting flies the majority of research has focused on the adult stage as a consequence of their economic impact and importance in disease transmission. However, the predicted sheer biomass of the larvae as gauged from the adult numbers means that the larvae are likely to be very important cohorts in the ecology of aquatic and swamp communities. However, despite a number of attempts to study larvae and to recover them from their habitat little is known of larval densities and their population dynamics (Kettle and Lawson 1952, Linley and Davies 1971, Aussel and Linley 1993, Blackwell *et al.* 1994a, 1999, Kline and Axtell 1977, Lardeux and Ottenwaelder 1997, Schmidtmann *et al.* 2000, Carpenter *et al.* 2001).

A detailed understanding of larval habitat requirements would be important in applying control methods to the larval stages. Correlations have been made for a few species between key characteristics of habitat and presence of larvae: for example in *C. impunctatus* Goetghebuer, the Scottish biting midge, larvae live in wet grasslands and are associated particularly with plants and mosses such as *Juncus* spp. and *Sphagnum* spp., high levels of organic matter and water content and high redox potential (Kettle 1961, Blackwell *et al.* 1994a, 1999, Carpenter *et al.* 2001).

The normal daily temperature range within the bog is not great and the larvae are within the top 5cm. However, during the overwintering period when the surface is covered with ice and snow the larvae are at lower levels in the soil and are active in temperatures as low as 2-4 °C. In both these studies a combination of botanical classification, soil type and larval sampling were carried out; Hribar (1989) provides a comprehensive review of larval sampling methods. A detailed study (Schmidtmann *et al.* 2000) of the varied soil characteristics across the USA in which larvae of the BTV vector complex *C.variipennis* complex exist has defined significant relationships between high organic content and high levels of phosphate and nitrate as found in livestock waste with *C. variipennis* (Coquillet) and *C. sonorensis* (Wirth & Jones) which is involved with the distribution of BTV. These components were at very low levels in the salt lakes where *C. occidentalis* (Wirth & Jones) is found. In addition *C. sonorensis* was found in wastewater lagoons from dairies and saline springs in regions that would otherwise have been inhabited by *C. variipennis*. These findings suggest that salinity plays an overriding role in the distribution of this species complex. The ability of Ceratapogonid larvae to exploit a large diversity of moist, brackish soils, rich in organic matter has no doubt contributed to the success of the group: it is worth being reminded that such a range of habitats include farmyard manure for *C. nubeculosus* (Meigen) and *C. imicola* Kieffer, elephant dung for *C. loxodontis* and rotting fruit for *C. tutti-fruitti*. Dung heaps, muddy puddles in farm tracks, moist ground around watering holes and streams are all yielding *Culicoides* larvae on farms in the UK.

Adults

Culicoides are some of the smallest blood feeding flies measuring between 1 to 2 mm wing span; despite their smallness they have a serious impact on humans and warm-blooded vertebrates as vectors of disease and nuisance insects. They are a serious biting nuisance to man, their bites in susceptible horses cause acute dermatitis (sweet itch) and some species transmit filarial worms, protozoa and viruses affecting humans and other vertebrates. However, their major importance insofar as vectors of disease is concerned is their transmission of virus to animals and man: more than 50 viruses have been isolated from *Culicoides* species and a number of these viruses are of huge international importance in that they cause the 'OIE' notifiable diseases BT of sheep and cattle and African horse sickness (AHS) of equids (Mellor *et al.* 2000).

Host location and blood feeding, which is by adult females only, involves host recognition, activation of host seeking behaviour, host location and feeding to engorgement (see also Sutcliffe 1987). As far as humans are concerned there is a plethora of stories relating to those individuals, and their particular characteristics, which result in them being pestered and/or bitten heavily compared with others who are bitten less. The same is true for animals and while most are attacked infrequently a subset can be very heavily attacked (Knols *et al.* 1995, Agelopoulos *et al.*1999, Kelly 2001, Jensen *et al.* 2004). Indeed Woolhouse *et al.* (1997) have estimated for vectors transmitting malaria and leishmaniasis that, on average, 20% of the host population contributes 80% of the net transmission potential. This patchiness in utilising only a proportion of potential hosts in a group by blood feeders is of great interest. Kelly (2001) suggests that feeding success is the prime driver for such host-discrimination; individuals that exhibit successful host defence behaviours are able to disrupt feeding especially to repletion. For defensive behaviours of cattle in reducing the load of predatory flies see Torr and Mangwiro (2000). The blood feeder is thus driven to feed on the most susceptible in a group: the evolutionary pressure is for vectors to evolve strategies to discriminate and feed on the least defensive hosts' (Kelly 2001).

There have been a number of studies (Humphreys and Turner 1973, Koch and Axtell 1979, Schmidtmann *et al.* 1980) into host preferences shown by *Culicoides* using animal baited traps; a large variety of wild and domestic/stock animals were used in the different studies ranging through quail, chicken, turkeys, rats, raccoons, opossum, goats, sheep, calves and ponies. Similarly, odours collected from different host species such as waterbuck, cattle, sheep, red deer and pony, result in differential trap catches of *C. impunctatus* females (Mands *et al.* 2004). It is well known that *Culicoides* vectors of BT feed equally well on cattle and sheep and increase incidence of disease in sheep in this way through the reservoir cattle hosts. Overall for the many species of midge monitored the main finding is that the larger host animals are attacked most frequently; no difference in midge preference for wild or domesticated animals has been detected. ELISA analyses of blood meals show that *C. imicola* feeds on more than one host in some feeds with horses, sheep, and cows predominating (Brown, Mordue (Luntz), Mordue and Venter unpublished observations). Analyses of the blood meals of wild caught *C. impunctatus* showed that large mammals such as cow and deer were preferred to rabbits and mice (Blackwell *et al.* 1994b). Bhasin *et al.* (2001) in experiments to show the responsiveness of *C. impunctatus* to CO_2 found maximum effect at levels of CO_2 equivalent to the respiratory output of large cows. Tsetse and stable flies are also attracted to the higher CO_2 output of individual cattle related to their mass (Torr *et al.* 2006).

Host preferences have been shown to be influenced by differential attractiveness between individual potential hosts of the same species that may signal non-appropriateness. The avoidance of an inappropriate host may occur to prevent or minimise time-wasting, energy-depleting journeys to inappropriate hosts, a poor quality blood meal, or to avoid the high risk of succumbing to the most host-defensive individuals (Kelly 2001). Non-host avoidance can also be seen in tsetse flies, *Glossina pallidipes* Austen and *G. m. morsitans* Westwood that preferentially feed on the host species ox and buffalo, compared with the non-host species waterbuck and impala (Gikonyo *et al.* 2002). This occurs despite them being able to perceive volatile chemicals from both host and non-host species. The authors suggest it could be attributable to an allomonal blend consisting of lactone and 2-ketones, produced by the non-host (waterbuck) that are responsible for the differences. Similarly, the head fly, *Hydrotaea irritans* (Fallén) and the horn fly, *Haematobia irritans* (L.) are differentially attracted to individual cattle of the same species within a herd (Steelman *et al.* 1993, Agelopoulos *et al.* 1999, Birkett *et al.* 2004). It has been suggested that the less attractive individuals produce chemicals that 'mask' or interfere with their attractiveness and it was shown that by placing less attractive individuals into an attractive herd, the numbers of flies on attractive cattle can be significantly reduced (Jensen *et al.* 2004). Linalool, naphthalene and propyl butanoate were shown to be repellent, whilst high concentrations of 6-methyl-5-hepten-2-one and 1-octen-3-ol reduced fly loads on heifers and were suggested to be 'non-host' or 'masking' chemicals and could be exploited to develop a push-pull control strategy for livestock (Birkett *et al.* 2004). Since these chemicals are also known to be found in human skin emanations (Bernier *et al.* 2000), they may have a similar role for haematophagous insects and man. Two human-specific compounds, 7-octenoic acid, (*E*) and (*Z*)-3-methyl-2-hexenoic acid are known to reduce trap catches of *An. gambiae* Giles when presented with human odour in odour baited entry traps (OBET's) and are consequently believed to also be masking compounds (Costantini *et al.* 2001). However, they have not been quantified on differentially attractive human individuals.

Compared to the information available for mosquitoes and blackflies there is little data available on the preferential biting sites of midges (see Townley *et al.* 1984). These authors have studied in detail landing and engorging sites of midges in relation to sweet itch. The majority of midges feeding to completion did so on the mane and lower legs; areas along the dorsal mid-line were also preferential engorgement sites. Mid-line and mane sites are areas for sweet-itch, however, the lower leg areas that were also extensively fed upon show no sweet-itch lesions. The propensity for sweet-itch manifestation is very likely due to areas of dorsal skin reacting more severely to *Culicoides* attack; moreover as discussed above some horses prone to sweet itch may be more prone to midge attack than others.

Midges as vectors of disease

The vast majority of *Culicoides* are obligate blood feeders seeking out mammals (including man) and birds. However, adult males do not blood feed. Females fly principally at dawn and dusk to seek a blood meal, a mate or oviposition site.

The main candidates for novel bluetongue vectors in Europe are member species of the *C. obsoletus* and *C. pulicaris* complexes, which are abundant and widespread in Palearctic regions particularly in farmland habitats. The northward expansion of *Culicoides imicola*, the old world vector of bluetongue, has allowed the virus to 'jump' into more Palaearctic species of midge via the biting of viraemic animals. Circumstantial evidence of the involvement of *C. obsoletus* and *C. pulicaris* complexes in BTV transmission in Europe is strong. The spatial and temporal distributions of the complexes coincide with BT outbreaks in Sicily (Torina *et al.* 2004, Purse *et al.* 2005), BTV is present in pools of wild-caught insects (Caracappa *et al.* 2003, De Liberato *et al.* 2005, Savini *et al.* 2005), and BTV replication to likely transmissible titres occurs in both the *C. obsoletus* and *C. pulicaris* complexes under laboratory conditions (Carpenter *et al.* 2006).

The member species of the *C. obsoletus* and *C. pulicaris* complexes are only identifiable with great difficulty on the basis of morphological traits. While this methodology has sufficed for studies in insect systematics, in an epidemiological context it is inadequate. Analyses of *Culicoides* morphology require specialised and specific knowledge that only a few researchers in the world possess, and these capacities cannot be transferred to new personnel when a new epidemiological threat arises unexpectedly. The members of the *C. obsoletus* complex are only easily distinguishable using traits expressed in adult males, and these are doubly irrelevant in an epidemiological context. Male *Culicoides* are not attracted to traps in sufficient numbers to aid species identification and play no part in virus transmission. In some cases, even the most divergent traits available are only semi-diagnostic, e.g. *C. pulicaris* (L.) and *C. punctatus* (Meigen) in the *C. pulicaris* complex overlap for wing morphology (Lane 1981). In addition where it now is important to be able to work year round and to identify vectors of disease in the overwintering larval stage when midges can only be identified to family, a universal identification tool is absolutely essential.

Species identification

Current research on *Culicoides* vector species is creating a knowledge base that promises to be useful for minimising the risk of further bluetongue epidemics in Europe and containing such outbreaks as do occur. A key resource for all these activities is clearly an expanded taxonomy of the *C. obsoletus and C. pulicaris* complexes based on species-diagnostic traits that researchers with generic skills can readily analyse, and that are expressed in epidemiologically relevant life stages that are easy to sample in large numbers.

Molecular identification of vector species

Global trends in biodiversity loss and disease vector emergence, coupled with the increasing rarity of traditional taxonomists, have led to the methodological framework known as 'DNA barcoding' (Hebert *et al.* 2003) or 'DNA taxonomy' (Tautz *et al.* 2003) being proposed as more efficient means of pursing the systematics and identification of animal species. For reasons highlighted elsewhere in the present volume, insect vectors are considered an especially important target for developing the improved taxonomies that DNA-based approaches promise (Besansky *et al.* 2003). In its technically simplest form, DNA barcoding uses partial DNA sequences of COI as raw data, and species differences are inferred according to two alternative criteria (1) an arbitrary threshold of 10-fold higher sequence divergence between members of different species than among conspecifics or (2) a clustering of COI clades from each species derived from a variety of phenetic or phylogenetic tree-building analyses.

Studies involving DNA analysis by the polymerase chain reaction (PCR) provides species-diagnostic traits for a range of *Culicoides* species relevant to BT epidemiology. Members of the *C. imicola* species complex from southern Africa form distinct lineages based on RAPD markers (Sebastiani *et al.* 2001) or sequences of the mitochondrial cytochrome oxidase gene (COI) (Linton *et al.* 2002, Nolan *et al.* 2004), and specimens of *C. imicola* from the Mediterranean basin and South Africa cluster in the same group according to phylogenetic analysis of COI (Dallas *et al.* 2003). Members of the *C. obsoletus* complex are more reliably identifiable using species-specific markers in COI (Pagès *et al.* 2005) or in intergenic transcribed spacer of ribosomal DNA (Gomulski *et al.* 2005, Perrin *et al.* 2006) than using morphological traits. Species-diagnostic PCR assays based on COI sequences can be used to identify four members of the *C. obsoletus* species complex and five members of the *C. pulicaris* species complex found in the UK (Figures 1 and 2, Nolan *et al.* 2007).

The effectiveness of DNA barcoding for species identification is being assessed in various dipteran groups, including potential disease vectors. Positive indications come from two studies of

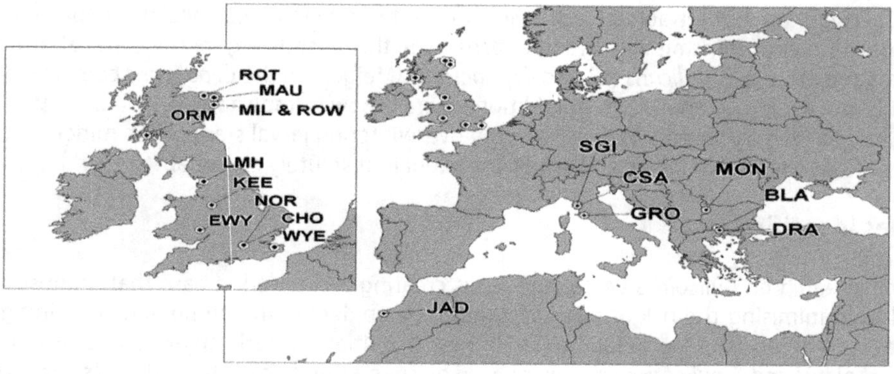

Figure 1. Locations of the sampling sites in the UK and Europe of Culicoides *species collected in 2004/05. UK; Chobham, Ewyas, Keele,Maud, LowMoorhead, Milltimber, Normandy, Ormsary, Rothiemay, Rowett, Wye. Morocco; El Jadida. Italy; San Giuliano, Grosetto, Colle Salvete, Bulgaria; Blagoevgrad, Montana. Greece; Drama. (Taken from Nolan* et al. *2007. Phylogenetic separation and development of molecular identification assays for members of the* Culicoides obsoletus *and* Culicoides pulicaris *species complexes, implicated vectors of bluetongue in Europe. Veterinary Microbiology (2007) doi:10.1016/j.vetmic.2007.03.019).*

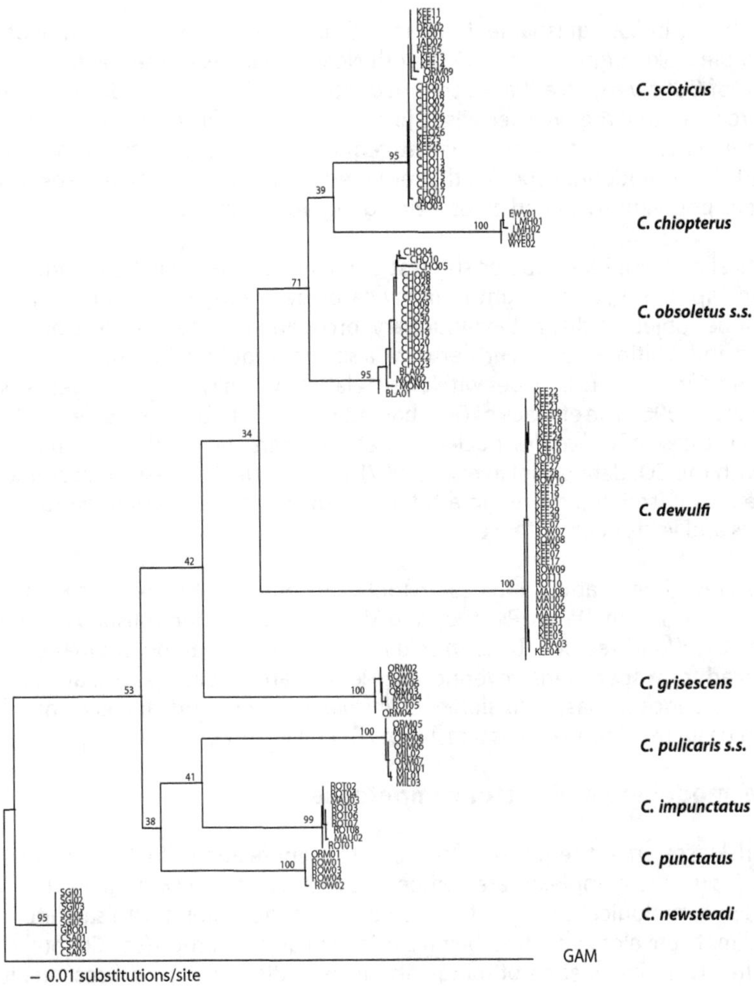

Figure 2. Phylogenetic relationships among COI haplotypes of 127 midges of the Culicoides obsoletus *and* Culicoides pulicaris *species complexes. Maximum likelihood tree constructed using a TrN substitution model with a proportion of invariable rates and a γ-distributed rate. Numbers on the nodes represent bootstrap values (10,000 replicates) obtained under distance criterion of a maximum likelihood setting with fast stepwise addition, 4-5 morphologically identified male specimens were used in the case of the* C. obsoletus *complex, in the case of the* C. pulicaris *complex all specimens used were morphologically identified females. (Taken from Nolan* et al. *2007 Phylogenetic separation and development of molecular identification assays for members of the* Culicoides obsoletus *and* Culicoides pulicaris *species complexes, implicated vectors of bluetongue in Europe. Veterinary Microbiology (2007) doi:10.1016/j.vetmic.2007.03.019).*

mosquitoes: 53 species in 9 genera from eastern Canada (Cywinska *et al.* 2006) and 63 species in 15 genera from India (Kumar *et al.* 2007). Interspecific divergence in COI averages 20-fold (Canada) or 8-fold (India) higher than intraspecific divergence in these groups, with only one pair of species in each study being indistinguishable. More mixed results emerge from a study of two genera of big-headed flies (Skevington *et al.* 2007): both Nearctic clades of the genus *Nephrocerus* had very low levels of COI divergence that overlapped intraspecific variation. Both studies used 'total evidence' approaches that require specialised joint analyses of morphometric and molecular data. In contrast, researchers of vector-borne disease require a methodology that is accessible to non-specialists and that can identify species that are members of closely-related species complexes and that exist as populations spread across wide geographic ranges.

Other theoretical and empirical studies suggest that using single-gene barcoding to distinguish closely-related Dipteran species might in many cases be ineffective. Theoretically, the mtDNA genome might be subject to distinct evolutionary forces and therefore have its own evolutionary history (Ballard and Whitlock 2004). Furthermore, a species requiring identification might overlap in its mtDNA and nuclear gene lineages with close relatives when species divergence is sufficiently recent, (Meier *et al.* 2006). The efficacy of DNA barcoding needs to be assessed empirically for each focal group. In the case of *Culicoides*, nuclear genetic markers such as rITS1 (Figure 3) show good congruence with the COI data of Nolan *et al.* (2007), and its effectiveness for vector identification will clearly be maximised when used in a total-evidence context in combination with data on morphometrics and vector competence.

As well as species identification, molecular tools can be used to great effect to investigate population structure. Nolan, Dallas, Piertney and Mordue (Luntz) (unpublished observation) have shown with the Old World vector of BT *C. imicola* that there are 27 haplotypes in the 272 base pairs of COI sequenced from specimens covering the Mediterranean basin and that their occurrence provide strong evidence of mass incursion of *C. imicola* from the south and east into the northern Mediterranean basin most probably as a result of climate change.

Surveillance, modelling and vector competence

An adequate risk assessment requires both accurate identification of which members of the *C. obsoletus* and *C. pulicaris* complexes are competent vectors and knowledge of their geographic distributions. Epidemiological surveys of *Culicoides* BTV vectors along with satellite imagery and modelling of climate envelopes are now being carried out across Europe (see Chapter 6 for details). Such surveys involving knowledge of midge abundance, distribution and seasonality, livestock distribution, farm habitats and meteorological data will provide advance notice of high risk areas within Europe. Coupled with this approach are laboratory studies of vector competence (Institute of Animal health (IAH), Pirbright, Surrey) whereby freshly collected adult female *Culicoides* are fed under biosecure conditions on blood containing BTV and incubated for viral replication and subsequent BTV detection, isolation and titre measurement. Key to understanding species identity is the application of DNA-based species identification methods on the same samples to unambiguously identify those species with high vector competence. Such work is currently being undertaken in the UK using the CO1 methods of Nolan *et al.* (2007). This assay is a one tube 'multiplex' reaction that produces diagnostic amplicon bands for each species in a speedy low-cost high throughput assay.

Monitoring and control

Culicoides midge control has been little researched in recent years. Although insecticide treatments can be used to control other haematophagous insects, they have limited efficacy for adult

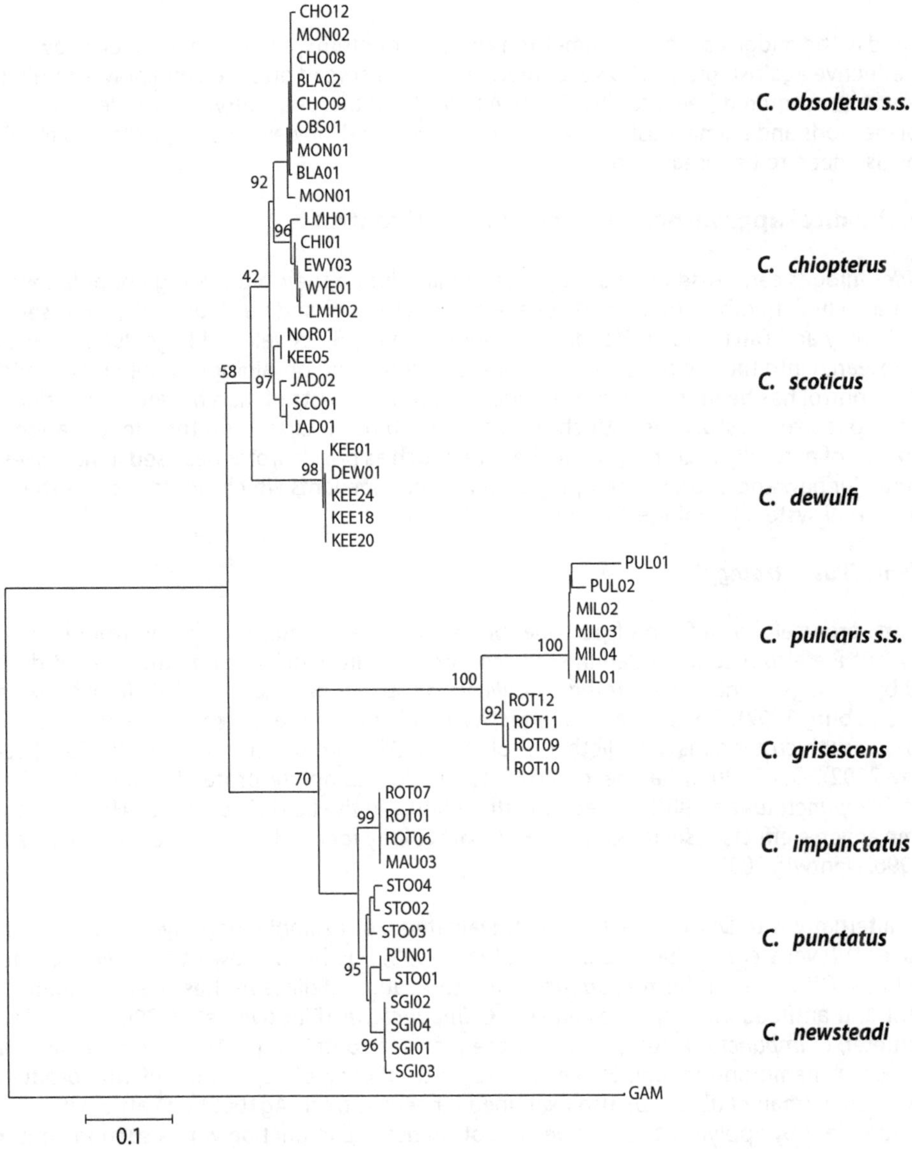

Figure 3. Phylogenetic relationships among ribosomal DNA ITS1 haplotypes of 47 midges of the Culicoides obsoletus *and* Culicoides pulicaris *species complexes. Neighbour joining tree constructed using the Tamura-Nei distance measure. Numbers on the nodes represent bootstrap values (10,000 replicates).*

Culicoides populations (Linley and Davies 1971). Additionally, insecticide use on larvae and the modification of extensive larval habitats (e.g. coastal wetlands) is impractical and environmentally unsound for many areas around the globe and often provides short-term control only (Linley 1976). Work is now starting however to study the efficacy of currently available treatments for sheep and cattle midge control. Deltamethrin and cypermethrin pour-ons have been shown to be highly effective against midges, however there is a lack of spread on sheep and new formulations of products are required (see also Chapter 7). A logical and thorough survey of modern veterinary control methods and animal husbandry protocols is required, however new approaches to midge control also need to be researched.

Semiochemical approaches to monitoring and control

Culicoides midges can cause great annoyance to man due to their bites. Midge populations can occur in such high numbers that recreational activities, tourism and outdoor work can be severely limited (Linley and Davies 1971, Hendry and Godwin 1988, Blackwell and Page 2003, Cilek *et al.* 2003). Research into the use of semiochemicals (behaviour and physiology modifying chemicals) in midge control has been triggered by midges as nuisance pests of man however the principles may be applied to livestock also. Much recent research has focussed on the identification and exploitation of naturally occurring semiochemicals such as host kairomones used in host seeking behaviour, for new and improved trapping systems and repellents which would provide the basis for a push-pull system for midge control.

Repellents (Push strategy)

Repellents are useful as a form of personal protection against nuisance biting from *Culicoides* midges, but are also recommended in some countries for livestock during outbreaks of disease caused by pathogens that are vectored by *Culicoides* spp. midges, e.g. *C. imicola* (Braverman and Chizov-Ginzburg 1997). One of the most effective repellents on the market for haematophagous insects is one which contains *N, N*, diethyl-n-toluamide DEET; in concentrations of 10–90% (Fradin and Day 2002). DEET (20%) has been shown to provide complete protection for 6 to 7 hours against *C. impunctatus* and 98% protection after 8 hours in the field (Trigg 1996). However, there are some adverse effects associated with DEET, particularly for children (Selim *et al.* 1995, Wantke *et al.* 1996, Nentwig 2003).

Natural alternatives to DEET have also been evaluated. For example, bog myrtle, *Myrica gale* L. (Myricaceae) (Evans *et al.* 1996, Stuart and Stuart 1998) has been shown to be repellent to *C. impunctatus*. Oil of neem, from *Azadirachta indica* A. Juss (Meliaceae) has been evaluated for repellent and antifeedant properties against *C. impunctatus* (Blackwell *et al.* 2004). In a Y-tube olfactometer, *C. impunctatus* females chose the arm not containing ≥ 1% of the repellent and fed less on a membrane feeder. However, the field efficacy of repellency of this product is unknown. Braverman *et al.* (2000) also examined a neem extract (Ag1000) on cattle with a 4.5% active ingredient by applying it to a polyester-cotton net in conjunction with a suction light trap plus CO_2. This extract reduced the trap catches by 34%. Other repellents tested in this way by Braverman *et al.*, included Lice free, Mosi-guard (with 50% *Eucalyptus maculata* var. *citriodora*) Hook extract and DEET. A 66% reduction in trap catches was found with DEET. Relative to DEET, Lice free repellent, and Mosi-guard reduced trap catches by 15 and 39% respectively (Braverman *et al.* 2000). However, when Mosi-guard was tested against *C. imicola* in the field using a black light trap covered with impregnated mesh, three times as many midges were caught on the treated traps than the controls. This could highlight the differential efficacy of repellents against different species of biting Diptera (Braverman *et al.* 1999), and that the repellent is attractive in the absence of host odour to *C. imicola*. Although the authors describe their method as promising for field evaluation of repellents (before animal testing), field tests on live animals is the definitive method which must eventually be applied to evaluate the efficacy and longevity of repellents. The eucalyptus-based insect repellent, *p*-menthane diol (PMD) has also been investigated as a human repellent against *C. impunctatus* and was found to provide complete protection up to 7 hours and 98% protection up to 8 hours. This matched the efficacy of DEET which was used as a positive control (Trigg 1996). In the laboratory, PMD prevents *C. sonorensis*, the primary vector of bluetongue virus in North America (Holbrook *et al.* 2000) from feeding on treated human arms for up to 9 h (Trigg and Hill 1996).

Some protection can also be provided to humans by lotions and oils applied to the skin, such as Avon Skin-So-Soft ®, although it is thought that these substances provide a barrier that traps the midges, rather than repelling them *per se* (Magnon *et al.* 1991). This product can give around 71% repellency against *C. furens* (Poey) compared with 98% provided by 75% DEET.

Insect repellents like DEET are normally applied to skin, clothing, or other surfaces that discourage insects from landing. However, the concept of having a repellent that works at a distance is receiving much attention. This idea of 'spatial repellency' (also termed 'attraction inhibition') could prevent insects from coming into an area such as a home, stable or barn without the need for direct application. Whilst DEET gives good contact repellency, in laboratory trials it has been shown not to display any spatial repellency to *Aedes aegypti* (L.) mosquitoes (Kline *et al.* 2005, Bernier *et al.* 2005). However, other compounds such as nepetalactone (the major constituent of catnip oil) linalool and dehydrolinalool do exhibit spatial repellency against *Ae. aegypti* mosquitoes (Kline at al. 2005, Bernier *et al.* 2005). Insecticides have also been shown to exhibit spatial repellency. For example, a paper strip impregnated with metofluthrin, a newly synthesised pyrethroid, reduces catches of *Anopheles* spp. and *Culex* spp. mosquitoes when used along side cow and human-baits. Laboratory and field trials demonstrated that the mosquitoes were affected by airborne metofluthrin vapour and not by contact with the strip. Spatial repellency caused by insecticides is thought to be due to high knockdown activity and a sub lethal effect (Kawada *et al.* 2004). Presently spatial repellents have not been investigated or developed for *Culicoides* midges against man or farm animals, however, the concept is certainly appealing considering the great numbers that can be found around a host.

Traps (Pull strategy)

Many different trap designs for insects exist and have been tested over the years, each with varying degrees of efficacy (see Chapter 19). Simple boards or tubes, which are covered in a sticky material, will retain all sizes of insects that have flown or been blown onto them and are therefore not selective. Light traps and suction traps catch a greater number of insects but are also, like sticky traps, non-selective and skilled taxonomists are required to undertake the laborious job of sorting through the catch. Traps baited with behaviour modifying chemicals, such as pheromones, are now being used extensively to monitor and sometimes to regulate populations of insect pests. For haematophagous insects traps have been developed primarily for monitoring mosquito populations. Recently, because of environmental needs to reduce the use of chemical insecticides and the increased resistance of mosquitoes to available pesticides, interest in the use of traps for control is increasing (see Kline 1999). However, with the exception of tsetse flies, very little is known of the impact of traps on the population dynamics of blood feeding flies. The success of baited traps and treated targets, in reducing tsetse populations (Vale 1993) has engendered serious interest in applying such technologies to other biting flies. Until recently only 2 trap types, or limited variations on their design, were used. These were CDC (Centre for Disease Control) trap and a light trap (Service 1993) both with fans for sucking insects attracted by light into a holding vessel. Present approaches are now focused on improved design to reduce the catch of non-target insects by the use of traps such as CFG (Counter Flow Geometry; Figure 4) and the incorporation of host-derived kairomones such as CO_2 and/or 1-octen-3-ol in the trap (Kline 1994, 1999).

One trap, that is now commercially available, called the Mosquito magnet (for example the American Biophysics Corporation PRO) has received attention in recent years and is an effective way of trapping *Culicoides* species midges (Cilek and Kline 2002, Cilek and Hallmon 2005). These traps provide heat, moisture and CO_2 through the catalytic combustion of propane (Cilek and Hallmon 2005). In coastal northwest Florida, much work has been done to evaluate the efficacy and consistency for this trap to reduce midge populations in a residential area (Cilek *et al.* 2003). *C. mississippiensis*, *C. furens* and *C. melleus* were caught in the greatest numbers when using CO_2 (500 ml/min) and 1-octen-3-ol, 3-n-propylphenol and 4-methylphenol (in a 2:1:8 ratio at 5.4mg/h) as bait. The combination of CO_2 plus the above compounds work synergistically, however, the traps failed to consistently reduce the populations. This was demonstrated by the variable weekly reduction which ranged from 4.2 to 85.3% (Cilek and Hallmon 2005). Other studies in Scotland have shown a similar pattern. Continuous trapping of midges in a domestic garden using three Mosquito Magnet traps emitting CO2 (500ml/min) and 1-octen-3-ol (5 mg/h) over a 4 week period failed to reduce the numbers of midges caught (Mordue(Luntz) unpublished observation) Such trapping systems do not reduce populations of midges where larval habitat is extremely extensive as in Scotland, or traps can become overwhelmed by the extremely high numbers of host-seeking

Figure 4. Willem Takken and Simon Carpenter preparing a counter flow geometry (CFG) odour baited trap for Culicoides *collection in Argyllshire Scotland, August 2000.*

adult females (Day *et al.* 2001, Cilek *et al.* 2003). However, when midge populations decline to less than 500 midges per trap a reduction may be achieved (Cilek *et al.* 2003). In such a domestic setting, the number of traps needed and their placement is crucial for effective control. Currently it seems that a high number of traps are required – more than one per backyard in some areas (Cilek *et al.* 2003). Trap design can also affect the species of midge caught in traps. For example, the ABC PRO trap baited with CO_2 plus 1-octen-3-ol, 3-n-propyl-phenol and 4-methylphenol caught significantly more *C. mississippiensis* Hoffman than a CFG trap baited with the same stimuli. In contrast, the CDC traps caught significantly more *C. barbarosa* than the ABC traps (Cilek and Kline 2002).In Scotland the addition of hexane extracts of hair of water buffalo (*Bubalus bubalis* L.) to CO_2 and 1-octen-3-ol bait in Mosquito Magnet Pro traps increased trap catches by 262% with greatest increases for *C. impunctatus* and *C. pulicaris* (Mands *et al.* 2004).

Pheromones, such as aggregation and sex pheromones, can provide a species-specific method of attracting and trapping insects. Pheromones are known to play a role in *Culicoides* midge ecology. For example, virgin female *C. nubeculosus* adults release a volatile sex pheromone that attracts males and stimulates copulation (Kremer *et al.* 1979). Females emit maximum levels of pheromone 24-48 h post emergence (Kremer *et al.* 1979, Ismail 1982, Mair and Blackwell, 1998), and this decreases with increasing age or after females take a blood meal (Kremer *et al.* 1979, Ismail 1982). Electroantennogram (EAG) studies on male *C. nubeculosus* have shown that volatile emissions from live virgin females are detected only by male antennal receptors (Mair and Blackwell 1998). Odours from mated females also produce smaller but significant EAG responses in males, suggesting that pheromone release continues after mating but at a reduced level (Mair and Blackwell 1998). The site of emission of the pheromone in virgin females is from the ventral side of the abdomen via atrichial areas which have a close association with haemolymph oenocytes (Ismail and Kremer 1983, Ismail and Zachary 1984). Release or dispersal of the pheromone may be associated with a characteristic grooming behaviour, including wing flutter which precedes mating. Preliminary work suggests this pheromone could be the hydrocarbon n-heptadecane, however, further work is required to confirm the activity of this compound (Mordue (Luntz) and Mordue 2003).

Female Scottish biting midges, *C. impunctatus*, are known to utilise an aggregation or recruitment pheromone that is attractive to other females and that may also be used during host-location (Blackwell *et al.* 1994b, 1996). However, this pheromone has not yet been identified. A contact sex pheromone, which stimulates copulation, has been identified for *C. melleus* (Coquillet) and comprises 2-methyldocosane, 8-methyldocosane and 9-methyltricosane (Linley and Carlson 1978).

Despite the evidence for the presence of pheromones in *Culicoides* midges and the identification of several putative components, these compounds have not been exploited to any extent in control technologies. Further work is required to identify more *Culicoides* pheromones (e.g. for *C. impunctatus*) and test the feasibility of using these compounds in trapping systems for monitoring or control.

The expectations for trapping and impact on midge populations are somewhat different from those for crop pests. For nuisance midges such as *C. impunctatus* in Scotland and *C. furens* and *C. hollensis* (Melander and Brues) in the eastern United States reduction in population density in relatively small domestic scenarios such as around dwelling houses etc to midge densities that can be tolerated is the aim. The trapping out for crop pests or for tsetse that requires reduction of 90% of the population to achieve control is not required and may not be achievable because of immigration from nearby sites. For midge vectors of disease where bites from viraemic insects must be avoided the challenge is great and an integrated pest management strategy is required utilising pesticides, repellents, semiochemicals, alteration to larval habitats and changed husbandry practises e.g. to maintain animals indoors at night when at most risk from midge bites.

Insecticides can be used in conjunction with odour-baited trapping systems. For example, insecticide treated fabric targets baited with CO_2 and 1-octen-3-ol can reduce local populations of *C. furens* in Florida and island populations in the Bahamas (Day *et al.* 2001). Similarly, Kline and Lemire (1998) demonstrated that mosquito populations can be reduced by using a single line barrier of either attractant-baited traps or cloth targets impregnated with the pyrethroid insecticide lambdacyhalothrin.

Although odour baited traps and repellents for *Culicoides* midge control are currently commercially available, further work is required to improve these technologies as both have had varying degrees of success. For livestock protection on farms much less research has been done and new approaches are needed. An additional problem in trapping techniques for monitoring is that odour baited traps cannot compete with a natural host and therefore their efficacy is debatable. However, by creating a scenario whereby odour-baited traps are used in combination with repellents, thus making the host less preferred, a 'push-pull' control strategy could be developed (Wanzala *et al.* 2004). In essence, future programmes should be developed that depend upon this technology: the 'push' away from potential hosts by the use of repellents augmented by the 'pull' into traps baited with specific attractants and localised toxins to kill the trapped midges.

The natural differential response of haematophagous insects to human hosts is thought to be a result of a combination of attractive and repellent olfactory stimuli (Costantini *et al.* 1998, Bernier *et al.* 2002) and can thus been described as a natural 'push-pull system' (Miller and Cowles 1990, Costantini *et al.* 1998). Although there is much evidence for the existence of chemicals causing behavioural movements towards a host that are involved in the 'pull' element of this system, such as CO_2 and 1-octen-3-ol, most evidence for the 'push' element suggests that the chemicals are inhibitors, rather than repellents (Costantini *et al.* 2001). Ultimately, this natural 'push-pull system' could be exploited and an artificial push-pull control strategy could be developed.

Future work

Future research on the molecular identification of European *Culicoides* in the context of the BT threat should focus on three areas. First, given the uncertainties about species discrimination outlined above, the effectiveness of single-gene DNA barcoding needs to be fully assessed in the full range of closely-related species present. Second, it is impossible to imagine using DNA barcoding alone to analyse the huge numbers of midges typically collected in field traps without developing and validating high-throughput methods such as those developed by Nolan *et al.* (2007). By analogy with the high-throughput genotyping methods used in human genetics (e.g. Yu *et al.* 2006), one promising approach is to analyse DNA extracted from pooled samples of hundreds or thousands of insects using species-diagnostic real-time PCR assays. DNA assays with a smaller physical scale than PCR, such as DNA mass spectrometry (Tost and Gut 2006), might also be suitable. Ideally, each sample would correspond to the yield of a single trapping session or trapping locality, and the outputs would be estimates of species proportions in each sample. This would obviate the need for time-consuming identification at the level of individual insects in routine monitoring, although individual analysis would be possible as well, e.g. in combined species identification-vector competence assays. Third, research on vector competence needs to move from phenotype to genotype, i.e. away from the highly specialised oral susceptibility assays that for reasons of biosecurity can only be done in a single licensed laboratory and towards genome mapping, transcription profiling and proteomics as these methodologies become easier to apply to insect vectors, especially vectors that cannot be maintained in laboratory culture.

If vector capacity is stably stratified among species then disease risk will be identifiable and predictable using the methods outlined in this chapter to analyse insect vectors. If it is unstable or more homogeneous among species, such that most or all species will at any given time contain some populations of competent vectors, then vector competence becomes ubiquitous, disease risk becomes more and more unpredictable, and the scientific value of analysing vectors becomes questionable. It will be especially important to assess the sensitivity of vector competence to systematic trends in the physical environment, such as species comparisons from North to South of the UK, to gain understanding of stability of vector competence within the same species in different climatic envelopes.

To elucidate fully the epidemiology of vector-borne diseases such as BT, surveillance and risk assessment of the vector requires to be carried out (see Chapter 6). This has been already carried out with the old world vector, *C. imicola* and work in this area is being applied to predictive modelling for novel more palaearctic species of midge. Governmental organisations in Europe (e.g. SEERAD in Scotland) are beginning to support such work for emerging midge vectors in northern Europe involving systematic trapping of the vectors for presence, distribution and abundance of *Culicoides* species and their potential as vectors of animal diseases. Such surveys will yield important data that requires to be linked together with all relevant research (vector haplotypes, vector abundance and seasonality, larval habitats, climate change, wind patterns and virus serotype) in order to understand and predict the risk of BT entering and establishing itself in more northerly parts of Europe. Coupled with this much more research is required on integrated pest management strategies for vector control including the use of semiochemicals such as pheromones, host kairomones, host allomones for use in push-pull strategies in vector control.

Acknowledgements

Work on molecular taxonomy and semiochemicals of midges by the authors had been funded by EU, Defra, BBSRC & Scottish Enterprise Proof of Concept. W. Mordue is thanked for scientific input & help with the manuscript.

References

Agelopoulos N, Birkett MA, Hick AJ, Hooper AM, Pickett JA, Pow EM, Smart LE, Smiley DWM, Wadhams LJ and Woodcock CM (1999) Exploiting semiochemicals in insect control. Pest Sci 55: 225-235.

Aussel JP and Linley JR (1993) Activity of *Culicoides furens* larvae (Diptera: Ceratogoponidae). J Med Entomol 31: 774-789.

Ballard JWO and Whitlock MC (2004) The incomplete natural history of mitochondria. Mol Ecol 13: 729-744.

Bernier UR, Kline DL, Barnard DR, Schreck CE and Yost RA (2000) Analysis of human skin emanations by gas chromatography-mass spectrometry. 2. Identification of volatile compounds that are candidate attractants for the yellow fever mosquito (*Aedes aegypti*). Anal Chem 72: 747-756.

Bernier UR, Kline DL, Schreck CE, Yost RA and Barnard DR (2002) Chemical analysis of human skin emanations: comparison of volatiles from humans that differ in attraction of *Aedes aegypti* (Diptera: Culicidae). J Am Mosq Control Assoc 18: 186-195.

Bernier UR, Furman KD, Kline DL, Allan SA and Barnard DR (2005) Comparison of contact and spatial repellency of catnip oil and *N,N*-Diethyl-3-methylbenzamide (Deet) against mosquitoes J Med Entomol 42: 306-311

Besansky NJ, Severson DW and Ferdig MT (2003) DNA barcoding of parasites and invertebrate disease vectors: what you don't know can hurt you. Trends Parasitol 19: 545-546.

Bhasin A, Mordue(Luntz) AJ and Mordue W (2001) Field studies on the efficacy of host odour baits for the biting midge, *Culicoides impunctatus* in Scotland. Med Vet Entomol 15: 147-156.

Birkett MA, Agelopoulos N, Jensen V, Jespersen MB, Pickett JA, Prijs J, Trapman HTGJ, Wadhams LJ and Woodcock CM (2004) The role of volatile semiochemicals in mediating host location and selection by nuisance and disease-transmitting cattle flies. Med Vet Entomol 18: 313-322.

Blackwell A, Dyer C, Mordue (Luntz) AJ, Wadhams LJ and Mordue W (1996) The role of 1-octen-3-ol as a host-odour attractant for the biting midge, *Culicoides impunctatus* Goetghebuer, and interactions of 1-octen-3-ol with a volatile pheromone produced by parous female midges. Physiol Entomol 21: 15–19.

Blackwell A, Young MR and Mordue W (1994a) The microhabitat of *Culicoides impunctatus* (Diptera: Ceratopogonidae) larvae in Scotland. Bull Entomol Res 84: 295-301.

Blackwell A, Mordue (Luntz) AJ and Mordue W (1994b) Identification of blood meals of the Scottish biting midge, *Culicoides impunctatus*, by indirect enzyme-linked immunosorbent assay (ELISA). Med Vet Entomol 8: 20-24.

Blackwell A, Dyer C, Mordue (Luntz) AJ, Wadhams LJ and Mordue W (1994c) Field and laboratory evidence for a volatile pheromone produced by parous females of the Scottish biting midge, *Culicoides impunctatus*. Physiol Entomol 19: 251-257.

Blackwell A, Lock KA, Marshall B, Boag B and Gordon SC (1999) The spatial distribution of larvae of *Culicoides impunctatus* biting midges. Med Vet Entomol 13: 362-371.

Blackwell A and Page SJ (2003) Biting Midges and Tourism in Scotland. In J. Wilks and S.J. Page eds, Managing Tourist Health and Safety in the N ew Millenium: Global Perspectives Advances in Tourism Research Series. Elsevier Science, Oxford pp177-196.

Blackwell A, Evans KA, Strang RHC and Cole R (2004) Toward development of neem-based repellents against the Scottish Highland biting midge *Culicoides impunctatus*. Med Vet Entomol 18: 449-452.

Braverman Y and Chizov-Ginzburg A (1997) Repellency of synthetic and plant-derived preparations for *Culicoides imicola*. Med Vet Entomol 11: 355-360.

Braverman Y, Chizov-Ginzburg A and Mullens BA (1999) Mosquito repellent attracts *Culicoides imicola* (Diptera: Ceratopogonidae). J Med Entomol 36: 113-115.

Braverman Y, Wegis MC and Mullens BA (2000) Response of *Culicoides sonorensis* (Diptera: Ceratopogonidae) to 1-octen-3-ol and three plant-derived repellent formulations in the field. J Am Mosq Control Assoc 16: 158-163.

Caracappa S, Torina A, Guercio A, Vitale F, Calabro A, Purpari G, Ferrantelli V, Vitale M and Mellor PS (2003) Identification of a novel bluetongue virus vector species of *Culicoides* in Sicily. Vet Rec 153: 71-74.

Carpenter S, Mordue (Luntz) AJ and Mordue W (2001) Oviposition in *Culicoides impunctatus* under laboratory conditions. Entomol Exp Appl 101:123-129.

Carpenter S, Lunt HL, Arav D, Venter GJ and Mellor PS (2006) Oral susceptibility to bluetongue virus of *Culicoides* (Diptera: Ceratopogonidae) from the United Kingdom. J Med Entomol 43: 73-78.

Cilek JE and Kline DL (2002) Adult biting midge response to trap type, carbon dioxide, and an octenol-phenol mixture in northwestern Flordida J Am Mosq Control Assoc 18(3): 228-231.

Cilek JE and Hallmon CF (2005) The effectiveness of the mosquito magnet ® trap for reducing biting midge (Diptera: Ceratopogonidae) populations in coastal residential backyards J Am Mosq Control Assoc 21(2): 218-221.

Cilek JE, Kline DL and Hallmon CF (2003) Evaluation of a novel removal trap system to reduce biting midge (Diptera: Ceratagoponidae) populations in Florida backyards. J Vector Ecol 28: 23-30.

Costantini C, Sagnon N, DellaTorre A, Diallo M, Brady J, Gibson G and Coluzzi M (1998) Odor-mediated host preferences of West African mosquitoes, with particular reference to malaria vectors. Am J Trop Med Hyg 58: 56-63.

Costantini C, Birkett MA, Gibson G, Ziesmann J, Sagnon NF, Mohammed HA, Coluzzi M and Pickett JA (2001) Electroantennogram and behavioural responses of the malaria vector *Anopheles gambiae* to human-specific sweat components. Med Vet Entomol 15: 259-266.

Cywinska A, Hunter FF and Hebert PDN (2006) Identifying Canadian mosquito species through DNA barcodes. Med Vet Entomol 20: 413-424.

Dallas JF, Cruickshank RH, Linton YM, Nolan DV, Patakakis M, Braverman Y, Capela R, Capela M, Pena I, Meiswinkel R, Ortega MD, Baylis M, Mellor PS and Mordue AJ (2003) Phylogenetic status and matrilineal structure of the biting midge, *Culicoides imicola*, in Portugal, Rhodes and Israel. Med Vet Entomol 17: 379-387.

Day JF, Duxbury CV, Glasscock S and Paganessi JE (2001) Removal trapping for the control of biting midge populations. In: Carlson DB, Carroll JD Jr, Cordero A, Rey JR, Taylor DS (eds.) 4th Workshop on Saltmarsh Management and Research. Tech Bull Fla Mosq Control Association. 2000 October 24-27; Vero Beach, FL. Fort Myers, FL: Florida Mosquito Control Association. 15-16.

De Liberato, C, Scavia G, Lorenzetti R, Scaramozzino P, Amaddeo D, Cardeti G, Scicluna M, Ferrari G and Autorino GL (2005) Identification of *Culicoides obsoletus* (Diptera: Ceratopogonidae) as a vector of bluetongue virus in central Italy. Vet Rec 156: 301-304.

Evans KA, Blackwell A, Deans SG and Simpson MJ (1996) The repellent properties of *Myrica gale* to haematophagous insect pests of man In: Widley KB (ed.) Proceedings 2nd International Conference of Insect Pests of Urban Environment Heriot-Watt University Edinburgh Scotland, Wheatons Exeter pp 427-430.

Fradin MS and Day JF (2002) Comparative efficacy of insect repellents against mosquito bites. New Engl J Med 347: 13-18.

Gikonyo NK, Hassanali A, Njagi PGN, Gitu PM and Midiwo JO (2002) Odor composition of preferred (buffalo and ox) and nonpreferred (waterbuck) hosts of some savanna tsetse flies. J Chem Ecol 28: 969-981.

Gomulski LM, Meiswinkel R, Délecolle JC, Goffredo M and Gasperi G (2005) Phylogenetic relationships of the subgenus Avaritia Fox, 1955 including *Culicoides obsoletus* (Diptera, Ceratopogonidae) in Italy based on internal transcribed spacer 2 ribosomal DNA sequences. Syst Entomol 30: 619-631.

Hebert PDN, Cywinska A, Ball SL and De Waard JR (2003). Biological identifications through DNA barcodes. Proc R Soc Lond B 270: 313-321.

Hendry G and Godwin G (1988) Biting midges in Scottish forestry: a costly irritant or a trivial nuisance? Scottish Forestry 42: 113-119.

Holbrook FR, Tabachnik WJ, Schmidtmann ET, McKinnon CN, Bobian RJ, Grogan WJ (2000) Sympatry in the *Culicoides variipennis* complex (Diptera: Ceratopogonidae): a taxonomic reassessment. J Med Entomol 37: 65-76.

Hribar LJ (1989) A review of methods for recovering biting midge larvae (Diptera: Ceratogoponidae) from substrate samples. J Agric Entomol 7(1): 71-77.

Humphrey J and Turner EC (1973) Blood-feeding activity of female *Culicoides* (Diptera: Ceratopogonidae). J Med Entomol 10: 79-83.

Ismail MT (1982) Factors affecting pheromone secretion and oviposition by fertilized females of *Culicoides nebeculosus* (Diptera: Ceratogoponidae). J Insect Physiol 28: 835-840.

Ismail MT and Kremer M (1983) Determination of the site of pheromone emission in the virgin females *Culicoides nubeculosus* Meigen (Diptera: Ceratogoponidae). J Insect Physiol 29: 221-224.

Ismail MT and Zachary D (1984) Sex pheromones in *Culicoides nubeculosus* (Diptera: Ceratogoponidae): possible sites of production and emission. J Chem Ecol 10: 1385-1398.

Jensen KMV, Jespersen JB, Birkett MA, Pickett JA, Thomas G, Wadhams LJ and Woodcock CM (2004) Variation in the load of the horn fly, *Haematobia irritans*, in cattle herds is determined by the presence or absence of individual heifers. Med Vet Entomol 18: 275-280.

Kawada H, Maekawa Y, Tsuda TY and Takagit M (2004) Laboratory and field evaluation of spatial repellency with metofluthrin-impregnated paper strip against mosquitoes in Lombok island, Indonesia. J Am Mosq Control Assoc 20: 292-298.

Kelly DW (2001) Why are some people bitten more than others? Trends Parasitol 17: 578-581.

Kettle DS and Lawson JWH (1952) The early stages of the British biting midges *Culicoides* Latreille and related genera. Bull Entomol Res 43: 421-467.

Kettle DS (1961) A study of the association between moorland vegetation and breeding sites of *Culicoides* (Diptera: Ceratogoponidae). Bull Entomol Res 52: 381-411.

Kline, D.L. 1994 Olfactory attractants for mosquito surveillance and control. J Am Mosq Control Assoc.10:280-296

Kline DL (1999) Comparison of two American Biophysics mosquito traps: the professional and a new counterflow geometry trap. J Mosq Cont Assoc. 15:276-282.

Kline DL and Axtell RC (1977) Distribution of *Culicoides hollensis*, *C. furens* and *C. bermudensis* in relation to plant cover in a North Carolina salt marsh (Diptera: Ceratopogonidae). J Med Entomol 13: 545-552.

Kline DL and Lemire GF (1998) Evaluation of attractant-baited traps/targets for mosquito management on Key Island Florida, USA. J Vector Ecol 23: 171-185.

Kline DL, Bernier UR, Posey KH and Barnard DR (2005) Olfactometric Evaluation of Spatial Repellents for *Aedes aegypti*. J Med Entomol 40: 463-467.

Knols BGJ, de Jong R and Takken W (1995) Differential attractiveness of isolated humans to mosquitoes in Tanzania. Trans R Soc Trop Med Hyg 89: 604-606.

Koch HG and Axtell RC (1979) Attraction of *Culicoides furens* and *Culicoides hollensis* (Diptera, Ceratopogonidae) to animal hosts in a salt-marsh habitat. J Med Entomol 15: 494-499.

Kremer M, Ismail MT and Rebholtz C (1979) Detection of a pheromone released by the females of *Culicoides nebeculosus* (Diptera: Ceratogoponidae) attracting the males and stimulating copulation. Mosq News 39: 627-631.

Kumar NP, Rajavel AR, Natarajan R and Jambulingam P (2007) DNA barcodes can distinguish species of Indian mosquitoes (Diptera: Culicidae). J Med Entomol 44: 1-7.

Lane RP (1981) A quantitative analysis of wing pattern in the *Culicoides pulicaris* species group (Diptera, Ceratopogonidae). Zool J Linn Soc 72: 21-41.

Lardeux FJR and Ottenwaelder T (1997) Density of larval *Culicoides belkini* (Diptera: Ceratogoponidae) in relation to physicochemical variables in different habitats. Ann Entomol Soc Am 34: 387-395.

Linley JR (1976) Biting midges of mangrove swamps and salt marshes (Diptera, Ceratopogonidae). In: Cheng L (ed.) Marine Insects. North Holland Publishing Company, Amsterdam, pp 335-376.

Linley JR and Carlson DA (1978) A contact mating pheromone in the biting midge,*Culicoides melleus*. J Insect Physiol 24: 423–427.

Linley JR and Davies JB (1971) Sandflies and tourism in Florida and the Bahamas and Caribbean area. J Econ Entomol 64: 264-278.

Linton YM, Mordue (Luntz) AJ, Cruickshank RH, Meiswinkel R, Mellor PS and Dallas JF (2002) Phylogenetic analysis of the mitochondrial cytochrome oxidase subunit I gene of five species of the *Culicoides imicola* species complex. Med Vet Entomol 16: 139-146.

Magnon GJ, Robert LL, Kline DL and Roberts LW (1991) Repellency of two DEET formulations and Avon Skin-So-Soft ® against biting midges (Diptera: Ceratopogonidae) in Honduras. J Am Mosq Cont Assoc 7: 80-82.

Mair J and Blackwell A (1998) Effect of multiple mating on the mating behaviour of *Culicoides nubeculosus* (Diptera: Ceratogoponidae). J Med Entomol 35: 996-1001.

Mands V, Kline DL and Blackwell A (2004) *Culicoides* midge trap enhancement with animal odour baits in Scotland. Med Vet Entomol 18: 336-342.

Meier R, Shiyang K, Vaidya G and Ng P (2006) DNA barcoding and taxonomy in Diptera: a tale of high intraspecific variability and low identification success. Syst Biol 55: 715-728.

Mellor PS, Boorman J and Baylis M (2000) *Culicoides* biting midges: Their role as arbovirus vectors. Ann Rev Entomol 45: 307-340.

Miller JR and Cowles RS (1990) Stimulo-deterrent diversion - a concept and its possible application to onion maggot control. J Chem Ecol 16: 3197-3212.

Mordue (Luntz) AJ and Mordue W (2003) Biting midge chemical ecology. Biologist 50: 159-162.

Nentwig G (2003) Use of repellents as prophylactic agents. Parasitol Res 90 (Suppl. 1): S40-48

Nolan DV, Carpenter S, Barber J, Mellor PS, Dallas JF, Mordue (Luntz) AJ and Piertney SB (2007) Rapid diagnostic PCR assays for members of the *Culicoides obsoletus* and *Culicoides pulicaris* species complexes, implicated vectors of bluetongue virus in Europe. Vet Microbiol 124: 82-94.

Nolan DV, Dallas JF and Mordue (Luntz) AJ (2004) Molecular taxonomy and population structure of a *Culicoides* midge vector. Vet Ital 40: 352–359.

Pagès N and Sarto V Monteys I (2005) Differentiation of *Culicoides obsoletus* and *Culicoides scoticus* (Diptera: Ceratopogonidae) based on mitochondrial cytochrome oxidase subunit I. J Med Entomol 42: 1026-1034.

Perrin A, Cêtre-Sossah C, Mathieu B, Baldet T, Délecolle JC and Albina E (2006) Phylogenetic analysis of *Culicoides* species from France based on nuclear ITS1-rDNA sequences. Med Vet Entomol 20: 219-228.

Purse BV, Mellor PS, Rogers DJ, Samuel AR, Mertens PPC and Baylis M (2005) Climate change and the recent emergence of bluetongue in Europe. Nat Rev Microbiol 3: 171-181.

Savini G, Goffredo M, Monaco F, Di Gennaro A, Cafiero MA, Baldi L, De Santis P, Meiswinkel R and Caporale V (2005) Bluetongue virus isolations from midges belonging to the Obsoletus complex (Culicoides, Diptera: Ceratopogonidae) in Italy. Vet Rec 157: 133-139.

Schmidtmann ET, Bobian RJ and Belden RP (2000) Soil chemistries define aquatic habitats with immature populations of the *Culicoides variipennis* complex (Diptera: Ceratopogonidae). J Med Entomol 37: 58-64.

Schmidtmann ET, Abend JF and Valla ME (1980) Nocturnal blood-feeding from pastured calves by the Ceratopogonid midge, *Culicoides Venustus*, in New York State. Mosq News 40: 571-577.

Sebastiani F, Meiswinkel R, Gomulski LM, Guglielmino CR, Mellor PS, Malacrida AR and Gasperi G (2001) Molecular differentiation of the Old World *Culicoides imicola* species complex (Diptera, Ceratopogonidae), inferred using random amplified polymorphic DNA markers. Mol Ecol 10: 1773-1786.

Selim SRE, Hartnagel G, Osimitz K, Gabriel L and Schoenig GP (1995) Absorption, metabolism and excretion of n,n-diethyl-m-toluamide following dermal application to human volunteers. Fundam Appl Toxicol 25: 95-100.

Service MW (1993) Mosquito Ecology: Field Sampling Methods 2nd Edition. Chapman and Hall London.

Skevington JH, Kehlmaier C and Stahls G (2007) DNA barcoding: mixed results for big-headed flies (Diptera: Pipunculidae). Zootaxa 1423: 1-26.

Steelman CD, Gbur EE, Tolley G and Brown AH (1993) Individual variation within breeds of beef-cattle in resistance to horn fly (Diptera: Muscidae). J Med Entomol 30: 414-420.

Stuart AE and Stuart CLE (1998) A microscopic slide test for the evaluation of insect repellents as used with *Culicoides impunctatus*. Entomol Exp Appl 89: 277-280.

Sutcliffe JF (1987) Distance orientation of biting flies to their hosts.Insect Sci Appl 8: 611-616.

Tautz D, Arctander P, Minelli A, Thomas RH and Vogler AP (2003) A plea for DNA taxonomy. Trends Ecol Evol 18: 70-74.

Torina A, Caracappa S, Mellor PS, Baylis M and Purse BV (2004) Spatial distribution of bluetongue virus and its *Culicoides* vectors in Sicily. Med Vet Entomol 18: 81-89.

Torr SJ and Mwangiro TN (2000) Interactions between cattle and biting flies: effects on the feeding rate of tsetse. Med Vet Entomol 14: 400-409.

Torr SJ, Mangwiro TN and Hall DR (2006) The effects of host physiology on the attraction of tsetse (Diptera: Glossinidae) and Stomoxys (Diptera: Muscidae) to cattle. Bull Entomol Res 96: 71-84.

Tost J and Gut IG (2006) DNA analysis by mass spectrometry - past, present and future. J Mass Spectrom 41: 981-995.

Townley P, Baker KP and Quinn PJ (1984) Preferential landing and engorging sites of the *Culicoides* species landing on a horse in Ireland. Equine Vet J 16: 117-120.

Trigg JK (1996) Evaluation of a Eucalyptus-based repellent against *Culicoides impunctatus* (Diptera: Ceratogoponidae) in Scotland. J Am Mosq Control Assoc 12: 329-330.

Trigg JK and Hill N (1996) Laboratory evaluation of a *Eucalyptus*-based repellent against four biting arthropods. Phytother Res 10: 313-316.

Vale GA (1993) Development of baits for tsetse flies (Diptera: Glossinidae) in Zimbabwe. J Med Entomol 30: 831-842.

Wantke F, Focke M, Hemmer W, Gotz M and Jarisch R (1996) Generalised urticaria induced by a diethyltoluamide-containing insect repellent in a child. Contact Dermatitis 35: 186-187.

Wanzala W, Sika NFK, Gule S and Hassanali A (2004) Attractive and repellent host odours guide ticks to their respective feeding sites Chemoecology 14: 229-232

Woolhouse MEJ, Dye C, Etard JF, Smith T, Charlwood JD, Garnett GP, Hagan P, Hii JLK, Ndhlovu PD, Quinnell RJ, Watts CH, Chandiwana SK and Anderson RM (1997) Heterogeneities in the transmission of infectious agents: Implications for the design of control programs. Proc Nat Acad Sci USA 94: 338-342.

Yu AR, Geng HF and Zhou XR (2006) Quantify single nucleotide polymorphism (SNP) ratio in pooled DNA based on normalized fluorescence real-time PCR. BMC Genomics 7: 143.

Surveillance, protection and control

19. Monitoring systems for adult insect pests and disease vectors

Yu Tong Qiu, Jeroen Spitzen, Renate C. Smallegange and Bart G. J. Knols

Abstract

In this chapter, we evaluate a selection of monitoring techniques for dipterous pests and disease vectors. Some of the monitoring systems can also be used as control measures to reduce nuisance biting or the chance of a disease outbreak. In Europe, resting catches, ovitraps, and CDC light traps are widely applied methods to monitor populations of mosquitoes. Recent developments in monitoring techniques in the USA and Germany have resulted in sophisticated traps such as the Mosquito Magnet Liberty and the BG-sentinel trap. Midges are successfully monitored with the Onderstepoort trap. Sandflies are monitored by light traps, sticky materials and human bait collection. Adding carbon dioxide to the majority of trapping devices generally enhances the capture of mosquitoes, midges and sandflies, whereas the effects of other odours are more species-specific. Sticky materials are commonly used to monitor houseflies, stable flies, and sheep blowflies. In addition, odour-baited traps have been developed to be able to monitor flies more species-specifically. Although there is a dire need for a universal trapping device that can be used to monitor diverse pests and thus the risk of invasions, dispersal, and dynamics of populations, no such device exists today and is unlikely to be forthcoming due to species-specific kairomones and behavioural responses towards traps. This hinders continuous monitoring efforts needed to study emerging pests and vector-borne diseases.

Keywords: monitoring, trapping, Diptera, pests, disease, vectors, Europe

Introduction

The introduction of a pathogen in a competent population of disease vectors in a specific area often goes unnoticed until a disease outbreak occurs. In spite of its eradication in the USA, locally transmitted malaria has been reported sixty-three times since 1957; a recent event being the outbreak of malaria in southern Florida (Filler *et al.* 2006). West Nile Virus was only detected in New York City after five patients developed severe disease in a short time (Asnis *et al.* 2000, 2001). Likewise, the introduction of Usutu virus in Vienna, Austria, went unnoticed until blackbird (*Turdus merula* L.) mortality became unusually high (Weissenböck *et al.* 2002). Although such incidents (i.e. the introduction of a vector or an infected human or animal host) can never be completely prevented, it would be desirable that the risk of such outbreaks could be better predicted and anticipated through regular or continuous intensive monitoring of potential vector populations. Examining the presence, or the likelihood, that a particular vector or pathogen will or has invaded a new area, however, can be complex (Lounibos 2002). Even in the presence of intensive monitoring, accidental introductions of vectors (e.g. by aircraft or other modes of transport (Tatem *et al.* 2006, van der Weijden *et al.* 2007)) may lead to small populations for which the sensitivity of monitoring devices is simply too low. Moreover, the introduced species may also not be responsive to the devices in use. Monitoring of vector population density and dynamics over time is often costly and labour intensive, which frequently results in *ad hoc* activities and absence of longitudinal studies. As such, identification of a potential threat and the onset of monitoring schemes is often following and not preceding an outbreak.

In this chapter, we describe and examine existing monitoring techniques for adult insect pests and disease vectors. We have limited ourselves to the Dipteran species that are described elsewhere in this volume. We give an outline of the techniques used in Europe and review recent developments elsewhere which have the potential to be useful for future application in Europe. Much of this information is summarised in Table 1.

Mosquito monitoring

One hundred and five different species of the Culicidae family have been described in Europe (http://www.faunaeur.org (Accessed 15 September 2007)). Most species only cause nuisance to humans and animals but a few can transmit diseases that have been or still are a serious threat to human and animal health. With the exception of Turkey (see Chapter 3), malaria has been eradicated in all of Europe (Bruce-Chwatt and De Zulueta 1980), and the likelihood of its return to previously endemic areas is considered negligibly small (Hay *et al.* 2004, see also Chapter 2). Arboviruses, like West Nile Virus (WNV) and Chikungunya, however, are a major cause for concern as has been witnessed by the recent outbreak of Chikungunya in northern Italy (see Chapter 10). Such events boost the search for novel methods to monitor vector populations, in this case *Aedes albopictus* (Skuse) (e.g. Facchinelli *et al.* 2007).

There are varying approaches for sampling adult mosquito populations (Service 1993). The responsiveness of mosquitoes to any given method is strongly dependent on their behaviour and/or physiological make-up. One can target emerging adults, resting adults, host-seeking individuals, or gravid females that are about to oviposit. Each physiological state requires a different approach and success rates vary between species and sexes. Certain sampling methods may selectively collect specific groups or age cohorts of a mosquito population, and sampling results may not necessarily represent and thus reflect the true age structure of the entire population. Because only female mosquitoes feed on humans and other animals and are thus responsible for transmission of disease, they have almost exclusively been the target of most monitoring systems. Only when male mosquitoes respond to mammalian odour in pursuit of virgin females (e.g. *Aedes diantaeus* Dyar and Knab (Jaenson 1985) and *Mansonia africana* (Theobald) (McIver *et al.* 1980)) do they end up in trapping devices, and only sound traps have been developed specifically to trap males (Ikeshoji 1982, 1986). In the following, we describe the most commonly used monitoring tools for adult female mosquitoes.

Sampling resting adults

Mosquitoes spend a considerable amount of time resting in man-made or natural shelters. Depending on the species, their resting places vary from human dwellings, animal stables, burrows, vegetation, hollow trees, and various crevices. For sampling resting adults indoors, either mouth - or battery-operated aspiration tubes or spray sheet collection can be used, although application of the latter method is limited in Europe due to regulatory restrictions on insecticide usage (EC 2005). Most mosquito species rest outside human houses and indoor sampling is therefore only suitable for so-called endophilic species that spend considerable periods of their gonotrophic cycle or overwintering period indoors. Outdoor sampling of resting adults in natural shelters is not very efficient because individuals can be scattered over a large area and may be difficult to locate. The efficiency of outdoor collections can be improved by creating artificial resting structures (e.g. resting boxes; WHO 1975) that attract mosquitoes, from which they can subsequently be collected. Some of these artificial devices have been proven to be effective in sampling certain

Table 1. An overview of commonly used monitoring tools for adult dipterous pests and disease vectors.

System	Photograph	Bait	Placement	Power	Suitability [1]					Advantages	Disadvantages	Website
					Cul[2]	Cer[2]	Phl[2]	Mus[2]	Cal[2]			
Aspirator		None	In hand	No	+++	-	+			Simple, inexpensive	Labour intensive; variation in personal skill	www.bioquip.com
Resting box		None	Ground	No	++	-				Simple, inexpensive	Not effective, trapped insects escape easily	
CDC Light traps		None or CO_2 (dry ice or compressed gas)	1.5m	6V battery	+++	++	+++			Light weight	Trapped mosquitoes damaged by fan; collects non-target insects	www-rci.rutgers.edu/~insects/cdctrap.htm; www.johnwhock.com
New Jersey Light trap		None or CO_2	1.5m	AC	+++	+	+			Runs for a long period of time without tending	Collect non-target insects; not usable in places without electricity	www.rci.rutgers.edu/~insects/njtrap.htm; www.johnwhock.com

Table 1. Continued.

System	Photograph	Bait	Placement	Power	Suitability [1]					Advantages	Disadvantages	Website
					Cul[2]	Cer[2]	Phl[2]	Mus[2]	Cal[2]			
Human baiting collection		Human	Ground	No	+++	-	+			Effective, provides direct human-vector contact rate	Risk of disease transmission; depending on skill and variation in individual attractiveness	
Animal baited traps		Blood hosts of vectors (horses, birds or dogs)	Ground or canopy	No	+++	-	+			Helpful if aim is e.g. ornithophilic species	Labour intensive; variation in attractiveness; ethical clearance needed with respect to animal welfare	
Mosquito Magnet traps		CO_2 (combustion of propane), humid air, heat, 1-octen-3-ol, Lurex3	Ground	AC or propane-powered	+++	++				High efficacy	Expensive; propane needed; often malfunctioning after low outside temperature and/or after rain storms	www.mosquitomagnet.com
Counterflow Geometry traps (model MM-X)		CO2, human odours, synthetic odours, 1-octen-3-ol, Lurex3	15-50cm	6 or 12V battery	+++	-				Light weight; can test effects of odour baits without CO2	Commercially not available	

BG-Sentinel trap		BG-lure, CO_2	Ground	12V battery	+++	-				Light weight; high trapping efficacy; washable; collapsible	Not very easy to connect to CO2	www.bg-sentinel.com/
Sticky ovitrap		Water or infusion	Ground	No	++	-				Inexpensive, collects both eggs and adults, effective for several Culicine species	Daily check needed	
Gravid trap		Water or infusion	Ground	No	++	-				Specific for Culicines	Only gravid females trapped	www.johnwhock.com
Onderste poort		None	1.5m	AC	-	+++	+			Specific for midges		
Sticky traps		None or liver and sodium sulphide	0.5-2.5m	No	-		+++	+++	+++	Simple, inexpensive, non-toxic, gives species information	Not species specific; inaccurate; time-consuming; accumulation of dust	www.olsonproducts.com; www.catchmaster.com; www.agrisense.co.uk; www.farnam.com

Table 1. Continued.

System	Photograph	Bait	Placement	Power	Suitability [1]					Advantages	Disadvantages	Website
					Cul[2]	Cer[2]	Phl[2]	Mus[2]	Cal[2]			
Spot card		None	Near resting sites	No				+		Simple, inexpensive	Inaccurate; time-consuming; gives no information on species; sex; physiological age	
Scudder grid		None	Ground	No				+		Simple, inexpensive	Identification difficult; variation due to weather; placement; time and personal skill	
Jug traps		Attractive fluid	1m	No				+++		Simple, inexpensive	Needs to be checked regularly; efficiency depends on colour and toxicant	www.farnam.com
Nzi trap		1-octen-3-ol, acetone, phenols	Ground	No				++		Catches few non-target species	Only outdoors; large	www.vestergaard-frandsen.com; www.nzitrap.com

Lucitrap		Lucilure	0.5-1m	No					-	Species-specific	Not effective in Europe	www.bioglobal.com.au
Agrilure Blowfly Trap		Freeze dried liver	1-1.5m	No					+++	Non-toxic, easy to use and maintain	Needs to be checked regularly; formulation, packaging and delivery of bait not favourable	www.agrimin.com

[1] +-+++: suitable-highly suitable and/or frequently used for monitoring this insect group; -: not suitable; blank: no information available.

[2] Cul: Culicidae; Cer: Ceratopogonidae; Phl: Phlebotominae; Mus: Muscidae; Cal: Calliphoridae.

mosquito species (see review by Service 1993). Instead of sampling resting adults, however, monitoring methods that lure and intercept flying mosquitoes by using attractants are often more effective and therefore more extensively used. Several sampling methods for the latter purpose are discussed below.

Human bait collection (HBC)

The human bait collection (HBC) method is one of the principal sampling methods in monitoring mosquito populations. Humans wait for mosquitoes to land on their exposed legs and catch them with an oral aspirator. Because this method uses a human as bait, it is most effective in catching unfed female anthropophilic mosquito species, which are often vectors of human diseases. Data from HBC, expressed as number of mosquito bites per human per day, can be used to estimate biting rates and infection rates. Factors influencing the reliability of the data collected using the human bait collection method include: (1) weather conditions such as temperature, wind speed, rain, and air pressure; (2) the circadian activity of mosquitoes and seasonal shift; (3) the location of the bait collection; (4) the preferential biting behaviour of mosquito species; and (5) temporal and inter-individual variation in 'attractiveness' of the fieldworkers in combination with their catching skills (Smith 1956, Curtis *et al.* 1987, Service 1993). As this method directly measures the intensity of (potentially infectious) mosquito bites it has long been considered the golden standard in mosquito monitoring (Mboera 2005). A serious drawback of the HBC is the (sometimes considerable) risk of infection for the person acting as an odour bait, which makes this method ethically unsatisfactory. The search for safe and reliable alternatives to replace the HBC method is ongoing. Mathenge *et al.* (2002) designed an exposure-free bednet trap ('Mbita' trap) and a semi-field study revealed that this Mbita trap caught 43.2% and 89.7% of the human landing catch in either competitive or non-competitive conditions, respectively. Field evaluation of this method in comparison with the HBC and CDC light trap (see below) in western Kenya yielded satisfactory results (Mathenge *et al.* 2005). However, a field study in Madagascar showed a much lower efficacy of the Mbita trap compared to the HBC method, possibly due to the zoophilic nature of *Anopheles arabiensis* Patton in that setting (Laganier *et al.* 2003). New trapping techniques are usually compared with the HBC to evaluate their value as an alternative (Costantini *et al.* 1998, Williams *et al.* 2006).

Animal-baited traps

In arbovirus surveillance, animals that are major or intermediate hosts are often used as bait to attract and lure mosquitoes into traps. The animals that are used as bait can either be protected or exposed to mosquito bites. When large animals such as horses, sheep or dogs are used as baits, they are kept in large netting cages; examples are the Shannon trap, Magoon, or Bates trap (Service 1993). A gap of ~5-20 cm is left open at the bottom of the trap to allow mosquitoes to enter the cage. Following a blood meal most mosquitoes can be found resting inside and often near the ceiling of the cage. Smaller hosts, like sentinel birds are kept in smaller netting cages that are connected to mosquito collecting chambers (often using the lobster-trap principle to retain mosquitoes) on one or both sides. Chickens, pigeons, and house sparrows are often-used sentinel species because they are preferred blood-hosts of the major WNV vectors such as *Culex pipiens* L. (Deegan *et al.* 2005, Balenghien *et al.* 2006, Darbro and Harrington 2006). A classical design is the Lard-Can trap (Lepore *et al.* 2004, Bellamy and Reeves 1952, Deegan *et al.* 2005).

Horse- and bird-baited traps are often used in WNV surveillance because horses are one of the dead-end hosts and birds are the intermediate hosts of the virus. By using horse- and bird-baited traps, as well as HBC, Balenghien *et al.* (2006) studied the blood-host preference of French mosquito species to evaluate their importance in the transmission of WNV. The height at which animal-baited traps are suspended has been found to influence their capture efficacy greatly: pigeon-baited Lard-Can traps placed at canopy-level (7-9 m) caught significantly greater numbers of *Cx. pipiens* and *Cx. restuans* Theobald than those placed at ground level (Deegan *et al.* 2005). The same was found for *Cx. pipiens* when carbon-dioxide baited CDC miniature light traps (see below) were used in the northeastern USA (Drummond *et al.* 2006, Andreadis and Armstrong 2007). In most cases, ethical clearance is needed when animals are being used as bait in Europe.

Light traps

Light traps are effective tools for sampling mosquitoes because many species are attracted to or disoriented by light. Light traps have a fan to suck mosquitoes flying close to the light into a collection container. Mosquito populations caught with light traps are biased by species and physiological state. Light traps are generally not useful in trapping diurnal mosquitoes. It has been reported that mosquitoes respond to light cyclically and most mosquitoes caught in light traps are host-seeking mosquitoes. A general disadvantage of light traps is that a broad range of non-target insects are caught, particularly when the traps are used outdoors. In addition, the fan often mutilates the insects caught, which then complicates species determination. Finally, due to the strong air current, mosquito survival is always very low and dried dead specimens are more difficult to dissect. Various types of light traps are used to sample mosquitoes. The following two types are often used as standard methods in mosquito surveillance programmes:

New Jersey light trap. The New Jersey (NJ) light trap (John W. Hock Company, Gainesville, FL, USA) was designed about 40 years ago and is still a commonly used monitoring tool (Service 1993, Mulhern 1942). The trap contains a 25 W white light bulb and a ventilator to attract and suck the mosquitoes into a catching jar. The light and the fan are powered by 120-220 Volt mains. Trapped mosquitoes are killed with a toxic agent. The efficacy of the NJ light trap is affected by position, suspension height, other nearby light sources, the lunar cycle, temperature, humidity, and wind. It is advised that the NJ light trap be operated during four successive nights, turning on the light before sunset and off just after sunrise. This type of trap is suitable for investigating the presence of species in an area and is commonly used in longitudinal surveys to estimate mosquito abundance (http://www.johnwhockco.com/products/1112.htm (Accessed 15 September 2007)) although its sensitivity has been considered lower than the CDC miniature light trap against certain species in California (Reisen *et al.* 1999, 2000, 2002).

CDC miniature light trap. The Centers for Disease Control (CDC) light trap (John W. Hock Company, Gainesville, FL, USA) was designed in 1962 for short-term surveillance of mosquito vectors of arboviruses (Sudia and Chamberlain 1962). Similar to the New Jersey light trap, a CDC trap is often equipped with white light and carries a motorised fan producing a down draft to collect mosquitoes in a capture net. The role of the light varies for different species and sometimes odour-baited CDC light traps catch similar numbers with or without the light bulb (Becker *et al.* 1995, Mboera *et al.* 2000). The lightweight CDC trap is powered by a 6-V battery. Its portability and the possibility it offers for trapping live mosquitoes makes the CDC trap the most commonly used mosquito monitoring system. The efficacy of the CDC trap is enormously increased when baiting it with carbon dioxide, either produced by dry ice (Newhouse, 1966) or from a cylinder (Parker *et al.* 1986).

The height at which a CDC trap is suspended affects the mosquito species composition of the catch. A recent study carried out in Maryland, USA, compared CDC traps suspended at four different heights. The results showed that CDC traps placed at 0.5 or 1.5 m caught higher numbers of the majority of mosquito species trapped, including *Aedes vexans* Meigen, *Anopheles bradleyi* King, *An. crucians* Wiedemann, *Coquillettidia perturbans* (Walker), *Culex salinarius* Coquillett, *Ae. sollicitans* (Walker), *Ae. taeniorhynchus* (Wiedemann), than when placed at 3m and 5m (Shone *et al.* 2006). Conversely, the highest capture of the WNV vector *Cx. pipiens* was found when CO_2 -baited CDC or light traps were placed in the tree canopy (~7.6 m) (Andreadis and Armstrong 2007).

In the tropics, the CDC trap is often placed next to a human sleeping under a bednet. The light bulb is often not used, as it does not seem to increase the catch of targeted anthropophilic mosquito species, including *An. gambiae* Giles, *An. arabiensis* Patton, *An. funestus* Giles, and *Cx. quinqufasciatus* Say (Mboera *et al.* 1998, Costantini *et al.* 1998, Mathenge *et al.* 2005). In some settings, like for malaria vectors in Tanzania, a good correlation between CDC light trap catches and HBC was found (Lines *et al.* 1991), where three CDC traps placed besides occupied bednets could replace two people performing an indoor HBC. This correlation was less clear in other studies (e.g. Hii *et al.* 2000) and depends also on the type of data analysis performed on catches that compare different methods (Smith 1995, see also http://www.sti.ch/fileadmin/user_upload/Pdfs/trapmeth.pdf (Accessed 15 September 2007)).

Odour-baited traps

Kairomones originating from the skin or exhaled breath of hosts are the major cues that guide host-seeking mosquitoes to find their blood meals (Takken and Knols 1999). Mosquito traps baited with host or artificial odour blends therefore increase the likelihood of catching female mosquitoes. Although odour-baited traps are effective in sampling specific mosquito species, the bait used may not be effective for other species present in the same area because of differences in host preference.

Carbon dioxide (CO_2) is a major compound in animal breath and is used by mosquitoes to locate hosts (Rudolfs 1922, Mboera and Takken 1997); therefore this compound is one of the most effective baits and is most commonly used in mosquito traps. Carbon dioxide is normally produced either using dry ice, combustion of propane or provided from gas cylinders. When using cylinders, the release rate of CO_2 can be regulated and is usually set at 100-1000 ml/min (animal breath equivalent: 250-350 ml/min (Gillies and Wilkes 1968, Mboera *et al.* 2000). Increasing the CO_2 flow rate extends the attractive distance (McIver and McElligott 1989).

Numerous studies have shown that the effectiveness of trapping devices is improved when baited with CO_2 (Mboera and Takken 1997). Studies in Northeast USA revealed that adding CO_2 to a CDC light trap or an AB flicker trap increased the number of *Ae. canadensis* (Theobald), *Ae. vexans*, *Cq. perturbans*, *Cx. salinarius*, and anopheline mosquitoes (Vaidyanathan and Edman 1997, Shone *et al.* 2006). Shone *et al.* (2003) also found that the WNV vector *Ae. albopictus* was more attracted to CDC Fay-Prince traps baited with CO_2 than traps without it. Hundreds to thousands of compounds are emitted by mosquito blood-hosts (Takken and Knols 1999, Bernier *et al.* 2000, Curran *et al.* 2005, 2007). Some of these compounds are used as kairomones by mosquitoes in their search for blood.

Several mosquito kairomones have become commercially available for baiting mosquito traps. One of the most widely used compounds is 1-octen-3-ol, which is a typical odour of herbivorous mammals (Hall *et al.* 1984), and the olfactory system of several mosquito species is highly sensitive to this compound (Grant and O'Connell 1996, Lu *et al.* 2007, Takken and Kline 1989). When carbon dioxide is used together with 1-octen-3-ol, an increase in trapping efficacy and a synergistic effect was reported in attracting various mosquito species (Kline *et al.* 1991, Kemme *et al.* 1993, Shone *et al.* 2006). Burkett *et al.* (2001) found that adding 1-octen-3-ol to a Mosquito Magnet trap (see below) baited with CO_2 trapped more *Ae. vexans* and *Cx. tritaeniorhynchus* in Korea. In several other studies, however, 1-octen-3-ol did not show any additional effect to CO_2 when added to a CDC light trap, AB flicker trap or MM-X trap (Vaidyanathan and Edman 1997, Shone *et al.* 2003, 2006). It has also been found that adding 1-octen-3-ol to CO_2 baited traps decreased the number of *Cx. pipiens* (Burkett *et al.* 2001) and *Cx. quinquefasciatus* (Russell and Ritchie 2004). The only other compound currently manufactured as mosquito attractant is Lurex3™ which contains L(+) lactic acid (Acree *et al.* 1968) and 2-hydroxy propionic acid. This blend is considered more suitable for trapping *Ae. albopictus* when combined with 1-octen-3-ol compared to this latter compound alone (Hoel *et al.* 2007).

Many other compounds have been used as baits in mosquito traps and include, amongst others, acetone, ammonia, dimethyl disulfide and carboxylic acids (Mboera *et al.* 2000, Kröckel *et al.* 2006, Njiru *et al.* 2006, Qiu *et al.* 2007). None of these compounds are currently available on the market as mosquito attractants.

Mosquito Magnet traps. The former American Biophysics Cooperation (ABC, Rhode Island, USA) developed several types of odour-baited Mosquito Magnet traps utilising counterflow technology, in which an outgoing air flow carries chemical lures that attract mosquitoes to the vicinity of the trap and an incoming air flow draws mosquitoes into a collecting chamber. These traps (www.mosquitomagnet.com (Accessed 16 September 2007)), are baited with either CO_2 or other odour lures, such as 1-octen-3-ol or Lurex3™ (Hoel *et al.* 2007). Carbon dioxide is provided through combustion of propane gas, which generates supposedly attractive heat and moisture at the same time (Kline 1999).

Commercially available ABC counterflow traps are Mosquito Magnet® Defender, Mosquito Magnet® Liberty (MM-L), Mosquito Magnet® Liberty PLUS and Mosquito Magnet® Pro (MM-Pro). The traps are easy to use and can operate without human interference for 2-3 weeks. However, there are often reports of malfunctioning (US CPSC 2006, Frans Jacobs personal communication) especially during cold weather conditions or after a heavy rain storm.

One of the counterflow traps that can be combined with cylinder generated CO_2 is the MM-X counterflow geometry trap. This type of trap is designed for research purposes and not available commercially. In a field assessment, this trap was found to capture higher numbers of *An. crucians* Wiedemann than propane-powered traps (Kline 2002). In a screen-house semi-field study in Kenya, MM-X traps captured more than 90% of the released female *An. gambiae s.s.* mosquitoes when baited with a blend of foot odour, ammonia and CO_2 (Njiru *et al.* 2006). In a field study in the Gambia, it was found that CO_2 is essential for the attractiveness of MM-X traps to various groups of mosquitoes (Qiu *et al.* 2007). The MM-X trap was found to be more effective in capturing *Cx. quinquefasciatus* and *Cx. tarsalis* Coquillet than two other types of propane-powered Mosquito Magnet traps (Cooperband and Cardé 2006). The MM-Pro trap uses a thermoelectric generator to power the fans and this trap is used to reduce mosquito populations, especially *Cx.*

quinquefasciatus in high-density areas (Kline *et al.* 2006). The drawbacks of such sophisticated traps are the difficulty of repairing malfunctioning traps, their high cost, and dependence on the availability of propane gas tanks.

Most commercially-available ABC counterflow traps utilise propane combustion to generate CO_2 (Kline 2002). Some of these ABC traps were compared with other types of traps in Korea. It was found that two types of Mosquito Magnet traps (MM-Pro baited with 1-octen-3-ol and MM-X baited with CO_2 or CO_2 + 1-octen-3-ol), trapped, in total, equal numbers of mosquitoes as a human-baited Shannon trap and more than an ABC light trap with or without light, baited with CO_2 or 1-octen-3-ol, or the combination of the two. The MM-X traps caught similar numbers of *An. lesteri* Baisas and Hu, *An. sinensis* Wiedemann and *An. yatsushiroensis* Miyazaki as the MM-Pro traps, whereas more *Ae. vexans* were caught with MM-X traps. Interestingly, it was found that the numbers of the WNV vector, *Cx. pipiens*, were strongly decreased when 1-octen-3-ol was added to any of the traps (Burkett *et al.* 2001). The same authors also reported that the counterflow traps had better performance in attracting anopheline and aedine mosquitoes than a light trap baited with CO_2, but no difference was found for culicines (Burkett *et al.* 2001). A field evaluation of mosquito traps in Sweden reported that an American Biophysics trap, CF2000 (alias MM-Beta-1), using propane to generate CO_2, attracted more species and more male mosquitoes than a CDC trap baited with dry ice (Blackmore and Dahl 2002). Cooperband and Cardé (2006) compared the trapping efficiency of four CO_2-baited traps (Encephalitis Virus Surveillance trap, MM-L, MM-F and MM-X) for *Cx. quinquefasciatus* and *Cx. tarsalis* that had been released in a field wind tunnel. The MM-X showed the highest trapping efficiency for both mosquito species. Flight behaviour was compared when mosquitoes entered the vicinity of the traps and apparently this became tortuous downwind for the traps that performed poorly, and only so upwind of the MM-X trap.

BG-Sentinel trap. Recently, the BG-Sentinel (BGS) trap has been developed by BioGents GmbH (Regensburg, Germany) for monitoring the yellow fever and dengue vector, *Ae. aegypti* (L.) It consists of a collapsible pop-up container with white gauze on the upper side, and a sucking air flow generated by a 12V DC-powered fan drawing the mosquitoes into a catching bag. The fan also produces an ascending air current carrying odours released from a BG-lure, a blend of ammonia, lactic acid and hexanoic acid, out of the container via the white gauze. Releasing CO_2 is not necessary to catch *Ae. aegypti* with the BGS trap (http://www.bg-sentinel.com/ (Accessed 16 September 2007)). It was found in a surveillance of *Ae. aegypti* in Brazil that the BGS trap captured more male and female mosquitoes than the Mosquito Magnet Liberty baited with CO_2. The trapping efficacy of the BGS baited with BG-lure was similar to an MML baited with CO_2 and BG-lure (Kröckel *et al.* 2006). The same study also showed that the BGS trap has a similar trapping efficiency as the HBC method.

A field study in Australia showed that the BGS trap has a strong visual component in attracting diurnal yellow fever mosquitoes (Williams *et al.* 2006a). A recent study in the USA showed that the BGS trap is more effective in capturing the dengue and Chikungunya fever vector *Ae. albopictus* than, among others, the CDC trap. Adding CO_2 increased its efficacy significantly (Meeraus *et al.* 2007). Aside from *Aedes* species,the BG-Sentinel trap is also claimed to capture high numbers of culicines such as *Cx. quinquefasciatus* and *Cx. pipiens* (http://www.bg-sentinel.com/bilder/BG-Sentinel_Manual_Addition_of_CO2.pdf (Accessed 16 September 2007)).

Traps for gravid females

Sticky ovitraps. Ovitraps are mainly used for collecting eggs of day-flying, container-inhabiting mosquito species such as the yellow fever, dengue and Chikungunya fever vectors *Ae. aegypti* and *Ae. albopictus*. The traps consist simply of a dark-coloured container with water or grass/hay infusion (Isoe *et al.* 1995) in which a wooden paddle wrapped in light-coloured material, such as germination paper, is placed. Adult mosquitoes are attracted to the container to lay eggs. Species identification of eggs collected in ovitraps can usually only be done after the eggs have developed into the larval or adult stage. Another disadvantage of this trap is that the number of eggs does not accurately reflect the adult density.

In order to be able to trap adult females, sticky ovitraps are made by placing a black sticky card on the side wall of the ovitrap container. Adult mosquitoes collected with the sticky ovitraps can be used for species identification and for virus detection (Bangs *et al.* 2001, Ritchie *et al.* 2003). Ovitraps need to be checked and refilled regularly (at least once a week but preferably more frequently) to remove the eggs and adults. The effectiveness of a sticky ovitrap depends on its position relative to houses, height, shade and type of infusion used (Williams *et al.* 2006, Russell and Ritchie 2004, Ritchie *et al.* 2003, Trexler *et al.* 1998). Sticky oviptraps are effective in capturing culicine mosquitoes, especially *Cx. quinquefasciatus, Ae. aegypti* (Ritchie *et al.* 2003, Ritchie *et al.* 2004) and *Ae. albopictus* (Facchinelli *et al.* 2007).

Gravid traps. Gravid female mosquitoes are attracted to odours emanating from oviposition sites (Trexler *et al.* 1998, Chadee *et al.* 1993, Mboera *et al.* 2000). Special traps have been developed to attract blood-fed females. The CDC Gravid traps A and B are especially effective in attracting culicine mosquitoes (Reisen *et al.* 1999, DiMenna *et al.* 2006, Allan and Kline 2004). In general, larger traps capture more mosquitoes, whereas the effect of trap colour seems to be species specific (Rapaport *et al.* 2005, Allan and Kline 2004). Various infusions are used as bait in gravid traps, resulting in different effects on the capture efficacy of the traps (Jackson *et al.* 2005). It was found that hay or grass infusion or marshy water were highly attractive for *Cx. pipiens, Cx. restuans* and *Cx. quinquefasciatus* (Jackson *et al.* 2005, DiMenna *et al.* 2006, Burkett-Cadena and Mullen, 2007, Allan *et al.* 2005). The Mosq-ovitrap has been developed by Lin *et al.* (2005) for both trapping gravid females and collecting eggs of *Ae. albopictus*. The trap appeared to be more effective in collecting eggs of *Ae. albopictus* than a CDC ovitrap. The trapping efficacy increased when the Mosq-ovitrap was placed at ground level instead of at a height of 1 or 1.5m (Lin *et al.* 2006).

Monitoring biting midges

Midges are nuisance insects because females take blood meals from humans and livestock. The occurrence of large numbers of biting midges in coastal areas reduces the number of visits by tourists, which can have a huge negative impact on the local economy (see Chapter 18). Moreover, some midge species are vectors of African horse sickness and bluetongue virus. For neither disease is there an effective treatment. In Europe, the epidemic area of bluetongue disease has moved 800 km north from the Mediterranean Sea in the past 10 years, raising serious public concern (Purse *et al.* 2005, 2006). Recently it has become clear that the vector responsible for the prevalence of the bluetongue virus (BTV) in northern Europe was not the Afro-Asian BTV vector *Culicoides imicola* Kieffer, but most likely members of the *C. obsoletus* (Meigen) complex (Mehlhorn *et al.* 2007, Purse *et al.* 2006, Savini *et al.* 2005, Cagienard *et al.* 2006, Chapter 7).

Light traps

Various designs of light traps are used for monitoring the population dynamics of biting midges. The Onderstepoort light trap is one of the most effective midge traps and widely used in midge research. A surveillance of the BTV vectors with Onderstepoort light traps revealed the most abundant species in Bulgaria to be *C. obsoletus* (Purse *et al.* 2006). The catching efficacy of light traps is largely affected by the type of light used, as the preference for certain light wavelengths is species-specific. A study in South Africa showed that *C. imicola* and other South African *Culicoides* species prefer Onderstepoort traps with black light to those with white light (Venter and Hermanides 2006). Surveillance in Germany revealed that ultraviolet-light traps caught much higher numbers of *Culicoides* species, including *C. obsoletus,* than traps with incandescent light (Mehlhorn *et al.* 2007). Bishop *et al.* (2004, 2006) compared light traps equipped with incandescent light to those with different colours of light-emitting diodes (LEDs) in Australia. These studies showed that the main vector of livestock viruses, *C. brevitarsis* Kieffer, was preferentially attracted to green LEDs compared to incandescent light, white, blue or ultraviolet LEDs, whereas other minor *Culicoides* species preferred LEDs emitting light in the ultraviolet region. One of the advantages of using UV LEDs is that it attracts fewer Lepidoptera than common UV lamps (Bishop *et al.* 2006).

Odour baited traps

The effectiveness of midge traps can be improved with odour baits. Carbon dioxide is the most important chemical cue that attracts biting midges to their blood hosts (see Chapter 18). When baited with dry ice, a CDC trap (without light) was more attractive to the Californian *Culicoides* species *C. variipennis sonorensis* Wirth and Jones than without CO_2 (Gerry and Mullens 1998). A synergistic effect was found for CO_2 and 1-octen-3-ol (Ritchie *et al.* 1994, Kline *et al.* 1994). Adding heat or light to the combination of these two compounds augmented the catch of biting midges (Kline and Lemire 1995, Ritchie *et al.* 1994).

Several traps baited with CO_2 and 1-octen-3-ol are used as a method to reduce biting midge populations, e.g. the Mosquito Magnet®, Midgeater® or Midg-it (Cilek *et al.* 2003, Chapter 18, http://www.midgeater.biz/index.htm (Accessed 16 September 2007)). Cilek and Kline (2002) studied the effectiveness of three different types of traps (CDC, MM-Pro, and MM-X) baited with a synthetic mixture (4:1:8 mixture of 1-octen-3-ol: 3-n-propylphenol: 4-methylphenol) with or without CO_2 in Florida, USA. They found that the mixture alone did not increase the catch, whereas a synergistic effect was found between CO_2 and the mixture for two local *Culicoides* species (*C. mississippiensis* Hoffman and *C. barbosai* Wirth and Blanton). The synergistic effect was only found when using MM-Pro or MM-X traps, but not for CDC traps.

Natural odour blends are also used in odour-baited traps for midges. A study in Scotland on the attraction of *C. impunctatus* Goetghebuer showed a synergistic effect of CO_2 and either acetone or 1-octen-3-ol in combination with the odour of cow urine (Bhasin *et al.* 2001). In another field study in Scotland, a hexane extract of buffalo hair was found to increase the number of *C. impunctatus* caught with a Mosquito Magnet® Pro™ (Mands *et al.* 2004). It is interesting to note that a mosquito repellent, the oil of *Eucalyptus maculata* var. *citriodora*, was found to be attractive to the BTV vector *C. imicola*, whereas two other mosquito repellents, DEET and Meliaceae product AG1000 were repellent to this species (Braverman *et al.* 1999).

Additional remarks

Because midges are small insects, about 1.5 mm, they can easily go through the mesh used in mosquito traps, and therefore ultra-fine mesh should be used instead. Biting midges are most active at dawn and at dusk until the early night, the highest number of midges usually being caught in the evening (Blackwell 1997, Bishop *et al.* 1995, Kettle *et al.* 1998).Traps for BTV vector surveillance should be placed in the vicinity of large livestock holdings with sheep, cattle, goats or horses. One should keep in mind that during bluetongue outbreaks, livestock holders are often obliged to use insecticides which can influence trap catches drastically.

Monitoring phlebotomine sandflies

Phlebotomine sandflies are a reason for serious health concerns in southern Europe (see Chapter 5). Because climate conditions are changing and the distribution of these sandflies is extending northwards, health authorities must be more alert to the spread of these vectors in the foreseeable future (Kovats *et al.* 1999).

In comparison with, e.g. mosquitoes, there is little known about the biology of these tiny dipterans, with a body length seldom exceeding 3 mm (Killick-Kendrick 1999). This section will discuss successful monitoring techniques used for sandflies belonging to the genera *Phlebotomus*, which are able to transmit leishmaniasis in Europe (Killick-Kendrick 1999, Chapter 5).

Techniques for adult sandfly monitoring can be categorised as either methods for sampling the resting population or for sampling the foraging insects at night. The chosen methodology depends on the aim of the study and the characteristics of the habitat. Males are not a vector of leishmaniasis, as they do not need a blood-meal for reproduction. Catching male sandflies, however, can be very useful for identification purposes. More males are caught using certain sampling techniques than others. Collecting resting adults using mouth or battery-operated aspirators can be time-consuming. The advantage is that both sexes are as likely to be caught and that the insects will stay intact and are therefore more easily identifiable compared to insects caught with other methods (e.g. sticky traps).

Both males and females can be caught with sticky traps. This kind of trap is inexpensive and can be easily stored. Most often, castor oil is used to impregnate sheets of paper to construct sticky traps. Alexander (2000) states that the viscosity of castor oil is low in habitats with a relatively high humidity, which reduces the efficiency of the traps. Substances with a greater viscosity are available, but this makes the handling of the insects that are stuck to the trap more difficult. Identification of fragile sandflies can be difficult when using sticky traps.

Light traps are most often used for collecting active sandflies. Light itself seems only attractive over a short range. Adding CO_2 increases the sampling area for sandflies (and mosquitoes) (Gillies 1980). CDC light traps have been used successfully for the collection of *Phlebotomus sergenti* Parrot and *P. papatasi* (Scopoli) in a study by Volf *et al.* (2002). In this study, traps were placed in or near occupied houses, together with numerous sticky traps. The CDC traps were more successful in catching sandflies than the sticky traps. Furthermore, the light traps caught more males than females.

A disadvantage of using light traps that are equipped with a fan is that the insects can get damaged, which makes identification difficult. Another disadvantage can be that other, larger insects are also caught and that part of the catch gets eaten by these before the catch can be immobilised (Alexander 2000). Some traps, like the Onderstepoort light trap have been fitted with a large-sized mesh netting to prevent larger insects from being trapped.

Recent studies in Greece and Turkey showed successful collection results using light traps, sticky traps and aspirators (Ivovic *et al.* 2007, Toprak and Özer 2007). In Greece, a 5-year collection programme of sandflies was conducted using CDC miniature light traps, Onderstepoort light traps and sticky traps (Ivovic *et al.* 2007). Additionally, indoor collections were made with mouth or electric aspirators. A total of 8,688 sandflies were collected but it was unclear from this study which of the techniques used was most effective. Although Onderstepoort traps are designed for midge (*Culicoides*) sampling, they appear to be suitable for sandfly collections as well.

Light traps were most effective for nine of the 13 phlebotomine species collected in the study in Turkey (Toprak and Özer 2007). Aspirators ranked second in this study, in which a total of almost 30 thousand sandflies were caught over a 2-year period. The sticky traps (impregnated with castor oil) were least effective. Human bait collections can be very effective in catching host-seeking sandflies. However, for reasons outlined above, this method is not recommended. A safe method is the collection of adults with aspirators from an animal-baited trap as described by Montoya-Lerma and Lane (1996, see review by Alexander 2000).

Monitoring house and stable flies and sheep blowflies

The housefly, *Musca domestica* L. (Diptera: Muscidae), is one of the most important pests world-wide with respect to hygiene. The flies are not only a nuisance, irritating people and animals and leaving regurgitation and faecal spots on surfaces, but they also are vectors of pathogens that may cause serious diseases in humans and animals (see Chapter 16). The stable fly, *Stomoxys calcitrans* (L.) (Diptera: Muscidae), causes production loss in cattle due to painful bites, especially on the lower legs, and fatigue from trying to avoid the biting flies (Campbell *et al.* 2001, Mullens *et al.* 2006, Taylor and Berkebile 2006). In addition, stable flies may transmit pathogens (Hamilton *et al.* 2003, Förster *et al.* 2007, Scoles *et al.* 2005, Mramba *et al.* 2007). The blowfly, *Lucilia sericata* (Meigen) (Diptera: Calliphoridae), is an economically important ectoparasite of sheep, causing cutaneous myiasis. This fly species is found in several areas in northern Europe (Martínez-Sánchez *et al.* 2007).

Monitoring methods of house flies mainly focus on the adult fly. Techniques used are spot cards, scudder grids, baited jug traps, and sticky ribbons, tapes, cards, pyramids and cylinders (Pickens and Miller 1987, Axtell and Arends 1990, Axtell 1999, Anonymous 2000, 2003, Geden and Hogsette 2001, Jacobs *et al.* 2003, Williams 2003, Geden 2005, 2006, Kaufman *et al.* 2005, Lole 2005). These methods are inexpensive to use and do not require a power source.

With spot cards, housefly activity indoors is measured by counting the number of faecal and regurgitation spots at 1-7 day intervals. Although commonly used because it is easy, this method cannot be used outdoors; it is time-consuming, does not give any information on species, sex or physiological status, and its accuracy is doubtful because one individual fly can make several spots. Besides, results depend largely on correct placement of the cards (Geden *et al.* 1999, Geden and Hogsette 2001, Jacobs *et al.* 2003, Geden 2005).

The scudder grid method can be used both indoors and outdoors, is quick and easy but identification of the flies landing on the grill is difficult and the counts depend largely on weather conditions, position of the grill, time of day and the person performing the observational counts (Geden 2005, Lole 2005).

A jug trap is made by cutting holes in the upper part of the sides of plastic (milk) jugs, and placing a bait at the bottom of the jug. The holes allow house flies, which are attracted to the bait, to enter the trap, where they are caught and killed. Jug trap catches depend largely on their colour and the toxicant used (Geden and Hogsette 2001). A similar principle is used in the recently introduced housefly trap, the Flybrella (Carlson and Hogsette 2007), whose efficiency has yet to be tested in practice. Other traps that have primarily been developed as killing devices for house flies, such as electrocuting traps emitting light in the ultraviolet range, are, besides having other disadvantages (see Chapter 16), usually not suitable for monitoring purposes.

Sticky traps are used to monitor both house flies and stable flies. These traps are non-toxic, inexpensive and easy to use. They can give information on species, population density, distribution, dispersal, sex ratio, physiological age and a photographic record can be made. However, they can also be messy and time-consuming to handle, are usually not species-specific, fly cadavers can be consumed by scavengers and accumulation of dust can reduce the capture potential (Geden *et al.* 1999, Kaufman *et al.* 2001, 2005, Cornelius 2002, Jacobs *et al.* 2003, Lole 2005, Winpisinger *et al.* 2005).

A recent comparison of five adhesive traps in the USA revealed that two BiteFree prototype traps (Farnam Co., Phoenix AZ, USA) were highly efficient at catching stable flies outdoors (Taylor and Berkebile 2006). Coroplast sticky cards (Coroplast, Great Pacific Enterprises Inc., Vancouver BC, Canada) were found to catch more stable flies than alsynite sticky cards and cylinders. White, grey, red, and green Coroplast sticky cards performed better than blue and black ones (Beresford and Sutcliffe 2006).

Geden (2005) showed that in the USA, the Terminator (Farnam Co., Phoenix AZ, USA) and the Victor Fly Magnet (Woodstream Co., Lititz PA, USA) jug trap, both commercially available, are effective in monitoring outdoor populations of house flies, contrary to a sticky cylinder. Combining the attractants of these two jug traps increased their attractiveness. A low-cost, widely available alternative attractant is molasses, the odour composition of which has recently been described (Quinn *et al.* 2007). The baited strip QuikStrike trap (Wellmark International, Schaumburg IL, USA) was found, when sheltered, to be very promising for monitoring purposes of house flies outdoors.

Tsetse flies are successfully caught using traps made of blue and black cloth (Mihok *et al.* 2007). One of these traps, the Nzi, has also been tested for its effectiveness to trap stable and house flies and to find out whether blue is also attractive for these fly species. The Nzi trap appeared to be less efficient for stable flies than adhesive traps (Taylor and Berkebile 2006), but this may depend on its placement (Gilles *et al.* 2007). For stable flies, blue fabric enhanced the attractiveness of alsynite, whereas alsynite was more attractive to house flies than blue fabric (Geden 2006, Taylor and Berkebile 2006). On Florida beaches, where stable flies can be a serious biting pest of humans, stable flies were found to prefer landing on blue and red adhesive boards rather than on orange and white ones (Cilek 2003). The number of stable flies caught in Nzi traps in Canada increased when 1-octen-3-ol was added as an odour attractant (Mihok *et al.* 2007).

Monitoring of *Lucilia sericata* Meigen in sheep pastures in Europe usually takes place with sticky targets baited with liver and sodium sulphide, as developed by Wall *et al.* (1992) and shown to be effective in the field by Smith and Wall (1998). A trap (LuciTrap®, Bioglobal Pty Ltd., Wacol Qld, Australia) combined with a blend of attractants (butyric acid, 2-mercaptoethanol, indole and 20% sodium sulphide solution, known as Lucilure®) mimicking the odour of fleece rot, animal carcasses, urine and faeces, has been developed for the Australian sheep blowfly, *L. cuprina* (Wiedemann) (Armstrong *et al.* 2005), which is morphologically and ecologically similar to *L. sericata*. Although positive results have been achieved for catching *L. cuprina* in Australia and South Africa, this trap was not effective in trapping *L. sericata* in Hungary: sticky targets baited with liver and sodium sulphide caught 500-1500 times more flies than the LuciTrap® (Hall *et al.* 2003). Recently, Broughan and Wall (2006) tested a newly developed trap, baited with freeze-dried liver, in Britain, showing that this trap may contribute to an effective control of blowfly strike. A modified version of this trap is now commercially available in Britain and is known as Agrilure® Blowfly Trap (Agrimin Ltd., North Lincolnshire, UK). This device can be used as a monitoring system of *L. sericata* in large sheep flocks.

Concluding remarks

In the above sections we have provided an overview of commonly used methods for monitoring dipteran pests and vectors of disease. We have not tried to be exhaustive but have merely attempted to give some insight into basic operating principles of these methods, alongside their advantages and problems (Table 1). We conclude that, in spite of growing needs for monitoring tools for European pest and vector species, there is a much greater effort to develop, manufacture and market such tools (both in the professional and home segments) outside Europe, predominantly in the USA. Therefore, most monitoring tools are merely evaluated against European species but were not specifically designed for these. Secondly, we observe that most of the commonly used methods are more or less suitable for collecting mosquitoes, but that for other insect families (Muscidae, Calliphoridae) the scala of options remains rather limited. A third conclusion is the observation that trapping efficiency varies dramatically for different species for which data were published, and that this is largely caused by variations in trap placement, operating times, and odour baits used. Although it is difficult to achieve, standardised operating procedures across various countries would provide more meaningful data that would also enable better comparative analyses of monitoring tools.

Little is known about the threshold population densities below which the monitoring efficiency of the various tools drops to zero. Particularly in cases whereby the precise spatial distribution of species is needed, such information will be highly valuable. Moreover, such knowledge will be essential in case local eradication of an invasion is needed.

The search for efficient monitoring tools for important pest and vector species is further influenced by the need to minimise environmental impact and trapping of non-target organisms. Most methods in use today are far from optimal in that regard, necessitating the need to develop more species-specific tools. We conclude that highly efficient and species-specific trapping methods can only be developed through a detailed understanding of the behavioural ecology (and in particular the host-seeking process of females) of the target species. With growing concerns over the (re)establishment of vector-borne diseases in Europe such research is urgently called for.

References

Acree F Jr, Turner RB, Gouck HK, Beroza M and Smith N (1968) L-Lactic acid: a mosquito attractant isolated from humans. Science 16: 1346-1347.

Alexander B (2000) Sampling methods for phlebotomine sandflies. Med Vet Entomol 14: 109-122.

Allan SA and Kline D (2004) Evaluation of various attributes of gravid female traps for collection of *Culex* in Florida. J Vector Ecol 29: 285-294.

Allan SA, Bernier UR and Kline DL (2005) Evaluation of oviposition substrates and organic infusions on collection of *Culex* in Florida. J Am Mosq Control Assoc 21: 268-273.

Andereadis TG and Armstrong PM (2007) A two-year evaluation of elevated canopy traping for *Culex* mosquitoes and West Nile virus in an operational surveillance program in the Northeastern United States. J Am Mosq Control Assoc 23: 137-148.

Anonymous (2000) Pest Management Recommendations for Poultry. A Cornell and Penn State Cooperative Extension Publication, Pennsylvania State University. 16 pp.

Anonymous (2003) The Spot Card Technique for sampling *Musca domestica*. Report of Santa Clara County Vector Control District. 18 pp.

Armstrong B, Knights G, Urech R, Green P and Ward M (2005) Sheep parasites. The LuciTrap sheep blowfly trapping system. In: Department of Primary Industries and Fisheries, Australia, Queensland.

Asnis DS, Conetta R, Teixeira AA, Waldman G and Sampson BA (2000) The West Nile Virus outbreak of 1999 in New York: the Flushing Hospital experience. Clin Infect Dis 30: 413-418.

Asnis DS, Conetta R, Waldman G and Teixeira AA (2001) The West Nile virus encephalitis outbreak in the United States (1999-2000): from Flushing, New York, to beyond its borders. Ann NY Acad Sci 951: 161-171.

Axtell RC (1999) Poultry integrated pest management: status and future. Integr Pest Manag Rev 4: 53-73.

Axtell RC and Arends JJ (1990) Ecology and management of arthropod pests of poultry. Ann Rev Entomol 35: 101-126.

Balenghien T, Fouque F, Sabatier P and Bicout DJ (2006) Horse-, bird-, and human-seeking behavior and seasonal abundance of mosquitoes in a West Nile virus focus of southern France. J Med Entomol 43: 936-946.

Bangs MJ, Tan R, Listiyaningsih E, Kay BH and Porter KR (2001) Detection of dengue viral RNA in *Aedes aegypti* (Diptera: Culicidae) exposed to sticky lures using reverse-transcriptase polymerase chain reaction. J Med Entomol 38: 720-724.

Becker N, Zgomba M, Petric D and Ludwig M (1995) Comparison of carbon dioxide, octenol and a host-odour as mosquito attractants in the Upper Rhine Valley, Germany. Med Vet Entomol 9: 377-380.

Bellamy RE and Reeves WC (1952) A portable mosquito bait-trap. Mosq News 12: 256-258.

Beresford DV and Sutcliffe JF (2006) Studies on the effectiveness of Coroplast sticky traps for sampling stable flies (Diptera: Muscidae), including a comparison to Alsynite. J Econ Entomol 99: 1025-1035.

Bernier UR, Kline DL, Barnard DR, Schreck CE and Yost RA (2000) Analysis of human skin emanations by gas chromatography/mass spectrometry. 2. Identification of volatile compounds that are candidate attractants for yellow fever mosquito (*Aedes aegypti*). Analytical Chem 72: 747-756.

Bhasin A, Mordue AJ and Mordue W (2001) Field studies on efficacy of host odour baits for the biting midge *Culicoides impunctatus* in Scotland. Med Vet Entomol 15: 147-156.

Bishop AL, Bellis GA, McKenzie HJ, Spohr LJ, Worrall RJ, Harris AM and Melville L (2006) Light trapping of biting midges Culicoides spp. (Diptera: Ceratopogonidae) with green light-emitting diodes. Austr J Entomol 45: 202-205.

Bishop AL, McKenzie HJ, Barchia IM and Harris AM (1995) The effects of habitat on the distribution of *Culicoides brevitarsis* Kieffer (Diptera, Ceratopogonidae) during its resting phase. Austr J Zool 43: 531-539.

Bishop AL, Worrall R, Spohr LJ, McKenzie HJ and Barchia IM (2004) Response of Culicoides spp. (Diptera: Ceratopogonidae) to light-emitting diodes. Austr J Entomol 43: 184-188.

Blackmore MS and Dahl C (2002) Field evaluation of a new surveillance trap in Sweden. J Am Mosq Control Assoc 18: 152-157.

Blackwell A (1997) Diel flight periodicity of the biting midge *Culicoides impunctatus* and the effects of meteorological conditions. Med Vet Entomol 11: 361-367.

Braverman Y, Chizov-Ginzburg A and Mullens BA (1999) Mosquito repellent attracts *Culicoides imicola* (Diptera: Ceratopogonidae). J Med Entomol 36: 113-115.

Broughan JM and Wall R (2006) Control of sheep blowfly strike using fly-traps. Vet Parasitol 135: 57-63.

Bruce-Chwatt LJ and De Zulueta J (1980) The rise and fall of malaria in Europe, A historico-epidemiological study. Oxford University Press, Oxford.

Burkett-Cadena ND and Mullen GR (2007) Field comparison of Bermuda-hay infusion to infusions of emergent aquatic vegetation for collecting female mosquitoes. J Am Mosq Control Assoc 23: 117-123.

Burkett DA, Lee WJ, Lee KW, Kim HC, Lee HI, Lee JS, Shin EH, Wirtz RA, Cho HW, Claborn DM, Coleman RE and Klein RA (2001) Light, carbon dioxide, and octenol-baited mosquito trap and host-seeking activity evaluations for mosquitoes in a malarious area of the Republic of Korea. J Am Mosq Control Assoc 17: 196-205.

Cagienard A, Griot C, Mellor PS, Denison E and Stark KDC (2006) Bluetongue vector species of *Culicoides* in Switzerland. Med Vet Entomol 20: 239-247.

Campbell JB, Skoda SR, Berkebile DR, Boxler DJ, Thomas GD, Adams DC and Davis R (2001) Effects of stable flies (Diptera: Muscidae) on weight gains of grazing yearling cattle. J Econ Entomol 94: 780-783.

Carlson DA and Hogsette JA (2007) FLYBRELLA: A Device to attract and kill house flies. J Econ Entomol 100: 483-487.

Chadee DD, Lakhan A, Ramdath WR and Persad RC (1993) Oviposition response of *Aedes aegypti* mosquitoes to different concentrations of hay infusion in Trinidad, West-Indies. J Am Mosq Control Assoc 9: 346-348.

Cilek JE (2003) Attraction of Colored PLASTICIZED CORRUGATED Boards to Adult Stable Flies, *Stomoxys calictrans* (Diptera: Muscidae). Florida Entomol 86: 420-423.

Cilek JE and Kline DL (2002) Adult biting midge response to trap type, carbon dioxide, and an octenol-phenol mixture in northwestern Florida. J Am Mosq Control Assoc 18: 228-231.

Cilek JE, Kline DL and Hallmon CF (2003) Evaluation of a novel removal trap system to reduce biting midge (Diptera: Ceratopogonidae) populations in Florida backyards. J Vector Ecol 28: 23-30.

Cooperband MF and Cardé RT (2006) Orientation of Culex mosquitoes to carbon dioxide-baited traps: flight manoeuvres and trapping efficiency. Med Vet Entomol 20: 11-26.

Cornelius ML (2002) Sticky traps. In: D. Pimentel (ed.) Encyclopedia of pest management. Marcel Dekker, Inc., New York, pp 795-797.

Costantini C, Sagnon NF, Sanogo E, Merzagora L and Coluzzi M (1998) Relationship to human biting collections and influence of light and bednet in CDC light-trap catches of West African malaria vectors. Bull Entomol Res 88: 503-511.

Curran AM, Rabin SI, Prada PA and Furton KG (2005) Comparison of the volatile organic compounds present in human odor using SPME-GC/MS. J Chem Ecol 31: 1607-1619.

Curran AM, Ramirez CF, Schoon AA and Furton KG (2007) The frequency of occurrence and discriminatory power of compounds found in human scent across a population determined by SPME-GEMS. J Chromatogr B Analyt Technol Biomed Life Sci 846: 86-97.

Curtis CF, Lines JD, Ijumba J, Callaghan A, Hill N and Karimzad MA (1987) The relative efficacy of repellents against mosquito vectors of disease. Med Vet Entomol 1: 109-119.

Darbro, J.M. and L.C. Harrington (2006) Bird-baited traps for surveillance of West Nile mosquito vectors: Effect of bird species, trap height, and mosquito escape rates. J Med Entomol 43: 83-92.

Deegan CS, Burns JE, Huguenin M, Steinhaus EY, Panella NA, Beckett S and Komar N (2005) Sentinel pigeon surveillance for West Nile virus by using lard-can traps at differing elevations and canopy cover classes. J Med Entomol 42: 1039-1044.

DiMenna MA, Bueno R, Parmenter RR, Norris DE, Sheyka JM, Molina JL, LaBeau EM, Hatton ES and Glass GE (2006) Comparison of mosquito trapping method efficacy for West Nile virus surveillance in New Mexico. J Am Mosq Control Assoc 22: 246-253.

Drummond CL, Drobnack J, Backenson PB, Ebel GD, Kramer LD (2006) Impact of trap elevation on estimates of abundance, parity rates, and body size of *Culex pipiens* and *Culex restuans* (Diptera: Culicidae). J Med Entomol 43: 177-184.

EC (2005) Commission Regulation (EC) No 1048/2005 of 13 June 2005 amending Regulation (EC) No 2032/2003 on the second phase of the 10-year work programme referred to in Article 16(2) of Directive 98/8/EC of the European Parliament and of the Council concerning the placing of biocidal products on the market. Off J Eur Comm L178, 0001-0098.

Facchinelli L, Valerio L, Pombi M, Reiter P, Costantini C, Della Torre A (2007) Development of a novel sticky trap for container-breeding mosquitoes and evaluation of its sampling properties to monitor urban populations of *Aedes albopictus*. Med Vet Entomol 21: 183-195.

Filler SJ, MacArthur JR, Parise M, Wirtz R, Eliades MJ, Dasilva A, Steketee R (2006) Locally acquired mosquito-transmitted malaria: a guide for investigations in the United States. MMWR Recomm Rep 55: 1-9.

Förster M, Klimpel S, Mehlhorn H, Sievert K, Messler S and Pfeffer K (2007) Pilot study on synanthropic flies (e.g. *Musca, Sarcophaga, Calliphora, Fannia, Lucilia, Stomoxys*) as vectors of pathogenic microorganisms. Parasitol Res 101: 243-246.

Geden CJ (2005) Methods for monitoring outdoor populations of house flies, *Musca domestica* L. (Diptera: Muscidae). J Vector Ecol 30: 244-250.

Geden CJ (2006) Visual targets for capture and management of house flies, *Musca domestica* L. J Vector Ecol 31: 152-157.

Geden CJ and Hogsette JA (2001) Research and extension needs for integrated pest management for arthropods of veterinary importance. In: Proceedings of a Workshop in Lincoln, Lincoln, Nebraska.

Geden CJ, Hogsette JA and Jacobs RD (1999) Effect of airflow on house fly (Diptera: Muscidae) distribution in poultry houses. J Econ Entomol 92: 416-420.

Gerry AC and Mullens BA (1998) Response of male *Culicoides variipennis sonorensis* (Diptera: Ceratopogonidae) to carbon dioxide and observations of mating behavior on and near cattle. J Med Entomol 35: 239-244.

Gilles J, David JF, Duvallet G, De La Rocque S and Tillard E (2007) Efficiency of traps for *Stomoxys calcitrans* and *Stomoxys niger niger* on Réunion Island. Med Vet Entomol 21: 65-69.

Gillies MT (1980) The role of carbon-dioxide in host-finding by mosquitoes (Diptera, Culicidae) - a review. Bull Entomol Res 70: 525-532.

Gillies MT and Wilkes TJ (1968) A comparison of the range of attraction of animal baits and of carbon dioxide for some West African mosquitoes. Bull Entomol Res 59: 441-456.

Grant AJ and O' Connell RJ (1996) Electrophysiological responses from receptor neurons in mosquito maxillary palp sensilla. In: Bock GR and Cardew G (eds.) Olfaction in mosquito host interactions Ciba Foundation Symposium 200, Wiley, Chichester, New York, pp 233-248.

Hall DR, Beevor PS, Cork A, Nesbitt BF and Vale GA (1984) 1-Octen-3-ol. A potent olfactory stimulant and attractant for tsetse isolated from cattle odours. Insect Sci Applic 5: 335–339.

Hall MJR, Hutchinson RA, Farkas R, Adams ZJO and Wyatt NP (2003) A comparison of Lucitrapsandreg; and sticky targets for sampling the blowfly *Lucilia sericata*. Med Vet Entomol 17: 280-287.

Hamilton JV, Lehane MJ and Braig HR (2003) Isolation of *Enterobacter sakazakii* from midgut of *Stomoxys calcitrans*. Emerg Infect Dis 9:1355-1356.

Hay SI, Guerra CA, Tatem AJ, Noor AM and Snow RW (2004) The globel distribution and population at risk of malaria: past, present, and future. Lancet Infect Dis 4: 327-336.

Hii JLK, Smith T, Mai A, Ibam E and Alpers MP (2000) Comparison between anopheline mosquitoes (Diptera: Culicidae) caught using different methods in a malaria endemic area of Papua New Guinea. Bull Ent Res 90: 211-219.

Hoel DF, Kline DL, Allan SA and Grant A (2007) Evaluation of carbon dioxide, 1-octen-3-ol, and lactic acid as baits in Mosquito Magnet Pro traps for *Aedes albopictus* in north central Florida. J Am Mosq Control Assoc 23: 11-17.

Ikeshoji T (1982) Attractive sounds for autochemosterilization of the male mosquitoes. Jpn J Sanit Zool 33: 41–49.

Ikeshoji T (1986) Distribution of the mosquitoes, *Culex tritaeniorhynchus,* in relation to disposition of sound traps in a paddy field. Jpn J Sanit Zool 37: 153–159.

Isoe J, Beehler JW, Millar JG and Mulla MS (1995) Oviposition responses of *Culex tarsalis* and *Culex quinquefasciatus* to aged Bermuda grass infusions. J Am Mosq Control Assoc 11: 39-44.

Ivovic V, Patakakis M, Tselentis Y and Chaniotis B (2007) Faunistic study of sandflies in Greece. Med Vet Entomol 21: 121-124.

Jackson BT, Paulson SL, Youngman RR, Scheffel SL and Hawkins B (2005) Oviposition preferences of *Culex restuans* and *Culex pipiens* (Diptera: Culicidae) for selected infusions in oviposition traps and gravid traps. J Am Mosquito Control Assoc 21: 360-365.

Jacobs RD, Hogsette JA and Millar RW (2003) Using sticky cards to monitor fly populations in poultry houses. In: Series of Animal Science Department, Florida Cooperative Extension Service, Institute of Food and Agricultural Sciences, University of Florida.

Jaenson TGT (1985) Attraction to mammals of male mosquitoes with special reference to *Aedes diantaeus* in Sweden. J Am Mosq Control Assoc 1: 195–198.

Kaufman PE, Rutz DA and Frisch S (2001) Sticky traps for large scale house fly (Diptera: Muscidae) trapping in New York poultry facilities. J Agric Urban Entomol 18: 43-49.

Kaufman PE, Rutz DA and Frisch S (2005) Large sticky traps for capturing house flies and stable flies in dairy calf greenhouse facilities. J Dairy Sci 88: 176-181.

Kemme JA, van Essen PHA, Ritchie SA and Kay BH (1993) Response of mosquitoes to carbon dioxide and 1-octen-3-ol in southeast Queensland, Australia. J Am Mosq Control Assoc 9: 431–435.

Kettle DS, Edwards PB and Barnes A (1998) Factors affecting numbers of *Culicoides* in truck traps in coastal Queensland. Med Vet Entomol 12: 367-377.

Killick-Kendrick R (1999) The biology and control of phlebotomine sand flies. Clin Dermatol 17: 279-289.

Kline DL (1999) Coamparison of two American Biophysics mosquito traps: The professional and a new counterflow geometry trap. J Am Mosq Control Assoc 15: 276-282.

Kline DL (2002) Evaluation of various models of propane-powered mosquito traps. J Vector Ecol 27: 1-7.

Kline DL and Lemire GF (1995) Field evaluation of heat as an added attractant to traps baited with carbon dioxide and octenol for *Aedes taeniorhynchus*. J Am Mosq Control Assoc 11: 454-456.

Kline DL, Hagan DV and Wood JR (1994) *Culicoides* responses to 1-octen-3-ol and carbon-dioxide in salt marshes near Sea Island, Georgia, USA. Med Vet Entomol 8: 25-30.

Kline DL, Patnaude M and Barnard DR (2006) Efficacy of four trap types for detecting and monitoring *Culex* spp. in north central Florida. J Med Entomol 43: 1121-1128.

Kline DL, Wood JR and Cornell JA (1991) Interactive effects of 1-octen-3-ol and carbon dioxide on mosquito (Diptera: Culicidae) surveillance and control. J Med Entomol 28:254–258.

Kovats RS, Haines A, Stanwell-Smith R, Martens P, Menne B. and Bertollini R (1999) Climate change and human health in Europe. BMJ 318: 1682-1685.

Kröckel U, Rose A, Eiras AE and Geier M (2006) New tools for surveillance of adult yellow fever mosquitoes: Comparison of trap catches with human landing rates in an urban environment. J Am Mosq Control Assoc 22: 229-238.

Laganier R, Randimby FM, Rajaonarivelo V and Robert V (2003) Is the Mbita trap a reliable tool for evaluating the density of anopheline vectors in the highlands of Madagascar? Malar J 2: 42.

Lepore TJ, Pollack RJ, Spielman A and Reiter P (2004) A readily constructed lard-can trap for sampling host-seeking mosquitoes. J Am Mosq Control Assoc 20: 321-322.

Lin LF, Lu WC, Cai SW, Duan JH, Yi JR, Deng F, Chen Q and Chen XG (2005) The design and efficacy observation of new Mosq-ovitrap for monitoring of vector of Dengue fever. Chin J Vector Biol Control 16: 26-28.

Lin LF, Cai SW, Li RB, Lu WC and Chen Q (2006) Effect of spatio-temporal factors on Mosq-ovitrap trapping efficacy for *A. albopictus* surveilliance. Acta Parasitol Med Entomol Sin 13: 25-27.

Lines JD, Curtis CF, Wilkes TJ and Njunwa KJ (1991) Monitoring human-biting mosquitoes (Diptera: Culicidae) in Tanzania with light-traps hung beside mosquito nets. Bull Entomol Res 81: 77-84.

Lole, MJ (2005) Nuisance flies and landfill activities: an investigation at a West Midlands landfill site. Waste Manag Res 23: 420-428.

Lounibos LP (2002) Invasions by insect vectors of human disease. Ann Rev Entomol 47: 233-266.

Lu T, Qiu YT, Wang GR, Kwon JY, Rutzler M, Kwon HW, Pitts J, van Loon JJA, Takken W, Carlson JR and Zwiebel JL (2007) Odor coding by maxillary palp neurons of the malaria vector mosquito *Anopheles gambiae*. Current Biol 17: 1-12.

Mands V, Kline DL and Blackwell A (2004) *Culicoides* midge trap enhancement with animal odour baits in Scotland. Med Vet Entomol 18: 336-342.

Martínez-Sánchez A, Smith KE, Rojo S, Marcos-García MA and Wall R (2007) Geographic origin affects larval competitive ability in European populations of the blow fly, *Lucilia sericata*. Entomol Exp Appl 122: 93-98.

Mathenge EM, Killeen GF, Oulo DO, Irungu LW, Ndegwa PN and Knols BG (2002) Development of an exposure-free bednet trap for sampling Afrotropical malaria vectors. Med Vet Entomol 16: 67-74.

Mathenge EM, Misiani GO, Oulo DO, Irungu LW, Ndegwa PN, Smith TA, Killeen GF and Knols BGJ (2005) Comparative performance of the Mbita trap, CDC light trap and the human landing catch in the sampling of *Anopheles arabiensis*, *An. funestus* and culicine species in a rice irrigation in western Kenya. Malar J 4: 7.

Mboera LEG (2005) Sampling techniques for adult Afrotropical malaria vectors and their reliability in the estimation of entomological inoculation rate. Tanzan Health Res Bull 7: 117-124.

Mboera LEG and Takken W (1997) Carbon dioxide chemotropism in mosquitoes (Diptera: Culicidae) and its potential in vector surveillance and control. Rev Med Vet Entomol 85: 355-368.

Mboera LEG, Kihonda J, Braks MAH and Knols BGJ (1998) Short report: influence of centers for disease control light trap position, relative to a human-baited bednet, on catches of *Anopheles gambiae* and *Culex quinqufasciatus* in Tanzania. Am J Trop Med Hyg 59: 595-596.

Mboera LEG, Knols BGJ, Braks MAH and Takken W (2000) Comparison of carbon dioxide-baited trapping systems for sampling outdoor mosquito populations in Tanzania. Med Vet Entomol 14: 257-263.

Mboera LEG, Takken W, Mdira KY and Pickett JA (2000a) Sampling gravid *Culex quinquefasciatus* (Diptera: Culicidae) in Tanzania with traps baited with synthetic oviposition pheromone and grass infusions. J Med Entomol 37: 172-176.

Mboera LEG, Takken W and Sambu EZ (2000b) The response of *Culex quinquefasciatus* (Diptera: Culicidae) to traps baited with carbon dioxide, 1-octen-3-ol, acetone, butyric acid and human foot odour in Tanzania. Bull Entomol Res 90: 155-159.

McIver SB and McElligott PE (1989) Effects of release rates on the range of attraction of carbon-dioxide to some Southwestern Ontario mosquito species. J Am Mosq Control Assoc 5: 6-9.

McIver SB,Wilkes TJ, GilliesMT (1980) Attraction to mammals of male *Mansonia* (Mansonioides) (Diptera: Culicidae). Bull Entomol Res 70: 11–16.

Meeraus W, Johnson J and Arias J (2007) Field comparison of novel and gold standard traps for collecting Aedes albopictus in northern Virginia. Presentation at the 32nd Annual Conference of the Mid-Atlantic Mosquito Control Association http://www.bg-sentinel.com/bilder/Meeraus_et_al_2007_Field_comparison_of_novel_and_gold_standard_traps_Ae_albopictus_Northern_Virginia.pdf

Mehlhorn H, Walldorf V, Klimpel S, Jahn B, Jaeger F, Eschweiler J, Hoffmann B and Beer M (2007) First occurrence of *Culicoides obsoletus*-transmitted Bluetongue virus epidemic in Central Europe. Parasitol Res 101: 219-228.

Mihok S, Carlson DA and Ndegwa PN (2007) Tsetse and other biting fly responses to Nzi traps baited with octenol, phenols and acetone. Med Vet Entomol 21: 70-84.

Montoya-Lerma J and Lane RP (1996) Factors affecting host preference of *Lutzomyia evansi* (Diptera: psychodidae), a vector of visceral leishmaniasis in Colombia. Bull Entomol Res 86: 43-50.

Mramba F, Broce AB and Zurek L (2007) Vector competence of stable flies, *Stomoxys calcitrans* L. (Diptera: Muscidae), for *Enterobacter sakazakii*. J Vector Ecol 32: 134- 139.

Mulhern TD (1942) New Jersey mechanical trap for mosquito surveys. New Jersey Agric Exp Stn Circ 421, 8 pp.

Mullens BA, Lii KS, Mao Y, Meyer JA, Peterson NG and Szijj CE (2006) Behavioural responses of dairy cattle to the stable fly, *Stomoxys calcitrans*, in an open field environment. Med Vet Entomol 20: 122-137.

Newhouse VF, Chamberlain RW, Johnson JG and WD Sudia (1966) Use of dry ice to increase mosquito catches of the CDC miniature light trap. Mosq News 26: 30-35.

Njiru BN, Mukabana WR, Takken W and Knols BGJ (2006) Trapping of the malaria vector *Anopheles gambiae* with odour-baited MM-X traps in semi-field conditions in western Kenya. Malar J 5: 39.

Parker M, Anderson AL and Slaff M (1986) An automatic carbon-dioxide delivery system for mosquito light trap surveys. J Am Mosq Control Assoc 2: 236-237.

Pickens LG and Miller RW (1987) Techniques for trapping flies on dairy farms. J Agricul Entomol 4: 305-313.

Purse BV, Mellor PS, Rogers DJ, Samuel AR, Mertens PPC and Baylis M (2005) Climate change and the recent emergence of bluetongue in Europe. Nature Rev Microbiol 3: 171-181.

Purse BV, Nedelchev N, Georgiev G, Veleva E, Boorman J, Denison E, Veronesi E, Carpenter S, Baylis M and Mellor PS (2006) Spatial and temporal distribution of bluetongue and its Culicoides vectors in Bulgaria. Med Vet Entomol 20: 335-344.

Qiu YT, Smallegange RC, ter Braak CJF, Spitzen J, van Loon JJA, Jawara M, Milligan P, Galimard AM, van Beek TA, Knols BGJ and Takken W (2007) The attractiveness of MM-X traps baited with human or synthetic odor to mosquitoes (Diptera: Culicidae) in The Gambia. J Med Entomol, in press.

Quinn BP, Bernier UR, Geden CJ, Hogsette JA and Carlson DA (2007) Analysis of extracted and volatile components in blackstrap molasses feed as candidate house fly attractants. J Chromatogr A 1139: 279–284.

Rapaport AS, Lampman RL and Novak RJ (2005) Evaluation of selected modifications to CO_2 and infusion-baited mosquito traps in Urbana, Illinois. J Am Mosq Control Assoc 21: 395-399.

Reisen WK, Boyce K, Cummings RC, Delgado O, Gutierrez A, Meyer RP and Scott TW (1999) Comparative effectiveness of three adult mosquito sampling methods in habitats representative of four different biomes of California. J Am Mosq Control Assoc 15: 24-31.

Reisen WK, Eldridge BE, Scott TW, Gutierrez A, Takahashi R, Lorenzen K, DeBenedictis J, Boyce K, Swartzell R (2002) Comparison of dry ice-baited centers for disease control and New Jersey light traps for measuring mosquito abundance in California. J Am Mosq Control Assoc 18: 158-163.

Reisen WK, Meyer RP, Cummings RFand Delgado O (2000) Effects of trap design and CO_2 presentation on the measurement of adult mosquito abundance using Centers for Disease Control-style miniature light traps. J Am Mosq Control Assoc 16: 13-18.

Ritchie SA, Vanessen PHA, Kemme JA, Kay BH and Allaway D (1994) Response of biting midges (Diptera, Ceratopogonidae) to carbon-dioxide, octenol, and light in Southeastern Queensland, Australia. J Med Entomol 31: 645-648.

Ritchie SA, Long S, Hart A, Webb CE and Russell RC (2003) An adulticidal sticky ovitrap for sampling container-breeding mosquitoes. J Am Mosq Control Assoc 19: 235-242.

Ritchie SA, Long S, Smith G, Pyke A and Knox TB (2004) Entomological investigations in a focus of dengue transmission in Cairns, Queensland, Australia, by using the sticky ovitraps. J Med Entomol 41: 1-4.

Rudolfs W (1992) Chemotropism of mosquitoes. Bull New Jersey Agric Exp Stn Circ, No. 367.

Russell RC and Ritchie SA (2004) Surveillance and behavioral investigations of *Aedes aegypti* and *Aedes polynesiensis* in Moorea, French Polynesia, using a sticky ovitrap. J Am Mosq Control Assoc 20: 370-375.

Savini G, Goffredo M, Monaco F, Di Gennaro A, Cafiero MA, Baldi L, De Santis P, Meiswinkel R and Caporale V (2005) Bluetongue virus isolations from midges belonging to the *Obsoletus* complex (*Culicoides*, Diptera: ceratopogonidae) in Italy. Vet Rec 157: 133-139.

Scoles GA, Broce AB, Lysyk TJ and Palmer GH (2005) Relative efficiency of biological transmission of *Anaplasma marginale* (Rickettsiales: Anaplasmataceae) by *Dermacentor andersoni* (Acari: Ixodidae) compared with mechanical transmission by *Stomoxys calcitrans* (Diptera: Muscidae). J Med Entomol 42: 668-675.

Service MW (1993) Mosquito Ecology: field sampling methods. Elsevier Applied Science, London and New York.

Shone SM, Ferrao PN, Lesser CR, Glass GE and Norris DE (2003) Evaluation of carbon dioxide- and 1-octen-3-ol-baited Centers for Disease Control Fay-Prince traps to collect *Aedes albopictus*. J Am Mosq Control Assoc 19: 445-447.

Shone SM, Glass GE and Norris DE (2006) Targeted trapping of mosquito vectors in the Chesapeake Bay Area of Maryland. J Med Entomol 43: 151-158.

Smith A (1956) The attractiveness of an adult and child to *Anopheles gambiae*. East Afr Med J 33: 409-410.

Smith KE and Wall R (1998) Suppression of the blowfly *Lucilia sericata* using odour-baited triflumuron-impregnated targets. Med Vet Entomol 12: 430-437.

Smith T (1995) Proportionality between light trap catches and biting densities of malaria vectors. J Am Mosq Control Assoc 11: 377-378.

Sudia WD and Chamberlain RW (1962) Battery-operated light trap: An improved model. Mosq News 22: 126-129.

Takken W and Kline DL (1989) Carbon dioxide and 1-octen-3-ol as mosquito attractants. J Am Mosq Control Assoc 5: 311-316.

Takken W and Knols BGJ (1999) Odor-mediated behavior of Afrotropical malaria mosquitoes. Annu Rev Entomol 44: 131-157.

Tatem AJ, Hay SI and Rogers DJ (2006) Global traffic and disease vector dispersal. Proc Nat Acad Sci USA 103: 6242-6247.

Taylor DB and Berkebile D (2006) Comparative efficiency of six stable fly (Diptera: Muscidae) traps. J Econ Entomol 99: 1415-1419.

Toprak S and Ozer N (2007) Distribution of sand fly (Diptera: Psychodidae) species and efficiency of capturing methods in Sanliurfa Province, Turkey. J Med Entomol 44: 23-28.

Trexler JD, Apperson CS and Schal C (1998) Laboratory and field evaluations of oviposition responses of *Aedes albopictus* and *Aedes triseriatus* (Diptera: Culicidae) to oak leaf infusions. J Med Entomol 35: 967-976.

U.S. Consumer Product Safety Commission (2006) American Biophysics Corp. announce recall of accessory used to clean Mosquito Magnet®. http://www.usrecallnews.com/USARecall2/USARecall2FDAArchive109.html (Accessed 16 September 2007).

Vaidyanathan R and Edman JD (1997) Sampling methods for potential epidemic vectors of eastern equine encephalomyelitis virus in Massachusetts. J Am Mosq Control Assoc 13: 342-347.

Van der Weijden W, Leewis R and Bol P (2007) Biological globalisation – Bio-invasions and their impacts on nature, the economy and public health. KNNV Publishing, Utrecht.

Venter GJ and Hermanides KG (2006) Comparison of black and white light for collecting *Culicoides imicola* and other livestock-associated *Culicoides* species in South Africa. Vet Parasitol 142: 383-385.

Volf P, Ozbel Y, Akkafa F, Svobodova M, Votypka J and Chang KP (2002) Sand flies (Diptera: Phlebotominae) in Sanliurfa, Turkey: Relationship of *Phlebotomus sergenti* with the epidemic of anthroponotic cutaneous leishmaniasis. J Med Entomol 39: 12-15.

Wall R, Green CH, French N and Morgan KL (1992) Development of an attractive target for the sheep blowfly *Lucilia sericata*. Med Vet Entomol 6: 67-74.

Weissenböck H, Kolodziejek J, Url A, Lussy H, Rebel-Bauder B and Nowotny N (2002) Emergence of Usutu virus, an African mosquito-borne Flavivirus of the Japanese encephalitis virus group, central Europe. Emerg Infect Dis 8: 652-656.

World Health Organisation (1975) Manual on Practical Entomology in Malaria: Part II - Methods and Techniques. WHO Offset Publications No. 13, Geneva.

Williams CR, Long SA, Russell RC and Ritchie SA (2006) Field efficacy of the BG-sentinel compared with CDC backpack aspirators and CO2-baited EVS traps for collection of adult *Aedes aegypti* in Cairns, Queensland, Australia. J Am Mosq Control Assoc 22: 296-300.

Williams CR, Long SA, Russell RC and Ritchie SA (2006) Optimizing ovitrap use for *Aedes aegypti* in Cairns, Queensland, Australia: Effects of some abiotic factors on field efficacy. J Am Mosq Control Assoc 22: 635-640.

Williams RE (2003) Livestock and Poultry. Control of poultry pests. In: Purdue University Cooperative Extension Service.

Williams RE (2003) Livestock & Poultry. Control of poultry pests. Purdue University Cooperative Extension Service. 8p. http://www.ces.purdue.edu/extmedia (Accessed 25 August 2007).

Winpisinger KA, Ferketich AK, Berry RL and Moeschberger ML (2005) Spread of Musca domestica (Diptera: Muscidae), from Two Caged Layer Facilities to Neighboring Residences in Rural Ohio. J Med Entomol 42: 732-738.

20. Personal protection against European disease vectors

Niels O. Verhulst, Christopher F. Curtis and Nigel Hill

Abstract

Apart from being a nuisance, bites from insects and other arthropods can transmit a variety of diseases. When one is out of doors application of a repellent to the skin is the appropriate form of personal protection. N,N-diethyl-3-methyl benzamide (deet) is the most commonly used repellent and is generally considered more effective than any other repellent currently on the market. However, in some trials, recently developed repellents (KBR 3023, IR 3535, PMD) have shown effectiveness and persistence levels equivalent to deet. When there is a risk of encountering ticks, tucking trousers into socks and treating these with permethrin is recommended to minimise the chance of contracting tick-borne diseases. Where biting occurs indoors and at night, screening of doors and windows, use of pyrethroid treated bednets and of vaporising mats, which emit a volatile pyrethroid, are also appropriate protective measures. Buzzers and vitamin B_1 are completely ineffective as insect repellents.

Keywords: mosquitoes, sand flies, ticks, repellents

Prevention of biting nuisance and disease transmission by arthropods in Europe

Biting insects and ticks can be a considerable source of annoyance in the warmer parts and seasons in Europe. Thus, many people attempt to protect themselves by using repellents on skin or clothing, insecticide aerosols, vaporising mats or coils, bednets (which may be treated with insecticide) and screening of windows and doors of houses. The motive for taking such precautions may be more than mere annoyance in individuals who show severe allergic reaction to bites (Peng and Simons 2004).

In the past, malaria (transmitted by *Anopheles* mosquitoes), was of major importance in Europe until the eradication of transmission between about 1900 and 1960 (De Zulueta 1998, Snowden 2006). Eradication in each country of southern Europe was mainly achieved by organised campaigns of drainage of *Anopheles* breeding sites and indoor residual spraying with dichloro-diphenyl-trichloroethane (DDT). In addition, personal protection, in the form of improvement in housing, including screening of doors and windows, which made houses less attractive and accessible to *Anopheles*, played a significant role. There are no animal reservoirs of the *Plasmodium* species which cause human malaria, but nowadays thousands of people per year acquire malaria infections in tropical or sub-tropical countries and arrive as 'imported' cases in Europe (Jelinek *et al.* 2002) (See Chapters 2 & 3). One must consider the risk of local *Anopheles* mosquitoes becoming infected after biting an imported malaria case. Rising temperatures will enhance the likelihood of *Plasmodium* completing its development in an infected mosquito so that onward transmission may occur (MacDonald 1957). In former Soviet Central Asia (which is part of the WHO European Region), there has been a resurgence of malaria (Sabatinelli *et al.* 2001) because health services are no longer available to everyone, which results in untreated cases arriving from Afghanistan and elsewhere that remain untreated and able to infect local mosquitoes. However, in Italy, despite its hot summers, there have been very few cases of local transmission of malaria in recent decades apparently because of prompt and effective treatment of imported cases by the widely available health service (Romi *et al.* 2001). Members of the *Anopheles maculipennis* complex remain potential vectors of *P. vivax* malaria in Europe if they encounter an untreated human case of malaria (See Chapter 3). These mosquitoes bite at night and so personal protection measures in houses would be applicable. *An. plumbeus* can be considered a potential vector if it bites an untreated imported case of *P. falciparum* to which (unlike members of the *An. maculpennis* complex) it is susceptible (Eling *et al.* 2003). Two cases of *P. falciparum* were considered to have been transmitted inside a German hospital by *An. plumbeus* from an imported malaria case, who had not yet been cured, to two other patients (Krüger *et al.* 2001). In Germany, this mosquito species has spread from its well-known breeding site in tree-holes to disused manure-pits on dairy farms and subsequently have become a major pest (Becker *et al.* 2003). Because it bites in daylight (Blacklock and Carter 1920), personal protection against this species should emphasise the use of repellents on skin more than protection at night and indoors by house screening, bednets or vaporising mats.

Another mosquito-borne disease, due to West Nile virus, has caused approximately 800 deaths since 1999 in North America (Kramer *et al.* 2007, Reisen and Brault 2007) and epidemics have also occurred in the Old World, for example in Volgograd, Russia (Platonov *et al.* 2001). The reservoir of the virus is in birds and, of the numerous possible vector species, members of the *Culex pipiens* complex are the most important (See Chapter 8). In Europe, *Cx. p. pipiens* tends to bite birds and *Cx. p. molestus* bites humans. However, Curtotti *et al.* (2007) have detected a few *Cx. p. molestus* biting birds and it appears that these mosquitoes could act as bridge-vectors, transmitting the virus from birds to humans. *Culex p. molestus* larvae are typically found in underground sites such as flooded basements of buildings which can become breeding sites, even in European winters (White and Chase 1980). They bite after dark, so house screening, bednets and vaporising mats are appropriate methods of personal protection.

In southern Europe, *Leishmania* pathogens are transmitted by *Phlebotomus* sand flies from dogs to humans. The normal human immune system generally prevents symptoms, but immuno-compromised HIV patients may develop visceral leishmaniasis (Berenguer *et al.* 1989)(see Chapter 5). The use of insecticidal dog collars to kill sand flies which might otherwise pick up *Leishmania* from the dogs is mentioned in the section on treated clothing while the use of insecticidal bednets against sand flies is referred to in the section on the use of nets against mosquitoes.

Culicoides biting midges are a serious nuisance in some parts of Europe. For example, in the Scottish highlands in the summer, the use of insect repellents by outdoor workers and tourists is essential. Members of this genus are also the vectors of Blue Tongue Virus (BTV), a serious disease of sheep, which has spread into southern Europe in recent years and even further north in 2006, reaching the Netherlands (See Chapter 7). Most *Culicoides* species bite outside in daylight and darkness.

There are two existing endemic tick-borne diseases in parts of Europe: Tick Borne Encephalitis (TBE) and Lyme Disease (Dumpis *et al.* 1999, Parola and Raoult 2001) (See Chapters 11 & 12). These are a significant risk to those who enter forests and brush against long grass where ticks are likely to crawl up under clothing and bite. The vaccines available to prevent TBE are highly effective and personal protection by vaccination against TBE in highly endemic areas is recommended (Dumpis *et al.* 1999, Rendi-Wagner 2004). An American vaccine against Lyme Disease is available (De Silva *et al.* 1996), but would probably not be effective in Europe since it is directed against American strains of the pathogen *Borrelia burgdorferi s.s.*[14] A European vaccine should be directed against all the different pathogen strains present in Europe (Wilske *et al.* 1996, Derdáková and Lenčáková 2005). Appropriate use and treatment of clothing with repellents can reduce the risk of tick bites and is discussed below.

Repellents on skin

Repellents applied on skin to repel biting insects and other arthropods are probably older than the existence of the human race. Many mammals and birds rub themselves with millipedes that discharge benzoquinones, which repel insects and ticks (Weldon *et al.* 2003, Carroll *et al.* 2005, Moore and Debboun 2007). Capuchin monkeys (*Cebus capucinus*) rub themselves with plants such as *Citrus, Piper* and *Clematis* (Mary 1996), which may have a repellent effect on insects. The same plant species are used by indigenous people of the New World.

Repellents of synthetic or natural origin, applied to skin, are an important line of defence for Europeans against arthropod bites. Insect repellents are relatively cheap, work immediately, and are easy to carry. However, some repellents are irritant to some people (Fradin 1998). They are volatile and require re-application every few hours, persistence is reduced at higher temperatures, and they may be washed off by sweat or rain. Application of a repellent before going to sleep would not last through the night and, for overnight protection, a bednet is recommended.

Deet

N,N-diethyl-3-methylbenzamide (DEET) is equally or more effective than any other repellent currently on the market. It is used by hundreds of millions of people and in a variety of formulations including sticks, aerosols, lotions and gels. It was first used by the U.S. Army in the 1940s and registered for public use in the 1950s.

Several concentrations of deet are available. However those above 50% do not notably improve the efficacy (Buescher *et al.* 1983) and those below 50% have been shown to be effective against a range of arthropods including; mosquitoes (Frances *et al.* 1996, Fradin and Day 2002, Costantini *et al.* 2004, Takken 2004), ticks (Carroll *et al.* 2004), midges (Perich *et al.* 1995), black flies (Robert *et al.* 1992) and chiggers (Lerdthusnee *et al.* 2003). A dose of 0.8 mg/cm^2 of 50% deet can protect for up to ten hours against mosquito bites (Takken 2004). It is recognised that one of the disadvantages of deet is that it can damage certain types of hard plastics, synthetic fabrics and leather (Fradin and Day 2002). Careless usage may lead to damage to watch faces, camera lenses etc. We tested several types of watch faces and, with some of them, prolonged contact with deet made them sticky to the touch. In combination with sunscreen, deet acts as a solvent and reduces the Sun Protection Factor (Montemarano *et al.* 1997).

A study with radio-labelled deet, showed that 3 to 8% of the applied dose was absorbed through human skin (Selim *et al.* 1995). Literature on cases of toxic reactions due to dermal absorption is scarce. Toxicity in adults is mainly dermatological. Some people develop blisters, burning and erythema, often in the antecubital and popliteal fossa (Frances 2007a). If deet is ingested, severe

[14] By 1998, the US Food and Drug Administration (FDA) had approved a new vaccine against Lyme disease called LYMErix. The vaccine was not 100 percent effective, however, and the FDA still recommended using other preventive measures. In 2002, the manufacturer of LYMErix announced that the vaccine would no longer be available commercially.

toxic reactions can occur, in a few cases resulting in death. A report on a toxic reaction in a child after ingestion of a low dose of deet (Petrucci and Sardini, 2000) has led to restriction of the concentration of deet in commercial products to a maximum of 50% in most countries. Although adverse effects of deet have been reported, the chances of these are small if the repellent is applied and used sensibly in accordance with the registered label. In 1998, the U.S. Environmental Protection Agency (EPA) concluded in a Reregistration Eligibility Decision that deet would generally not cause unreasonable risk to humans or the environment (EPA 1998). However, the EPA also required additional labelling regarding the use of deet on children. The US Public Health Service has made a strong recommendation in favour of the use of deet repellent as a protective measure against mosquito vectors of West Nile virus.

KBR 3023 and IR 3535

In recent years, the synthetic products KBR 3023, (1-)(1-methyl-propxycarbaryl) – 2 – (2 hydroxy-ethyl) piperidine, also known as Picardin) and IR 3535, (2[N-hydroxy-N-acetyl] amino proprionic acid ethyl ester) have been developed and are now the active ingredients of many insect repellent products. KBR 3023 has little or no plasticising effect (Nentwig 2002) which is a serious disadvantage of deet. KBR 3023 and IR 3535 are approximately equivalent to deet in repellent activity against culicine mosquitoes when freshly applied and repel over 6-8 hours, as shown in Table 1 (Yap *et al.* 1998, Barnard *et al.* 2002). In a field trial in Burkina Faso, comparing KBR 3023, IR 3535 and deet, the repellent activity against *An. gambiae s.l.*, 10 hours after application, was consistently higher for KBR3023 (Costantini *et al.* 2004). Additional data reviewed by Puccetti (2007) confirm that the efficacy of deet and IR3535 is comparable against *Aedes* and *Culex* mosquitoes, but for *Anopheles*, deet showed longer-lasting repellency. Additional data, reviewed by Frances (2007b), on IR 3535 confirmed that its repellency is as good or better than comparable deet formulations.

Favourable toxicity data, on KBR 3023 and IR 3535 are reviewed respectively by Frances (2007b) and Puccetti (2007). In contrast to the earlier described adverse effect of deet on a certain type of

Table 1. Mean and range of repellency (in %) of four compounds over 6-8 hours after application.

Data from Barnard *et al.* (2002) on *Ochlerotatus taeniorhynchus*	Product			
	25% DEET	25% KBR3023	25% IR3535	40% PMD
Mean repellency over 6 hrs	94.8% [ab]	97.5% [a]	88.6% [b]	89.2% [b]
Range from 1st to last hour	100%→77.0%	100%→90.0%	100%→81.0%	100%→60.0%
Data of Yap *et al.* (1998) on *Aedes* and *Culex*	*Ae. albopictus*		*Cx. quinquefasciatus*	
	20% Deet	20% KBR3023	20% Deet	20% KBR3023
Mean repellency over 8 hrs	84.9% [a]	94.4% [a]	98.5% [b]	98.5% [b]
Range from 1st to last hour	100%→66.3%	99.4%→88.3%	100%→90.1%	100%→98.6%

Note: % repellency defined as difference between numbers of bites without and with the repellent, divided by number of bites without repellent and multiplied by 100. For each species in the table, data not sharing the same superscript letter differ statistically significantly at $P<0.05$.

watch face, we found that prolonged exposure of the watches faces to KBR 3023 or IR 3535 had no negative effect.

Plant derived repellents

Many plant species are repellent against insects, no doubt as a defence mechanism which has evolved against phytophagous insects and biting arthropods (Moore *et al.* 2007b). In developing countries, people often rely on locally-available plant derived repellents. Nowadays, natural repellents are attracting increasing interest among people in developed countries who do not like synthetic repellents. Large numbers of plant-derived essential oils have recently been tested in studies for their repellent activities, with promising results in some cases (Tawatsin *et al.* 2001, Trongtokit *et al.* 2005, Amer and Mehlhorn 2006). However, in most cases their efficacy is much lower than that of synthetic repellents such as deet. One should be sceptical about the effectiveness of commercial products labelled as 'natural repellents' unless and until they have been shown to be effective in controlled tests. Plant derived repellents generally contain essential oils extracted from plants by steam distillation. Because these plant oils have a high vapour pressure, their longevity is generally much lower than that of deet (Moore *et al.* 2007b). Citronella and PMD (*p*-menthane-3,8-diol) from the lemon eucalyptus plant are moderately to very effective and are available in commercial products.

Citronella

Cymbopogon is a genus of grasses with a citrus-like odour. Common names are lemon grass or citronella grass. Citronella oil is predominantly derived from *Cymbagon nardus* or *C. citratus* and, when used as a repellent, is generally diluted in ethanol. A laboratory study by Tawatsin *et al.* (2001) showed that volatile oils from *C. winterianus* with 5% vanillin could repel *Ae. aegypti*, *An. dirus* and *Cx. quinquefasciatus* for up to eight hours. In a study with 38 essential oils, *C. nardus* was one of the most effective, with 2 hours of complete protection (Trongtokit *et al.* 2005). The persistence of repellency with citronella was found to be markedly less than that of deet (Trigg and Hill, 1996, Cockcroft *et al.* 1998). A field study in Bolivia with *C. citratus* in ethanol showed 78% and 91% protection for 3 h against *An. darlingi* and *Mansonia spp.* (Moore 2005). The current European Biocide Directive requires that all active ingredients of biocidal products are reviewed. Producers and formulators had to either identify or notify their existing active substances to the European Chemicals Bureau (EC 1998). Citronella was not identified or notified as an active ingredient and therefore does not appear on the list of active ingredients which should be reviewed (EC 2005). Because Citronella is not included in the review programme, it is forbidden to sell products with citronella in the European Union since 1 September 2006 (EC 2005).

PMD (p-menthane-3,8-diol)

PMD (*p*-menthane-3,8-diol) is an insect repellent generally derived from oil extracted from the lemon eucalyptus plant (*Eucalyptus maculate citriodon*). It has been used in China as an insect repellent, where it is called *Quwenling* (data of Lu Bao-lin and Li Zu-zi in Curtis 1990). The main ingredient of *Quwenling* is PMD, but it also contains a mix of terpenoids (Strickman 2007). PMD has eight possible stereoisomers, of which four are found when PMD is derived from oil extracts (Barasa *et al.* 2002). Each of these isomers can also be synthesised chemically. No difference in repellency was found between these stereoisomers (Barasa *et al.* 2002).

Several studies showed that repellents containing PMD can be very effective in providing protection against biting insects for several hours and are comparable to deet and other synthetic repellents (See Table 1 and Trigg 1996, Govere *et al.* 2000). PMD is more persistent than other plant-based products such as citronella (Trigg, 1996). It is effective against most biting insects including mosquitoes (*Anopheles*, *Culex* and *Aedes*), midges (*Culicoides*), ticks (*Ixodes*) and stable flies (*Stomoxys calcitrans*) (Trigg and Hill, 1996, Gardulf *et al.* 2004).

It has been decided that the European Union Biocides Directive (EC 1998) applies to repellents as well as insecticides. Negotiations are continuing about whether extensive toxicological testing on animals would be required to clear PMD for continued commercial sale. Small companies are unlikely to fund such testing and it is feared that in the absence of the required information on safety of PMD that it might therefore disappear from the market.

Impact of insect repellents on skin against malaria

Although insect repellents on skin have long been valued against nuisance and biting mosquitoes, evidence that deet and PMD products are effective against malaria has only recently been obtained (Rowland *et al.* 2004, Hill *et al.* in press). These data are from Asia and Latin America where much of the biting by malaria vectors is at hours when people have not yet gone to bed and may be out of doors (Harris *et al.* 2006). Where bites by malaria vectors are received indoors or in bed, bednets, house screening, vaporising mats etc. (discussed later) are more effective and an application of a skin repellent before going to sleep will not last the whole night. It should also be noted that repellents tend to divert biting from one person to another (Moore *et al.* 2007a) rather than killing the vector mosquito, as can happen with an insecticide treated net.

Insecticide Treated Nets (ITNs)

In a review of the history of bednets, Lindsay and Gibson (1988) describe the first recorded bednet use in the 6th century BC. Ross (1910) noted that most malaria vectors bite late at night and recommended bednets as protection against them. Bednets are widely used in the summer in Europe against night-biting nuisance insects. By treating a bednet with a pyrethroid, such as permethrin, deltamethrin, alphacypermethrin or lambdacyhalothrin, a chemical barrier is added to the incomplete physical barrier provided by the net. The pyrethroid deposit drives away or kills insects landing on the net and thus improves it by: (a) reducing the likelihood of insects finding any holes large enough to allow insect entry into the net (Curtis *et al.* 1996) or insects biting through the net of part of the body touching the net; (b) reducing the likelihood of diversion of insects from someone with a net to another person in the same house without one (Lines *et al.* 1987), (c) where ITNs are used by a whole community, so many insects are killed after being attracted to contact nets that the number living long enough for *Plasmodium* (or other pathogens) to reach the infective stage are substantially reduced (Magesa *et al.* 1991, Maxwell *et al.* 1999b). There is now abundant evidence that ITNs are effective in control of diseases transmitted by vectors which bite indoors at night, especially malaria (Alonso *et al.* 1991, Lengeler and Snow 1996, Lengeler 2004), and also the forms of leishmaniasis transmitted by sand flies which bite indoors (Reyburn *et al.* 2000). Nowadays, ITNs (together with Indoor Residual Spraying) are the cornerstones in the Roll Back Malaria programme launched by the World Health Organisation for tropical countries (WHO 2002). If malaria transmission again becomes a threat in Europe, the use of ITNs should be actively encouraged. Until recently, it has been necessary to re-treat ITNs after they have been washed about five times. However, there are now several brands of long lasting insecticidal nets being manufactured with an insecticide deposit which is wash-proof (Graham *et al.* 2005, Maxwell *et al.* 2006) and there are kits with which a wash-proof pyrethroid deposit can be applied to any net (Graham *et al.* 2005, Yates *et al.* 2005).

Until recently, tests for pyrethroid resistance showed that even high frequency of resistance genes did not prevent effective use of ITNs (Henry *et al.* 2005). However, N'Guessan *et al.* (2007) found a population of *An. gambiae* in Benin with resistance of a kind which seriously interferes with the effectiveness of ITNs and Indoor Residual Spraying (IRS) using pyrethroids.

The World Health Organisation considers the use of nets treated with a pyrethroid to be safe, even for young children (Zaim *et al.* 2000, WHO 2005). However, for a few days after re-treatment of nets with certain pyrethroids, users may experience some skin or nasal irritation (Maxwell *et al.* 1999b).

Impregnated clothing

Ticks wait on grass and other plants, attach themselves to the legs of passing animals or humans and move slowly to other parts of the body to obtain blood meals. The pathogens causing TBE and Lyme Disease may be transmitted during the taking of these meals. Tucking trousers into socks and treating both with the pyrethroid permethrin is an effective and convenient means of protecting against ticks when one is going to walk through long grass or forest.

Of the pyrethroids mentioned earlier in connection with bednet treatment, permethrin is the least likely to irritate skin in close contact with treated clothing. Most pyrethroids are not volatile and therefore remain effective against arthropods which contact them for far longer than the volatile repellents used on the skin. As with treatment of nets, it is possible to use formulations of permethrin which make the deposit resistant to many washings (Faulde *et al.* 2003, Faulde and Uedelhoven 2006).

There has been much interest among military medical authorities in the treatment of uniforms with permethrin to reduce mosquito biting through uniforms (Schreck *et al.* 1984, Scholdt *et al.* 1988, Deparis *et al.* 2004, Khoobdel *et al.* 2006). Permethrin-treated clothing is also becoming available for the civilian market. However, it should be noted that the majority of mosquito bites are not through clothing but on exposed skin. Permethrin-treated clothing gives a rate of protection against biting on exposed skin adjacent to the clothing which is practically negligible (e.g. C.F. Curtis, unpublished observation of mosquitoes taking full blood meals on an exposed arm within a few centimetres of the sleeve of one of the commercially available treated shirts). Schreck *et al.* (1984) have emphasised that a skin repellent needs to be used in conjunction with permethrin treated clothing.

Where dogs are reservoirs of *Leishmania* it has been found that, after fitting dogs with collars containing a pyrethroid, the insecticide diffuses across the dog's skin and kills sand flies attempting to bite (Killick-Kendrick *et al.* 1997). In Iran, the long-term consequence of community-wide use of pyrethroid-treated dog collars was a reduced risk of *Leishmania* infection in children (Gavgani *et al.* 2002).

House construction and screening

Improvement of rural houses which rendered them less attractive as mosquito resting places, was an important factor in the decline of malaria in the first half of the 20th century in the USA (Boyd 1926). Studies in Europe and the USA of the medical history of people living in houses with screened doors and windows proved that such screening reduced malaria risk (Manson 1900, USPHS 1947, Snowden 2006). The very marked difference in incidence of dengue in northern Mexico and nearby areas in the southern USA is considered to be largely due to the better quality of housing and more time spent indoors in the USA (Reiter *et al.* 2003). Attempts are now being made to find affordable ways of upgrading rural, tropical houses to make them less attractive to mosquitoes (Lindsay *et al.* 2003).

Winter time breeding of *Cx. p. molestus* in the flooded basement of a building in London was eliminated by repairing a leaking sewer pipe (White and Chase 1980). In Zanzibar town, *Culex* breeding in flooded basements was quickly and permanently eliminated by application of floating layers of polystyrene beads (Maxwell *et al.* 1999a) and this method may be useful in Europe.

Apart from vaccination in attempting to protect against Blue Tongue Virus (BTV), bringing livestock indoors, where one of the vectors, *Culicoides imicola,* will not enter, is one of the prevention methods being attempted (Mellor and Wittmann 2002). However, *C. bolitinos* is willing to enter animal sheds to bite (P. Mellor, *pers.comm.*) and members of the BTV-vector taxonomic groups *C. obsoletus* and *C. pulicaris* are readily found biting outdoors and indoors (see Chapter 7).

Repellent vaporising mats, mosquito coils and insecticide aerosols

Vaporising mats heated electrically are widely available as household mosquito repellents/insecticides. By vaporising a volatile pyrethroid compound, such as bio-allethrin, the device can kill mosquitoes or repel them from a room for about 10 hours (Curtis and Hill, 1988). Mosquito coils consist of a synthetic pyrethroid or natural pyrethrum, a binder and organic filler, which will smoulder when ignited. Mosquito coils can give off insecticidal smoke for about 8 hours and are commonly used in tropical countries. In a comparison with vaporising mats, Curtis and Hill (1988) showed that the latter were markedly better at preventing *Anopheles* biting than were coils, especially in a draught. Lawrance and Croft (2004) concluded that there is no evidence that burning mosquito coils reduce malaria risk significantly, but there is consistent evidence that burning coils inhibit nuisance biting by various mosquito species.

There has been concern about inhaling insecticidal vapour or smoke from vaporising mats or mosquito coils. More pyrethroid is likely to be absorbed through the lungs than is taken up by skin contact with impregnated clothing or bednets (Barlow *et al.* 2001). Smoke from a mosquito coil contains heavy metals, allethrin and, in some cases, octachlorodipropyl ether (Liu *et al.* 1987, Krieger *et al.* 2003), which is a potent lung carcinogen. A lengthy exposure to the smoke may well have adverse effects (Liu *et al.* 1987).

If screened windows and doors are opened during daytime, mosquitoes may enter and rest inside rooms. An insecticide aerosol is useful to kill these insects in the evening when doors and windows are closed. If these are not screened, an aerosol will not be of much use as its effects are non-residual and more mosquitoes are likely to enter throughout the night.

Buzzers

Electronic buzzers are sold with the claim that they repel mosquitoes because the sonic frequencies produced by the devices mimic the sound of the male wing beats, or the sound of predators (Cabrini and Andrade, 2006). However, Rasnitsyn *et al.* (1974) and twelve subsequent laboratory or field studies in the 1970s and 80s found no difference in biting rate when the electronic buzzers were switched on or off. More recent studies have confirmed this conclusion (Foster and Lutes 1985, Cabrini and Andrade 2006, Ahmad *et al.* 2007). Claims that an ultrasonic device would repel cockroaches could also not been substantiated (Ahmad *et al.* 2007). Two companies were prosecuted and found guilty under the Trade Descriptions law of the U.K. for making unsubstantiated claims in the advertising of buzzers (Curtis 1994).

Vitamin B_1

Many people believe that taking doses of vitamin B_1 makes one emit an odour repellent to mosquitoes. However, laboratory tests (Khan *et al.* 1969, Curtis 1986) and field trials in the Norwegian arctic (Brunn 1964) and in Liberia (Maasch 1973) found no such repellency.

Acknowledgements

We thank Jeroen Spitzen and Dr. Yu Tong Qiu for their comments on this chapter.

References

Ahmad A, Subramanyam B and Zurek L (2007) Responses of mosquitoes and German cockroaches to ultrasound emitted from a random ultrasonic generating device. Entomol Exp Appl 123: 25-33.

Alonso PL, Lindsay SW, Armstrong JRM, Conteh M, Hill AG, David PH, Fegan G, Francisco AD, Hall AJ, Shenton FCCK and Greenwood B (1991) The effect of insecticide-treated bed nets on mortality of Gambian children. Lancet 337: 1499-1502.

Amer A and Mehlhorn H (2006) Repellency effect of forty-one essential oils against *Aedes*, *Anopheles*, and *Culex* mosquitoes. Parasitol Res 99: 478-490.

Barasa SS, Ndiege IO, Lwande W and Hassanali A (2002) Repellent activities of stereoisomers of p-menthane-3,8-diols against *Anopheles gambiae* (Diptera: Culicidae). J Med Entomol 39: 736-741.

Barlow SM, Sullivan FM and Lines J (2001) Risk assessment of the use of deltamethrin on bednets for the prevention of malaria. Food Chem Toxicol 39: 407-422.

Barnard DR, Bernier UR, Posey KH and Xue RD (2002) Repellency of IR3535, KBR3023, para-menthane-3,8-diol, and deet to black salt marsh mosquitoes (Diptera: Culicidae) in the Everglades National Park. J Med Entomol 39: 895-899.

Becker N, Zgomba M, Petric D, Dahl C, Boase C, Lane J and Kaiser A (2003) Mosquitoes and their control Kluwer Academic / Plenum Publishers, New York.

Berenguer JJ, Moreno SS, Cercenado EE, Bernaldo de Quirós JJC, García de la Fuente AA and Bouza EE (1989) Visceral leishmaniasis in patients infected with human immunodeficiency virus (HIV). Ann Intern Med 111: 129-132.

Blacklock B and Carter HF (1920) Observations on *Anopheles (coelodiazesis) plumbeus*, Stephens, with special reference to its breeding-places, occurrence in the Liverpool district, and possible connection with the spread of malaria. Ann Trop Med Parasitol 13: 421-444.

Boyd MF (1926) The influence of obstacles unconsciously erected against anophelines (Housing and Screening) upon the incidence of malaria. Am J Trop Med 6: 157-160.

Brunn J (1964) Thiamin som middel mot mygg. Nordiske Med 71: 650-651.

Buescher MD, Rutledge LC, Wirtz RA and Nelson JH (1983) The dose-persistence relationship of deet against *Aedes aegypti*. Mosq News 43: 364-366.

Cabrini I and Andrade CFS (2006) Evaluation of seven new electronic mosquito repellers. Entomol Exp Appl 121: 185-188.

Carroll JF, Kramer M, Weldon PJ and Robbins RG (2005) Anointing chemicals and ectoparasites: effects of benzoquinones from millipedes on the lone star tick, *Amblyomma americanum* J Chem Ecol 31: 63-75.

Carroll JF, Solberg VB, Klun JA, Kramer M and Debboun M (2004) Comparative activity of deet and AI3-37220 repellents against the ticks *Ixodes scapularis* and *Amblyomma americanum* (Acari: Ixodidae) in laboratory bioassays. J Med Entomol 41: 249-254.

Cockcroft A, Cosgrove J and Wood R (1998) Comparative repellency of commercial formulations of deet, permethrin and citronellal against the mosquito *Aedes aegypti*, using a collagen membrane technique compared with human arm tests. Med Vet Entomol 12: 289-294.

Costantini C, Badolo A and Ilboudo-Sanogo E (2004) Field evaluation of the efficacy and persistence of insect repellents DEET, IR3535, and KBR 3023 against *Anopheles gambiae* complex and other Afrotropical vector mosquitoes. Trans R Soc Trop Med Hyg 98: 644-652.

Curtis CF (1986) Fact and fiction in mosquito attraction and repulsion. Parasitol Today 11: 316-318.

Curtis CF (1990) Appropriate technology in vector control. Boca Raton: CRC.

Curtis CF (1994) Anti-mosquito buzzers, advertising and the law. Wing Beats 1994: 10-12.

Curtis CF and Hill N (1988) Comparison of methods of repelling mosquitoes. Entomol Exp Appl 49: 175-179.

Curtis CF, Myamba J and Wilkes TJ (1996) Comparison of different insecticides and fabrics for anti-mosquito bednets and curtains. Med Vet Entomol 10: 1-11.

Curtotti A, Fonseca DM, Curtis CF, Ismáil N, Davies D and Malcolm CA (2007) Can British *Culex pipiens* s.l. be a bridge vector for the transmission of WNV to humans? Eur Mosq Bull 22: 31-32.

De Silva AAM, Telford SSR, Brunet LLR, Barthold SSW and Fikrig EE (1996) *Borrelia burgdorferi* OspA is an arthropod-specific transmission-blocking Lyme disease vaccine. J Exp Med 183: 271-275.

De Zulueta J (1998) The end of malaria in Europe: an eradication of the disease by control measures. Parassitologia 40: 245-246.

Deparis X, Frere B, Lamizana M, Leroux NGRF, Lefevre P, Finot L, Hougard JM, Carnevale P, Guillet P and Baudon D (2004) Efficacy of permethrin-treated uniforms in combination with DEET topical repellent for protection of French military troops in Côte d'Ivoire. J Med Entomol 41: 914-921.

Derdáková M and Lenčáková D (2005) Association of genetic variability within the *Borrelia burgdorferi sensu lato* with the ecology, epidemiology of Lyme borreliosis in Europe. Ann Agr Environ Med 12: 165-172.

Dumpis UU, Crook DD and Oksi JJ (1999a) Tick-borne encephalitis. Clin Infect Dis 28: 882-890.

EC (1998) Directive 98/8/EC of the European Parliament and of the Council of 16 February 1998 concerning the placing of biocidal products on the market. Off J Eur Comm L123: 0001-0063.

EC (2005) Commission Regulation (EC) No 1048/2005 of 13 June 2005 amending Regulation (EC) No 2032/2003 on the second phase of the 10-year work programme referred to in Article 16(2) of Directive 98/8/EC of the European Parliament and of the Council concerning the placing of biocidal products on the market. Off J Eur Comm L178, 0001-0098.

Eling W, van Gemert GJ, Akinpelu O, Curtis J and Curtis CF (2003) Production of *Plasmodium falciparum* sporozoites by *Anopheles plumbeus*. Eur Mosq Bull 15: 12-13.

EPA (1998) Reregistration Eligibility Decision: DEET. In: United States Environmental Protection Agency, pp 1-119.

Faulde MK, Uedelhoven WM and Robbins RG (2003) Contact toxicity and residual activity of different permethrin-based fabric impregnation methods for *Aedes aegypti* (*Diptera: Culicidae*), *Ixodes ricinus* (*Acari: Ixodidae*), and *Lepisma saccharina* (*Thysanura: Lepismatidae*). J Med Entomol 40: 935-941.

Faulde MK and Uedelhoven W (2006) A new clothing impregnation method for personal protection against ticks and biting insects. Int J Med Microbiol 296: 225-229.

Foster WA and Lutes KI (1985) Tests of ultrasonic emissions on mosquito attraction to hosts in a flight chamber. J Am Mosq Control Assoc 1: 199-202.

Fradin MS (1998) Mosquitoes and mosquito repellents: A Clinician's Guide. Ann Intern Med 128: 931-940.

Fradin MS and Day JF (2002) Comparative efficacy of insect repellents against mosquito bites. N Engl J Med 347: 13-18.

Frances SP (2007a) Efficacy and safety of repellents containing deet. In: M. Debboun, S.P. Frances & D. Strickman (eds.), Insect Repellents, Principles, Methods and Uses. CRC Press, Boca Raton, pp 311-326.

Frances SP (2007b). Picaridin. In: M. Debboun, S.P. Frances & D. Strickman (eds.), Insect Repellents, Principles, Methods and Uses. CRC Press, Boca Raton, pp 337-340.

Frances SP, Klein TA, Hildebrandt DW, Burge R, Noigamol C, Eikarat N, Sripongsai B and Wirtz RA (1996) Laboratory and field evaluation of DEET, CIC-4, and AI3-37220 against *Anopheles dirus* (Diptera: Culicidae) in Thailand. J Med Entomol 33: 511-515.

Gardulf A, Wohlfart I and Gustafson R (2004) A prospective cross-over field trial shows protection of lemon eucalyptus extract against tick bites. J Med Entomol 41: 1064-1067.

Gavgani ASM, Hodjati MH, Mohite H and Davies CR (2002) Effect of insecticide-impregnated dog collars on incidence of zoonotic visceral leishmaniasis in Iranian children: a matched cluster randomised trial. Lancet 360: 374-379.

Govere J, Durrheim DN, Baker L, Hunt R and Coetzee M (2000) Efficacy of three insect repellents against the malaria vector *Anopheles arabiensis*. Med Vet Entomol 14: 441-444.

Graham K, Kayedi MH, Maxwell C, Kaur H, Rehman H, Malima R, Curtis CF, Lines JD and Rowland MW (2005) Multi-country field trials comparing wash-resistance of PermaNet™ and conventional insecticide-treated nets against anopheline and culicine mosquitoes. Med Vet Entomol 19: 72-83.

Harris AF, Marias-Arnez A and Hill N (2006) Biting times of *Anopheles darlingi* in the Bolivian Amazon and implications for controlling malaria. Trans Roy Soc Trop Med Hyg 100: 46-47.

Henry MC, Assi SB, Rogier C, Dossou-Yovo J, Chandre F, Guillet P and Carnevale P (2005) Protective efficacy of lambda-cyhalothrin treated nets in *Anopheles gambiae* pyrethroid resistance areas of Côte d'Ivoire. Am J Trop Med Hyg 73: 859-864.

Hill N, Lenglet A, Arnez AM and Carneiro (2007). Plant-based insect repellent used in combination with insecticide treated bed nets give greater protection against malaria than treated nets alone in areas of early evening biting vectors: a double-blind, placebo controlled clinical trial in the Bolivian Amazon. BMJ (in press).

Jelinek T, Schulte C, Behrens R, Grobusch MP, Coulaud JP, Bisoffi Z, Matteelli A, Clerinx J, Corachán M, Puente S, Gjørup I, Harms G, Kollaritsch H, Kotlowski A, Björkmann A, Delmont JP, Knobloch J, Nielsen LN, Cuadros J, Hatz C, Beran J, Schmid ML, Schulze M, Lopez-Velez R, Fleischer K, Kapaun A, McWhinney P, Kern P, Atougia J, Fry G, Da Cunha S and Boecken G (2002) Imported falciparum malaria in Europe: Sentinel surveillance data from the European network on surveillance of imported infectious diseases. Clin Infect Dis 34: 572-576.

Khan AA, Maibach HI, Strauss WG and Fenley WR (1969) Vitamin B1 is not a systemic mosquito repellent in man. Trans St Johns Hosp Dermatol Soc 55: 99-102.

Khoobdel M, Shayeghi M, Vatandoost H, Rassi Y, Abaei MR, Ladonni H, Mehrabi Tavana A, Bahrami SH, Najafi ME, Mosakazemi SH, Khamisabadi K, Azari Hamidian S and Akhoond MR (2006) Field evaluation of permethrin-treated military uniforms against *Anopheles stephensi* and 4 species of *Culex* (Diptera:Culicidae) in Iran. J Entomol 3: 108-118.

Killick-Kendrick R, Killick-Kendrick M, Focheux C, Dereure J, Puech MP and Cadiergues MC (1997) Protection of dogs from bites of phlebotomine sandflies by deltamethrin collars for control of canine leishmaniasis. Med Vet Entomol 11: 105-111.

Kramer LD, Li J and Shi P-Y (2007) West Nile virus. Lancet Neurol 6: 171-181.

Krieger RI, Dinoff TM and Zhang X (2003) Octachlorodipropyl ether (S-2) mosquito coils are inadequately studied for residential use in Asia and illegal in the United States. Environ Health Perspect 111: 1439-1442.

Krüger A, Rech A, Su XZ and Tannich E (2001) Short communication: Two cases of autochthonous *Plasmodium falciparum* malaria in Germany with evidence for local transmission by indigenous *Anopheles plumbeus*. Trop Med Int Health 6: 983-985.

Lawrance CE and Croft AM (2004) Do mosquito coils prevent malaria? A systematic review of trials. J Trav Med 11: 92-96.

Lengeler C and Snow RW (1996) From efficacy to effectiveness: insecticide-treated bednets in Africa. Bull World Health Organ 74: 325-332.

Lengeler C (2004) Insecticide-treated bed nets and curtains for preventing malaria. Cochrane Database Syst Rev CD000363.

Lerdthusnee K, Khlaimanee N, Monkanna T, Mungviriya S, Leepitakrat W, Debboun M and Coleman RE (2003) Development of an *in vitro* method for the evaluation of candidate repellents against *Leptotrombidium (Acari: Trombiculidae)* chiggers. J Med Entomol 40: 64-67.

Lindsay SW and Gibson ME (1988) Bednets revisited - Old idea, new angle. Parasitol Today 4: 270-272.

Lindsay SW, Jawara M, Paine K, Pinder M, Walraven GEL and Emerson PM (2003) Changes in house design reduce exposure to malaria mosquitoes. Trop Med Int Health 8: 512-517.

Lines JD, Myamba J and Curtis CF (1987) Experimental hut trials of permethrin-impregnated mosquito nets and eave curtains against malaria vectors in Tanzania. Med Vet Entomol 1: 37-51.

Liu WK, Wong MH and Mui YL (1987) Toxic effects of mosquito coil (a mosquito repellent) smoke on rats: I. Properties of the mosquito coil and its smoke. Toxicol Lett 39: 223-230.

Maasch HJ (1973) Outdoor studies on the mosquito repellent effect of vitamin B 1. Freilanduntersuchungen zur muckenabweisenden Wirkung von Vitamin B 1. Z Tropenmed Parasitol 24: 119-122.

MacDonald G (1957) The epidemiology and control of malaria. Oxford University Press, London.

Magesa SM, Wilkes TJ, Mnzava AEP, Njunwa KJ, Myamba J, Kivuyo MDP, Hill N, Lines JD and Curtis CF (1991) Trial of pyrethroid impregnated bednets in an area of Tanzania holoendemic for malaria vector population. Effects on the malaria vector population. Acta Trop 49: 97-108.

Manson P (1900) Experimental proof of the mosquito-malaria theory. Lancet I: 923-925.

Mary B (1996) Fur rubbing: Use of medicinal plants by capuchin monkeys (*Cebus capucinus*). Am J Primatol 38: 263-270.

Maxwell CA, Mohammed KK, Kisumku UU and Curtis CF (1999a) Can vector control play a useful supplementary role against bancroftian filariasis? Bull World Health Organ 77: 138-143.

Maxwell CA, Myamba J, Njunwa KJ, Greenwood BM and Curtis CF (1999b). Comparison of bednets impregnated with different pyrethroids for their impact on mosquitoes and on re-infection with malariaafter clearence of pre-existing infections with chlorproguanil-dapsone. Trans Roy Soc Trop Med Hyg 93: 4-11.

Maxwell CA, Myamba J, Magoma J, Rwegoshora RT, Magesa SM & Curtis CF (2006). Tests of Olyset nets by bioassay and in experimental huts. J Vector Borne Dis 43: 1-6.

Mellor PS and Wittmann EJ (2002) Bluetongue Virus in the Mediterranean Basin 1998-2001. Vet J 164: 20-37.

Montemarano AD, Gupta RK, Burge JR and Klein K (1997) Insect repellents and the efficacy of sunscreens. Lancet 349: 1670-1671.

Moore SJ (2005) A methodology for developing plant-based biocides against *Anopheles* mosquitoes. London School of Hygiene and Tropical Medicine. London University, London.

Moore SJ and Debboun M (2007) History of insect repellents. In: M. Debboun, S.P. Frances & D. Strickman (eds.), Insect repellents, Principles, Methods and Uses. CRC Press, Boca Raton, pp 3-29.

Moore SJ, Davies CR, Hill N and Cameron MM (2007a). Are mosquitoes diverted from repellent-using individuals to non-users? Results of a field study in Bolivia. Trop Med Int Health 12: 532-539.

Moore SJ, Lenglet A and Hill N (2007b). Plant-based insect repellents. In: M. Debboun, S.P. Frances & D. Strickman (eds.), Insect repellents, Principles, Methods and Uses. CRC Press, Boca Raton, pp 275-304.

N'Guessan R, Corbel V, Akogbéto M and Rowland M (2007) Reduced efficacy of insecticide-treated nets and indoor residual spraying for malaria control in pyrethroid resistance area, Benin. Emerg Infect Dis 13: 199-206.

Nentwig GG (2002) Bayrepel® (KBR 3023), a new mosquito repellent: from laboratory synthesis to a worldwide commercial product. In: European Conference on Travel Medicine. Florence, pp 85.

Parola P and Raoult D (2001 Tick-borne bacterial diseases emerging in Europe. Clin Microbiol Infect 7: 80-83.

Peng Z and Simons FER (2004) Mosquito allergy: Immune mechanisms and recombinant salivary allergens. Int Arch Allergy Immunol 133: 198-209.

Perich MJ, Strickman D, Wirtz RA, Stockwell SA, Glick JI, Burge R, Hunt G and Lawyer PG (1995) Field-evaluation of 4 repellents against *Leptoconops americanus* (Diptera, Ceratopogonidae) biting midges. J Med Entomol 32: 306-309.

Petrucci N and Sardini S (2000) Severe neurotoxic reaction associated with oral ingestion of low-dose diethyltoluamide-containing insect repellent in a child. Pediatr Emerg Care 16: 341-342.

Platonov AE, Shipulin GA, Shipulina OY, Tyutyunnik EN, Frolochkina TI, Lanciotti RS, Yazyshina S, Platonova OV, Obukhov IL, Zhukov AN, Vengerov YY and Pokrovskii VI (2001). Outbreak of West Nile virus infection, Volgograd Region, Russia, 1999. Emerg Infect Dis 7: 128-132.

Puccetti C (2007) IR 3535 (Ethyl Butylacetylaminopropionate). In: M. Debboun, S.P. Frances & D. Strickman (eds.), Insect repellents, Principles, Methods and Uses. CRC Press, Boca Raton.

Rasnitsyn SP, Alekseev AN, Gornostaeva RM, Kupriyanova YS, Potapov AA and Razumova OV (1974) Negative results of testing of sound generator prototypes designed for mosquito repellence (Russian). Meditsinskaya Parazitologiya i Parazitarnye Bolezni 43: 706-708.

Reisen, W and Brault AC (2007) West Nile virus in North America: perspectives on epidemiology and intervention. Pest Manag Sci 63: 641-646.

Reiter P, Lathrop S, Bunning M, Biggerstaff B, Singer D, Tiwari T, Baber L, Amador M, Thirion J, Hayes J, Seca C, Mendez J, Ramirez B, Robinson J, Rawlings J, Vorndam V, Waterman S, Gubler D, Clark G and Hayes E (2003) Texas lifestyle limits transmission of dengue virus. Emerg Infect Dis 9: 86-89.

Rendi-Wagner P (2004) Risk and Prevention of tick-borne encephalitis in travelers. J Trav Med 11: 307-312.

Reyburn H, Ashford R, Mohsen M, Hewitt S and Rowland M (2000) A randomized controlled trial of insecticide-treated bednets and chaddars or top sheets, and residual spraying of interior rooms for the prevention of cutaneous leishmaniasis in Kabul, Afghanistan. Trans Roy Soc Trop Med Hyg 94: 361-366.

Robert LL, Coleman RE, Lapointe DA, Martin PJS, Kelly R and Edman JD (1992) Laboratory and field evaluation of five repellents against the black flies *Prosimulium mixtum* and *P. fuscum* (Diptera: Simuliidae). J Med Entomol 29: 267-272.

Romi R, Sabatinelli G and Majori G (2001) Could malaria reappear in Italy? Emerg Infect Dis 7: 915-919.

Ross R (1910) The prevention of malaria. Murray, London.

Rowland M, Downey G, Rab A, Freeman T, Mohammad N, Rehman H, Durrani N, Reyburn H, Curtis C, Lines J and Fayaz M (2004) DEET mosquito repellent provides personal protection against malaria: A household randomized trial in an Afghan refugee camp in Pakistan. Trop Med Int Health 9: 335-342.

Sabatinelli G, Ejov M and Joergensen P (2001) Malaria in the WHO European Region (1971-1999). Euro surveill 6: 61-65.

Scholdt LL, Schreck CE, Qureshi A, Mammino S, Aziz A and Iqbal W (1988) Field bioassays of permethrin treated uniform and a new extended duration repellent against mosquitoes in Pakistan. J Am Mosq Contr Assoc 4: 233-236.

Schreck CE, Haile DG and Kline DL (1984) The effectiveness of permethrin and deet, alone or in combination, for protection against *Aedes taeniorhynchus*. Am J Trop Med Hyg 33: 725-730.

Selim S, Hartnagel REJ, Osimitz TG, Gabriel KL and Schoenig GP (1995) Absorption, metabolism, and excretion of N,N-diethyl-m-toluamide following dermal application to human volunteers. Toxicol. Sci. 25: 95-100.

Snowden FM (2006) The conquest of malaria, Italy, 1900-1962. Yale University Press.

Strickman D (2007) PMD (p-Menthane-3,8-Diol) and Quwenling. In: M. Debboun, S.P. Frances & D. Strickman (eds.), Insect repellents, Principles, Methods and Uses. CRC Press, Boca Raton, pp 347-351.

Takken W (2004) Onderzoek naar de werkzaamheid van zeven verschillende muggen-repellentia. In: Entomologie, Wageningen Universiteit, Dept. Plantwetenschappen, Wageningen.

Tawatsin A, Wratten SD, Scott RR, Thavara U and Techadamrongsin Y (2001) Repellency of volatile oils from plants against three mosquito vectors. J Vector Ecol 26: 76-82.

Trigg JK (1996) Evaluation of a Eucalyptus-based repellent against *Anopheles* spp. in Tanzania. J Am Mosq Contr Assoc 12: 243-246.

Trigg JK and Hill N (1996) Laboratory evaluation of a eucalyptus-based repellent against four biting arthropods. Phytother Res 10: 313-316.

Trongtokit Y, Rongsriyam Y, Komalamisra N and Apiwathnasorn C (2005) Comparative repellency of 38 essential oils against mosquito bites. Phytother Res 19: 303-309.

United States Public Health Service (1947) Malaria control in impounded water [by the] United States Public Health Service and [the] Tennessee Valley Authority, Health and Safety Department. Washington, U.S. Govt. Print. Off., 1947.

Weldon PJ, Aldrich JR, Klun JA, Oliver JE and Debboun M (2003) Benzoquinones from millipedes deter mosquitoes and elicit self-anointing in capuchin monkeys (Cebus spp.). Naturwissenschaften 90: 301-304.

White GB and Chase CA (1980) Eradication of a localized outbreak in London of the pest mosquito *Culex molestus*. Trans Roy Soc Trop Med Hyg 74: 685.

WHO (2002) Scaling up insecticide-treated netting programmes in Africa; a strategic framework for coordinated national action. In: Technical Support Network for Insecticide-Treated Materials. Roll Back Malaria. World Health Organization, Geneva, Switzerland.

WHO (2005) Safety of pyrethroids for public use. World Health Organization Pesticide Evaluation Scheme (WHOPES).

Wilske B, Busch U, Fingerle V, Jauris-Heipke S, Preac Mursic V, Rössler D and Will G (1996) Immunological and molecular variability of OspA and OspC. Implications for Borrelia vaccine development. Infection 24: 208-212.

Yap HH, Jahangir K, Chong ASC, Adanan CR, Chong NL, Malik YA and Rohaizat B (1998) Field efficacy of a new repellent, KBR 3023, against *Aedes albopictus* (SKUSE) and *Culex quinquefasciatus* (SAY) in a tropical environment. J Vector Ecol 23: 62-68.

Yates A, N'Guessan R, Kaur H, Akogbeto M and Rowland M (2005) Evaluation of KO-Tab 1-2-3®: a wash-resistant 'dip-it-yourself' insecticide formulation for long-lasting treatment of mosquito nets. Malar J 4: 52.

Zaim M, Aitio A and Nakashima N (2000) Safety of pyrethroid-treated mosquito nets. Med Vet Entomol 14: 1-5.

21. Mosquito control in Europe

Norbert Becker and Marija Zgomba

Abstract

The history of mosquito control goes back to ancient times and was first mentioned in the western hemisphere by Herodotus 450 years BC. However, effective control started with the discovery of the role of mosquitoes in pathogen and parasite transmission at the turn of the 20th century. In the following decades, effective insecticides used in integrated control strategies were discovered. With the foundation of the European Mosquito Control Association (EMCA) in 2000, a forum was created for intensive exchange of knowledge in the field of sound mosquito control. Soon after its foundation, several hundred members joined the EMCA from 22 countries, including more than 20 European abatement districts, which were founded between 1958 and 2001. More than 2.3 million hectares of larval habitats such as flood plains, rice fields and salt marshes must be treated for mosquito control regularly. In most programmes the bio-insecticides *Bacillus thuringiensis israeliensis* and *B. sphaericus* are used, which provide effective control in an environmentally-benign manner.

Keywords: history, mosquito control, *Bacillus thuringiensis israeliensis*, *B. sphaericus*, Europe, association

Introduction

No other organisms have influenced the socio-economic development of humans more than mosquitoes. It is therefore not suprising that humans have sought means to control mosquitoes since prehistoric times. Herodotus (480 BC) was the first in the western world who commented on mosquito control while he was in Egypt. He described that Egyptians had two ways of avoiding mosquito bites. In the Nile delta they slept under fishing nets, and in Upper Egypt on the top of buildings, where wind speeds are high enough to prevent mosquito access. In early days, humans used inorganic materials to combat insect pests. The Sumerians apparently used sulphur compounds to control insects well before 2500 BC. The Chinese, in 1200 BC, used plant-derived fumigants as well as mercury and arsenic compounds as insecticides (Retnakaran *et al.* 1985). Plinius the Elder, in his *Historia Naturalis* (70 AD), includes a summary of pest control practices extracted from the Greek literature of the preceding 200-300 years. Most of the methods and materials, often based on superstition, were useless.

Interest in insects increased with the discovery of the microscope, and Gutenberg's printing press made it possible for Johannes Wonneke to publish the first drawing of a mosquito, in 1480 in Frankfurt, in his *Ortus Sanitatis*, one of the first publications on Medical Entomology. The term 'mosquito' is derived from the Latin name 'mosca'. European explorers called them 'muscetos', later in English 'mosquitoes'.

In the 16th and 17th century, natural philosophers wondered if there was something that could be done to prevent these 'noyous iniuries', and in doing so initiated the era of modern mosquito control. A landmark publication in entomology is the 'Memoires pour servir à l'Histoire des Insects' by Reaumur which was published between 1734-1742. Reaumur wrote, 250 years ago, 'Mosquitoes are certainly our declared enemies, and very annoying enemies at that. But they are enemies good for us to know. If we give them even the least bit attention we find ourselves forced to admire them and to admire even the instrument they bleed us with, for this it is only necessary to examine their structure. Furthermore, throughout their lives they offer facts adequate to please any of those minds which are curious about the marvels of nature.' This view remains valid until today.

Boccone (1697) and Buxbaum (1728) were the first to mention that Asians used powder ground from the heads of local pyrethrum species (*Pyrethrum carneum* and *P. roseum*) flowers against insects.

Between 1800 and 1850, pyrethrum, a mixture of lime and sulphur, phosphorous paste, and rotenone were used. Around 1870, a range of secondary plant products (e.g. nicotine, derris and quassia) were recommended. With these products, and the availability of the arsenic compound Paris green and kerosene emulsion as insecticides, the era of professional pesticide application had begun.

The history of mosquito control in modern times can be divided into the era before and after the discovery of the role of mosquitoes in pathogen and parasite transmission by a handful European and American researchers at the turn of the 20th century (Boyd, 1949). At the beginning of the twentieth century, mosquito control was based on two basic facts and two basic tools: (a) mosquitoes act as vectors of disease agents and water is a natural larval habitat, and (b) tools to fight larvae are the shovel and kerosene. Between 1900 and 1942, mosquito control was mostly based on this approach.The first recorded use of a synthetic organic insecticide, dinitro-*o*-cresol, occurred in 1892 and by the 1930s a range of such compounds had been discovered but remained very limited in use (Cremlyn 1978). From the 1920s onwards, the increasing potential of insecticides as tools for insect control led to their growing predominance as an approach to insect control (Becker *et al.* 2003).

In 1939, the first insecticide of the chlorinated hydrocarbon group, dichloro-diphenyl-trichloroethane, known as DDT, was introduced. This was soon followed by the development of organophosphates and in the early 1950s carbamates were created. In the era of chemical control agents, during 1960 - 1970s, the discovery and development of photostable pyrethroids led to profound changes in the practice of insecticide use. The adverse effects of sometimes indiscriminate and excessive use of DDT were dramatically portrayed by the publication of Rachel Carson's 'Silent Spring' (1962). As a result of the problems with chlorinated hydrocarbons, alternative methods were sought. These alternatives include insecticides developed as a result of basic entomological research on metabolic disrupters, moulting inhibitors and behaviour-modifying compounds. Highly potent inhibitors of chitin synthesis in the insect integument were developed in the early 1970s. The discoveries that insect juvenile hormones regulate many developmental functions in insects and that they are unique to arthropods, initiated the synthesis of juvenoids in the early 1980s for selective insect control. As with other alternative strategies for insect control, the full potential of juvenoids in selective and environmentally-sound insect control has not been fully realised to date.

Since the 1960s, various genetic control techniques such as translocations, cytoplasmatic incompatibility, meiotic drive and sterile hybrids have been investigated, but none of these have been implemented on a large scale. It is also important to note the extensive research on the evaluation and development of biological control agents such as protozoa, fungi, nematodes, and other parasites and pathogens. So far none of these agents have reached the stage of large-scale use in vector control programmes (Becker *et al.* 2003) with the exception of *Bacillus thuringiensis israeliensis* (*Bti*) and *B. sphaericus* (*Bs*).

Both *Bti* and *Bs* produce protein toxins during sporulation that are assembled into crystals referred to as the parasporal inclusion body. These proteins are highly toxic to mosquito larvae. Whereas *Bti* toxins kill mosquito and black fly larvae, and at higher dosages also larvae of some other nematoceran flies (e.g. Chironomidae, Sciaridae, Tipulidae), *Bs* has a more narrow host range and is only active against mosquitoes, in particular against larvae of *Culex* species and *Anopheles gambiae* Giles. The only other insects killed by *Bs*, at higher dosages, are Psychodidae larvae. The high potential of *B. sphaericus* lies not only in its spectrum of efficacy but also in its ability to persist in nature under certain conditions, which means that long-term control can be achieved (Becker *et al.* 2003). The time-span between retreatments can thus be extended and personnel costs reduced. This opens up the possibility of a successful and cost-effective control of *Culex* species, particularly of *Cx. quinquefasciatus* Say, the most important vector of lymphatic filariasis that breeds primarily in highly polluted water bodies in urban areas.

It was shown that the selectivity of the bacilli derives from a variety of factors: (1) the protein crystals (inactive protoxin) must be ingested by the target insect, and this depends on its feeding habits, (2) proteases must then convert the protoxin into biologically active toxins in the alkaline milieu of the target insect midgut, and (3) the toxins must then bind to a cell surface receptor (glycoprotein) of the larval midgut epithelial cells. This disturbs the osmoregulatory mechanisms of the cell membrane, which causes swelling and bursting of the midgut cells. Non-target-organisms do not activate protoxin into toxin, or remain undamaged because of a lack of specific receptors.

These two microbial control agents were rapidly developed with the support of industry, universities, and national and international organisations such as the World Health Organisation (WHO). Following extensive safety tests and environmental impact studies, the bacilli were quickly put into use. This rapid exploitation was aided by a series of useful properties of the bacterial control agents. In addition to the relative ease with which they can be mass-produced, bacterial control agents are highly efficient, environmentally safe, easy to handle, stable when stored, cost-effective, and suitable for integrated control programmes based on community participation. The risk of resistance, especially when *Bti* is used, is much lower compared to conventional chemical insecticides (Goldberg and Margalit 1977, Becker and Margalith 1993).

Whereas products of *Bti* were introduced into mosquito and blackfly control programmes at the beginning of the 1980s, *Bs* found its way into mosquito control programmes by the beginning of the 1990s.

The best strategies and techniques for mosquito control should be integrated, well suited to achieving the objectives of the programmes, and at the same time they should preserve the ecological balance. This approach has gained much support, not only among scientists, but also among the general public in the last 20 years.

The European Mosquito Control Association (EMCA)

With the foundation of the European Mosquito Control Association (EMCA) in 2000, a forum was created for intensive exchange of knowledge in the field of sound mosquito control. The overall objective of the EMCA is to promote mosquito control and related subjects in the broadest sense, and to disseminate relevant information to its members and others in Europe and neighbouring regions. The specific aims of the EMCA are: (1) to improve of the life quality of people by reducing the mosquito nuisance level; (2) the promotion of co-operation and participation in mosquito control projects in Europe and developing countries; (3) the promotion of the study, prevention and control of mosquito-borne diseases; (4) the development and implementation of new materials, techniques and tools for mosquito control; (5) the enhancement of control measures based on Integrated Pest Management principles by favouring methods with low toxicological profile and low environmental impact; (6) the preservation of wetland biodiversity by use of selective materials and methods for mosquito control and promoting participation in management of natural areas; (7) the specification, development and standardisation of rules of 'good laboratory and field practice' in testing and application of insecticides, sampling methods, monitoring techniques, treatment procedures and other control measures; (8) the organisation of educational and training courses, visits and staff exchanges between the members of the association in order to achieve better professional skills in accordance with EMCA aims; (9) the promotion of general public and authorities awareness about mosquito control and related subjects; (10) active support for expansion of research, evaluation and expertise; and (11) in general, the stimulation of information and knowledge exchange in the broadest sense in accordance with the objectives of the association: e.g. publishing activities and the organisation of events, exhibitions and meetings. Soon after its foundation, 22 countries joined EMCA, including more than 20 European mosquito abatement districts that were founded between 1958 and 2001 and deal with more than 2.3 million hectares of controlled breeding sites such as flood plains, rice fields and salt marshes. Bi-annual workshops are organised that study important practical subjects related to modern mosquito control operations.

In Europe, the mosquito species most commonly causing nuisance problems are the so-called floodwater mosquitoes, such as *Aedes vexans* (Meigen) or *Ochlerotatus sticticus* (Meigen) that occur in regularly-flooded river valleys. Other important species are the snow-melt mosquitoes, e.g. *Oc. communis* (de Geer), *Oc. punctor* (Kirby) and *Oc. hexodontus* (Dyar) in swampy woodlands and tundra areas, the halophilous species *Oc. caspius* (Pallas) and *Oc. detritus* (Haliday), which breed mainly in the shallow lagoons found along the coasts of southern Europe, Asia Minor and North Africa, or the rock-pool mosquito *Oc. mariae* (Sergent & Sergent) found along parts of the Mediterranean rocky coasts. *Culex pipiens molestus* Forskål, which is known as the 'house mosquito' because of its presence close to human settlements, is a noticeable nuisance in temperate zones.

Implemenation of Control Programmes in Europe

The control programmes are implemented along the following steps:

1. Baseline collection of entomological and ecological data which includes:
   a. the occurrence of mosquito species with special regard to their vector capacity and/or nuisance potential;
   b. monitoring of the population dynamics of the most important mosquito species in relation to abiotic (e.g. climatic factors, such as rainy and dry seasons, or temporary flooding periods) and biotic conditions (the assessment of predators of the mosquito immature stages, which can assist in forecasting the development of mosquito populations); and
   c. information about mosquito migration, biting and resting habits to define the intervention area.

2. Mapping and individual numbering of significant breeding site (especially when larvicides are used).
3. Selection of appropriate tools and application systems to attack the target organisms.
4. Assessment of the optimal dosage to ensure a cost-effective operation.
5. Design of the control strategy based on results obtained in the above studies.
6. Training of field staff.
7. Governmental application formalities.
8. Intensive public relations activities.

Modern techniques such as Geographic Information Systems (GIS), have now become widely used by professionals in research, government and industry when dealing with spatial data. A GIS consists of computer hardware, software, and geographical data, designed to efficiently capture, store, update, manipulate, analyse and display all forms of geographically referenced information. Modern information technology allows for the integration of GIS systems with database technology, and with digital mobile field data collection systems supported by a Global Positioning System (GPS). The ability to link information about where things are, with the information about what things are like, provides the user with a much improved understanding of spatial phenomena and their relationships. GIS and information technology can greatly improve the survey, logistics and documentation of mosquito control operations.

Routine mosquito control programmes in Europe

Croatia

In Croatia, 50 species of mosquitoes have been recorded, of which 25 occur in the Pannonian plains. The major nuisance species is *Ae. vexans*, which is mainly associated with *Oc. sticticus, Ae. cinereus* Theobald, and *An. messeae* Falleroni that breed in the inundation areas of the rivers Sava, Drava and Danube. In the mountain areas (Gorski Kotar, Lika, Velebit, North Dalmatia) species such as *Oc. nigrinus* (Eckstein), *Oc. punctor, Oc. geniculatus* (Olivier), and *An. maculipennis* Meigen occur in moderate numbers.In the Mediterranean area, 20 mosquito species have been identified so far, the most important ones being *Cx. p. pipiens* (the main nuisance species in human settlements), *Oc. caspius, Oc. mariae, Oc. leucomelas* (Meigen) and *Culiseta annulata* (Schrank). Until the 1950s malaria was transmitted in Dalmatia by *An. sacharovi* Favr and *An. labranchiae* Falleroni.

Mosquito control operations are usually organised and financed at local levels (cities, towns, villages) such as in Osijek, Vukovar, Slavonski Brod, Vinkovci, Županja, Bilje, Beli Manastir, Pula, Poreč, Umag, Split, and Zagreb. Regional initiatives are aimed at merging the ongoing programmes into larger operational units. Aerial spraying against adult mosquitoes is restricted by the Croatian government. For ground application, malathion, the carbamate Baygon, and the pyrethroids Icon® (lambdacyhalothrin) and AquaReslin® (permethrin) are used. More and more larval control is based on the use of *Bti*. In recent years ice granules containing *Bti* toxins have been applied over several thousands of hectares.

Czech and Slovak Republics

In the Czech Republic, the Elbe Lowland (along a ca. 150 km stretch of the Elbe River from the town of Melnik to Kradec Kralove) and the Morava Lowland (the banks of the Morava river and its tributaries from the town of Olomouc to Breclav, ca. 200 km^2), have a total population of ca. 500,000 people in need of effective mosquito control. In the Slovak Republic, Rye Island – north bank (ca. 70 km) of the Danube river from Bratislava to Komarno (the 'Mosquito City'), the inundation areas of the Vah river and the lowlands of East Slovakia (inundated by the Bodrog and Latorica rivers) all experience mosquito problems. In the Elbe and Morava lowlands, the snow-melt mosquito *Oc. cantans* (Meigen) and the floodwater mosquitoes, *Ae. vexans*, *Oc. sticticus* and *Ae. rossicus* Dolbeshkin, the latter in southern Moravia and Rye Island only, are the dominant nuisance species causing great discomfort to inhabitants, and serious economic damage by their impact on tourism and property values. In some regions, severe mosquito nuisance may make agricultural or forestry work temporarily impossible, and in some localities (natural deer parks) it may cause serious loss of game.

The anautogenic *Cx. p. pipiens* Linnaeus is the main mosquito pest in households in warmer parts of both countries. Females of the autogenic *Cx. p. pipiens* biotype *molestus* are an important mosquito pest in the major cities. They breed inside buildings (e.g. in flooded cellars) and bite inhabitants all year round.

Abundant breeding places of *Anopheles* mosquitoes are widespread in southern Moravia, Southern and East Slovakia, but no endemic malaria cases have occurred since its eradication in 1952.

Until 1989, aerial adulticiding (with organophosphates) or granular application of various larvicides (organophosphates or *Bti*) was carried out in selected areas. Since 1989, only small-scale adulticiding using pyrethroids has been carried out. Field larviciding is practically non-existent. In 1997 (after catastrophic floods in Moravia), tens of thousands of hectares were aerially treated by cold aerosol containing pyrethoids (permethrin). Autogenic *Cx. molestus* breeding sites (inside houses) are removed by drainage practices (if technically possible) or treated with organophosphate larvicides. Anautogenic domestic mosquitoes (*Cx. p. pipiens*) are not controlled at all.

France

Organised mosquito control has a long history in France, which reflects the severe problems caused by these insects. Mosquito control in the areas of France in which mosquitoes cause the greatest health and nuisance problems, is the responsibility of five independent organisations. EID Méditerranée (Entente Interdepartementale pour la Démoustication), based in Montpellier, covers the French part of the Mediterranean basin, EID Ain Isère-Rhône-Savoie, based in Chindrieux, covers the Rhone-Alps region, EID Atlantique, based in St-Crépin, covers the Atlantic coast, two abatement district in the Alsace protect the human population on the French side of the Upper Rhine Valley and one mosquito control organisation is active in Paris.

EID Méditerranée was founded in 1958, and has now grown into a respected institution in southern France, as the development of tourism and economic growth in that region would not have been possible without the activities of this organisation. The EID Atlantique was established in 1963, the EID Rhone-Alps in 1966, the Alsace region started mosquito control activities in 1984, and the Paris district in 1992. Initially sustained by the central French government until the 1980s, the control operations are now paid for by the local Départments and communities as well as by regional administrations.

Most effort is directed at larval control, either through physical methods, management of wetlands and taking into account the whole fluvial hydrosystem, or by use of larvicides that are effective against larval instars in specific habitats. Correlation of larval abundance and water fluctuation, followed by ecological maps and dominant vegetation of the sites, provide the basis for implementation of larval control. From the mid 1980s, chemical insecticides were partly replaced with *Bti*. and *Bs*. By 1997 some of the organisations, such as EID Rhone-Alpes and in the Alsace region, based their larval control exclusively on *Bti* and *Bs*.

Each French mosquito district has its own particular environmental conditions for mosquito development, which therefore requires specific strategies. The Mediterranean organisation has developed control programmes for 350,000 ha from Marseille to the Spanish border, the Rhone-Alps carries out mosquito control across 250,000 ha, the Atlantic organisation covers 100,000 ha, Alsace approximately 8,000, and the Paris district 2,000 ha. In addition to flood areas and/or wetlands, numerous *Culex* generations may be produced in urban areas, in a variety of breeding sites, for example water containers. These urban breeding sites require a different treatment. Some are physically eliminated, while others are sprayed with conventional larvicides or biological control agents (*Bs* and *Bti*).

Germany

The control of mosquitoes in Germany has a long history. In the 1920s and 1930s breeding sites were treated with petroleum oils (Becker and Ludwig 1983). During the 1950s and 1960s adulticides were used. In the early 1970s, the mosquito population was extremely high because of frequent fluctuations of the water level of the river Rhine. The people in villages in the Rhine valley could not spend much time outdoors without being seriously attacked by biting mosquitoes; there was an attack rate of more than 500 mosquitoes per minute. As a consequence of this distressing situation, 44 towns and communities on both sides of the river Rhine formed a jointmosquito control programme, the KABS (Kommunale Aktionsgemeinschaft zur Bekämpfung der Stechmückenplage e.V.) which was founded in 1976. Currently, 100 cities and municipalities along a 310 kilometer stretch of the Upper Rhine River, with a total population of 2.5 million people, have joined forces to control mosquitoes, mainly *Ae. vexans* over a breeding area of some 600 km^2 of the Rhine's flood-plain. The budget of the programme is approximately € 2.4 million per year.

The overall concept used is integrated biological control (IBC), based on the protection of humans against mosquitoes and conservation of biodiversity. When considering the ecosystem web, the KABS aims to only target the niche occupied by mosquitoes without causing any collateral damage to non-target species. This goal can only be achieved when biological control methods are used. Conservation and the encouragement of natural predator species is an important part of the programme. Therefore, microbial and biological methods are integrated with environmental management (e.g. improving of the canal system for regulation of the water level and provision of permanent habitats for aquatic predators such as fish).

For almost three decades, *Bti* and *Bs* have been successfully used in Germany as biological control agents against floodwater mosquitoes (e.g. *Ae. vexans*) and *Culex* mosquitoes (e.g. *Cx. p. pipiens* biotype *molestus*). Over 2,500 km^2 of breeding areas have been treated with *Bti*, resulting in an annual reduction of mosquito densities by more than 95%.

The flood plains of the river Rhine are usually inundated two to 12 times each summer. The extent of the flooding depends on snow-melt in the Alps and on rainfall, and constant monitoring of the water flow in the Rhine and in the flood plains is needed. During flooding, *Ae. vexans* and other floodwater mosquito larvae hatch within minutes or hours at temperatures exceeding 8 °C. Before control measures can be conducted, the larval density and the larval stages are checked by means of 10 sample scoops at representative breeding sites, in order to justify control efforts and to establish the correct dosage and the best formulation to be applied. One day after application, water samples are taken from the reference breeding sites to check mosquito density and to thereby evaluate the efficacy of the treatment.

According to the extent of the flooding, 10-20% of the potential breeding areas of 600 km^2 must be treated regularly by up to 400 collaborators of the KABS. For treating first and second larval instars, 200 g of powder formulation (3,000 ITU/mg) or 0.5 l of liquid concentrate (1,200 ITU/mg) are dissolved in 9-10 l of filtered pond water for each hectare treated, and applied by knapsack sprayers. For deeper breeding sites or when late instar larvae are present, the dosage is doubled.

During the most severe floods, usually one third of the area must be treated with *Bti* granules, which are dispensed with the aid of a helicopter (at 10-20 kg/hectare). From 1981 to 2006, 150 tonnes of *Bti* powder and fluid concentrate and almost 1,500 tonnes of *Bti* ice and sand granules mixed with quartz sand, vegetable oil and *Bti* powder, were used, to treat more than 1,500 km^2 of breeding area. Since 1997, the more cost-effective *Bti* ice granules have been applied by helicopter over thousands of hectares of floodwater mosquito breeding sites.

Control of urban mosquito species is mainly carried out at household level. To assist with this, KABS provides information on the biology of *Cx. p. pipiens* biotype *molestus* and the appropriate control measures. Culinex® tablets have been particularly successful. They kill *Culex* larvae in water containers over a period of several weeks. In drainage systems and large cesspools with eutrophic water bodies, *Bs* is applied, as a liquid or powder formulation, against *Culex* larvae. Each year, about one million Culinex-*Bti/Bs* tablets are successfully applied against *Cx. p. pipiens*, especially in rainwater containers.

Monitoring the programme: Some 5% of the KABS budget is invested in monitoring mosquito populations, mosquito resistance and environmental impact. All studies carried out to date have shown that the introduction of *Bti* and *Bs* has reduced the numbers of nuisance mosquitoes to a tolerable level, but that the diversity and quality of the ecosystem as a whole has not been affected.

a. Monitoring mosquito abundance: To monitor the abundance of adult mosquitoes, a sufficient number of traps are placed at comparable sites throughout the entire inundation area. These are monitored twice a month from April to September. On each occasion, the mosquito density is sampled during one night by means of carbon dioxide baited light-traps. Catches in areas where no control measures have been undertaken serve as reference points (100% of the mosquito population) for catches from areas being treated, in order to determine the success of control operations (mortality rate in percent). It has been shown that the widespread application of *Bti* since 1981 has successfully averted mass incidences of mosquitoes. Naturally, these control measures have had an extremely positive reception among the local people.
b. Monitoring the environmental impact: it has been essential to document the environmental impact of *Bti* and *Bs* applications, in order to provide a scientific basis for rebutting the

arguments commonly brought against mosquito control by its opponents. Before large-scale application of microbial control agents was undertaken, the most important members of various aquatic groups (*Cnidaria* to *Amphibia*) were screened in the laboratory and in small-scale field trials for their susceptibility to microbial control agents. This study showed that in addition to mosquitoes (Diptera: Culicidae) and black flies (Diptera: Simuliidae), only few species of midges (Diptera: Chironomidae) were affected by *Bti*. For the most part, these midges were much less susceptible to *Bti* than the target organisms. *Bs* is toxic to an even narrower range of insects: certain mosquito species, such as *Culex* species, are highly susceptible, *Aedes/Ochlerotatus* species are much less susceptible, and black fly larvae as well as other insects (with the exception of certain Psychodidae) and non-target organisms are not susceptible. The development of insects in treated and untreated water is regularly monitored using emergence traps. The occurrence and abundance of insects in treated areas is assessed by regular light trap catches. All investigations have shown that while the numbers of *Aedes/Ochlerotatus* mosquitoes are drastically reduced, all other insects continue to develop in the water and, as winged adults, provide a food resource for birds, amphibians and bats.

c. Resistance monitoring**:** Mosquito populations are checked at regular intervals for the development of resistance. No resistance has been detected after 10 years of treatment with *Bti* (Becker and Ludwig 1993). To prevent resistance to *Bs* developing in *Culex*, *Bs* and *Bti* are used in rotation.

Greece

One of the largest mosquito control projects in Europe, as far as treatment areas are concerned, is the project related to rice field mosquito control in the area of Thessaloniki (a city of 1.2 million inhabitants in northern Greece). The project started in 1997, and consists of larviciding with the organophosphate temephos, using three helicopters. GIS data on mosquito larval productivity gathered systematically (by 30 technicians for 18,000 ha of rice cultures) for individual parcels of the rice fields (15 ha each on an average) exhibit very distinct spatial-temporal patterns for the years 1998-2001 (depending on the different water management practices). In addition to the project in Thessaloniki, control activities were recently implemented in the areas of the prefecture of Serres, Chalkidiki (14 municipalities), Larissa (two municipalities), Kavala (threee municipalities), Pieria (11 municipalities), Imathia (three municipalities), Pella (seven municipalities), and Lamia. Altogether, more than 100,000 ha of breeding sites are treated with temephos (Abate 1SG, Abate 50 EC), *Bti* (Vectobac and Teknar), diflubenzuron (IGR, DU-DIM 10WP). *Bacillus thuringiensis israeliensis* is mainly used in ecologically sensitive areas.

In seven out of the 50 Departments across Greece, there is currently an active network of 46 mosquito monitoring stations. Twenty of the 50 mosquito species found in Greece are monitored weekly, including the three main anthropophilic species *Oc. caspius*, *An. sacharovi* and *Cx. p. pipiens*. Quantitative data over time provide considerable insight into the effectiveness of the abatement programme. Thus, in the Thessaloniki area, the overall average nuisance during sunset in the month of July was approximately 60 bites/15 min in 1996 (baseline year, before the project implementation) versus two bites/15 min in 2001.

Hungary

There are two major regions in which abundant mosquito populations occur in Hungary, the regular flooding areas along the Danube, the Tisza river and their tributaries, and the shores and marsh area around Lake Balaton and Lake Velence. The temperate Central European climate offers favourable conditions for the development of large mosquito populations. So far there have been 44 species recorded in the country. *Aedes/Ochlerotatus* species are prolific in all breeding sites, and in addition, *Coquillettidia* species are abundant at Lake Balaton in the thickly reeded marsh areas. *Culex* species are frequently abundant in villages.

The area of inundation and other potential breeding sites along the rivers and lakes is estimated to be about 2,500 km^2. The country's nature reserves are connected to the rivers, and during overflows, they become ideal breeding sites for mosquitoes. In such habitats, larval samples may contain 1,000-2,000 larvae per liter. Along the river Tisza the area has been uniquely endowed with more than 170 ox-bow lakes. Since most of these lakes are situated in the country's nature reserves, the lakes are an important destination for eco-tourists.

Over recent decades, many breeding sites in Hungary have disappeared while new ones have been created. Mosquito abundance has increased in many areas, apparently due to the deterioration of water ecosystems and formation of new environments favourable for breeding, development and host availability.

Current mosquito control in Hungary is carried out by several private abatement or pest control organisations, which perform this activity in conjunction with urban pest control. Recently, attempts have been made to introduce novel technologies and strategies that involve regional cooperation between municipalities and public health authorities. Mapping of breeding sites and monitoring of all developing mosquito stages has high priority. Some organisations are working with various strategies to improve the formulations available for mosquito control. However, at present, the main products used are either organophoshates or pyrethroids, and to a lesser extent some IGRs and products based on *Bti*.

The Hungarian Mosquito Control Association was founded in 2001 and in 2002 the major companies dealing with mosquito control joined forces to implement control activities based on environmentally safe biological control tools in order to reduce the use of chemical insecticides.

Italy

As in most European countries, organised mosquito control activities in Italy suffered from the lack of general interest during the post malaria-eradication period 1950-1980. In this period, programmes were no longer centrally coordinated and mainly focused on adult mosquito control. Following improved socio-economic conditions, the demand for mosquito control on a large scale started in northern Italy in the middle of the 1980s. At the same time, two main factors determined a new way to implement mosquito control: the general awareness on the ecological impact of traditional mosquito control methods and the discovery and rapid market availability of selective microbial agents active against mosquito larvae. Therefore, several programmes were implemented mainly based on regional and local public financial support.

One of the first programmes to be established, in 1987, was in Bologna province, an irrigated agricultural area of about 900 km^2 in the plains of the river Po. From its onset major attention was devoted to larval control with occasional adult control support in defined areas. As a result of specific research and monitoring activities, the control programme has adopted larviciding as its sole control technique. In the campaigns against the two major target species *Oc. caspius* and *Cx.*

pipiens, 95% of the total larvicide products used is based on *Bti*, while the remaining 5% consists of temephos, which is still used in catch basins only.

According to data collected weekly by the adult mosquito monitoring programmes using carbon-dioxide baited (CO_2) traps at 23 fixed stations, and from surveys based on public opinion, the results are highly satisfactory. A tolerance threshold based on CO_2 trap catches has been developed.

Breeding sites have been mapped and coded, and their characteristics digitalised using appropriate software. This enables field technicians to conduct both ecological surveys and control operations, updating the info on a daily basis. These data may be used for daily reports, and can be highly valuable when a cost/benefit analysis of the programme is to be made.

One particular achievement of this programme has been the development of a special larvicide applicator. This device consists of a remote-controlled hydraulic arm fitted with larvicide nozzles and a camera, which enables a single operator to gain access and view, and then to treat difficult breeding sites in urban areas and in dense vegetation. The use of this device has enabled the amount of product used to be optimised and the overall efficacy increased.

Breeding habitats are treated with *Bti* at intervals of 5 to 7 days, depending on water temperature. The mosquito season of the region usually requires 20-24 treatments in order to control *Cx. pipiens* in a satisfactory manner.

In the region of Comacchio, a tourist resort in the Po delta and a natural, protected area which includes 47 km of coast with an expected yearly presence of 500,000 tourists, the mosquito control programme was started in 1991. Its strategy is still based on a combination of both larvicides and adulticides. Adult mosquito densities are estimated by three-weekly collections using CO_2 traps, and these have indicated that the current measures have achieved a reduction in mosquito biting densities of 95% compared to pre-intervention densities. The main target species of this region are, in order of importance, *Oc. caspius*, *Cx. pipiens.* biotype *molestus*, *Oc. detritus*, and *Cx. modestus* Ficalbi. Breeding sites have been digitally mapped and stored on MapInfo since the beginning of the project, and are currently being transferred for use with ArcView software.

Since 1996, following the approval of the Regional Law n.75/1995, several programmes were started in the Piedmont region, which rapidly became the most organised region in Italy. Here the guidelines adopted as mandatory to obtain the regional financial support require standard mapping of the breeding sites, storage of data in software specifically developed inside ArcView, adoption of larval control operations only, and weekly monitoring of adult populations using CO_2-baited traps.

The total area under protection is 273,058 ha, with a resident population of 1,722,766.

More programmes are expected to be established in the near future particularly in Central and southern Italy to protect revenues from tourism, and in northern Italy to reduce *Aedes albopictus* (Skuse) which is rapidly becoming the main nuisance species in this country (see Chapter 14).

Poland

In the past, mosquito control activities in Poland were strongly connected with the need to reduce the numbers of anophelines, because of their role as malaria vectors. Malaria was endemic in several areas in Poland during the years following World War I. In 1921, 52,965 malaria cases were recorded, mostly in the eastern and south-eastern parts of the country. Anti-malarial action, organised since 1925 in the areas where endemic foci existed, involved not only health care and medical treatment, but also control measures against mosquitoes with Paris Green (Szata 1997). After World War II, malaria was present on the Baltic coast, in the Mazury-region (north-eastern Poland) and near Warsaw. A serious outbreak of the disease in the northern part of the country in the 1940s and 1950s necessitated anopheline mosquito control. In the years 1948-1951 the control measures against both larvae and adult mosquitoes in the region of Gdanska Bay and Zulawy were undertaken as part of the anti-malarial campaign. The measures consisted of residual treatment of resting places outside and inside of buildings, space spraying, and treatment of larval breeding places with DDT. Physical modification of the environment (drainage, levelling of ground, and filling in small ponds) was also included.

In 1952 in Podgrodzie resort (near Szczecin) the first large-scale operation against mosquitoes took place, because of their presence in enormous numbers. Screening surveys carried out at the beginning of July, before the intervention started, showed the presence of anophelines (mostly *An. messeae*) and *Cx. p. pipiens* mosquitoes. Buildings (outdoors and indoors), forest areas near the resort, marshes, ponds, parts of Szczecinski Bay, were all treated with DDT (Brodniewicz 1953). In 1956-58 on Karsiborz Island in the Szczecinski Bay, control measures against *Oc. communis* and *Oc. annulipes* (Meigen) were undertaken. DDT was used in the forest environment and indoors in residential houses (Bosak *et al.* 1961). Upon completion of this operation, resulting in significant reduction of mosquito populations, similar control was carried out in 1958 and 1959 in other resorts in the Szczecinski Bay. However, it was pointed out that, although control actions were evaluated as very efficient, in several areas the dose of DDT applied exceeded many times that recommended by WHO, and negative impacts on non-target organisms were observed (Krzeminski and Brodniewicz 1969). As a result, large-scale mosquito control actions in natural biotopes were no longer recommended, but abatement operations in small, selected areas in tourist resorts were continued, mostly directed against nuisance mosquitoes. For chemical treatments chlorinated hydrocarbons were used: DDT, BHC, Dieldrin, and in several cases pyrethroids (Krzeminski and Brodniewicz 1969).

In the 1960s, however, malaria was eradicated in Poland and the measures taken against mosquitoes were usually aimed at reducing their numbers to a level at which they no longer caused a biting nuisance, especially during periods of massive and synchronised emergence.

In the 1970s and 1980s, according to fragmentary data, mosquito control in Poland was focused on adults only and was implemented with residual insecticides in selected places when necessary. The chemicals used are mostly pyrethroid insecticides: permethrin, deltamethrin, cypermethrin, and also etofenprox. Being relatively inexpensive, products containing the organophosphate DDVP were also applied.

The flood that affected Central Europe in the summer of 1997 resulted in mass breeding of mosquitoes all over the flooded areas, including Poland, where mosquito control was carried out on a large scale in six of the 26 provinces affected by floods, situated mainly along the Odra river. In total, 10,000 ha were treated, partially by aerial spraying from helicopters and partially by spraying from the ground. Mainly adulticides were used: AquaReslin Super (permethrin); Trebon 10 S.C. (etofenprox); K-Othrine flow (deltamethrin) and Alfasep Super Kill (cypermethrin) (Gliniewicz *et al.* 1998).

It must be pointed out, that the catastrophic 1997 floods and the subsequent huge mosquito problems increased public interest in mosquito control. As a result, large-scale control activities have been re-initiated in several regions. Since 1998, mosquitoes have been controlled in the city of Wroclaw, where each year mass emergence of nuisance mosquitoes causes problems. The control measures have been directed against larvae and adults using Simulin (cont. *Bti*) and Dimilin 25 GR (diflubenzuron) and the adulticides AquaReslin Super and Trebon Mega S.C. (etofenprox). According to surveillance data, the main nuisance species in Wroclaw are *Cx. p. pipiens*, *Oc. caspius*, *Oc. sticticus*, *Ae. vexans*, *Oc. cantans*, *Oc. leucomelas* and *Cs. annulata*. Approximately 1,600 ha have to be treated. The efficiency of control was 85% in 1998 and 70% in 1999 (Rydzanicz *et al.* 2000). Since 2001, on behalf of the local authorities of Wroclaw, regular treatments with *Bti* (Vectobac WDG and Vectobac 12AS) are conducted in fields irrigated for sewage purification resulting in a reduction rate of mosquito populations of 95% in the treated areas (Rydzanicz, personal communication).

The town of Szczecin and the islands of Uznam (town Swinoujscie) and Karsiborz also experienced problems in the 1990s resulting from the mass occurrence of nuisance mosquitoes, and mosquito control operations have therefore continued. In Szczecin only adults were controlled (mainly with Trebon Mega 10 S.C.), while in Swinoujscie and Karsiborz, both *Bti* (Simulin) as larvicide and etofenprox (Trebon Mega 10 S.C.) as adulticide were used.

In the town of Gorzow Wielkopolski, situated in the western part of Poland, a mosquito control programme was started in 2001. The treated area includes 1,250 ha of the city area, suburbs and marshy grounds on river banks. Adult mosquitoes are controlled with Trebon Mega S.C., while larval breeding places are treated with *Bti* agent (Simulin), applied aerially by helicopters and from the ground with knapsack sprayers. Local authorities have described the programme as being highly effective.

In 2000, a mosquito abatement programme based on *Bti* was started in Krynica Morska, a tourist resort, located on Vistula-Spit, in an area of protected landscape. Mosquito breeding sites in temporary and stable water pools were identified and mapped. Qualitative and quantitative studies on the mosquito fauna showed that *Oc. cantans* was the predominant species (84,7%), followed by *Oc. communis* (Kubica-Biernat *et al.* 2000). The control measures were begun in April 2000. Vectobac 12 AS and Vectobac TP - *Bti* preparations - were used as larvicides. The efficiency of control was about 99% (with Vectobac 12 AS) and 97% (with Vectobac TP) (Kubica-Biernat *et al.* 2000).

Serbia

The province of Vojvodina has been subject to continuous mosquito control for the last 30 years. Although other areas, such as the Yugoslavian Adriatic coast, also provide favourable mosquito breeding conditions, a consistent approach to mosquito control in these parts of the country has not been developed so far.

The province of Vojvodina lies in the northernmost part of Yugoslavia, covering a surface of about 20,000 km^2. The southeastern part of the Pannonian Plain bears several plateaus and hills ranging between 150 and 600 m above sea level, but the remaining area is flat, with an average elevation of 80 m above sea level. Three large rivers enrich it: the Danube, Tisza and Sava, and several smaller ones with a total length in Vojvodina of 597 km. Several protected areas and swamps are connected to the river lowlands and hilly regions on the river banks. All of these protected areas provide very productive breeding sites for a great variety of mosquito species. The climate is temperate Central European, characterised by large fluctuations in air temperature. The annual mean temperature is 11 °C. The average precipitation varies from 550 to 670 mm per year.

The number of mosquito species recorded in Vojvodina is 32, out of the 51 known in Yugoslavia. *Aedes/Ochlerotatus* mosquitoes, whic comprise more than 90% of trap catches and almost 100% of human bait catches, are dominant in the floodplains of the river Danube. The number of females landing to feed on one forearm can exceed 300 mosquitoes in 5 minutes. The lower parts of the plains are occasionally flooded. Each increase of the Danube water level of between 250-650 cm at the Novi Sad measuring station triggers hatching of *Ae. vexans*, *Oc. sticticus*, *Ae. cinereus*, *Ae. rossicus* and *An. messeae* larvae in huge numbers. Many extensive marshes, ox-bow-swamps and a complex system of irrigation channels are present, which provide numerous and extensive mosquito breeding sites. Deep depressions of the old river arms, which contain water the whole year round, show typical zones of submerged, floating and emergent hydrophyte vegetation. The main species in the last two biotopes are *An. messeae*, *An. maculipennis* s.s., *Cx. modestus* and *Cx. p. pipiens*. The banks of the rivers, ox bows and marshes are usually lined by willow, poplar and alder trees. Brambles, various shrubs and climbers make a thick coverage of the ground and an excellent daytime resting-place for the hordes of floodwater mosquitoes. The stagnant and slowly flowing water on the alkaline and saline soils, with a relatively high concentration of dissolved salts, offer good conditions for *An. atroparvus* Van Thiel, *Oc. caspius* and *Oc. dorsalis* (Meigen). On the Danube floodplains, the most abundant *Aedes/Ochlerotatus* species are *Ae. vexans* and *Oc. sticticus*, which occur in very large numbers two to five times per year, depending on the pattern of fluctuation of the Danube river. The network of irrigation channels has been neglected for decades. It is blocked by aquatic vegetation and cannot collect and drain water quickly enough to prevent the formation of puddles and marshes suitable for mosquito development. Under such conditions, the previously mentioned species of *Anopheles* and *Culex* mosquitoes are most dominant. Saline waters close to several spas are breeding sites of *Oc. caspius*, *Oc. dorsalis* and *An. atroparvus*. Marshes with water are rich in *Coquillettidia richiardii* (Ficalbi). Urban environments up to 10 km from major rivers are often invaded by masses of dispersing *Aedes/Ochlerotatus* females. Inhabitants are also continually bitten by *Culex p. pipiens*, which are very abundant where *Aedes/Ochlerotatus* are not present.

Malaria was eradicated from the region after World War II. The last cases of endemic *Plasmodium vivax* malaria were recorded in 1955, close to Nis in southern Serbia, in the vicinity of rice fields that were breeding sites of *An. maculipennis* s.s. and *An. messeae* (Guelmino, 1956). In the last decade, the average number of imported malaria cases was about 40 per year. Owing to global climatic changes influenced by El Niño and La Niña activity, warming of the Pannonian Plain climate is predicted. The average number of mosquitoes of the *An. maculipennis* complex in the Koviljski nature reserve has increased over the last few years, and may increase the risk of re-emergence of malaria in Vojvodina province (see Chapter 3). Apart from malaria, several surveys of Tahyna virus have been conducted in Serbia. The highest percentage of positive sera was found in the southern part of the river Tisza lowlands (Gligic and Adamovic 1976).

Mosquito control in Vojvodina has been organised since 1976 under the umbrella of the Provincial government. The Faculty of Agriculture of the University of Novi Sad was the founder and organiser of the control programme until 1985, when insecticide purchase and pest control organisations (PCO) were transferred to trained personnel in the local mosquito abatement district. The Faculty of Agriculture remained involved in education, monitoring and the evaluation of insecticide properties, dosages and the effectiveness of control measures. Since 1980, when *Bti* was first introduced as a larvicide in parallel with organophosphates and IGR compounds, *Bti* has been the subject of continuous research and application programmes at various scales. Year after year, various *Bti* formulations have been tested, but not used on a large scale in Vojvodina mainly because of economic constraints. From 1993 onwards, the municipality of Novi Sad has started to support the 'Environmentally friendly approach in mosquito control' project. The project's main goal is to provide a base for rational biological mosquito control, mainly by using *Bti* products. Monitoring of larval and adult floodwater and house mosquitoes was established, together with mapping of their breeding sites. Development of software supporting the implementation of biological control of floodwater mosquitoes has been conducted as well. GIS software programmes have been designed to serve as a database for the information on physico-chemical characteristics of mosquito breeding sites, plant communities, limnofauna, egg, larval and adult densities at breeding sites, and speed and direction of female mosquito dispersal. The programme can also simulate different levels of flooding, give precise information about areas to be treated and record effectiveness of control measures.

A network of CO_2 baited traps has been established since 1994, which samples mosquitoes on a weekly basis from April to October. Additional traps are used temporarily in the region of Novi Sad municipality, particularly during periods of mosquito dispersal, or to evaluate control measures. Furthermore, 30 traps are continuously used in several localities of Vojvodina (Sombor, Subotica, and Pancevo) to assess the dispersal potential of the major nuisance species. As a result of project activities, since 1997 at least one generation of floodwater mosquitoes is controlled solely by *Bti* application. Other seasonal outbreaks of floodwater mosquitoes are suppressed by combined methods (*Bti* for larviciding and ULV pyrethroids/organophosphates for adulticiding) using air or ground equipment.

Spain

In Spain, organised mosquito control started in the early 1900s following the discovery of mosquitoes as vectors of malaria. Nevertheless, interest in mosquito studies ceased as soon as the disease was eradicated from this country in 1963 (Pletsch 1965).

Spain's economic development resulted in improved quality of life, and it was increasingly recognized by the public and by public administrators that mosquito nuisance should be considered as a limiting factor for the development of the country. This fact was especially important in towns and cities close to mosquito breeding sites, and in areas where tourism was the most important industry.

The first mosquito control service (MCS) was created in 1982, in Roses Bay and Ter River. In the following years, three more services were created under the direction of public administrations. In 1983, the MCS of the Baix Llobregat was created, followed by the MCS of Huelva in 1985, and the MCS of the Ebro Delta in 1991. Three of the MCS are situated in Catalonia, in the northeast of Spain, along 350 km of the Mediterranean coast, and the other is located in Andalusia, in the southwest of Spain.

All mosquito control services operate on the basis of integrated pest management focused on larval control, are completely independent, and are linked to different local public administrations. The first MCS of Roses Bay and Ter River were created in 1982 for two reasons: to limit mosquito nuisance and to demonstrate to the tourist industry that a salt marsh area can be compatible with human activities. This area lies between the Pyrenées and the Montgrí massif, forming a bay with swamps, salt marshes and lagoons. The area covers about 34,000 ha, of which 7,200 ha are natural mosquito breeding sites. The region also contains more than 700 ha of rice fields and about 7,000 septic tanks, in one of the most important tourist areas of Catalonia.

Ochlerotatus caspius, *Oc. detritus* and *Ae. vexans* are the most important species in natural breeding sites, while *An. atroparvus* is the major species of the rice fields. Natural areas are mostly swamps, salt marshes and pasture meadows with variable water levels due to rain, sea intrusions or manmade causes. This area, including a natural park, is exclusively treated by biological agents (*Bti*) while the most important *Cx. pipiens* breeding sites (the septic tanks dispersed over the residential areas) are treated with the insect growth regulator pyriproxyfen.

The MCS of the Baix Llobregat region east of Barcelona was founded in 1983. This abatement district must reduce the mosquito nuisance along the river area where important tourist resorts exist. The service is responsible for fifteen municipalities stretching south of Barcelona, including the airport. The area of the municipalities involved is 25,000 ha, including 9,000 ha of the river delta. The species causing the problem are mainly *Cx. pipiens* and *Oc. caspius*. Although nineteen mosquito species have been identified, the aforementioned are the most abundant. Natural areas of this region have been disappearing progressively due to pollution and construction activities in the vicinity of Barcelona. Only 340 ha are potential natural breeding sites. While the natural zones are only a problem where rainfall or sea intrusions occur, there is potentially continuous *Cx. pipiens* development between May and November in 290 km of polluted ditches in the district, and to a lesser extent even during the winter because of underground sites. Larval habitats of this species are treated with *Bs*.

The MCS of the Ebro Delta, covering seven municipalities and more than 32,000 ha, performs an extremely valuable service in Catalonia. Because of the dramatic nuisance caused in villages and touristic areas by enormous numbers of mosquitoes, it was inevitable that by 1991, a mosquito control service came into being. This delta is covered by more than 22,000 ha of rice fields and 10,000 ha of natural areas, which form the Natural Park created in 1983. The majority of problems are related to natural or artificial fluctuations of the water level that cause the appearance of *Oc. caspius*, *Oc. detritus*, *An. atroparvus* and *Cx. modestus*. *Culex pipiens* occurs in ditches and septic tanks in all villages spread across the delta.

The abundant larval mosquito population in natural areas is controlled by *Bti*. For the control of rice field mosquitoes, a buffer zone around each village is treated with temephos.

The region of Huelva established a control service in 1985, and covers an area of 11 municipalities with more than 130,000 ha. This region is located on the southwest coast of Spain and partially extends along the tidal salt marshes between Portugal and the Lower Guadalquivir River marsh. This natural area was drastically modified by the building of an industrial complex, destroying more than 400 ha of salt marshes. The pollution of the swamp water and the creation of physical barriers have further complicated the already difficult mosquito control on the salt marshes. Nowadays, 100% of the tidal marsh area is protected by environmental policy.

Ochlerotatus caspius and *Oc. detritus* are the most important species on the Huelva salt marshes. This area covers 16,000 ha, and is treated by both the conventional insecticide Temephos, and by the biological agent *B. thuringiensis israeliensis. Culex pipiens* breeding places in Huelva are mainly in rain ditches and gutters. In the city of Huelva, 60,000 gutters are treated each year with *Bs*, and the treatments are usually repeated on a bi-weekly basis.

Methods used by the Spanish MCS are based on an integrated control strategy. Some of the control problems are similar across all the MCS, but there are also important local differences resulting from landscape features, varying breeding habitats and, of course, different species. Services were established in areas where mosquito control was needed, and after the failure of several private companies. After the gradual introduction of methods appropriate to local needs, it has now been shown that control programmes as carried out by MCS is most probably the only effective approach.

Sweden

The first Swedish abatement district (Biologisk Myggkontroll-Nedre Dalälven, BMK-ND) was created in September 2000, as a response to many years of complaints about mosquito nuisance by the local population of the lower portion of the Dalälven river (Nedre Dalälven) in Central Sweden. In the initial phase, the seven municipalities with part of their population in the Nedre Dalälven area, as well as four regional governments, guaranteed financing of the project. Further financial support was provided by the Swedish government. A first estimate of the size of the potential mosquito breeding areas was 10 km^2. It was not possible, within a reasonable time frame, to use traditional ecological mapping for such a large and often inaccessible area. Thus, from the beginning, high resolution orthophoto maps in combination with direct DGPS were used to create precise maps of the location and size of breeding sites.

Approximately 10 km^2 of the Dalälven flood plain between the city of Avesta and the Sea of Bothnia provide temporary wetland areas. These areas are usually flooded one to two times during spring and summer. The extent of the flood depends on the snow melt in the Scandinavian mountain range and on rainfall in the very large catchment area for the River Dalälven. It is necessary to monitor the water level in the river and in the flood plain from the beginning of snow melt.

Some 24 different species of mosquitoes have been recorded in the Nedre Dalälven area, and the majority of species attack humans. Determining the seasonal and spatial distribution of mosquito populations is an integral component of providing solutions to periodic infestations of mosquitoes in the Nedre Dalälven area. Distribution patterns and seasonal dynamics of adult mosquitoes in this area have shown that the predominant species is *Oc. sticticus* (80%), followed by *Ae. rossicus* (8%) and *Ae. cinereus* (7%). These three species constitute the key nuisance species during spring and summer floods. In spring, *Oc. communis, Oc. punctor* and *Oc. intrudens* (Dyar) can also be found in high numbers in the wetlands. *Ochlerotatus sticticus* causes the most problems, due to its mass occurrence and dispersal ability. It feeds viciously on humans and almost prevents outdoor activities. Up to 60,000 mosquitoes per trap per night were recorded, showing that mosquito populations can be very large. The broad flight range of *Oc. sticticus* extends the affected area to about 100 km^2 or about 10 times the size of the wetland breeding areas. The seasonal dynamics of mosquitoes in the area are mainly influenced by the flooding regime of the Dalälven River, which is controlled by a large number of hydro-electric plants.

Malaria disappeared from Sweden 50 years ago; however, mosquito-borne viruses are present. Of the 24 mosquito species recorded in the area, five are involved in virus transmission: *Oc. communis* and *Oc. punctor* are potential vectors of Inkoo virus; *Cx. torrentium* Martini, *Cs. morsitans* (Theobald) and *Ae. cinereus* are the proven vectors of Sindbis virus causing Ockelbo disease. Four of these vector species were found frequently during surveys, and one isolate of Sindbis virus from *Ae. cinereus* has been found in the Nedre Dalälven area.

The control of *Ochlerotatus* and *Aedes* mosquitoes is based on the use of *Bti* products. As a prerequisite, the following activities were conducted in 2000 and 2001: (1) mapping of 1000 ha of potential breeding areas for floodwater mosquitoes, (2) governmental approval for the application of *Bti* by helicopter, (3) calibration of helicopter spraying equipment, (4) bi-weekly adult mosquito monitoring covering the whole Nedre Dalälven area, (5) training of the local fire-brigade to perform field work with larval sampling, (6) creation of a frequently-updated website to disseminate accurate information to the public (www.sandviken.se/mygg; Accessed 28 August 2007), and (7) performing 100% successful treatment against *Oc. sticticus* larvae in 70 ha of flooded wetlands.

All major non-profit organisations concerned about eventual negative impacts on non-target organisms were invited to discuss the development of the mosquito control operations in order to minimise ecological side effects. Several research projects were initiated to analyse the patterns of mosquito interaction with other faunal components in the wetland areas, as well as long-term studies of possible side-effects of *Bti* on non-target organisms.

Switzerland

Switzerland has several wetland conservation areas in the foothills on both sides of the Alps. They are located at the ends of valleys where the rivers have formed plains or deltas before flowing into lakes. These wetland areas are flooded periodically during the spring and summer months by snow melt and increasingly heavy precipitations along the rim of both sides of the Alps. Although relatively small in size, these wetland areas exhibit a unique fauna and flora, influenced by special climatic conditions. Many are under federal protection, such as the natural reserve Bolle di Magadino in the south of Switzerland, which is located along one of the main routes for migratory birds. Another area of interest, exhibiting an extraordinary biological diversity is situated at the upper end of the Lac de la Gruyère.

The frequently-flooded wetland zones have become major breeding places for floodwater mosquitoes such as *Ae. vexans* and *Oc. sticticus*. Adult mosquitoes with a flight range of 10 km or more have created a serious nuisance problem for the nearby residential areas and for tourism, which is a vital economic source of income for Switzerland.

Conservation of natural areas requires mutual acceptance and support by the local population as well as the authorities. In 1988, the first long-term mosquito control project in the plain of Magadino was initiated (Lüthy 2001). Because this mosquito control programme, using exclusively products based on *Bti*, was so successful, a second project followed in 1995 at the Lac de la Gruyère.

It is likely that other areas such as the natural reserve 'Les Grangettes' at the upper end of Lake Geneva will follow.

Reports are submitted regularly to working groups, which have been established within the two mosquito control projects. Members of the working groups represent the federal government, the cantons, the communities and tourist organisations. The main topics discussed within both working groups are the safety aspects and the economic feasibility of the mosquito control. Annual authorisation for mosquito control operations must be issued by the federal government.

Over the past decade, control practices have been refined and optimised. Although the intention is to promote an integrated approach, the use of *Bti* has so far been the main method of choice, because of its safety (WHO 1999) and commercial availability. Other strategies were either not efficient enough or would interfere negatively with the ecosystem.

Aerial application of *Bti* by helicopter has proven to be the only efficient method to reach the larval breeding sites. For this purpose, granular formulations of *Bti* are ideal for use in the flooded zones, which are predominantly covered by dense vegetation. The granular formulations have an excellent shelf life of 2-3 years. With enough stock on hand, the control authorities can react very quickly to eliminate mosquito breeding sites. Mapping of the breeding sites and timing are the most important elements for successful larviciding. Timing includes the climatic factors, the expected movements of the water levels in the wetland zones, and the availability of manpower and equipment.

Discussion

The examples of mosquito control activities described above demonstrate the important role of mosquitoes as nuisance, and vectors of pathogens in Europe (e.g. Chikungunya and West-Nile-Viruses). Governments and local communities are sufficiently aware of the problem to provide funding for such activities, although in several cases, citizens are being taxed to pay for the costs of the control operations. Most control activities are directed against the immature stages of the mosquitoes, and concern aquatic habitats which often serve as important, biologically-rich habitats sensitive to disturbance. For that reason, the large-scale use of environmentally-safe biocides such as *Bti* and *Bs* is an encouraging signal. However, it should be realised that the number of environmentally-acceptable products is limited, and in the inadvertent case of resistance, few alternative products are available. Therefore, research to broaden the number of effective, environmentally-sound anti-larval products should be encouraged. The impact of climate change might be favourable for expansion and increase of mosquito populations due to higher annual temperatures and precipitation. For those reasons the countries should be prepared for a continuous investment in Mosquito Control Organizations as they could act adequately in preventing transmission of diseases, implement sound mosquito control strategies assuring good health policy and environmentally friendly approach.

Acknowledgments

The authors thank all members of EMCA and KABS who contributed information to this chapter: Carlos Aranda, Romeo Bellini, Roger Eritja, Raul Escosa, Remi Foussadier, Jose Carlos Galvez, Aleksandra Gliniewicz, Raymond Gruffaz, Klaus Hoffmann, Achim Kaiser, Said Karch, Beata Kubica-Biernat, Oszkar Kufcsak, Christophe Lagneau, Elzbieta Lonc, Peter Lüthy, Jan Lundström, Eduard

Marques Mora, Jean-Pierre Mas, Enrih Merdic, Spiros Mourelatos, Odile Moussiegt, Dusan Petric, Françoise Pfirsch, Frantizek Rettich, Katarynza Rydzanicz, Martina Schäfer, Francis Schaffner, Asghar Talbalaghi, and Elzbieta Wegner.

References

Becker N and Ludwig HW (1983). Mosquito Control in West Germany. Bull Soc Vector Ecol 8: 85-93.

Becker N and Margalit J (1993) Control of Dipteran pests by *Bacillus thuringiensis*, in: *Bacillus thuringiensis:* Its uses and future as a biological insecticide. (P Entwistle, MJ Bailey, J Cory and S Higgs eds.), John Wiley & Sons, Ltd., Sussex, England.

Becker N and Ludwig M (1993) Investigations on possible resistance in *Aedes vexans* field populations after a 10-year application of *Bacillus thuringiensis israeliensis*. J Am Mosq Contr Assoc 9: 221-224

Becker N, Petrić D, Zgomba M, Boase C, Dahl C, Lane J and Kaiser A (2003). Mosquitoes and their control. Kluwer Academic Publishers, London, 497 pp.

Boccone P (1697) Museo di plante rare della Sicilia, Malta, Corsica, Italia, Piemonte, e

Germania. Venezia.

Bosak T, Dworzak Z, Golba J and OgoÕ ska A (1961) Zwalczanie plagi komarów na wyspie Karsibórz w osiedlach ludzkich i przyleglych terenach otwartych. Przeglad Epidemiologiczny 15: 59-66.

Boyd MF (1949) Malariology. A comprehensive survey of all aspects of this group of diseases from a global standpoint. By sixty-five contributors, edited by M. F. Boyd. Vol 1, 787 pp.

Brodniewicz A (1953) Dezynsekcja Miasteczka Dzieciecego w Podgrodziu kolo Szczecina. Warszawa.

Buxbaum JC (1728) Plantarum minus cognitarum centuria I. complectens plantas circa Byzantium and in oriente. Acad. scient. socium. Petropoli

Carson R (1962) Silent spring. Houghton Mifflin Co., Boston, 368pp.

Cremlyn R (1978) Pesticides. Preparation and mode of action. John Wiley and Sons, New York, 240 pp.

Gligic A and Adamovic ZR (1976) Isolation of Tahyna virus from *Aedes vexans* mosquitoes in Serbia. Mikrobiologija 13(2): 119–129.

Gliniewicz A, Krzeminska A and Sawicka B (1998) Mosquito control in Poland in the areas affected by flood in July/ August 1997. Acta Parasitol Port 5: 39.

Goldberg LH and Margalit J (1977) A bacterial spore demonstrating rapid larvicidal activity against *Anopheles sergentii, Uranotaenia unguiculata, Culex univittatus, Aedes aegypti* and *Culex pipiens*. Mosq News 37: 355-358.

Guelmino D (1956) The sequelle of physical and chemical changes in the biotops of anofelines larves. Recueil des travaux de l'Acad. Serbe des Sc. LII – Institut de recherches medicales. No. 2, 79-87.

Krzeminski J and Brodniewicz A (1969) Plaga komarów w uzdrowiskach i kąpieliskach na polskim wybrzezu Baltyku w swietle badan w ostatnim dwudziestoleciu. Balneologia Polska 14: 93-104.

Kubica-Biernat B, Gliniewicz A, Kowalska B and Stanczak J (2000) Development of the mosquito integrated biological control (IBC) program in the Vistula-Spit (Mierzeja Wislana) region, northern Poland. Proc. 13th Eur. SOVE Meeting, 24-29 Sept. 2000, Belek, Turkey, 217.

Lüthy P (2001) La lotta biologica control le zanzare alle Bolle di Magadino. In: Contributo alla conoscenza delle Bolle di Magadino (N Patocchi ed.): 139–145.

Pletsch D (1965) Informe sobre una mision efectuada en España en Septiembre-Noviembre de 1963 destinada a la certificación de la erradicación del paludismo. Revista de Sanidad e Higiene Pública. 7,8,9: 309–355.

Retnakaran A, Grenett J and Ennis T (1985) Insect Growth Regulators. In: Comprehensive insect physiology, biochemistry and pharmacology (GA Kerkut and LI Gilbert eds.), Vol 12, Pergamon Press, Oxford: 529-601.

Szata W (1997) Zimnica w Polsce. Przegl Epid 51: 177-183.

Rydzanicz K, Lonc E, Puzio A, Markiewicz M (2000) Proby integrowanej kontroli komarow na terenie miasta i okolic Wroclawia. Biuletyn PSPDDD 1/2000, 8-10.

World Health Organisation (1999) Microbial Pest Control Agent *Bacillus thuringiensis*, Environmental Health Criteria 217, IOMC, World Health Organization, Geneva, 105 pp.

Nature conservation, wildlife management and human activities as drivers

22. Changes in global scale land use and its implications for nature conservation and emerging vector-borne diseases[15]

Herbert H.T. Prins and Iain J. Gordon

Abstract

The European nature has been subject to man's impact for thousands of years, which led to major vegetation changes and, with it, major changes in the numerical abundance of wild animals. A number of these wild animals can be vectors of diseases, notably rodents, roe deer and lagomorphs and perhaps waterfowl and storks. The human population explosion of the 19th and 20th century led to a near-disappearance of wild nature and a near-eradication of the non-rodent vectors that can transfer diseases to people. Recently, nature is rebounding because in many parts of Europe land abandonment and depopulation of the countryside takes place, coupled to a strong nutrient deposition, yielding great and new possibilities for nature conservation. This leads to an increase in mammals and birds that depend on nutrient-rich conditions, possibly resulting in a strong built-up of tick populations that are the main vectors of diseases like Lyme disease or tick-borne encephalitis. Yet very little is known of the ecology of the mammals and birds that form an important linkage in the life cycle of the diseases. Such a lack of knowledge is irresponsible given the threat to human health and life.

Keywords: nature conservation, landscape, wildlife, environmental change, vector-borne disease

Introduction

In a remote hamlet in the Italian Apennines, the last local person, an aged woman, died in the spring of 2005. In that same village that summer we saw a roaming wild wolf. In 2006 Bruno the bear ran riot across the German countryside, turning a conservation success into a public relations nightmare. The wilderness is rebounding in the western world and with this rebound there are new possibilities for the well-being of vectors of vector-borne diseases, both vertebrate vectors such as mammals and birds and arthropods such as ticks or mosquitoes (Enserink and Vogel 2006, Prins and Gordon 2007). Few people realise the major changes that are taking place because more money is to be made out of the fear of climate change for which technological answers are given than out of the small but insidious changes in our landscape for which it has been proven that neither subsidies nor policies appear to give solutions. The media thus give little attention to what is happening in Europe's backyard.

The population of Japan was smaller on 31 December 2005 than it was on 1 January 2005. This highly developed and rich country is the first in which the population has started to shrink, but others, such as Italy and Germany, will follow soon (Anonymous 2006a, Population Reference Bureau 2005). In the coming forty years, over fifty countries will have lower human populations than today (Anonymous 2006a). This population contraction is most notable at the eastern flank of Europe: Russia's population is expected to fall by 22% between 2005 and 2050, that of Ukraine by a stunning 43% (Chamie 2005, Population Reference Bureau 2005). Apart from this population

[15] This chapter is partly based on Prins and Gordon (2007) and Gordon and Prins (2007).

decline there is the depopulation of the countryside. This has been going on for some fifty to seventy years already, and it will continue to happen in all temperate countries, from New Zealand to Argentina, Japan, Canada, the USA, and the whole of Europe and former USSR. Just as an example: in the past decade, some 13,000 of Russia's 155,000 villages were abandoned and 35,000 are now populated by 10 or fewer people, many of them old (LaFraniera 2003, Grammaticas 2004). Although the reduction in agricultural lands started in the 1980s (Pallot 1990), the loss between 1990 and 2001 in the Russian Federation has been close to 10% in land surface: since 1970 some 25 million hectares reverted to wild lands (Ioffe *et al.* 2004). The same abandonment of villages and reversal to wild land is happening in southern Europe (and in North America, Japan, southern South America, Australia and New Zealand). This rebound of wilderness may herald enormous possibilities for nature conservation in Europe and the rest of the highly developed world. Indeed, in its editorial comment of the 7th of January 2006, the Economist wrote *'During the second half of the 20th century, the global population explosion was the big demographic bogey ... Now that worry has evaporated, and this century is spooking itself with the opposite fear: the onset of demographic decline.... People should not mind, though. What matters for economic welfare is GDP per person .... The new demographics that are causing populations to age and to shrink are something to celebrate ...'* From a financial or economic point, this journal's rejoice may be correct, but here we draw attention to the fact that wildlife is coming back into contact with urban populations. This may be causing changes in the urbanites' perception of wildlife from one of cute and cuddly to one of a nuisance. We fear that the heralded celebration by the *Economist* will be marred by the onset of new diseases and the recurrence of old ones. Cases in point are Lyme disease, babesiosis and tick-borne encephalitis, all spread by ticks, or malaria spread by mosquitoes. Especially diseases that are spread by invertebrates like ticks, which are dependent for their own life cycle on the presence of wild animals, are of importance in this context, because with the rebound of wilderness there will be a significant increase in wildlife which are the natural blood hosts of these arthropod vectors.

Why are large parts of the countryside being abandoned, and does this have consequences for people's health? In this chapter we sketch some of the processes that take place and have taken place in Europe. It is a fascinating story, but there are some worrisome trends going on.

Landscape changes in Europe

The European landscape has been used for a very long time. Palaeolithic people were hunting and gathering already during the Pleistocene some 200,000 years ago and may have used fire as a tool to modify the landscape for hunting purposes. This has gone on ever since that time. The effect of such landscape management has been observed in Australia: non-agricultural people that had settled there some 40,000 years ago had been using fire to modify that landscape for an uncounted number of generations (e.g. Hughes 1987, Lewis 2002). During the Mesolithic period, the European landscapes may have undergone vegetation modification for production purposes (e.g. cultivation of hazel in the Mesolithic: Simmons *et al.* 1981). Perhaps at the time the vegetation modifications may not have been very noticeable to the untrained eye: indeed, it has taken decades of observation before it was realised that the 'primordial' Amazonian, Middle American and other rainforests have been strongly modified by local people (e.g. Noble and Dirzo 1997). Some changes are not even noticeable to trained ecologists and only become apparent by advanced mathematical analyses of remotely sensed images as was recently shown by Murwira (2003). Perhaps the European Mesolithic people may have had a lifestyle similar to that of the pre-European inhabitants of the temperate forest zones of North America; indeed when the first

European colonists came to what is now called New England on the eastern seaboard of the USA, the forests were so open and 'Arcadian' that settlers thought it had been created especially for them (Cronon 1983, Prins 1994, cf. Motzkin and Foster 2002, Rusell and Davis 2001).

European indigenous plants that had been shaped by natural selection reacted to the new forces, and new types of vegetation emerged (Prins 1998). Momentous changes started with the development of agriculture in the Neolithic; not only did plant communities become modified because the competitive interactions between the domesticated and native plant species altered, but land was cleared and alien species were introduced for food consumption or involuntarily. Many of these species are now considered part of the native flora and only palynological and archaeological research can reveal their non-local provenance and their alien roots (e.g. Godwin 1975, Knörzer 1971, 1975, Opravil 1978, Pennington 1969, Van Zeist 1980). More changes took place in the European landscape when early farmers adopted domesticated livestock from the Middle East and southern Russia (and perhaps even northern Siberia in case of the reindeer) (e.g. Legge 1996, Köhler-Rollefson 1996, Zeuner 1967). In much of Europe, the use of domestic grazers and browsers in the already modified landscapes led to a landscape that would have been unrecognisable to earlier generations. Superimposed on the changes brought about by climatic change, fire and felling, the impact of domesticated indigenous browsers and grazers caused forests to disappear first from the Zagros Mountains (e.g. Hole 1996) or in the Lebanon, and then across that whole range of landscapes of much of Europe (Prins 1998). The spread of farming (or farmers) from the Levant across Europe was 1 km per year (Cavalli-Sforza 1996) while the rate for pastoralism across Africa was 0.9 km per year (Prins 2000). In other words, the rate of spread of these changes was about 20 km per generation. People in Europe have thus been changing the landscape for a very long time. Deforestation may have forced rats into people's homes, thus spreading the plague to people, as presently still happens in parts of Africa (Laudisoit *et al.* 2007). Undoubtedly the distributions of tick species must have changed profoundly when forests disappeared, and grasslands formed at the hand of man (Randolph 2001). Thus, nature has been changing already for thousands of years and so has the interaction between animals, vector-borne diseases and people. This observation should not lead to complacency but to alertness, because it proves that old certainties are not necessarily good beacons for the future.

Over-utilisation of European natural resources and the eradication of wildlife

Often nature conservationists hark back to the end of the 19th century as if these were the golden days for nature (e.g. Olmig 2002). Where the indigenous peoples of Europe had eked out a meagre existence through sheer resource mining (i.e. extensive burning, removing of top-layers, forest felling, charcoal making, etc., removed nitrogen and phosphorus from many parts of the landscape), some plant species could survive and then had their halcyon days: parasitic plants such as *Orobanche* or *Cuscuta*, half-parasites as *Rhinanthus* or *Pedicularis*, or even plants that need insects as nitrogen source such as *Drosera* were thriving as evidenced by vegetation descriptions of the end of the 19th century. Other plants that thrived were hardy shrubs as *Erica*, *Calluna* or *Juniperus*, thorny shrubs, and of course all those plants that were poisonous or unpalatable, such as *Euphorbia* or *Oenanthe*. Vegetation descriptions from those days for different parts of Europe by plant ecologists such as Radde (1899) or Engler and Drude (1901) sketch a picture of a denuded landscape where only the hardiest plants could survive, where grazing by domestic stock was relentless. With the advance of artificial fertilisers, these vegetations became rarer and rarer again, while motorised ploughing enabled their destruction as well (e.g. McCollin *et al.* 2000). During those days at the end of the 19th century and first half of the 20th, natural grasslands have nearly been wiped off the face of the earth because they could be ploughed. For example, in Ukraine only about 4% of the steppes remained in the original state (Goriup 1998) and the figure for western Europe is even lower. In a country like the Netherlands the forest cover reached its absolute nadir when only about 2% of the country was clad in some form of forest (cf. Janssen and van de Westeringh 1983, Van der Woud 1987).

We know from many sources that wild predators were unyieldingly persecuted, game was unrelentingly pursued, nests of birds were ruthlessly denuded of their eggs (as is still done in some of the more remote parts of Europe), while migratory birds were shot at, caught in nets or with lime sticks. During those reputedly arcadian days mammals and bird numbers reached dangerously low numbers: wolf (*Canis lupus* L.), wild cat (*Felis sylvestris* L.) and lynx (*Lynx lynx* L.) were locally eradicated, red deer (*Cervus elaphus* L.) and wild boar (*Sus scrofa* L.) locally went extinct, eagles and other birds of prey, but also ravens (*Corvus corax* L.) and cranes (*Grus grus* L.), disappeared from the western part of the continent. Roe deer (*Capreolus capreolus* L.) survived at many places, but at extremely low numbers. Other wild ungulates even went extinct, such as the aurochs (*Bos primigenius* Heck) and the wild (forest) horse (*Equus caballus* L.) but also the European bison (*Bison bonasus* L.) reached the brink of extinction during the 19th century. Should any of these wild birds or animals have been vectors of diseases (see Table 1), the local Europeans had ensured that they did not pose any danger any more.

When artificial fertilisers were introduced at an enormous scale in a number of countries, agricultural productivity increased so much that overproduction took place. In the European Union and also the former Eastern bloc (but equally so in the USA, southern Africa, Japan, Australia and New Zealand) agricultural policies then started to distort markets at a massive scale in the 1960s till the present. World markets would have dictated a shift away from agriculture in these countries because unsubsidised agriculture became, too a large extent, not profitable if farmers in these areas had to provide their products at world market prices but had to pay labour costs as dictated by the local society's norms. Governments then started subsidising agriculture for two major reasons, namely, to maintain adequate national self-sufficiency and to guarantee farmers (and their dependent communities) a sufficiently high income so that they could continue to participate in society at large. This often had a spin-off that thus the seat of many a politician was secured too. These rich and highly-developed countries maintained a landscape geared towards maximum agricultural production and very high numbers of domestic cattle and sheep remained dominant in Europe. In these completely rationalised agricultural systems, there was no place left for wild animals, but also hardly for wild plants. The widespread use of pesticides that proved to be highly poisonous to the wild fauna resulted in a further decrease of fish and birds. At that time the conviction took hold that nature does not pose any danger to people in the West any more – nature became synonymous to beauty (e.g. Vera 1997).

Because of the increased income of the majority of the population, an ever-increasing number of people started developing feelings of loss concerning nature during the 1960s to 1980s. The amount of leisure time people had at their disposal increased quickly in the richer parts of Europe, and people started recreating at a vastly increased rate. Governments and NGOs reacted by purchasing lands at a greater scale than ever before for setting aside as 'nature'. These areas, however, were practically depopulated of fauna by that time (say, at the beginning of the 1980s), so the chances of encountering a mammal or bird which could have been carrying an arthropod vector of a disease were minimal. In the period between the 1930s and 1950s, malaria was eradicated from the richer parts of Europe and fleas carrying plague bacteria were effectively

Table 1. Human diseases in the temperate zones of the world for which wild animals and livestock have been implicated (based on Canadian Cooperative Wildlife Health Centre 2007, World Health Organisation 2006, William et al. *2002, Kruse* et al. *2004, Spratt 2004); the information is likely to be biased towards North America.*

	Disease	Causative agent	Vector	Region	Reservoir
I. Tick-borne diseases					
a. Viruses	Colorado tick fever	Orbivirus (fam. Reoviridae)	*Dermacentor andersoni* [1]	West USA, Canada	Golden-mantled ground squirrel; deer mouse; bushy-tailed woodrat; least chipmunk; Uinta chipmunk; Nuttal's cottontail rabbit; porcupine
	Rocky mountain spotted fever	*Rickettsia rickettsii, R. parkeri*	*Dermacentor variabilis*	Eastern North America	Perhaps Ground hog; mice; white-tailed deer
		Rickettsia rickettsii	*Dermacentor andersoni; Rhipicephalus sanguineus*	Western North America	Perhaps rabbits; large game
	Boutonneuse fever	*Rickettsia conorii, R. helvetica, R. mongolotimonae, R. aeschlimannii, R. slovaca*	*Rhipicephalus sanguineus, R. turanicus, Haemaphysalis punctata, H. sulcata*	Europe	Rodents, roe deer chamois, perhaps rabbit
	Siberian tick typhus	*Rickettsia helvetica, R. japonica, R. heilongjaiangensis*	*Ixodes holocyclus; Dermacentor marginatus, Haemaphysalis concinna, Rhipicephalus sanguineus*	Asia	?
	Queensland tick typhus	*Rickettsia australis, R. nonei*	*Ixodes holocyclus, I. tasmani, I cornuatus*	Australia	Bandicoots, rodents
	Powassan encephalitis	Flavivirus (Arbovirus)	*Ixodes cookei; I. marxi, I. spinipalpus; Dermacentor andersoni*	North America	Woodchuck, ground hog, snowshoe hare, Hudson red squirrel, grey squirrel, voles
			Haemaphysalis neumanni, Ixodes persulcatus, Dermacentor silvarum	Russia	Hare, marmot
	Loupin ill	Flavivirus	*Ixodes ricinus*	Old world	Red deer, red grouse
	Tick-borne encephalitis	Flavivirus	*Ixodes persulcatus, I. ricinus*	Old world	Rodents; wandering birds may spread the ticks

Table 1. Continued.

	Disease	Causative agent	Vector	Region	Reservoir
b. Protozoan	Babesiosis	*Babesia microti*	*Ixodes scapularis*	Eastern USA	White-footed deer mouse; meadow vole
			Ixodes pacificus	Western USA	
		Babesia divergens	*Ixodes ricinus, I. trianguliceps*	Europe	Cattle; perhaps deer
c. Other	Ehrlichiosis-like:				
	Ehrlichiosis	*Ehrlichia chaffeensis, Ehrlichia ewengii*	*Amblyomma americanum*	North America	White-tailed deer
	Granulocytic ehrlichiosis	*Anaplasma phagocytophilum*	*Ixodes scapularis*	Eastern USA, Canada	Deer, elk, rodents' perhaps birds
			Ixodes pacificus	Western USA, Canada	
	Sennetsu fever	*Neorickettsia sennetsu*	(Tremadode flatworms)	Japan, Malaysia	?
	Lyme disease	*Borrelia burgdorferi*	*Ixodes scapularis*	Eastern North America	White-footed deer mouse; Dusky-footed woodrat
			Ixodes pacificus	Western North America	California kangaroo rat
		Borrelia garinii / B. afzelii	*Ixodes ricinus*	Europe	Fieldmice; voles; roe deer
		B. sinica, B. japonica, B. tanukii, etc.	*Ixodes persulcatus*	Siberia, China, Japan	Field mice; voles; roe deer
	Tick-borne borreliosis	*Other Borrelia species (B. hermsiii, B. parkeri, B. venezuelensis, B. mazzottii, B. persica, B. duttonii, B. hispanica, B. crocidurae, B. merionesi, B. microti, B. dipodilli, B. caucasica, B. latyschewii)*	*Ornithodoros spp.*	Mediterranean, China, North America	(Perhaps) rodents

Table 1. Continued.

	Disease	Causative agent	Vector	Region	Reservoir
	Southern tick-associated rash illness	*Borrelia lonestari*	*Amblyomma americanum, Ixodes scapularis*	North America	White-tailed deer
	Tularaemia	*Franciscella tularensis*	*Amblyomma*, 5 spp. *Dermacentor, Haemaphysalis*, 5 *Ixodes* spp.	world wide	Rabbits, hares, beaver,
II. Mite-borne diseases					
	Chigger-borne rickettsiosis / Scrub typhus	Orientia (Rickettsia) tsutsugamushi	*Leptotrombidium deliense*, *Leptotrombidium* spp.	Australia, Asia,	Rodents
III. Mosquito-borne					
a. Viruses	Yellow fever	Yellow fever virus	*Aedes* spp., *Haemagogus* spp.	tropics	Monkeys, bush babies; mosquitos
	Dengue and Dengue Hemorrhagic fever	Arbovirus	*Aedes* spp.		Humans, perhaps some monkeys
	Japanese encephalitis	Japanese encephalitis virus	*Culex* spp.	East and South Asia	Night herons, egrets
	California encephalitis	14 different viruses	*Aedes* spp., *Anopheles spp., Culex* spp., *Culiseta spp., Psorophora* spp.	Europe, America, Africa, East Asia	Rodents and Lagomorphs
	Murray Valley encephalitis	arbovirus		Australia	Birds (stork family)
	Kunjin	arbovirus		Australia	Birds (stork family)
	Ross River	arbovirus	*Aedes, Anopheles, Coquillettidia, Culex, Mansonia, Tripteroides*	Australia	Kangaroo, wallaby (fruit bat)
	Barmah Forest	arbovirus		Australia	Brushtail possum
	less important Australian: Kokobera, Alfuy, Edge Hill, Stratford, Sindbis, Gan gan and Trubanaman			Australia	Macropodids, marsupials, herons, possums, birds

Table 1. Continued.

	Disease	Causative agent	Vector	Region	Reservoir
	Snowshoe Hare	arbovirus	*Aedes spp., Culex spp., Culiseta spp.*	North America	Snowshoe Hare, 4 carnivores, 3 ungulates, ruffed grouse, 7 rodents
	West Nile	Flavivirus	*Culex spp.*	North America, Africa	Birds
	Q-fever		*Coxiella burnetti, but also ticks Ornithodoros coriaceus*		Large game
	Eastern equine encephalitis	Alphavirus	*Aedes spp., Coquillettidia perturbans*	North and South America	Birds, mosquito: *Culiseta melanura*
	Venezuelan equine virus	Alphavirus	*Aedes spp., Culex portesi, Psorophora ferox*	North and South America	Rodents
	St. Louis encephalitis	Flavivirus	*Culex spp.*	North and South America	Birds
	Western equine virus (related in Europe: Sinbis virus) (also related Highlands J virus and Fort Morgan virus)	Alphavirus (Flavivirus)	*Culex spp., Culiseta spp.;* also *Aedes spp, Anopheles spp., Coquillettidia spp.*	North America	Passerine birds; also Richardson's ground squirrel, Snowshoe hare, garter snakes
b. Other	Malaria	*Plasmodium falciparum, P. malariae, P. ovale, P. vivax*	*Anopheles spp.*	tropics; in the past temperate	Humans, perhaps some monkeys
IV. Rodent-borne					
	Hanta haemorrhagic fever with renal syndrome	Hanta types:			
		Hantaan	see right-hand column	East Asia, Eastern Europe	*Apodemus agrarius*
		Seoul	see right-hand column	East Asia; seaports	*Rattus norvegicus*
		Dobrave/Belgrade	see right-hand column	Eastern Europe	*Apodemus agrarius*
		Puumala	see right-hand column	Scandinavia, Western Russia	*Clethrionomys glareolus*

Table 1. Continued.

Disease	Causative agent	Vector	Region	Reservoir
	Hanta types (continued)			
	New York	see right-hand column	Eastern USA	*P. leucopus*
	Black Creek Canal	see right-hand column	Florida	*Sigmodon hispidus*
	Bayou	see right-hand column	South-eastern USA	*Oryzomuys palustris*
	Juquitiba	see right-hand column	Brazil	Unknown
	Laguna Negra	see right-hand column	Paraguay and Bolivia	*Calomys laucha*
	Andes	see right-hand column	Argentina and Chili	*Oligoryzomys longicaudatus*
	Oran	see right-hand column	North-western Argentina	*O. longicaudatus*
	Lechiguanas	see right-hand column	Central Argentina	*O. flavescens*
V. Fruitbat-borne				
Hendra	Hendra virus (paramyxovirus)	see right-hand column	Australia, New Guinea	*Pteropus alecto, P. poliocephalus, P. scapulatus, P. conspiciilatus*
Nipah	Nipah virus (paramyxovirous)	see right-hand column	Australia, Malaysia	*Cynopterus brachyotis, Eonycteris spelaea, Pteropus hypomelanus, P. vampyrus*, and one non-fruit bat: *Scotophilus kuhlii*

[1] White-tailed deer is not a competent reservoir

combated too. The overwhelming majority of vector-borne diseases in Europe are since that time transmitted by ticks (Randolph 2001; Table 1), and because intermediary hosts were rare and vegetations were kept in highly productive states, chances of encountering ticks were low even for people who had the time for recreating in the countryside.

The last 30 years and the immediate future

Since the 1980s major changes in the non-urban landscapes of Europe and North America are taking place because arable- and livestock-based agriculture is no longer cost-effective; the countryside is becoming devoid of people who use the local resources, while an ever increasing number of people move to urban metropolises or peri-urban neighbourhoods (Figure 1). The same trend is visible in other developed countries, such as South Africa, Japan, USA, Thailand and Australia. Nowadays proportionally more and more people live in cities and towns: in Australia about 85% of the people are living in urban areas (Anonymous 2006a), in France it is more than 80% (Anonymous 2005b), in the most developed part of the world it is on average 76% and worldwide it is now 40% (United Nations 2006) or even 50% (United Nations Population Fund 2007 http://www.unfpa.org/swp/2007/english/introduction.html (Accessed 10 September 2007). Fewer people realise, however, that even the absolute numbers of people living in rural landscapes is decreasing and, as a result, the number of abandoned villages in France, Italy, Spain and Portugal is increasing. Even when British or Dutch pensioners take over property in these villages, the surrounding countryside remains unused. According to preliminary results of the first national census since 1989, more than half of Russia's 155,290 villages are abandoned or populated by 50 or fewer stragglers (Anonymous 2006b, cf. Kontorovich 2000). The same is happening in parts of the USA, especially Wyoming, Nebraska and Colorado (Anonymous 2005c), but also in the forested states of the eastern Atlantic seaboard (Flinn *et al.* 2005, Litvaitis 1993, Motzkin and Foster 2002, Ramankutty and Foley 1999). In 'Old Europe' (the European Union of 10 countries), the utilised agricultural area diminished by more than 3 million hectares between 1975 and 1987; between 1980 and 1990 this was 1,251,000 ha in Germany, 1,000,000 ha in France, and 307,000 ha in Italy

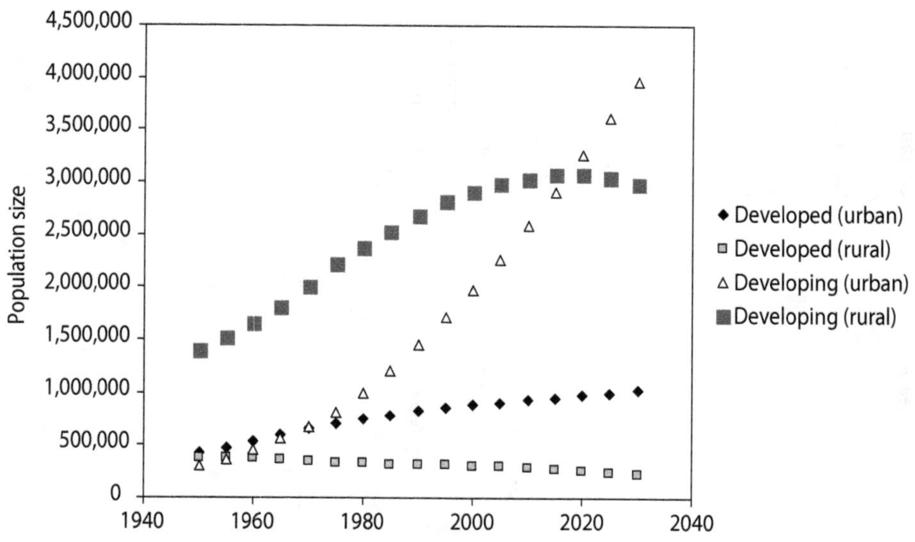

Figure 1. The development of the human population (×1000) in different parts of the world (United Nations 2003).

(European Commission data, http://epp.eurostat.ec.europa.eu/portal/page?_pageid=1090,30070682,1090_33076576and_dad=portaland_schema=PORTAL (Accessed 10 September 2007). The surface area of permanent meadows and grassland have strongly declined in 'Old Europe' too (between 1980 and 1990 1 million hectare; European Commission data). Also the number of domestic stock declines. Between 1991 and 2001, the number of cattle in western Europe stabilised at about 90 million head, in eastern and Central Europe it declined from 50 to 30 million head, and in the EECCA-countries (the 12 countries from the Caucasus up to Russia) it showed a massive decline from about 115 to 60 million head. Similar trends are reported for sheep and goats (Anonymous 2003; see also FAO 2006).

The reasons for this abandonment are not the same in every situation, but nearly always the ultimate cause is that more jobs and an easier life are available in the cities. Additional factors include lack of children, stock theft, lack of schools, hospitals or other services. Demography, market and neo-liberalism all lead to the same result: shops close, services disappear, jobs disappear in the villages but are offered in towns. With the ever-increasing costs of labour as compared to assets, farmers lay-off farm-hands and use machinery to do the work, which further decreases the local job markets. For necessary loans, farmers then get more and more dependent on banks, fodder producers or sellers of farm equipment, which makes them more vulnerable in times of adversity. Ultimately this frequently results in bankruptcy. Politicians have tried to stop this tide by doling out hundreds of billions of euros in subsidies to maintain the countryside in a productive setting. This has led to an ever-increasing cost for food for the urban people, and an ever-increasing call for reducing the toll barriers around 'Fortress Europe'. Agricultural subsidies have decreased already, and will decrease further. To save the last of the uneconomic farmers, governments provide subsidies for maintaining the countryside for the sake of biodiversity. Land abandonment leads to changes in the landscape to which wild grazers and browsers react. Yet, the decreasing numbers of domestic stock inhabiting the former extensive farms in developed countries offers opportunities for wild herbivores to exploit the unconsumed vegetation resource (Table 2). White-tailed deer (*Odocoileus virginianus* Boddaert) numbers, for example, have increased dramatically between 1980 and 2000 in many areas of North America (Doerr *et al.* 2001, Riley *et al.* 2002). The same happens in Europe with, for example, the number of Roe deer and Wild boar in Germany, France and other European countries (Federal Republic Germany 2003, Alban *et al.* 2005) spreading at an amazing rate even in a densely populated country as the Netherlands (T. van der Have, personal communication). Conservationists continue to emphasise the decline of nature instead of also mentioning these changes that are taking place (for example: European Mammal Assessment 2007), hardly realising that this rebound of nature takes place in real time and for all to see. Indeed, species associated with past land practices that typically were labour-intensive but capital-extensive and that led to resource mining in a fine-grained agricultural landscape, have disappeared and/or will decrease even further. Typical species in this context are the plant species referred to before such as those belonging to the genera *Erica*, *Calluna*, *Juniperus*, *Euphorbia* but also *Iris*, or *Gentiana*, and bird species associated with the landscapes in which these plants occur, such as Black grouse (*Tetrao tetrix* L.), European nightjar (*Caprimulgus europaeus* L.), Woodchat shrike (*Lanius senator* L.), Tree pipit (*Anthus trivialis* L.) or Stone curlew (*Burhinus oedicnemus* L.). But species that are favoured by nutrient-rich situations and quietude, such as herbivores ranging from voles to red deer, also increase in numbers.

Apart from economic drivers, sociological and demographic changes, and market distortions, there is a fourth insidious driver of change, namely climate change, especially the increasing concentrations of carbon dioxide in the atmosphere. This change may lead to changed competitive

Table 2. The effect of different scenarios in a changing European countryside on domestic herbivores and on wild herbivores (from Prins and Gordon 2007).

Scenario elements (CO_2-levels are assumed to increase in all possible scenarios)	Effect on **domestic herbivores** in the context of the ecology of grazers and browsers	Effect on **wild herbivores** in the context of the ecology of grazers and browsers
Rural areas without land abandonment or depopulation, while market distortions and farm subsidies do not allow much scope for increased land holding size due to mergers, and farmers rely heavily on further farm intensification	Very few browsers, if any, will find a place in this system. Animals will be kept indoors, and be fed with imported food and silage from intensively used (parts of) farms.	Small wild grazers, like migratory geese, can benefit but otherwise the landscape will become devoid of wild grazers and browsers.
Rural areas without land abandonment or depopulation, but cessation of market distortions and farm subsidies lead to a moderate increase in size of land holdings in places where it is economically profitable to merge. Cessation of agriculture in regions where it is no longer viable, but in regions where it is profitable most farms will further intensify	Market for novelty products can become important, such as lamas in Europe, and farmers may switch to game farming like in South Africa or New Zealand.	There will be an effect outside the region towards other regions: there abandonment will be promoted leading to woody encroachment and build-up of browser populations.
Land abandonment but no depopulation of rural areas, while market distortions and farm subsidies continue leading to much increased land holding size due to mergers even in areas where it does not make sense economically. Intensification is driven by subsidies even in places far removed from markets. There will be little land abandonment because this trend will be counteracted by subsidies.	Very few browsers, if any, will find a place in this system. The industry would like to keep animals indoors, and fed on imported food and silage from intensively used farms, but the public will demand 'romantic' landscapes with domestic stock kept outdoors.	Little opportunity for wild species in the landscape because abandonment is counteracted upon by government and farmers. Nature management will ask for woody species suppression by browsing, mowing and cutting. Increased options for small wild browsers; if wild grazers are available, they will be kept fenced.
Land abandonment but no depopulation of rural areas, but cessation of market distortions and farm subsidies lead to an increase in size of land holdings in places where it is economically profitable but cessation of agriculture where it is not. Intensification is not necessary on farms that stay in operation.	'Traditional' grazers (cattle, sheep) will be used on big farms, but market for novelty products could become important, and farmers may switch to game farming.	Reasonable opportunities for wild species in the landscape because pockets of land will be abandoned by agriculture. The opportunities are there for small and not so small browsers. Little scope for 20th century nature management. Increased options for large herds of wild grazers if they can be made economically; they will be kept fenced.

Table 2. Continued.

Scenario elements (CO_2-levels are assumed to increase in all possible scenarios)	Effect on **domestic herbivores** in the context of the ecology of grazers and browsers	Effect on **wild herbivores** in the context of the ecology of grazers and browsers
Depopulation of rural areas but without land abandonment, while market distortions and farm subsidies continue leading to increased land holding size due to mergers and further farm intensification.	Very few browsers, if any, will find a place in this system. Animals will be kept both indoors and outdoors, and be fed on imported food and silage from intensively used (parts of) farms.	Small wild grazers, like migratory geese, can benefit but otherwise the landscape will become devoid of wild grazers and browsers.
Depopulation of rural areas but without land abandonment, but cessation of market distortions and farm subsidies lead to a moderate increase in size of land holdings in places where it is economically profitable to merge. Cessation of agriculture where it is no longer viable, but on most farms intensification is necessary.	Very few browsers, if any, will find a place in this system. The industry would like to see animals kept indoors but mostly outdoors, and partly fed on silage from intensively used parts of the farms. Feedlot farming.	There will be a strong effect outside the region towards other regions: there abandonment will be promoted leading to woody encroachment and build-up of browser populations.
Land abandonment and depopulation of rural areas, while market distortions and farm subsidies continue to lead to increased land holding size due to mergers. Intensification is not necessary on farms that stay in operation. There will be little land abandonment because this trend will be counteracted by subsidies.	Very few browsers, if any, will find a place in this system. The industry would like to see animals kept indoors but mostly outdoors, and partly fed on silage from intensively used parts of the farms. Good options for mega-dairy projects, feedlot farming and extensive ranching	Small wild grazers, such as migratory geese, can benefit to a limited extent because of decreased fertiliser input. The landscape will get devoid of grazers and browsers, because farmer will suppress woody encroachment and will see game as a competitor for forage.
Land abandonment and depopulation of rural areas, but with cessation of market distortions and farm subsidies will lead to an increase in size of land holdings in places where it is economically profitable but cessation of agriculture where it is not. Intensification is not necessary on farms that stay in operation.	Domestic herbivore species richness may increase if there is a market for novelty products. Low fertilizer input, coupled with large farm sizes will require the suppression of woody encroachment. Burning and stumping necessary to maintain grazing lands.	Increasing place for wild herbivores. 'Wild game' tolerated if cost-effective. Place for wild herbivores on game ranches and private game parks. Burning necessary to encourage grazers. There will be a strong effect outside the region towards other regions: there abandonment will be promoted leading to woody encroachment and build-up of browser populations.

interactions between different plant groups; carbon dioxide is used in horticulture as an artificial 'fertiliser' and it allows plants to grow for more hours in the day so that more carbohydrates are produced. It appears as if a number of woody species surpassed a threshold of starch deposition in their roots, making them much more fire-resistant, because with sufficiently high carbohydrate reserves these woody species can resprout and grow quickly out of the fire-trapping zone. This in turn leads to an altered balance between grasses and woody species, favouring the latter (Bond *et al.* 2003) and causes changes in species composition and, in turn, creates decreasing opportunities for grazers and increased ones for browsers. In Europe, most wild ruminants are either mixed feeders or browsers. In a landscape that will thicken in areas where land abandonment or cessation of agriculture takes place, browsers and mixed feeders can thrive (Table 2). Also ticks may benefit form these changes.

Under a mode of production that encourages high efficiency, land abandonment and land use intensification are two sides of one coin. Wheat leads the field in world crops. In 2004, the worldwide area at which wheat was grown was about 220 million ha; for rice this was about 150 million ha, for maize some 140 million ha and for other cereals about 160 million ha again (soybeans 90, pulses 70, roots and tubers 50 million) (FAO 2006). Intensification is much easier to achieve in flat areas than in montane or steeply hilly areas (Anonymous 2005d). The number of farms decreases fast in OECD-countries. In Europe, most changes take place in mountainous areas (Batzing *et al.* 1997, MacDonald *et al.* 2000). For example in France the number of farms declined from about 880,000 in 1950 to 220,000 in 2005 (Anon 2005d). Poor, hilly areas in France, Italy, Spain and Portugal lost more than half their active population since the 1970s (EC Commission data, Torrano and Valderrabano 2004). In the north of the former East Germany, between 1991 and 1997 the number of people working in agriculture declined from 296,000 to 164,000 (Shaphiro and Stankevich 2000). In much of southern Europe bush encroachment is occurring widely because of land abandonment (e.g. Batzing *et al.* 1997, Bonazountas *et al.* 2005, Debussche *et al.* 1999, Preiss *et al.* 1997, Tabara *et al.* 2004; see also Haight *et al.* 2004, Finney 2005, O'Laughlin 2005). In Portugal, for example, rural depopulation and land abandonment has lead to a significant decline in agricultural land and low shrub and an increase in tall scrublands and forest, with an increase of 20-40% in fuel accumulation (shrubbery) at a landscape level (Moreira *et al.* 2001, see also Debussche *et al.* 1999, Tinner *et al.* 1998, Preiss *et al.* 1997, Valderrabano and Torrano 2000). Bush encroachment also takes place on a large scale in Ukraine (Kuemmerle *et al.* 2006). A shift away from fossil fuel to producing biofuels on these hilly terrains is not very likely, because the same reasons that led to the cessation of agriculture for food will lead to precluding agriculture for fuel, although politicians are ready again to use the old tool of market distortion by providing subsidies again.

The net results of these changes in land use have severe repercussions for the occurrence of wild animals that can act as vectors of vector-borne diseases. In Table 2 we present different scenarios (from Prins and Gordon 2007), focussing on the balance between woody species and grasses or herbs. Grazers and browsers react to these changes, but they can also be used to prevent changes in vegetation composition, or to facilitate them. Herbivores, in contrast to many other organisms, can act as landscape modifiers, but they can only do this to a limited extent. From Table 1 it is clear that most intermediary hosts or vectors are not the large herbivores or migratory geese: indeed, most are rodents. Very little research has been done on the relation between land use changes as highlighted in Table 2 and the ecology of all those herbivores, including the large ones. Incriminating those species as a threat to human health is not our intention, especially when it is known that many outbreaks of diseases in humans, poultry, pigs and other livestock can be ascribed to legal or illegal transport of humans, life animals or livestock products (e.g. Grootenhuis 2000, Food Standards Agency *et al.* 2007). However, because ticks belong to the most important vectors of the pathogenic infections for humans (Randolph 2001; see Table 1) and because ticks need small and large animals in the European landscape, we think that we need much more research into the ecology of these animals.

Relevance for vector-borne diseases

Depopulation and land abandonment is leading to enormous opportunities for wildlife to re-establish itself in the countryside, and for a new natural system to develop. In many of the areas that have been intensively used for decades or even centuries, wildlife was on the brink of extinction but now it is bouncing back. The number of Roe deer in Europe or White-tailed deer in North America have not been so high for centuries, and that is true for many others species as well. For example, the number of Elk (Moose: *Alces alces* L.) shot per year in Norway in 1950 was less than 5,000, but at present is more than 40,000 (Statistics Norway 2001), the central European Roe deer population now stands at some 15 million individuals (European Mammal Assessment 2007). The total number of Red deer in Germany is already close to 200,000 individuals of which some 60,000 per year are shot (European Mammal Assessment 2007). Generally speaking, though, estimates of population sizes are amazingly fragmentary or even close to absent (see the European Mammal Assessment 2007). At present, browsers and intermediate feeders are doing well in the increasingly forested landscapes of Europe, but an increase of indigenous large grazers is unlikely simply because most of them were exterminated in the past. To our knowledge Europe is more-or-less free of feral large herbivores (except in some small parts of the Spain, Portugal, United Kingdom, Ireland, Finland, Czech Republic; e.g. Equid Specialist Group 1996, S.E. van Wieren, personal communication) but nature-development programmes in various countries are introducing feral horses (*koniks*, but also Przewalski-horses) or semi-wild cattle.

Increased wildlife densities, especially of deer, can lead to the closure of the life cycle of tick-borne diseases. For example, *Borrelia* parasites causing Lyme disease needs small intermediary hosts such as mice or blackbirds (*Turdus merula* L.) but also large ones, such as roe deer or red deer. An increased number of deer may lead to an increased incidence of life-endangering parasitic diseases (Jensen and Jespersen 2005, Randolph 2004, Zavaleta and Rossignol 2004). Table 1 indicates that a relative small group of animals is responsible for most diseases: rodents and squirrels are perhaps the most important group, the second group consists of white-tailed deer (in North America) and Roe deer (in Europe). The third group in temperate areas are the Lagomorphs (Table 1). Finally, a number of bird species are of importance, although the role of waterfowl in the spread of influenza is still unclear. Nature conservation in Europe does not focus very much on the protection of rodents; birds, hares and rabbits, or wild deer species are much more important.

Too little is yet known about the relation between rodent densities and those of larger grazers although some work has been done already (e.g. Grant *et al.* 1982, Smit *et al.* 2001, Dekker 2007). In Europe knowledge on the effect of land abandonment on densities of large game species, song birds, and rodents is very limited. We think that with the trends we have sketched, there is an extremely urgent need to know much more about these issues: people's health and lives are at risk. It may be of great importance to know more about climate change, or nuclear fusion, or astronomy, but in the backyards of European citizens the world is changing fast without us knowing much about it. We assert that maintaining this lack of knowledge is irresponsible, and we express our dismay that government research funding agencies have repeatedly rejected research proposals in different countries to investigate this urgent matter. It is time they, and their political masters, wake up. Yet, also those studying Conservation Medicine miss this important new trend in the temperate zones. Recently, the Working Group on Land Use Change and Disease Emergence convened and ranked the most important land use changes in relation to new diseases (Patz *et al.* 2004); according to the group of experts the top five were agricultural development, urbanisation, deforestation, population movement and increasing population. For the developed world and for the temperate zones we maintain that this sequence should be different, namely, de-agrification, urbanisation, reforestation, population movement and decreasing population.

References

Alban L, Andersen MM, Asberg T, Boklund A, Fernandez N, Goldbach SG, Greiner M, Hojgaard A, Kramer-Schadt S, Stockmarr A, Thulke HH, Uttenthal A and Ydesen B (2005) Classical Swine Fever and Wild Boar in Denmark: a risk analysis. Danish Institute for Food and Veterinary Research (Ministry of Family and Consumer Affairs), Copenhagen.

Anonymous (2003) Europe's Environment: The Third Assessment. Environmental assessment report 10. European Environment Agency, Copenhagen.

Anonymous (2005a) Ears of plenty: the story of man's staple food. The Economist 24 December 2005, p. 26-30.

Anonymous (2005b) The Economist 24 December 2005, p. 75.

Anonymous (2005c) The poorest part of America: not here surely? The Economist 10 December 2005, p. 37-38.

Anonymous (2005d) Europe's farm follies: why the European Union remains its strange fondness for farm subsidies. The Economist 10 December 2005, p. 25-27.

Anonymous (2006a) http://earthtrends.wri.org/pdf_library/country_profiles/pop_cou_036.pdf (Accessed 10 September 2007)

Anonymous (2006b) http://www.cdi.org/russia/johnson/7246-17.cfm (Accessed 10 September 2007)

Batzing W, Perlik M and M. Dekleva (1997) Urbanization and depopulation in the Alps (with 3 coloured maps). Mtn Res Dev 16: 335-350.

Bennett J, van Bueren M and Whitten S (2004) Estimating society's willingnes to pay to maintain viable rural communities. Austr J Agric Res Econ 48: 487-512.

Bonazountas M, Kallidromitou D, Kassomenos PA and Passas N (2005) Forest fire risk analysis. Hum Ecol Risk Assess 11: 617 – 626.

Bond WJ, Midgley GF and Woodward WI (2003) The importance of low atmospheric CO2 and fire in promoting the spread of grasslands and savannas. Global Change Biol 9: 973-982.

Canadian Cooperative Wildlife Health Centre http://wildlife1.usask.ca/wildlife_health_topics/arbovirus/arbointro.php (Accessed 10 September 2007).

Cavalli-Sforza LL (1996) The spread of agriculture and nomadic pastoralism: insights from genetics, linguistics and archaeology. In: D.R. Harris (ed.) The Origins and Spread of Agriculture and Pastoralism in Eurasia, p. 51-69. UCL Press, London.

Chamie J (2005) 21st Century demographics: highs and lows. The Globalist 14 July 2005.

Cronon W (1983) Changes in the Land: Indians, Colonists, and the Ecology of New England. Hill and Wang, New York.

Debussche M, Lepart J and Devieux A (1999) Mediterranean landscape changes: evidence from old post cards. Global Ecol Biogeogr 8: 3-15.

Dekker JJA (2007) Rabbits, Refuges and Resources: how foraging of herbivores is affected by living in burrows. PhD thesis, Wageningen University, The Netherlands.

Doerr ML, McAninch JB and Wiggers EP (2001) Comparison of 4 methods to reduce white-tailed deer abundance in an urban community. Wildl Soc Bull 29: 1105 – 1113.

Engler A and Drude O (1901) Die Vegetation der Erde: die Heide Norddeutschlands. Engelmann, Leipzig.

Enserink M and Vogel G (2006) The carnivore comeback. Science 314, 746-749.

European Mammal Assessment (2007) http://ec.europa.eu/environment/nature/conservation/species/ema/findings.htm (Accessed 10 September 2007).

Equid Specialist Group 1996. Equus ferus. In: IUCN Red List of Theathened Species. www.iucnredlist.org. Accessed 10 September 2007.

FAO (2006) http://www.fao.org/es/ess/census/default.asp (Accessed 10 September 2007).

Federal Republic Germany (2003) Federal Republic of Germany's plan for combating and monitoring classical swine fever (CSF). http://ec.europa.eu/food/fs/ah_pcad/eradication/art45_germany.pdf (Accessed 10 September 2007)

Finney MA (2005) The challenge of quantitative risk analysis for wildland fire. Forest Ecol Manag 211: 97-108.

Flinn KM, Vellend M and Marks PL (2005) Environmental causes and consequences of forest clearance and agricultural abandonemnet in central New York, USA. J Biogeogr 32: 439-452.

Food Standards Agency, Health Protection Agency, Meat Hygiene Service and Defra (2007) Possible transmission of H5N1 avian influenza from imported turkey meat to the UK. http://www.food.gov.uk/multimedia/pdfs/birdfluinvest.pdf (Accessed 10 September 2007). See also http://www.defra.gov.uk/animalh/diseases/notifiable/disease/ai/pdf/epid_findings050407.pdf (Accessed 10 September 2007).

Godwin H (1975) History of the British Flora: A factual basis for phytogeography, 2nd ed. Cambridge Univ. Press, Cambridge.

Gordon I and HHT Prins (2007) Browsers and grazers in a changing world. In: Gordon I and Prins HHT (eds.) The Ecology of Browsing and Grazing. Ecological Studies Vol. 195. Springer, New York.

Goriup P (1998) The Pan-European Biological and Landscape Diversity Strategy: integration of ecological agriculture and grassland conservation. Parks 8: 37-46.

Grammaticas D (2004) Life ebbs away from Russian villages. BBC news 5 November 2004. http://news.bbc.co.uk/2/hi/europe/3984951.stm (Accessed 10 September 2007).

Grant WE, Birney EC, French NR and Swift DM (1982) Structure and productivity of grassland small mammal communities related to grazing-induced changes in vegetation cover. J Mammol 63: 248-260.

Grootenhuis JG (2000) Wildlife, livestock and animal disease reservoirs. In: Prins, HHT, Grootenhuis JG and Dolan TT (eds.) Conservation of Wildlife by Sustainable Use, pp. 81 – 113. Kluwer Academic, Boston.

Haight RG, Cleland DT, Hammer RB, Radeloff VC and Rupp TS (2004) Assessing fire risk in the wildland-urban interface. J Forestry 102: 41-48.

Hole F (1996) The context of caprine domestication in the Zagros region. In: Harris DR (ed.) The Origins and Spread of Agriculture and Pastoralism in Eurasia, p. 263-281. UCL Press, London.

Hughes R (1987) The Fatal Shore: a History of the Transportation of Convicts to Australia 1787 – 1868. Pan Books, London.

Ioffe G, Nefedova T and Zaslavsky I (2004) From spatial continuity to fragmentation: the case of Russian farming. Ann Assoc Am Geog 94: 913-943.

Janssen JCGM and van de Westeringh W (1983) Dat ging over zijn hout: overmatig gebruik van bossen in het zuiden van Limburg, van de Hoge Middeleeuwen tot in de 20ste eeuw. In: Studies over de sociaal-econmische geschiedenis van Limburg. Jaarboek van het Sociaal Historisch Centrum voor Limburg 28, 19 – 63 (Van Gorcum, Assen).

Jensen PM and Jespersen JB (2005) Five decades of tick-man interaction in Denmark: an analysis. Exp Appl Acarol 35: 131-146.

Knörzer KH (1971) Urgeschichtler Unkräuter im Rheinland: Ein Beitrage zur Entstehungsgeschichte der Segetalgesellschaften. Vegetatio 23: 89-111.

Knörzer KH (1975) Entstehung und Entwicklung der Grünlandvegetation im Rheinland. Decheniana 127: 195-214.

Köhler-Rollefson I (1996) The one-humped camel in Asia: origin, utilization and mechanisms of dispersal. In: Harris DR (ed.) The Origins and Spread of Agriculture and Pastoralism in Eurasia, p. 282-294. UCL Press, London.

Kontorovich V (2000) Can Russia resettle the Far East? Post-Communist Economies 12: 365-384.

Kruse H, Kirkemo AM and Handeland K (2004) Wildlife as source of zoonotic infections. Emerg Infect Dis 10: 2067 – 2072.

Kuemmerle V, Radeloff C, Perzanowski K and Hostert P (2006) Cross-border comparison of land cover and landscape pattern in Eastern Europe using a hybrid classification technique. Remote Sens Env 103: 449-464.

LaFraniera S (2003) A Survival Mission in Rural Russia: To keep farm alive, director keeps money and vodka away from workers. Washington Post, 25 July 2003. http://www.cdi.org/russia/johnson/7265-8.cfm (Accessed 10 September 2007).

Laudisoit A, Leirs H, Makundi RH, van Dongen S, Davis S, Neerinckx S, Deckers J and Libois R (2007) Plague and the human flea, Tanzania. *J Emerg Infect Dis* 13: 687-693.

Legge T (1996) The beginning of caprine domestication in Southwest Asia. In: Harris DR (ed.) The Origins and Spread of Agriculture and Pastoralism in Eurasia, p. 238-262. UCL Press, London.

Lewis D (2002) Slower than the Eye can See: Environmental Change in Northern Australia's Cattle Lands – a case study from the Victoria River District, Northern Territory. Tropical Savannas CRC, Darwin.

Litvaitis JA (1993) Response of early succesional vertebrates to historic changes in land-use. Conserv Biol **7**: 866-873.

MacDonald D, Crabtree JR, Wiesinger G, Dax T, Stamou N, Fleury P, Lazpita JG and Gibon A (2000) Agricultural abandonment in mountain areas of Europe: environmental consequences and policy response. J Environ Manag 59: 47-69.

McCollin D, Moore L and Sparks T (2000) The flora of a cultural landscape: environmental determinants of change revealed using archival sources. Biol Conserv 92: 249-263.

Motzkin G and Foster DR (2002) Grasslands, heathlands and shrublands in coastal New England: historical interpreations and approaches to conservation. J Biogeogr 29: 1569-1590.

Moreira F, Rego FC and Ferreira PG (2001) Temporal (1958 – 1995) pattern of change in a cultural landscape of northwestern Portugal: implications for fire occurrence. Landscape Ecol 16: 557-567.

Murwira A (2003) Scale matters: a new approach to quantify spatial heterogeneity for predicting the distribution of wildlife. PhD thesis, Wageningen University/ITC Enschede, The Netherlands.

Noble IR and Dirzo R (1997) Forests as human-dominated ecosystems. Science 277: 522-525.

Olmig K (2002) Landscape, Nature and the Body Politic: from Britain's Renaissance to America's New World. University of Wisconsin Press, Madisson.

O'Laughlin J (2005) Policies for risk assessment in federal land and resource management decisions. Forest Ecol Manag 211: 15-27.

Opravil E (1978) Synanthrope Pflanzengesellschaften aus der Burgwallzeit (8. - 10. Jh.) in der Tschechoslowakei. Ber Deutsch Bot Ges 91: 97-106.

Pennington W (1969) The History of the British Vegetation. English Univ. Press, London.

Pallot J (1990) Rural depopulation and the restoration of the Russian village under Gorbachov. Soviet Stud 42: 655-674.

Patz JA, Daszak P, Tabor GM, Aguirre AA, Pearl M, Epstein J, Wolfe ND, Kilpatrick AM, Foufopoulos J, Molyneux D, Bradley DJ (2004) Env Health Persp 112:1092–1098.

Population Reference Bureau (2005). 2005 World Population Data Sheet. http://www.prb.org/pdf05/05WorldDataSheet_Eng.pdf (Accessed 10 September 2007).

Preiss E, Martin JL and Debussche M (1997) Rural depopulation and recent landscape changes in a Mediterranean region: consequences for the breeding avifauna. Landscape Ecol 12: 51-61.

Prins HEL (1994) Children of Gluskap: Wabanaki Indians on the Eve of the European Invasion. In: Baker EW, Churchill EA, D'Abate RS, Jones KL, Konrad VA and Prins HEL (eds.) American Beginnings: Exploration, Culture and Cartography in the Land of Norumbega, p. 165-211. U. Nebraska Press, Lincoln,

Prins HHT (1998) The origins of grassland communities in northwestern Europe. In: Wallis de Vries MF, Bakker JP and van Wieren SE (eds.). Grazing and Conservation Management, p. 55 - 105. Kluwer Academic Publishers, Boston.

Prins HHT (2000) Competition between wildlife and livestock. In: Prins HHT, Grootenhuis JG and Dolan TT (eds.) Conservation of Wildlife by Sustainable Use, p.51 - 80. Kluwer Academic, Boston

Prins HHT and Gordon I (2007) Grazers and browsers in a changing world. In: Gordon I and Prins HHT (eds.). The Ecology of Browsing and Grazing. Ecological Studies Vol. 195. Springer, New York.

Radde G (1899) Grundzüge der Pflanzenverbreitung in den Kaukasusländern von der unteren Wolga über den Manytsch-Schneider bis zur Scheitelfläche Hocharmeniens. W. Englemann, Leipzig.

Ramankutty N and Foley JA (1999) Estimating historical changes in land cover: North American croplands from 1850 to 1992. Global Ecol Biogeogr 8: 381-396.

Randolph SE (2001) The Shifting Landscape of Tick-Borne Zoonoses: Tick-Borne Encephalitis and Lyme Borreliosis in Europe. Phil Trans R Soc Lond B 356: 1045-1056.

Randolph SE (2004) Evidence that climate change has caused 'emergence' of tick-borne diseases in Europe? Int J Med Microbiol 293: 5-15.

Riley SJ, Decker DJ, Enck JW, Curtis PD, Lauber TB and Brown TL (2002) Deer populations up, hunter populations down: implications for interdependence of deer and hunter population dynamics on management. Ecoscience 10: 455-461.

Rusell EWB and Davis RB (2001) Five centuries of changing forest vegetation in the Northeastern United States. Plant Ecol 155: 1-13.

Shaphiro S and Stankevich B (2000) Structural changes in agriculture in North Eastern Germany (in Russian). Vestsi Akademii Agrarnykh Navuk Respubliki Belarus 2: 40-43.

Simmons IG, Dimbley GW and Grigson C (1981) The Mesolithic. In: Simmons IG and Tooley MJ (eds.), The Environment in British Prehistory, p. 82 - 124. Duckworth, London.

Smit R, Bokdam J, den Ouden J, Olff H, Schot-Opschoor H and Schrijvers M (2001) Effects of introduction and exclusion of large herbivores on small rodent communities. Plant Ecol 155: 119-127.

Spratt DM (2004) *Homo sapiens*: purveyor of ecosystem health or of emerging wildlife diseases? Conf Proc Wildl Dis Assoc, Australasian Section.

Statistics Norway (2001) Forest and Uncultivated Land. http://www.ssb.no/english/subjects/01/sa_nrm/arkiv/nrm2001/kap04-en.pdf (Accessed 10 September 2007).

Tabara D, Sauri D and Cerdan R (2004) Forest fire risk management and public participation in changing socioenvironmental conditions: a case study in a Mediterranean region. Risk Anal 23: 249-260.

The Economist, Editorial, 7 January 2006 p. 12

Tinner W, Conedura M, Amman B, Gaggeler HW, Gedye S, Jones R and Sagesser B (1998) Pollen and charcoal in lake sediments compared with historically documented forest fires in southern Switzerland since AD 1920. Holocene 8: 31-42.

Torrano L and Valderrabano J (2004) Impact of grazing on plant communities. Spanish J Agric Res 2: 93-105.

United Nations (2003) Population Division of the Department of Economic and Social Affairs of the United Nations Secretariat, World Population Prospects: The 2004 Revision and World Urbanization Prospects: The 2003 Revision, http://esa.un.org/unpp (Accessed 10 September 2007).

United Nations (2006) World Urbanization Prospects, The 1999 Revision. http://www.prb.org/ (Accessed 10 September 2007).

Valderrabano J and Torrano L (2000) The potential for using goats to control *Genista scorpis* shrubs in European black pine stands. Forest Ecol Manag 126: 377-387.

Van der Woud A (1987) Het Lege Land: de ruimtelijke orde van Nederland 1798 – 1848. Meulenhoff Informatief, Amsterdam.

Van Zeist W (1980) Prehistorische cultuurplanten, onstaan, verspreiding, verbouw. In: Chamaulan M and Waterbolk HT (eds.), Voltooid Verleden Tijd?: Een hedendaagse kijk op de prehistorie, Intermediair, Amsterdam: 147-165.

Vera FWM (1997) Metaforen voor de Wildernis: eik, hazelaar, rund, paard. PhD thesis, Wageningen University, The Netherlands.

Williams ES, Yuill T, Artois M, Fischer J and Haig SA (2002) Emerging infectious diseases in wildlife. Rev Sci Tech Off Int Epiz 21: 139-157.

World Health Organization (2006) Geographical distribution of arthropod-borne diseases and their principal vectors. World Health Organization, Vector Biology and Control Division WHO, Geneva, Switzerland.

Zavaleta JO and Rossignol PA (2004) Community-level analysis of risk of vector-borne disease. Trans R Soc Trop Med Hyg 98: 610-618.

Zeuner FE (1967) Geschichte der Haustiere. Bayerische Landwirtschaftsverlag, München.

23. Wildlife and the risk of vector-borne viral diseases

Byron E. Martina and Albert D.M.E. Osterhaus

Abstract

In the past decades, several viruses have been identified that originated from animal reservoirs and pose a direct hazard to humanity. These include HIV, Sin Nombre virus, SARS coronavirus, and avian influenza viruses. Numerous longer known viruses are rapidly expanding their geographical range, with a corresponding increase in burden of human disease. These include several vector-borne viral diseases. In our globalising world, there are numerous, largely interrelated factors, which predispose for this increase. For the emergence of vector-borne viral diseases, changes in land-use, climate and human behaviour, modification of agricultural practices, increases in international and intercontinental travelling, and disruption of wildlife habitats have proven to be important in this respect. Virtually all these factors directly or indirectly influence the presence and density of reservoir hosts and vectors. Most vector-borne viral diseases originating from wild animal reservoirs are maintained in sylvatic and urban cycles. Here we discuss vector-borne viral diseases, which pose a risk to public health in Europe.

Keywords: Europe, viruses, ecology, epidemiology, wildlife, mosquito, ticks, birds, reservoir host

Introduction

The past century has seen an unprecedented increase in the number of emerging and re-emerging infectious diseases (EID) throughout the world, with serious implications for humans, domestic animals, and wildlife (Kuiken *et al.* 2003, Osterhaus 2001). This increase in EIDs was illustrated by the identification of novel pathogens such as Nipah virus, Ebola virus, hantavirus, SARS coronavirus, but is also used to denote the geographic expansion of diseases caused by 'old' pathogens, e.g. avian influenza viruses. Most of these infections are zoonoses, i.e. primarily infections of animals that can be transmitted directly to humans. The rise in EIDs is attributed to many factors, including human alteration of habitats, accelerated transportation of vectors and pathogens, and anthropogenic climate changes affecting the ecology of host, pathogen and vector. Most of the recognised EIDs in humans have their origin in domestic or wild animals. Most vector-borne diseases, which are diseases transmitted to humans by arthropod vectors such as mosquitoes and ticks, originated from a sylvatic cycle (a cycle between vector and animal host that is usually maintained in the jungle) and some are now maintained in an independent urban cycle, while wildlife is an important constituent of both cycles. However, most mosquito-borne diseases start from a sylvatic cycle and extend to peri-urban and urban areas, where much higher densities of humans and mosquitoes can promote outbreaks or epidemics. Animals are central to the transmission cycle of both mosquito and tick-borne infections and are defined here as animal reservoirs. If we are to predict the emergence of vector-borne diseases and design appropriate management strategies for their future containment, we need to understand the ecology and epidemiology of emerging pathogens. In this review, we discuss examples of vector-borne viral diseases emerging in Europe and factors that might affect their emergence.

Examples of emerging vector-borne viruses in Europe

Viruses belonging to the families *Flaviviridae, Bunyaviridae, Togaviridae, Reoviridae* and *Orthomyxoviridae* represent the most important causes of vector-borne diseases in Europe (Table 1).

Flaviviridae

The family *Flaviviridae* consists of three genera: *Flavivirus, Pestivirus* and *Hepacivirus*. Members of the genus *Flavivirus* represent a major class of vector-borne viruses of animals and humans. There are essentially four groups of flaviviruses: (1) tick-borne, (2) mosquito-borne, (3) tick- and mosquito-borne, and (4) non-vectored flaviviruses. The following flaviviruses circulate in Europe or pose a threat to unaffected European countries: tick-borne encephalitis virus (TBEV), Powassan virus (POWV), Louping ill virus (LIV), Omsk haemorrhagic virus (OHFV), West Nile virus (WNV), and Usutu virus (USUV).

TBEV was first discovered in 1937 in far-eastern Russia and has since been isolated in Asia and many European countries, particularly in Germany (DE), Croatia (HR), Czech Republic (CZ), Estonia (EE), Belarus (BY), Hungary (HU), Latvia (LV), Lithuania (LT), Poland (PL), Slovenia (SI), Slovakia (SK), Sweden (SE), Finland (FI), Norway (NO), Austria (AT), Switzerland (CH), and Russia (RU). The virus is absent in the UK (GB), Ireland (IE), Spain (ES), Portugal (PT), Italy (IT), Greece (GR), Turkey (TR), Cyprus (CY), Bulgaria (BG), Romania (RO), Yugoslavia (YU), Albania (AL), The Netherlands (NL), and Belgium (BE) (Bormane *et al.* 2004, Dumpis *et al.* 1999, Suss *et al.* 2004). TBEV affects thousands of people annually and has a significant impact on public health. TBEV is transmitted to humans mainly through tick bites, although infection through the consumption of raw goat's milk has also been reported (Gresikova *et al.* 1975, Kerbo *et al.* 2005). TBEV affects principally the nervous system, resulting in clinical features ranging from asymptomatic infections to more severe diseases including meningitis or meningoencephalitis (Ceroni 2006). TBEV is maintained in nature in a cycle involving ticks and wild vertebrate hosts, but also trans-ovarial and trans-stadial transmissions during the tick life cycle have been recognised as important mechanisms (Danielova *et al.* 2002) (Figure 1).

Although many tick species can sustain virus replication and transmit the virus to susceptible reservoir hosts under laboratory conditions, in Europe only *Ixodes ricinus* L., and in Russia only *Ixodes persulcatus* Schulze and *Haemaphysalis concinna* Koch play a significant role in virus transmission (Danielova 2002, Hudson *et al.* 2001, Korenberg 1994). TBEV has also been isolated sporadically from *Dermacentor* and *Hyalomma* species in Russia (Charrel *et al.* 2004). Most commonly, the virus is transmitted to vertebrate hosts by nymphal ticks, though feeding larvae can also infect a host. Reservoir hosts are small, wild-living rodents (e.g. *Apodemus flavicollis* Melchior), whereas humans and large animals such as deer or goats function as incidental hosts (Perkins *et al.* 2003). Two mechanisms of virus transmission to uninfected ticks are known: viremic and non-viremic (Charrel *et al.* 2004, Labuda *et al.* 1993, Randolph *et al.* 1999). Most small rodent species produce low viremia levels and this only for a short time, whereas ticks remain infected throughout their life cycle. Therefore, in the case of TBEV, many vertebrate hosts play a role in maintaining the infected tick population (Bakhvalova *et al.* 2006). The epidemiology and ecology of TBEV is partly dependent on the distribution of ticks and their seasonality of feeding activity. The patchy appearance of TBEV-endemic areas in Europe, despite the fact that *Ixodes ricinus* is the

Table 1. Medically important vector-borne diseases: geographic distribution, vector, and wildlife reservoir in Europe, and factors that affect the risk of emergence into a new geographic area.

Family	Geographic distribution	Main vector	Wildlife reservoir	Factors affecting disease emergence or incidence
Flaviviridae				
TBEV	DE, HR, CZ, EE, HU, LV, LT, PL, SI, SE, FI, AT, NO, CH, SK, BY, RU	*Ixodes ricinus* *Ixodes persulcatus* *Haemaphysalis concinna*	*Apodemus flavicollis* *Clethrionomys glareolus* *Apodemus sylvaticus* *Sorex araneus*	1. Climate change: temperature, rainfall 2. Availability of all vertebrate hosts 3. Wildlife translocation 4. Displacement of tick-infested rodents 5. Urbanisation and Human recreational behavior
POWV	RU	*Ixodes persulcatus* *Haemaphysalis neumanni* *Dermacentor silvarum*	*Marmota monax*	See TBEV
LIV	GB, IE, ES, GR, TR, NO	*Ixodes ricinus*	*Apodemus flavicollis* *Lagopus scoticus*	See TBEV
OHFV	RU	*Dermacentor reticulatus* *Dermacentor marginatus*	*Ondatra zibethicus*	See TBEV
WNV	RU, BY, CZ, PT, RO, HU, FR, IT, PT	*Culex pipiens* *Culex modestus*	*Corvus corone* *Passer montanus* *Passer domesticus*	1. Mosquito species population density 2. Bird population density 3. Climate change: higher temperature, more rainfall, heavy storms
USUV	AT, IT, HU, CH	*Culex pipiens*	*Turdus merula* *Strix nebulosa* *Hirundo rustica*	See WNV
Bunyaviridae				
CCHF	AL, BG, YU, GR, TR	*Hyalomma marginatum*	*Lepus europaeus*	See TBEV
Tahyna	NO, FR, DE, IT, RO, RU, ES, CZ, PL, PT	*Aedes vexans*	*Lepus europaeus* *Oryctolagus cunniculus*	See WNV
Inkoo	NO, RU, FI, SE	*Ochlerotatus punctor*	Not known	See WNV
Batai	NO, FI, RU, AT, HU, RO, PT	*Anopheles maculipennis*	Not known	See WNV
Lednice	CZ, RO	*Culex modestus*	Not known	See WNV

Table 1. Continued.

Family	Geographic distribution	Main vector	Wildlife reservoir	Factors affecting disease emergence or incidence
TSV	IT, ES, FR, GR, PT, CY	*Phlebotomus perniciosus ?* *Phlebotomus papatasi ?*	Not known	Climate change: temperature and humidity
UUKV	DE	*Phlebotomus perniciosus ?* *Phlebotomus papatasi ?*	Not known	Climate change: temperature and humidity
Togaviridae				
SINV	SE, FI, NO, GB, IT, RO, GR, YU, RU	*Culex torrentium* *Aedes cinereus*	*Turdus iliacus* *Turdus pilaris* *Parus caeruleus*	See WNV
Reoviridae				
TRBV	RU, CZ, IT	*Ixodes ricinus*	Not known	See TBEV
EYAV	CZ, FR, DE	*Ixodes ricinus*	*Oryctolagus cunniculus*	See TBEV
Orthomyxoviridae				
THOV	IT, PT	*Rhipicephalus appendiculatus*	Not known	See TBEV
DHOV	RU, PT	*Hyalomma marginatum*	*Lepus europaeus*	See TBEV

Germany (DE), Croatia (HR), Czech Republic (CZ), Estonia (EE), Belarus (BY), Hungary (HU), Latvia (LV), Lithuania (LT), Poland (PL), Slovenia (SI), Slovakia (SK), Sweden (SE), Finland (FI), Norway (NO), Austria (AT), Switzerland (CH), and Russia (RU), UK (GB), Ireland (IE), Spain (ES), Portugal (PT), Italy (IT), Greece (GR), Turkey (TR), Cyprus (CY), Bulgaria (BG), Romania (RO), Yogoslavia (YU), Albania (AL), Netherlands (NL), and Belgium (BE)

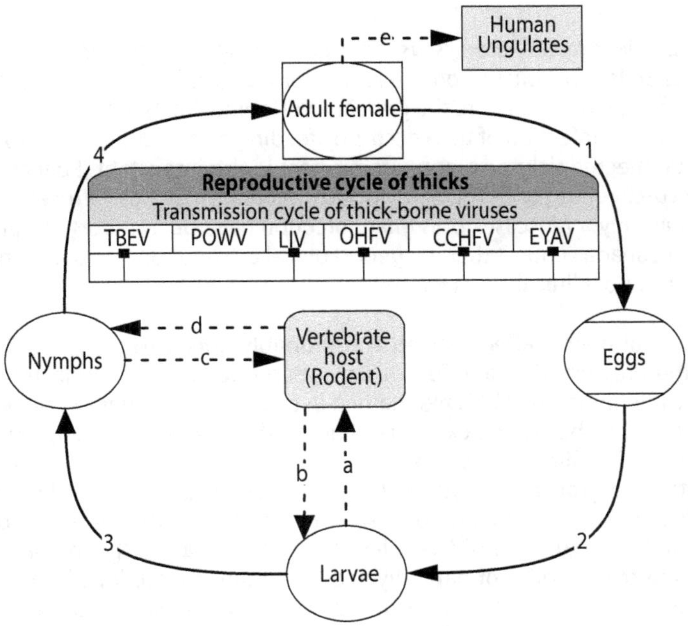

Figure 1. The transmission cycle of tick-borne viruses depends on survival, turn-over and reproductive rates of the tick (1-4). Each tick stage feeds on a different vertebrate host (a, c, e), which is necessary to develop to the next stage. The virus is trans-stadially maintained from one developmental stage to the next within a generation (steps 1 to 4). Some viruses are transmitted trans-ovarially (⫯) via eggs to larvae (steps 1 and 2), allowing infection of vertebrate hosts by larvae (a). Transmission of virus to uninfected ticks occurs during feeding on viremic host or during co-feeding on non-viremic vertebrate host (b, d). TBEV: Tick-borne Encephalitis Virus; POWV: Powassan Virus; LIV: Louping Ill Virus; OHFV: Omsk Hemorrhagic Fever Virus; CCHFV: Crimean-Congo Hemorrhagic Fever Virus; EYAV: Eyach Virus

dominant hard tick species across Europe, indicates that other ecological factors are important for TBEV emergence.

POWV was first isolated from a fatal case of encephalitis in the town of Powassan in Ontario, Canada, in 1958. The virus is also known to circulate in the coastal part of the United States, Mexico, South Dakota, and Russia (south-eastern Siberia and northeast of Vladivostok). It is believed that the majority of POWV infections are asymptomatic, but infection may also result in meningoencephalitis (Gholam *et al.* 1999), especially in children. The major route of POWV transmission to humans is through the bite of infected ticks (MMWR 2001). POWV is maintained in nature in a cycle involving ticks and wild rodents, with geographic differences in tick and vertebrate host species involved (McLean *et al.* 1961). For instance, in North America *Ixodes cookei* Packard, *Ixodes marxi* Banks, *Ixodes spinipalpus* Hadwen and Nuttall and *Dermacentor andersoni* Stiles may all be part of the transmission cycle, whereas in Russia, *Ixodes persulcatus* and *Haemaphysalis neumanni* Donitz are the principal vectors (Gritsun *et al.* 2003, Korenberg and Kovalevskii 1999). Although the woodchuck (*Marmota monax* L.) seems to be the principal host in North America, high infection rates (the proportion of a susceptible population that is infected) have also been observed in squirrels (*Tamiasciurus hudsonicus* Erxleben and *Sciurus carolinensis* Gmelin), chipmunks

(*Tamias striatus* L.), deer mice (*Peromyscus maniculatus* Wagner), voles (*Microtus* species), hares (*Lepus americanus* Erxleben), and raccoons (*Procyon lotor* L.) (Derby *et al.* 2007, Gritsun *et al.* 2003). *Apodemus* mice and *Microtus* voles are the principal vertebrate hosts in Russia (Gritsun *et al.* 2003). It is not known whether infection of ticks occurs by feeding on a viremic host only or if co-feeding of infected and uninfected ticks on a non-viremic host is also important (Ebel and Kramer 2004). In Russia, POWV co-circulates with TBEV and mixed infections with both viruses have been shown to occur. Although only a few cases of human infection have been reported, serosurveys in the United States and Canada suggest that the geographical distribution is wider than what would be expected on the basis of clinical case reports.

LIV is a flavivirus that mainly affects sheep and probably horses (Hyde *et al.* 2007). Variants of LIV circulate in the UK, Ireland, Spain, Turkey, Greece and Norway. LIV is transmitted to sheep by the tick *Ixodes ricinus*. In nature, LIV is maintained in a rodent-tick-rodent cycle with non-viremic transmission of the virus between ticks co-feeding on their vertebrate host, similar to what has been described for TBEV (Gilbert *et al.* 2000, Jones *et al.* 1997). A bird-tick-bird cycle has also been described with the red grouse (*Lagopus scoticus* Lath.) as the principal bird species (Hudson *et al.* 1995). Humans are also susceptible to infection with LIV, resulting in a neurological disease resembling the clinical picture of TBEV. Very few cases of tick-transmitted human infections have been reported, and the majority of naturally-acquired human infections have occurred mainly in individuals with occupational exposure such as veterinarians, sheep farmers, and laboratory workers.

OHFV was first isolated from the blood of a febrile patient living in the rural region of the Omsk district in Siberia, Russia. The natural vertebrate host for this virus is the water vole *(Arvicola terrestris* L.*)*, but the virus has adapted to the non-native muskrat *(Ondatra zibethicus* L.), which was introduced into Siberia from Canada in the late 1920s (Charrel *et al.* 2004). In contrast to the above-mentioned flaviviruses, OHFV is highly pathogenic for muskrats (Dunaev *et al.* 1975, Marenko *et al.* 1974), which may explain the infection rate variability observed in muskrat populations. OHFV circulates in a number of other animal species native to the endemic region, including many mammals, reptiles, and amphibians (Charrel *et al.* 2004). It is believed that the virus is transmitted to humans through a bite with infected ticks (*Dermacentor reticulatus* Hermann, *Dermacentor marginatus* Sulzer*)*. OHFV, like the other flaviviruses discussed above, can also be transmitted via the milk of infected goats or sheep (Reid *et al.* 1984, Woodall and Roz 1977). Furthermore, many reported infections in humans occurred during the hunting season and were related to direct contact with muskrats (Fedorova and Sizemova 1964). Human OHFV infection results in clinical symptoms quite different from those caused by TBEV infection (Banerjee *et al.* 1979, Borio *et al.* 2002). OHFV infection is associated mainly with capillary haemorrhages that include visceral haemorrhages in the nose, gums, uterus, and lungs and only occasionally neurological involvement is seen. The overlapping geographical range of the vector and muskrats, but the lack of OHFV emergence in new regions, underlines the complex ecology of OHFV emergence.

WNV was first isolated from a febrile patient in the West Nile district of Uganda in 1937. WNV has been shown to circulate in Africa, the Middle-East, North-America and in many European countries (Russia, Belarus, Czech Republic, Poland, Romania, Hungary, France, Italy, and Portugal) (Malkinson and Banet 2002). However, Europe has only had sporadic cases and outbreaks of WNV in humans and horses. In France, the first reported WNV outbreak affecting horses and humans occurred during the summer of 1962, followed by an outbreak among horses in 2000 (Malkinson and Banet 2002, Roehrig *et al.* 2002). This occurred in the wetland part of the Carmague region where diverse populations of migratory and resident birds share the habitat with large populations of mosquitoes. Birds, especially those belonging to the order Passeriformes (mainly members of the *Corvidae* family), are the natural host for WNV. The infection is transmitted to and maintained among birds by mosquitoes (*Culex pipiens* L. and *Culex modestus* Ficalbi). Studies performed in Egypt, Israel, and Romania suggest that WNV is maintained in two transmission cycles that can exist independently of each other (Figure 2) (Malkinson and Banet 2002). The sylvatic cycle occurs in or near wetlands where large numbers of migratory and resident birds amplify the virus, which is transmitted to other birds by ornithophilic/zoophilic mosquitoes. Bird species with a peri-domestic habitat then introduce the virus into an urban area where bridge vectors (e.g. *Culex pipiens*) and domestic birds (e.g. crows; *Corvus corone* L.) establish the urban (dryland) cycle (Jourdain *et al.* 2007). In endemic areas, the virus is amplified by continuous transmission between birds and mosquitoes and transmission is therefore increased where large numbers of mosquitoes are close to susceptible bird populations (Hayes *et al.* 2005). Humans and other mammalian species such as horses are classified as 'incidental' or 'dead end' hosts, as they do not play a role in maintaining transmission cycles. Humans become infected through infected bridge mosquitoes. The majority of human infections are subclinical and a small proportion develops meningoencephalitis, resembling that caused by TBEV infection. The outbreak of WNV in North America illustrates the potential emergence of vector-borne diseases when a virus is introduced in areas where the appropriate ecological conditions are present (Roehrig *et al.* 2002, Turell *et al.* 2002). The outbreak of WNV encephalitis in a flock of domestic geese in Hungary in the summer of 2003 and the recognition that three different pathogenic lineages circulate in that region are

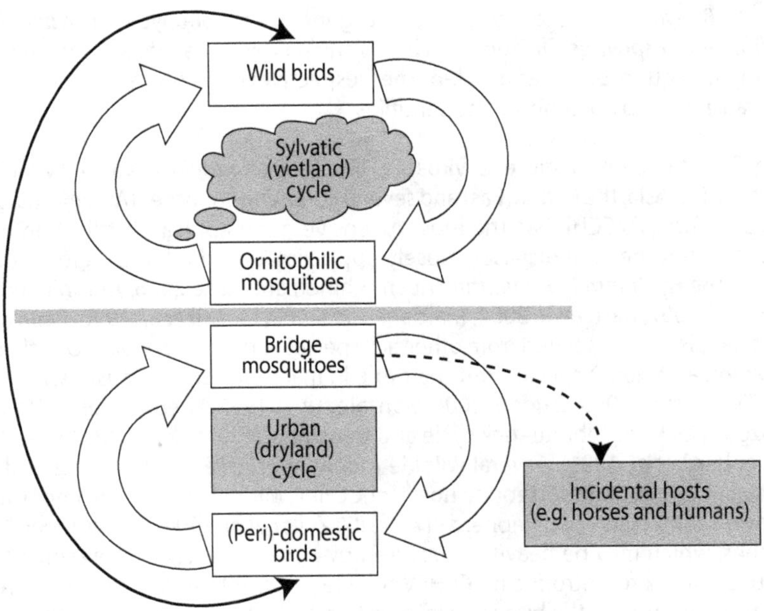

*Figure 2. West Nile virus transmission cycle. The sylvatic cycle is sustained by wild birds (migratory and resident) and ornithophilic/zoophilic mosquitoes. Bridge vectors present in both wetland and urban areas can transmit the virus to horses and humans (dashed line). Peri-domestic bird species such as house sparrow (*Passer domesticus *L.) and tree sparrow (*Passer montanus *L.) can function as introductory and amplifying hosts in urban settings (solid line). An urban cycle may be established and maintained independent of the sylvatic cycle.*

important events that should increase the level of alertness in Europe (Bakonyi *et al.* 2005, Bakonyi *et al.* 2006, Glavits *et al.* 2005).

USUV was first isolated from mosquitoes in South Africa in 1959 and named after a river in Swaziland. Until 2001, the virus had never been associated with severe or fatal diseases in animals or humans, and it had never before been observed outside tropical and subtropical Africa. Therefore, little is known about the ecology and epidemiology of USUV. In 2001, USUV was detected for the first time in Vienna, Austria, characterised by an epizootic among blackbirds (*Turdus merula* L.), and some fatal cases were also observed in great grey owls (*Strix nebulosa* Forster) and barn swallows (*Hirundo rustica* L.) (Chvala *et al.* 2004, Weissenböck *et al.* 2003, Weissenböck *et al.* 2002). The virus has since become established in Vienna and has spread from its initial area of emergence to large areas of the federal states of Lower Austria, Burgenland and Styria (Chvala *et al.* 2007). The emergence of USUV in central Europe was similar to the beginning of the WNV epidemic in the United States in 1999, except that most dead Austrian birds were found in urban areas, rather than in the natural wetland environment. USUV has now been detected in Hungary, Switzerland, and Italy, suggesting that USUV is spreading across Europe (see Chapter 9). Understanding how USUV will spread to unaffected areas in the coming years will be an important lesson for understanding flavivirus emergence in Europe. The lesson hidden in the USUV example should be that the introduction of a previously exotic flavivirus onto a new continent is a potentially serious threat for a wide range of avian and mammalian species, including humans.

Bunyaviridae

The family *Bunyaviridae* is divided into five genera: *Orthobunyavirus, Nairovirus, Phlebovirus, Hantavirus,* and *Tospovirus*. Orthobunyaviruses and nairoviruses that are known to occur in Europe are mainly mosquito-borne and tick-borne, respectively, in contrast to phleboviruses that are mainly transmitted by phlebotomine sandflies.

Crimean-Congo haemorrhagic fever virus (CCHFV; genus *Nairovirus*) is a fatal viral infection present in parts of Africa, Asia, the Middle East and several European countries (Albania, Bulgaria, Yugoslavia, Greece, and Turkey). CCHF has the most extensive geographical distribution of the medically-important tick-borne viral diseases, closely approximating the known global distribution of its tick vector, the *Hyalomma marginatum* Koch (Ahmeti and Raka 2006, Antoniadis and Casals 1982, Gozalan *et al.* 2007, Karti *et al.* 2004, Mardani *et al.* 2007, Morrill *et al.* 1990, Papa *et al.* 2002, 2004). The virus has also been isolated from other tick species, e.g. *Rhipicephalus, Ornithodoros, Boophilus, Dermacentor,* and *Ixodes* species, but their role in maintaining the transmission cycle in nature is unclear (Dohm *et al.* 1996, Ergonul 2006, Gonzalez *et al.* 1991, Shepherd *et al.* 1987). CCHFV perists in an enzootic tick–vertebrate–tick cycle and transmission to ticks occurs by feeding on a viremic host (Shepherd *et al.* 1991). Several wildlife rodent species, as well as hares and hedgehogs, are considered to be the main vertebrate hosts for CCHFV. Migratory birds are important for spreading CCHFV over a long distance (Shepherd *et al.* 1987, Zeller *et al.* 1994). In this regard, the importation of ostriches, which may be heavily parasitised by *Hyalomma* ticks, to breeding farms in western Europe poses a risk for introducing CCHFV-infected ticks into unaffected areas. Although human infections occur through the bite of an infected tick, infections are more commonly seen in people handling infected animals (Ergonul 2006). Infection with CCHFV may lead to the development of hemorrhagic manifestations, similar to those caused by other hemorrhagic fever viruses, reaching case-fatality rates up to 35%. The widespread geographic distribution of CCHFV, its ability to produce severe human disease with high mortality rates, and its classification as a bioterroristic agent (Borio *et al.* 2002, Bronze *et al.* 2002) make this virus an important human pathogen.

Other bunyaviruses having less human impact known to circulate in Europe include: Tahyna- (*Orthobunyavirus*), Batai- (*Orthobunyavirus*), Inkoo- (*Orthobunyavirus*), Lednice- (*Orthobunyavirus*), Toscana- (*Phlebovirus*), and Uukuniemi- (*Phlebovirus*) viruses (Lundstrom 1999, Niklasson and Vene 1996). Most orthobunyaviruses are transmitted by mosquitoes, mostly *Aedes* species (e.g. *Aedes trivittatus* Coq. and *Aedes albopictus* (Skuse)), but other species may also be competent. Tahyna virus infection causes fever and respiratory symptoms and sometimes also central nervous system involvement (Bardos *et al.* 1975, Bardos *et al.* 1980). It is present in most countries of central and southern Europe (Table 1). *Aedes vexans* Meigen and *Ochlerotatus caspius/dorsalis* Meigen are the most important vectors for Tahyna virus (Lundstrom 1999, Medlock *et al.* 2007). Batai virus (also known as the Calovo virus) was first isolated in Slovakia in 1960 and has subsequently been found in southern, central, and northern Europe (Francy *et al.* 1989) (Table 1). So far, infection with Batai virus has not been associated with clinical disease in humans. The primary vector for Batai virus appears to be mainly *Anopheles maculipennis* L. and *Anopheles claviger* Meigen, but the virus has also been isolated from several *Aedes* and *Culex* species (Brummer-Korvenkontio and Saikku 1975, Francy *et al.* 1989). Few studies have been conducted on the epidemiology and ecology of the Tahyna and Batai viruses. One study conducted in 1993, showed high sero-prevalence to Tahyna and Batai virus in roe deer (*Capreolus capreolus* L.), wild boar (*Sus scrofa* L.), brown hare (*Lepus europaeus* Pallas), and European rabbit (*Oryctolagus cunniculus* L.) (Bardos 1975, Simkova 1962b), suggesting a possible role of these animal species as vertebrate hosts. Antibodies to Tahyna virus hasve also been found in several birds (sparrows, ducks, swallows, and martins) (Simkova 1962a).

Inkoo virus is found in northern Europe, and is transmitted by *Ochlerotatus communis* De Geer and *Ochlerotatus punctor* Kirby. Despite the high seroprevalence observed in the population of Finland and Sweden (Brummer-Korvenkontio *et al.* 1973, Francy *et al.* 1989, Niklasson and Vene 1996), infection with Inkoo virus has not been associated with disease in humans. There is some evidence from a Russian study that Inkoo virus can cause encephalitis (Lundstrom 1999) in humans. The Lednice virus has been reported only in *Culex modestus* from former Czechoslovakia and Romania, with no reported transmission to humans (Lundstrom 1999, Malkova *et al.* 1972). The epidemiology and ecology of Inkoo and Lednice virus are largely unexplored. Specific sampling of potential vectors and bird species for virus detection and characterisation will help to elucidate the emerging risk of these orthobunyaviruses in Europe. It is noteworthy that re-assortment of bunyaviruses may lead to the emergence of more virulent variants, as was shown recently (Briese *et al.* 2006, Yanase *et al.* 2006).

Phleboviruses are mainly transmitted by phlebotomine sandflies (*Phlebotomus perniciosus* Newstead and *Phlebotomus papatasi* (Scopoli)) (Dohm *et al.* 2000), but can also be transmitted by mosquitoes (e.g. Rift valley fever virus; RVFV) and ticks. Phleboviruses are a major cause of encephalitis in humans in Mediterranean countries and have a major impact on livestock. The transmission cycle of phleboviruses in nature is largely unexplored. Three members of the genus *Phlebovirus* have been recognised to cause disease in humans: RVFV, Toscana virus (TSV) and Sandfly fever virus (SFV), whereas the Uukuniemi virus (UUKV; only detected in West Germany) has not been associated with human disease (Charrel *et al.* 2007, Di Nicuolo *et al.* 2005, Kuhn *et al.* 2005, Lundstrom 1999, Sanbonmatsu-Gamez *et al.* 2005). RVFV and SFV have not yet been detected in Europe, while TSV has been found in Italy, Spain, Portugal, France, Greece, Cyprus, Turkey, and Slovenia (Charrel *et al.* 2005). Infection with SFV results in encephalitis, especially in children (Dionisio *et al.* 2003), while TSV is a frequent cause of aseptic meningitis in children in central Italy (Valassina *et al.* 1998, 2000) and a rare cause of meningoencephalitis in the elderly (Dionisio *et al.* 2001, Kuhn *et al.* 2005).

Togaviridae

The family *Togaviridae* consists of two genera, the *Alphavirus* and the *Rubivirus*, with the genus *Alphavirus* containing all the vector-borne (mainly mosquito-borne) viruses. The Sindbis virus (SINV; *Alphavirus*), also known as Ockelbo virus, is a mosquito-borne virus found in Africa, Australia and Europe (Niklasson and Vene 1996). No other alphavirus has been reported in Europe. SINV is maintained in bird populations by ornithophilic mosquitoes (*Culex torrentium* Martini and *Culiseta morsitans* Theobald and to a lesser degree *Aedes* species). *Aedes cinereus* Meigen is the main vector for human transmission from viraemic birds to humans (Medlock *et al.* 2007). Several birds are considered amplifying hosts for SINV, such as redwing (*Turdus iliacus* L.), fieldfare (*Turdus pilaris* L.), blue tit (*Parus caeruleus* L.), chaffinch (*Fringilla coelebs* L.) and song thrush (*Turdus philomelos* Brehm). Human infections with SINV result in a disease characterised by polyarthritis, similar to Ross River virus and Chikungunya-virus induced polyarthritis (Laine *et al.* 2004).

Reoviridae

The family *Reoviridae* is divided into 11 genera and only members of the genus *Orbivirus*, *Coltivirus* and *Seadornavirus* are transmitted by vectors. Two members of this family are known to circulate in Europe, the Tribec virus (TRBV, genus *Orbivirus*), and the Eyach virus (EYAV, genus *Coltivirus*). Orbiviruses are vector-borne pathogens, transmitted by ticks, mosquitoes, and midges (Attoui *et al.* 2005). TRBV is known to circulate in Russia, former Czechoslovakia, and Italy and infection with this virus is associated with development of meningoencephalitis and polyradiculitis. EYAV circulates in several European countries, namely former Czechoslovakia, France and Germany (Attoui *et al.* 2005, Hassler *et al.* 2003). Patients infected with EYAV may develop meningoencephalitis, reminiscent of the disease seen in TBEV-infected individuals. Both viruses are transmitted by ticks, mainly *Ixodes ricinus*, although EYAV has also been isolated from *Ixodes ventalloi* Gil Collado (Chastel *et al.* 1984). The vertebrate hosts involved in the transmission cycle of both viruses are not clear, but the European rabbit (*Oryctolagus cunniculus*) and the European brown hare (*Lepus europaeus*) may be potential hosts for EYAV and TRBV (Attoui *et al.* 2005, Dobler *et al.* 2006). Whether autochthonous circulation in Europe has been established for both viruses is not known.

Orthomyxoviridae

The family *Orthomyxoviridae* consists of four genera: *Influenza virus A*, *Influenza virus B*, *Influenza virus C*, and *Thogotovirus*. The genus *Thogotovirus* consists of two tick-borne viruses, the Thogoto virus (THOV), which has six RNA segments and the Dhori virus (DHOV), which has seven RNA segments (Kuno *et al.* 2001). Influenza viruses are among the best-studied human pathogens, but little is still known about the tick-borne members of this family. THOV was first isolated in Thogoto forest near Nairobi, Kenya, and has since been found in several tick species and vertebrate hosts in Central Africa, Egypt, Middle-East, Italy and Portugal (Filipe and Calisher 1984, Srihongse *et al.* 1974). THOV has been associated with febrile encephalic disease in humans, including encephalitis and meningoencephalitis. DHOV was first isolated in 1971 from ticks collected from camels in India and the virus is now found in Russia, Egypt, and Portugal (Filipe and Casals 1979, L'Vov *et al.* 2002). THOV is transmitted by ticks (*Rhipicephalus appendiculatus* Neumann) and can be maintained intra-stadially (Jones *et al.* 1990, Kaufman and Nuttall 2003, Wang and Nuttall 2001). It is not clear which vertebrate hosts play a role in the transmission cycle of THOV in nature. DHOV has been isolated from hares (*Lepus europaeus*) and their ticks *Hyalomma marginatum* Koch in the middle zone of the Volga delta, in the Astrakhan region, suggesting a role for these species in the transmission cycle of DHOV (Butenko *et al.* 1987).

Factors contributing to the emergence of vector-borne diseases

The emergence of vector-borne diseases is dependent on several factors that influence the relationships between host, pathogen, and vector (Figure 3). Several reviews have specifically addressed the effect of these individual components and the risk for emergence of infectious diseases (Chan *et al.* 1999, Epstein 2001, Harvell *et al.* 1999, Hunter 2003, Koelle *et al.* 2005, Patz *et al.* 2005, Patz and Olson 2006, Sutherst 2004, Taylor 1997, Weiss and McMichael 2004). Although the aim of this review is not to describe all the factors that determine the emergence of vector-borne diseases in detail, we will briefly touch upon some of the components that can increase the risk for disease emergence.

Vector and reservoir components

The risk of an outbreak of a vector-borne disease is dependent, among other things, on factors that influence the density of the vector, the animal reservoir, and humans. As the density of vectors and reservoir hosts increases, the probability of human contact with infected vectors increases proportionally. Several factors are believed to influence the presence and density of vector populations.

Clearly, for the emergence of a vector-borne disease into a new geographical area, a competent vector must be present. Modern transportation facilitates movements of animals and merchandise,

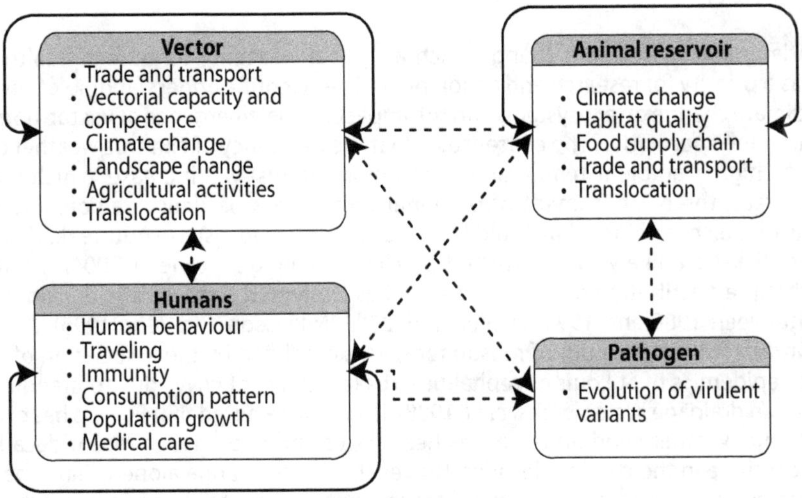

Figure 3. Interrelations between factors that influence the ecologic balance between vector, reservoir, host, and pathogen. Several factors can influence emergence of vectors and availability of an animal reservoir (boxes), or influence their density (loop).

thereby increasing the risk of introducing pathogens or vectors into new regions. This is exemplified by the introduction of exotic mosquito species in the United States (e.g. *Aedes albopictus*) (Foley *et al.* 1998, Francy *et al.* 1990) and Europe (Italy, e.g. *Aedes albopictus* and *Aedes atropalpus* Coquillet; and France, e.g. *Ochlerotatus japonica* Theobald) (Benedict *et al.* 2007, Knudsen *et al.* 1996). The increased international trade in exotic reptiles, which are frequently infested with ticks, also increases the chance of introducing exotic ticks to new geographic locations, as exemplified by the introduction of *Aponomma* and *Amblyomma* species into the United Kingdom (Burridge *et al.* 2002, Mertins and Schlater 1991, Pietzsch *et al.* 2006). Although most tick species found on reptiles are found in tropical regions and have particular preference for reptilian hosts, some of these species have the ability to establish a transmission cycle in more temperate zones (Reeves *et al.* 2006). Whilst the public health risk of reptile ticks is considered to be low, the fact that some tick species have been associated with human infections or have been shown to function as a reservoir for Russian spring-summer encephalitis virus underlines the risk of importation of exotic wild animals and the emergence of vector-borne diseases.

Ticks are usually found on a wide range of bird species, especially thrushes, sparrows and ground-foraging birds (Comstedt *et al.* 2006, Morshed *et al.* 2005, Schwan *et al.* 2007). Therefore, migratory birds may play an important role in disseminating ticks into previously uninfected areas. This is exemplified by data suggesting that seabirds from colonies in the southern and northern hemisphere can disperse *B. burgdorferi*-infected ticks trans-hemispherically (Reed *et al.* 2003). Additional work is warranted to explore the role of birds as vehicles for disseminating ticks across countries and continents.

Many mosquito and tick vectors have been shown to support the replication of viruses, but may not transmit the virus themselves and are therefore considered incompetent. Vector competence is influenced by intrinsic factors such as genetic make-up and host feeding preferences (Beerntsen *et al.* 2000). However, even if a particular vector is competent to transmit a pathogen, if not present temporally and spatially with the reservoir, the vector will not be suitable to start the transmission cycle.

The potential impact of climate change, such as global warming, on human health has been identified as a priority for research and action plans. The scientific understanding of the potential effects of climate change on ecosystems and the impact on the emergence of vector-borne disease remain scarce and speculative. The consensus is that the frequency of extreme weather conditions (fewer frosts, higher ambient temperatures and more floods) will increase, which will expand vegetation areas, the establishment of potential vectors in a particular region, and increased vector densities, all conditions that could favour a vector-borne disease outbreak. Cold-blooded vectors (e.g. ticks) are likely to be affected by climate change (Sutherst 2004). There is some evidence that the distribution of ticks in Sweden has expanded consistent with climate changes observed between 1980 and 1994 (Daniel *et al.* 2003, Niklasson and Vene 1996). Also extreme dryness may affect the behaviour of mosquitoes, as exemplified by the association of decreased rainfall with epidemics of St Louis encephalitis virus (a relative of WNV) due to breeding of *Culex pipiens* in urban drainage systems (Tasi *et al.* 1988). It should be noted that Europe has experienced several extreme weather conditions such as heat waves and floodings in recent decades, which is likely to continue in the coming decades. However, climate change alone will not be sufficient to affect the emergence of vector-borne diseases, but is likely to do so in concert with other ecological and demographical changes.

Other factors that may influence the ability of a vector to establish a transmission cycle in urbanised settings are change of landscape and agricultural practices. Increase in the world's human population drives intensification of agriculture and the need for expanding the number of farms. The latter is usually associated with deforestation and the need to create storage for drinking water and irrigation. These developments lead to the creation of potential breeding sites for mosquito vectors. In addition, deforestation leads to fragmentation of habitats, loss of biodiversity, and alteration of existing sylvatic cycles. For instance, yellow fever virus was maintained in a sylvatic transmission cycle involving monkeys and mosquitoes in both Africa and the Americas. Large outbreaks occurred when the transmission cycle was bridged in peri-urban and urban settings as a result of habitat fragmentation. In the past years, several cases of yellow fever have been reported in urban areas of South American countries, which are infested with *Aedes aegypti* (L.) (Gubler 2004, Petersen and Marfin 2005), the main vector for the virus. Therefore, it may only be a matter of time before urban yellow fever epidemics occur, with subsequent spread to other regions, with a pattern similar to the spread of dengue virus in the past (Bryant *et al.* 2003).

The reservoir population density is influenced by characteristics of the biotic environment such as habitat quality and their place in the food chain, which in turn are influenced by environmental factors. Therefore, the same factors that affect the vectors will affect the reservoir component. It is important to realise that although the geographical distribution of potential reservoir hosts may be indicative of the risk of disease emergence, the prevalence of infection within animal reservoirs and vector population(s) may be dynamic, varying among different biotic habitats. Much can be learned from ecological studies of rodent reservoirs involved in zoonotic diseases (Mills and Childs 1998).

Another way to influence the reservoir component is by the translocation of wildlife in biotic environments for conservation, agriculture, and hunting purposes. Translocation of wildlife animals in closed biotic environments introduces a risk of exposing resident wildlife species to 'new' pathogens, but may also establish an animal reservoir population that forms the basis for disease emergence (Woodford and Rossiter 1993).

Host and pathogen components

Human behaviour is another factor likely to influence the emergence of vector-borne diseases in humans. For example, in times of favourable weather, there is an increase in outdoor recreational activity, resulting in more exposed individuals. This has been shown to increase the incidence of diseases associated with tick and mosquito transmission. It should be emphasised that emergence of a vector-borne virus in a previously unaffected population becomes obvious only when outbreaks occur in animals or humans. Changes in host susceptibility through aging or immunosuppression can also contribute to the emergence of new pathogens and diseases (Harrus and Baneth 2005, Morris and Potter 1997, Ross *et al.* 2000, Van Loveren *et al.* 2000).

The majority of emerging diseases identified in the past decades are caused by viruses. The predilection of viruses, especially RNA viruses, to cause emerging infectious diseases can be explained by higher mutation rates and shorter generation times as compared to other pathogens such as Rickettsia and bacteria. Viruses able to infect multiple reservoir hosts in more than one taxonomic order seem to have a higher probability to cause disease in animals and humans. A possible explanation for this is that multi-host infection could increase the selection pressure for more virulent virus strains, better adapted to the reservoir host or to humans (Woolhouse *et al.* 2001).

Tick-borne encephalitis virus emergence analysed in relation to several factors

In the last decade, a marked increase of TBE cases has been observed in several European countries (Daniel *et al.* 2003) (Figure 4). The most dramatic changes in TBE-incidence were observed in Latvia, Lithuania, Estonia, Poland, Switzerland, Sweden and Finland, while incidences in Germany, Slovakia, and the Czech Republic have increased more modestly. In 1997, the first cases of TBE were seen in Norway (Skarpaas *et al.* 2004) and since then have increased steadily. In contrast to the general trend observed in Figure 4, the reported incidence in Austria has decreased over time. Several of the factors described above could have contributed to the incidence picture as depicted in Figure 4.

The development of *Ixodes ricinus* populations is dependent on the availability of three vertebrate hosts; in each stage of the development cycle, ticks feed once on a vertebrate host before they moult into the next stage. The stage-to-stage development may take weeks to months to complete and its rate is temperature dependent. Humidity also influences tick activity and survival. Extensive dryness or very low temperatures are limiting factors for tick development and activity.

It has been proposed that the relatively warmer weather in spring and autumn observed in Sweden has allowed longer activity seasons for both ticks and humans, which may have contributed to the spread of ticks and TBEV to new parts of Sweden (Johan *et al.* 2006), but may also explain the recent appearance of cases of TBE in Norway (Skarpaas *et al.* 2006, 2004). On the other hand, the downside effect of extreme high temperatures is an increase of the mortality rate of ticks (Randolph 2002). Increase in TBE incidence is directly associated with the infection rate within the tick population, which depends on the degree of seasonal overlap between larval and nymphal ticks. In order to increase the infection rate of ticks, the relatively few infected nymphs must transmit the virus to the more numerous larvae during the co-feeding on rodents, a process that is associated with ground temperature (Randolph 2002). Furthermore, it is known that the density of deer has an important impact on the tick population. No studies have been conducted today to show whether climate change significantly affects the abundance of the vertebrate hosts involved in the development cycle of *Ixodes ricinus*. Although decreases in TBE incidences may be ascribed to for instance an effective vaccination campaign against TBEV in Austria, or to extreme high temperatures in Slovenia, Croatia and Hungary, proper meta-analysis of the available data and multifactorial studies related to TBE incidence are lacking.

Human behaviour is another element that contributes to emergence of vector-borne diseases. Altered patterns in human behaviour, characterised by change in agricultural practices and expansion of working areas into tick-infested habitats usually results in an increase of vector-borne diseases (Randolph 2002). Alternatively, improvements of TBEV diagnostics, surveillance, vaccination, and more awareness of the risk factors related to recreational areas and the subsequent avoidance behaviour could all influence the incidence of TBEV in Europe.

Management strategies to prevent and deal with vector-borne diseases

As discussed above, when all ecological determinants are present, emergence of a vector-borne disease is determined by the relationship between vectors, wildlife animal reservoirs, and humans.

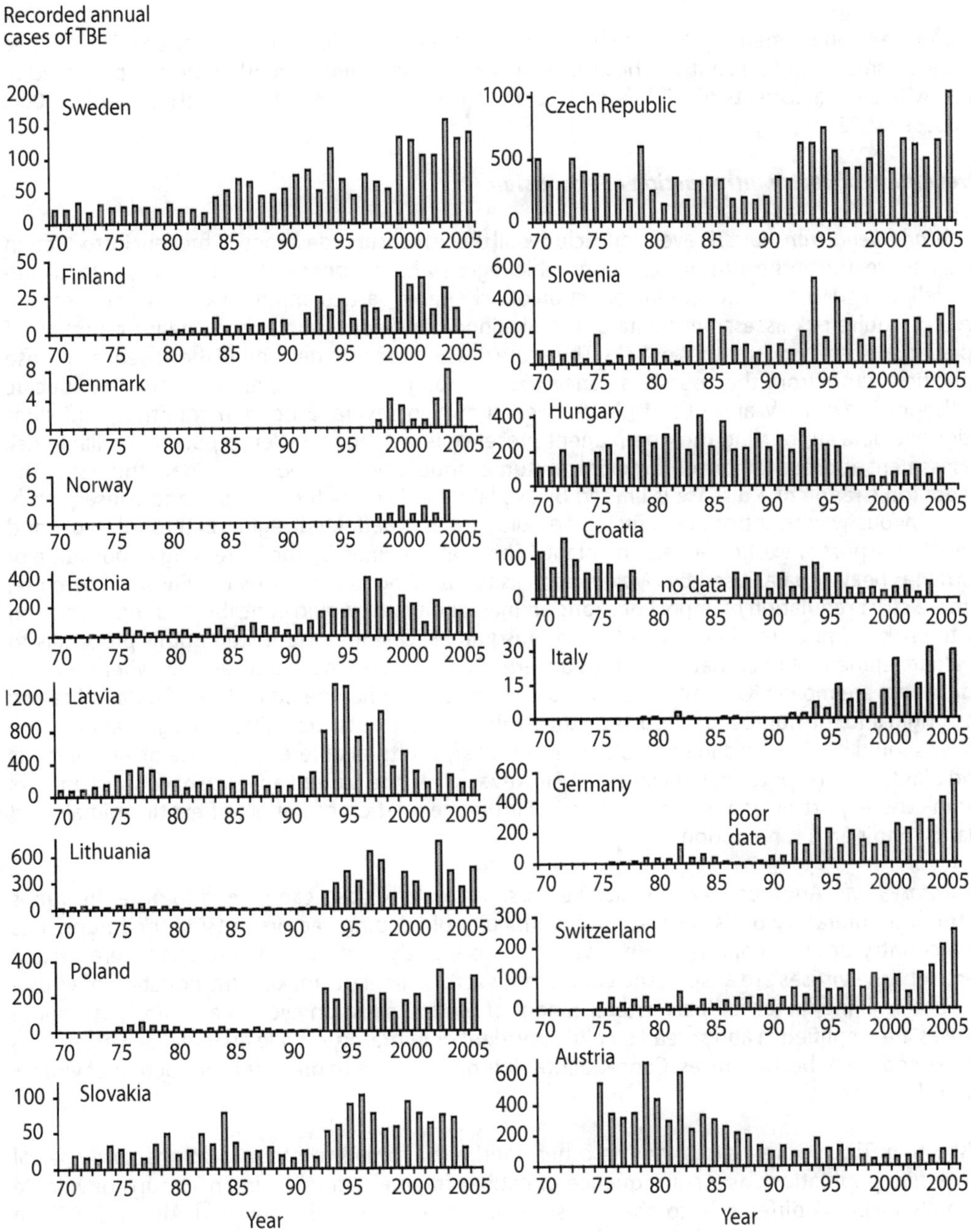

Figure 4. Changes in incidences of TBEV infections in several European countries (taken from Randolph and Šumilo (Chapter 11).

Therefore, management strategies should be designed to achieve (1) prevention, (2) real-time risk assessment, and (3) control. These aspects may be combined with other models proposed to deal with several aspects of wildlife disease and their emergence in humans (Heisey *et al.* 2006, Wobeser 2002)

Prevention of vector introduction or expansion

Legal and illegal transports: Prevention includes all those measures designed to prevent introduction of a new vector-borne virus into a geographic region where it does not occur. In order to design guidelines and rules to prevent introduction of ticks into a particular region, countries concerned would require risk assessment data of the likelihood and consequences of the introduction of specific species of mosquitoes or ticks. Therefore it is necessary to determine the absence of those specific vectors from the countries concerned. Although many competent vectors for specific pathogens (e.g. WNV and CCHFV) have been shown in several European countries, sufficient scientific data supporting their permanent presence is lacking. In order to pursue a reliable risk assessment approach, more entomological studies should be devoted to elucidate this issue. The geographic range of a disease is limited by availability of competent vectors, and consequently one obvious way to introduce infected vectors into an unaffected region is through legal and illegal transport of exotic animals and plants. This is exemplified by the increasing importation of ostriches heavily parasitised by *Hyalomma* ticks to breeding farms in Western Europe. Therefore, a more strict regulation of import of exotic animals and plants would contribute to the restriction of the risk of importing infected ticks and mosquitoes. One important step in this process is to evaluate epidemiological data from the concerned country and the countries from where animals and plants are moved. Quantitative risk assessment is often hampered by lack of data indicating the origin and the number of imported exotic animals and plants. In addition to legalisation, more effort should be deployed in understanding social and cultural differences in use of wild animals and plants and the potential risk involved. However, given the current absence of country borders for in- and export of goods and services in Europe, regulation of import of exotic animals and plants in this area is no option.

Migratory birds: Another way by which vectors, particularly ticks, can spread to other locations is through migratory birds, indicating the difficulty of keeping vectors outside the boundaries of a country or even continent. However, it is more likely that in many regions where certain vector-borne viruses are absent, the competent vectors circulate, making introduction of virus a key determining factor for local establishment of the transmission cycle. Many mosquito-borne viruses are amplified in and spread by birds (resident or migratory), while mammalian vertebrates spread most tick-borne viruses. Consequently, it is not possible to prevent introduction of viruses by birds.

Dispersion: Tick-borne viruses, on the other hand, may spread to other locations by dispersal of wildlife populations as a consequence of habitat fragmentation. Different groups of wildlife animals respond differently to the pressures of urbanisation (Bradley and Altizer 2007). In general, animals that required the ecological environment, disrupted by urbanisation, to survive (specialists) will disperse to other locations. Deforestation can even eliminate specialists from a region. In contrast, animals that do not require the special ecological conditions provided by the disrupted habitat will adapt and become abundant. As the survival conditions in the 'new' habitat deteriorate (decreased food sources) generalists migrate to more favourable areas. Therefore, strategies should be designed to manage urbanisation, by developing explicit risk assessments of habitat fragmentation. Creating a buffer zone between wildlife habitat and urban areas, e.g. deforested, uninhabited corridors, may be a strategy to reduce opportunities of contact between vector, animal reservoir, and humans.

Translocation: Translocation is defined as the movement of living organisms from one area for free release to another. Throughout the world, increasing numbers of native and exotic mammals, birds, and fish, some of them being endangered species, are translocated every year for sporting or reintroduction purposes. Potential health risks are associated with the translocation of wild animals in that they can introduce infectious agents and ectoparasites. This is exemplified by the introduction by elk (*Cervus elaphus* L.) of liver flukes (*Fascioloides magna* Bassi) to Italy, the abomasal nematodes *Spiculopteragia spiculoptera* Guschanskaia to Argentina and *Onchocerca cervipedis* Wher and Dikans to Poland (Corn and Nettles 2001). Several tick species have been recovered from elks, including *Ixodes ricinus, Ixodes persulcatus, Dermacentor silvarum* Olenev and *Dermacentor reticulatus* Fabr. (Corn and Nettles 2001). Therefore, when translocation of wild animals is considered, a protocol should be followed to reduce the risk of introduction of both pathogens and ectoparasites. Important aspects of such a protocol should include:

1. Evaluation of the health status of the animals to be translocated. In this regard, relevant samples of these animals should be collected and tested for evidence of specified infections by serologic, PCR, and/or culture techniques. Priority should be given to PCR and culture methods in order to detect the presence of active infection.
2. Quarantine measures should be implemented. Guidelines should be designed, which are specifically linked to the pathogen of interest. Although availability of sensitive and specific tests may reduce the need for long-term quarantine, it is recommended to implement quarantine measures to reduce the risk of introducing pathogens that were not included in the testing protocol.
3. Restriction of geographical areas for wildlife introduction. Designation of such areas requires knowledge of the kind of competent vectors present in that area and the pathogens that they may harbour. A quantitative risk assessment must be conducted based on epidemiological and ecological studies of the area designated for translocation of wild animals.
4. Introduction of ectoparasites can be prevented by treatment of animals with parasiticides. However, due to limited first-hand experience with these products on or in many vertebrate hosts, their efficacy and safety must be checked for each individual situation.

Dealing with global warming

As discussed above, climate change may lead to geographic range shifts and large changes in prevalence of vector-borne diseases because temperature may affect vector development, distribution, density as well as transmission rates. Much of the climate change we have observed to date has been attributed to human activity. This view has been contradicted by other sources (Keller 2007). Be it as it may, to reduce global warming we should consider to fundamentally transform the way we power our global economy, shifting away from an established legacy of unrestrained use of fossil fuels to more efficient and renewable sources of energy. Such a transformation would require society to engage in a converged effort to design and implement action plans to reduce gas emissions. This will be extremely difficult to achieve.

Real-time risk assessment

Once ecological convergence of vector, animal reservoir, and humans occur, incidental cases of human infections may occur. However, these isolated cases are usually missed, as the causative virus will not be incorporated in the differential diagnosis. Consequently, emerging viruses are only recognised during outbreaks in humans or animals. Although predicting precisely when, where, or how a virus will emerge is almost impossible, planning for it is essential, as exemplified by the pandemic threat of avian influenza viruses (Daems *et al.* 2005, Ducatez *et al.* 2006). Clearly, an adequate surveillance system is the cornerstone of any preparedness strategy. Therefore, setting up manageable surveillance systems could provide qualitative and quantitative information about the presence, distribution, and intensity of pre-defined vector-borne viruses. To improve the detection power of surveillance programmes and consequently their predictive value, it is important that local parties collaborate intensively to share and combine information. The type of surveillance system is likely to vary between geographical locations, depending on ecological conditions, list of viruses posing a threat to a particular region, and presence of specific vectors and hosts. Figure 5 summarises a multi-level surveillance system, consisting of different surveillance strategies, which may be implemented as a whole or a part thereof, depending on the need, specific expertise and resources available. Two surveillance strategies can be used to monitor for presence of virus in a particular geographic area; passive and active strategies. Passive strategies rely on the reporting of sick or dead animals for testing, whereas active strategies involve testing of samples of live animals. This combination of surveillance strategies has been used for detection of WNV in the USA. The same system could be adopted to set up a surveillance system for other mosquito- and tick-borne diseases.

Dead bird surveillance: Dead bird surveillance has proved to be the most sensitive method for detecting circulation of WNV in the USA. In the summer of 1999, WNV emerged in New York City, killing thousands of bird species pertaining to 17 orders. Especially members of the order Passeriformes, family *Corvidae,* have been shown to be highly susceptible to WNV infections. Dead bird surveillance consist of two components: reporting of sightings of dead birds and testing of samples of dead birds for presence of WNV by polymerase chain reaction (PCR), virus isolation, or antigen detection. The success of this method is dependent on the variability in detecting sick or dead animals, which is related to bird mobility, public awareness and interest to report bird mortality. Therefore, dead bird surveillance outside an outbreak situation is less sensitive when the goal is identifying the unknown.

Live bird surveillance: Live bird surveillance has not been used consistently to monitor WNV activity. This approach requires the capture and sampling of free-living wild birds to subsequently test them by PCR or serology. Alternatively, captive birds (e.g. chickens) have been used successfully as sentinels for detection of WNV circulation in the USA and UK (Buckley *et al.* 2006, Komar 2003). The advantage of using free-living birds over captive ones is that they move around, increasing the chance of getting infected. Screening of migratory birds would allow a better risk assessment of potential viruses being introduced into an unaffected area and therefore allow a more targeted approach towards preparedness. However, since free-living birds are unlikely to be recaptured, seroconversion will easily be missed, and a seropositive status may not provide information about the time point of infection, making this method less sensitive than the sentinel. Active surveillance of captive birds for a predefined list of mosquito-borne viruses such as WNV represents a powerful tool to monitor changes in a situation. It should be emphasised that an appropriate avian sentinel should meet the following criteria:

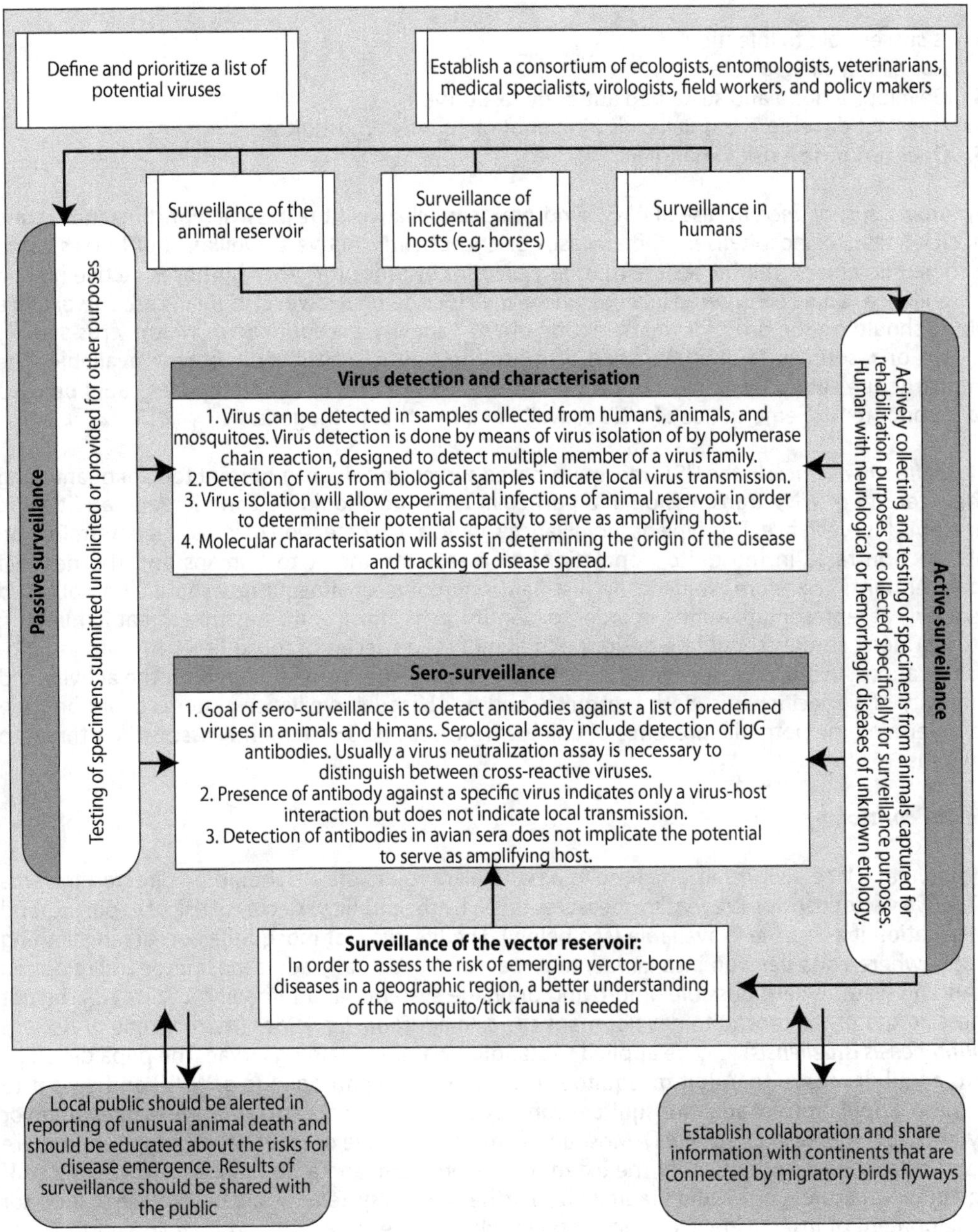

Figure 5. Surveillance flowchart depicting the methods used for passive and active surveillance strategies.

1. Is susceptible to infection.
2. Has a high survival rate.
3. Develops a good and sustained antibody response.
4. Does not develop viraemia levels high enough to infect mosquitoes.
5. Does not pose a risk to handlers.

Syndromic surveillance: In case of WNV, syndromic surveillance is the passive reporting and testing of sick horses or hospitalised human cases of encephalitis. Active surveillance of horses could also be implemented if horses are routinely bled for instance for other purposes. Active human surveillance can be considered in areas where WNV is endemic. However, human case surveillance alone should not be used for the detection of WNV activity, except in areas where virus activity is rare or resources to support avian-or mosquito based surveillance are not available. The infrastructure set up by many countries to screen blood for transfusion purposes could be used to monitor for the emergence of WNV as well as many other pathogens.

Mosquito surveillance: In the USA, mosquito-based surveillance is an important tool for quantifying the intensity of WNV transmission in a particular area and should be conducted in addition to dead-bird surveillance. This type of surveillance should allow detection of virus and determination of infection rates in mosquito populations to assess the threat to humans and the need of interventions. Therefore, statistically justifiable numbers of mosquitoes should be collected regularly at representative sites, in order to identify areas at risk and assess the threat to humans. Furthermore, studies should be designed to identify the species of mosquitoes present at high-risk locations. In analogy, tick-based surveillance could be performed to monitor the activity and intensity of a specific virus. Both mosquito- and tick-based surveillance systems could be used to determine the potential presence of unknown viruses and subsequently assess their threat to humans.

Vector control

Mosquitoes: Once a virus has emerged in a particular region, efforts should be directed towards effective vector control, preventive measures taken by the public to decrease risk of exposure, and vaccination if a vaccine is available (see below). The life cycle of mosquitoes occurs in standing water, where eggs develop into larvae, to pupae, and eventually adult mosquitoes. Eliminating standing water where possible, improving drainage systems, and removing debris can disrupt these stages of the mosquito development cycle. In addition, larvicides (methoprene or *Bacillus thuringiensis israeliensis*) may be applied to standing water-containing larvae. The pupa develops into an adult mosquito. Adult mosquitoes can acquire the pathogen (e.g. WNV) and pass it to humans. Populations of adult mosquitoes can be controlled by spraying adulticides (malathion or pyrethroids) (see also Chapter 21). Mosquito control (proactive or reactive) programmes require a surveillance system to identify the extent of the problem, and a permanent well-trained staff and infra-structure (Reisen and Brault 2007). Furthermore, a system should be in place to monitor development of insecticide resistance. Insects exhibit genetic variation in their susceptibility to insecticides and their use exerts a selection pressure that favours the survival of resistant strains. For instance, since the late 1940s, *Aedes aegypti* resistance to a range of insecticides has been documented from many parts of the world (Brown and Pal 1971, Chareonviriyahpap *et al.* 1999), affecting the control of viruses transmitted by this mosquito. Nevertheless, mosquito control has been shown to be an effective way to largely control WNV epidemics in the USA. Proactive mosquito control programmes focus on preventing human infection by suppressing mosquito populations below the levels needed to start an epidemic. Reactive control depends on the timely diagnosis and reporting of human and animal cases of WNV. In addition, the public is usually advised to take personal protective measures such as use of mosquito repellents, or wear long-sleeved clothing.

Ticks: Reducing the burden of tick-borne diseases can be accomplished by reducing the abundance of ticks. Basically, tick abundance can be controlled chemically or biologically. Although ticks are not insects, use of several insecticides has proven highly effective in reducing tick numbers. These insecticides can be delivered to the environment where ticks infest hosts and undergo diapause or can be applied directly to the host on which ticks feed. It is of paramount importance, however, that the nature and extent of undesirable side effects of insecticide use on other vertebrates and invertebrates are analysed before long-term use of these agents is considered. Targeting wildlife hosts of ticks with pesticides poses many challenges due to the variety of vertebrate hosts for many tick species and their ability to disperse. For instance, carbaryl and chlorpyrifos are effective agents for tick control, but are highly toxic for many vertebrates (George *et al.* 2004), while some synthetic pyrethroids are lethal to many invertebrates (Croft and Jung 2001, Eisler *et al.* 2003). These disadvantages of broadcast chemical insecticides have led to the search for biological alternatives. Biological control of ticks is based on the use of natural enemies of ticks such as birds or micro-organisms to reduce tick numbers (Samish 2000). Several biological systems seem effective under experimental conditions, but more systematic studies are needed to assess their usefulness. Overall, it is important that summer camps and other recreational areas are treated with anti-tick chemicals, as well as educating the population, particularly those working in yards and dealing with animals about the methods of protection against ticks.

Vaccination

Humans: Emergence of a vector-borne disease will be triggered by viruses to which the general population has little or no immunity. Therefore, vaccination of the susceptible population would be an obvious way to control outbreaks of vector-borne diseases. While the production and use of antivirals is an important option to consider, it is an impractical way to protect the population at large. However, with some exceptions, no effective or licensed vaccines are available against most of the medically important vector-borne viruses. Considering the relatively limited geographic distribution of most known vector-borne viruses and the risk of producing a vaccine for which no market exists, makes development and production of such vaccines unattractive to commercial companies. Therefore, in the interest of public health, incentive systems should be coordinated through stakeholder partnerships and where necessary and possible, intercontinental collaborations should be established.

Animal vaccination: Animal vaccination has been used as a measure to protect horses against for instance WNV. Mass vaccination of horses in the western USA resulted in a reduction in the numbers of equine WNV cases from 15,000 in 2002 to 1,200 in 2005 as the WNV epicentre moved into the west (Reisen and Brault 2007). Vaccination of animals with the aim of protecting humans is also a strategy worth considering. For example, vaccination of sheep against LIV could help reduce the number of infections occurring in animal handlers.

References

Ahmeti S and Raka L (2006) Crimean-Congo haemorrhagic fever in Kosova: a fatal case report. Virol J 3: 85.

Antoniadis A and Casals J (1982) Serological evidence of human infection with Congo-Crimean hemorrhagic fever virus in Greece. Am J Trop Med Hyg 31: 1066-1067.

Attoui H, Mohd Jaafar F, de Micco P and de Lamballerie X (2005) Coltiviruses and seadornaviruses in North America, Europe, and Asia. Emerg Infect Dis 11: 1673-1679.

Bakhvalova VN, Dobrotvorsky AK, Panov VV, Matveeva VA, Tkachev SE and Morozova OV (2006) Natural tick-borne encephalitis virus infection among wild small mammals in the southeastern part of western Siberia, Russia. Vector-Borne Zoon Dis 6: 32-41.

Bakonyi T, Hubalek Z, Rudolf I and Nowotny N (2005) Novel flavivirus or new lineage of West Nile virus, central Europe. Emerg Infect Dis 11: 225-231.

Bakonyi T, Ivanics E, Erdelyi K, Ursu K, Ferenczi E, Weissenbock H and Nowotny N (2006) Lineage 1 and 2 strains of encephalitic West Nile virus, central Europe. Emerg Infect Dis 12: 618-623.

Banerjee K, Gupta NP and Goverdhan MK (1979) Viral infections in laboratory persnnel. Indian J Med Res 69: 363-373.

Bardos V (1975) The role of mammals in the circulation of Tahyna virus. Folia Parasitol (Praha) 22: 257-264.

Bardos V, Medek M, Kania V and Hubalek Z (1975) Isolation of Tahyna virus from the blood of sick children. Acta Virol 19: 447.

Bardos V, Medek M, Kania V, Hubalek Z and Juricova Z (1980) [The clinical picture in Tahyna virus (California group) infections in children] Das klinische Bild der Tahyna-Virus (California-Gruppe) -- Infektionen bei Kindern. Padiatr Grenzgeb 19: 11-23.

Beerntsen BT, James AA and Christensen BM (2000) Genetics of mosquito vector competence. Microbiol Mol Biol Rev 64: 115-137.

Benedict MQ, Levine RS, Hawley WA and Lounibos LP (2007) Spread of the tiger: global risk of invasion by the mosquito Aedes albopictus. Vector-Borne Zoon Dis 7: 76-85.

Borio L, Inglesby T, Peters CJ, Schmaljohn AL, Hughes JM, Jahrling PB, Ksiazek T, Johnson KM, Meyerhoff A, O'Toole T, Ascher MS, Bartlett J, Breman JG, Eitzen EM, Jr., Hamburg M, Hauer J, Henderson DA, Johnson RT, Kwik G, Layton M, Lillibridge S, Nabel GJ, Osterholm MT, Perl TM, Russell P and Tonat K (2002) Hemorrhagic fever viruses as biological weapons: medical and public health management. JAMA 287: 2391-2405.

Bormane A, Lucenko I, Duks A, Mavtchoutko V, Ranka R, Salmina K and Baumanis V (2004) Vectors of tick-borne diseases and epidemiological situation in Latvia in 1993-2002. Int J Med Microbiol 293 Suppl 37: 36-47.

Bradley CA and Altizer S (2007) Urbanization and the ecology of wildlife diseases. Trends Ecol Evol 22: 95-102.

Briese T, Bird B, Kapoor V, Nichol ST and Lipkin WI (2006) Batai and Ngari viruses: M segment reassortment and association with severe febrile disease outbreaks in East Africa. J Virol 80: 5627-5630.

Bronze MS, Huycke MM, Machado LJ, Voskuhl GW and Greenfield RA (2002) Viral agents as biological weapons and agents of bioterrorism. Am J Med Sci 323: 316-325.

Brown AW and Pal R (1971) Insecticide resistance in arthropods. Public Health Pap 38: 1-491.

Brummer-Korvenkontio M and Saikku P (1975) Mosquito-borne viruses in Finland. Med Biol 53: 279-281.

Brummer-Korvenkontio M, Saikku P, Korhonen P and Oker-Blom N (1973) Arboviruses in Finland. I. Isolation of tick-borne encephalitis (TBE) virus from arthropods, vertebrates, and patients. Am J Trop Med Hyg 22: 382-389.

Bryant J, Wang H, Cabezas C, Ramirez G, Watts D, Russell K and Barrett A (2003) Enzootic transmission of yellow fever virus in Peru. Emerg Infect Dis 9: 926-933.

Buckley A, Dawson A and Gould EA (2006) Detection of seroconversion to West Nile virus, Usutu virus and Sindbis virus in UK sentinel chickens. Virol J 3: 71.

Burridge MJ, Simmons LA, Peter TF and Mahan SM (2002) Increasing risks of introduction of heartwater onto the American mainland associated with animal movements. Ann N Y Acad Sci 969: 269-274.

Butenko AM, Leshchinskaia EV, Semashko IV, Donets MA and Mart'ianova LI (1987) [Dhori virus--a causative agent of human disease. 5 cases of laboratory infection] Virus Dkhori--vozbuditel' zabolevaniia cheloveka. Piat' sluchaev laboratornoi infektsii. Vopr Virusol 32: 724-729.

Ceroni M (2006) Neurological manifestation of tick-borne encephalitis. Neurol Sci 27: 85.

Chan NY, Ebi KL, Smith F, Wilson TF and Smith AE (1999) An integrated assessment framework for climate change and infectious diseases. Environ Health Perspect 107: 329-337.

Chareonviriyahpap T, Aum-aung B and Ratanatham S (1999) Current insecticide resistance patterns in mosquito vectors in Thailand. Southeast Asian J Trop Med Public Health 30: 184-194.

Charrel RN, Attoui H, Butenko AM, Clegg JC, Deubel V, Frolova TV, Gould EA, Gritsun TS, Heinz FX, Labuda M, Lashkevich VA, Loktev V, Lundkvist A, Lvov DV, Mandl CW, Niedrig M, Papa A, Petrov VS, Plyusnin A, Randolph S, Suss J, Zlobin VI and de Lamballerie X (2004) Tick-borne virus diseases of human interest in Europe. Clin Microbiol Infect 10: 1040-1055.

Charrel RN, Gallian P, Navarro-Mari JM, Nicoletti L, Papa A, Sanchez-Seco MP, Tenorio A and de Lamballerie X (2005) Emergence of Toscana virus in Europe. Emerg Infect Dis 11: 1657-1663.

Charrel RN, Izri A, Temmam S, Delaunay P, Toga I, Dumon H, Marty P, de Lamballerie X and Parola P (2007) Cocirculation of 2 genotypes of Toscana virus, southeastern France. Emerg Infect Dis 13: 465-468.

Chastel C, Main AJ, Couatarmanac'h A, Le Lay G, Knudson DL, Quillien MC and Beaucournu JC (1984) Isolation of Eyach virus (Reoviridae, Colorado tick fever group) from Ixodes ricinus and I. ventalloi ticks in France. Arch Virol 82: 161-171.

Chvala S, Bakonyi T, Bukovsky C, Meister T, Brugger K, Rubel F, Nowotny N and Weissenbock H (2007) Monitoring of Usutu virus activity and spread by using dead bird surveillance in Austria, 2003-2005. Vet Microbiol 122: 237-245.

Chvala S, Kolodziejek J, Nowotny N and Weissenbock H (2004) Pathology and viral distribution in fatal Usutu virus infections of birds from the 2001 and 2002 outbreaks in Austria. J Comp Pathol 131: 176-185.

Comstedt P, Bergstrom S, Olsen B, Garpmo U, Marjavaara L, Mejlon H, Barbour AG and Bunikis J (2006) Migratory passerine birds as reservoirs of Lyme borreliosis in Europe. Emerg Infect Dis 12: 1087-1095.

Corn JL and Nettles VF (2001) Health protocol for translocation of free-ranging elk. J Wildl Dis 37: 413-426.

Croft BA and Jung C (2001) Phytoseiid dispersal at plant to regional levels: a review with emphasis on management of Neoseiulus fallacis in diverse agroecosystems. Exp Appl Acarol 25: 763-784.

Daems R, Del Giudice G and Rappuoli R (2005) Anticipating crisis: towards a pandemic flu vaccination strategy through alignment of public health and industrial policy. Vaccine 23: 5732-5742.

Daniel M, Danielova V, Kriz B, Jirsa A and Nozicka J (2003) Shift of the tick Ixodes ricinus and tick-borne encephalitis to higher altitudes in central Europe. Eur J Clin Microbiol Infect Dis 22: 327-328.

Danielova V (2002) Natural foci of tick-borne encephalitis and prerequisites for their existence. Int J Med Microbiol 291 Suppl 33: 183-186.

Danielova V, Holubova J, Pejcoch M and Daniel M (2002) Potential significance of transovarial transmission in the circulation of tick-borne encephalitis virus. Folia Parasitol (Praha) 49: 323-325.

Derby CD, Kicklighter CE, Johnson PM and Zhang X (2007) Chemical composition of inks of diverse marine molluscs suggests convergent chemical defenses. J Chem Ecol 33: 1105-1113.

Di Nicuolo G, Pagliano P, Battisti S, Starace M, Mininni V, Attanasio V and Faella FS (2005) Toscana virus central nervous system infections in southern Italy. J Clin Microbiol 43: 6186-6188.

Dionisio D, Esperti F, Vivarelli A and Valassina M (2003) Epidemiological, clinical and laboratory aspects of sandfly fever. Curr Opin Infect Dis 16: 383-388.

Dionisio D, Valassina M, Ciufolini MG, Vivarelli A, Esperti F, Cusi MG, Marchi A, Mazzoli F and Lupi C (2001) Encephalitis without meningitis due to sandfly fever virus serotype toscana. Clin Infect Dis 32: 1241-1243.

Dobler G, Wolfel R, Schmuser H, Essbauer S and Pfeffer M (2006) Seroprevalence of tick-borne and mosquito-borne arboviruses in European brown hares in Northern and Western Germany. Int J Med Microbiol 296 Suppl 40: 80-83.

Dohm DJ, Logan TM, Linthicum KJ, Rossi CA and Turell MJ (1996) Transmission of Crimean-Congo hemorrhagic fever virus by *Hyalomma impeltatum* (Acari:Ixodidae) after experimental infection. J Med Entomol 33: 848-851.

Dohm DJ, Rowton ED, Lawyer PG, O'Guinn M and Turell MJ (2000) Laboratory transmission of Rift Valley fever virus by *Phlebotomus duboscqi*, *Phlebotomus papatasi*, *Phlebotomus sergenti*, and *Sergentomyia schwetzi* (Diptera: Psychodidae). J Med Entomol 37: 435-438.

Ducatez MF, Olinger CM, Owoade AA, De Landtsheer S, Ammerlaan W, Niesters HG, Osterhaus AD, Fouchier RA and Muller CP (2006) Avian flu: multiple introductions of H5N1 in Nigeria. Nature 442: 37.

Dumpis U, Crook D and Oksi J (1999) Tick-borne encephalitis. Clin Infect Dis 28: 882-890.

Dunaev NG, Egorova LS and Fedorova TN (1975) [An experimental study of mixed tularemia and Omsk hemorrhagic fever infection in muskrats] Eksperimental'noe izychenie smeshannoi infektsii tuliaremii i omskoi gemorragicheskoi likhoradki u ondatr. Zh Mikrobiol Epidemiol Immunobiol: 108-112.

Ebel GD and Kramer LD (2004) Short report: duration of tick attachment required for transmission of powassan virus by deer ticks. Am J Trop Med Hyg 71: 268-271.

Eisler MC, Torr SJ, Coleman PG, Machila N and Morton JF (2003) Integrated control of vector-borne diseases of livestock--pyrethroids: panacea or poison? Trends Parasitol 19: 341-345.

Epstein PR (2001) Climate change and emerging infectious diseases. Microbes Infect 3: 747-754.

Ergonul O (2006) Crimean-Congo haemorrhagic fever. Lancet Infect Dis 6: 203-214.

Fedorova TN and Sizemova GA (1964) [Omsk Hemorrhagic Fever in Human Subjects and Muskrats During the Winter Period.] Zabolevaniia Omsko I Gemorragichesko I Likhoradko I Liude I I Ondatr V Zimni I Period. Zh Mikrobiol Epidemiol Immunobiol 41: 134-136.

Filipe AR and Calisher CH (1984) Isolation of Thogoto virus from ticks in Portugal. Acta Virol 28: 152-155.

Filipe AR and Casals J (1979) Isolation of Dhori virus from *Hyalomma marginatum* ticks in Portugal. Intervirology 11: 124-127.

Foley P, Hemsley C, Muller K, Maroske G and Ritchie S (1998) Importation of *Aedes albopictus* in Townsville, Queensland. Commun Dis Intell 22: 3-4.

Francy DB, Jaenson TG, Lundstrom JO, Schildt EB, Espmark A, Henriksson B and Niklasson B (1989) Ecologic studies of mosquitoes and birds as hosts of Ockelbo virus in Sweden and isolation of Inkoo and Batai viruses from mosquitoes. Am J Trop Med Hyg 41: 355-363.

Francy DB, Moore CG and Eliason DA (1990) Past, present and future of *Aedes albopictus* in the United States. J Am Mosq Control Assoc 6: 127-132.

George JE, Pound JM and Davey RB (2004) Chemical control of ticks on cattle and the resistance of these parasites to acaricides. Parasitology 129 Suppl: S353-366.

Gholam BI, Puksa S and Provias JP (1999) Powassan encephalitis: a case report with neuropathology and literature review. Can Med Assoc J 161: 1419-1422.

Gilbert L, Jones LD, Hudson PJ, Gould EA and Reid HW (2000) Role of small mammals in the persistence of Louping-ill virus: field survey and tick co-feeding studies. Med Vet Entomol 14: 277-282.

Glavits R, Ferenczi E, Ivanics E, Bakonyi T, Mato T, Zarka P and Palya V (2005) Co-occurrence of West Nile Fever and circovirus infection in a goose flock in Hungary. Avian Pathol 34: 408-414.

Gonzalez JP, Cornet JP, Wilson ML and Camicas JL (1991) Crimean-Congo haemorrhagic fever virus replication in adult Hyalomma truncatum and Amblyomma variegatum ticks. Res Virol 142: 483-488.

Gozalan A, Esen B, Fitzner J, Tapar FS, Ozkan AP, Georges-Courbot MC, Uzun R, Gumuslu F, Akin L and Zeller H (2007) Crimean-Congo haemorrhagic fever cases in Turkey. Scand J Infect Dis 39: 332-336.

Gresikova M, Sekeyova M, Stupalova S and Necas S (1975) Sheep milk-borne epidemic of tick-borne encephalitis in Slovakia. Intervirology 5: 57-61.

Gritsun TS, Nuttall PA and Gould EA (2003) Tick-borne flaviviruses. Adv Virus Res 61: 317-371.

Gubler DJ (2004) The changing epidemiology of yellow fever and dengue, 1900 to 2003: full circle? Comp Immunol Microbiol Infect Dis 27: 319-330.

Harrus S and Baneth G (2005) Drivers for the emergence and re-emergence of vector-borne protozoal and bacterial diseases. Int J Parasitol 35: 1309-1318.

Harvell CD, Kim K, Burkholder JM, Colwell RR, Epstein PR, Grimes DJ, Hofmann EE,

Lipp EK, Osterhaus AD, Overstreet RM, Porter JW, Smith GW and Vasta GR (1999) Emerging marine diseases--climate links and anthropogenic factors. Science 285: 1505-1510.

Hassler D, Oehme R, Kimmig P and Dobler G (2003) [Eyach virus: relative of the Colorado tick fever virus rediscovered in Baden-Wurttemberg] Eyach-Virus: Verwandter des Colorado-Zeckenfieber-Virus in Baden-Wurttemberg wiederentdeckt. Dtsch Med Wochenschr 128: 1874.

Hayes EB, Komar N, Nasci RS, Montgomery SP, O'Leary DR and Campbell GL (2005) Epidemiology and transmission dynamics of West Nile virus disease. Emerg Infect Dis 11: 1167-1173.

Heisey DM, Joly DO and Messier F (2006) The fitting of general force-of-infection models to wildlife disease prevalence data. Ecology 87: 2356-2365.

Hudson PJ, Norman R, Laurenson MK, Newborn D, Gaunt M, Jones L, Reid H, Gould E, Bowers R and Dobson A (1995) Persistence and transmission of tick-borne viruses: Ixodes ricinus and louping-ill virus in red grouse populations. Parasitology 111 Suppl: S49-58.

Hudson PJ, Rizzoli A, Rosa R, Chemini C, Jones LD and Gould EA (2001) Tick-borne encephalitis virus in northern Italy: molecular analysis, relationships with density and seasonal dynamics of *Ixodes ricinus*. Med Vet Entomol 15: 304-313.

Hunter PR (2003) Climate change and waterborne and vector-borne disease. J Appl Microbiol 94 Suppl: 37S-46S.

Hyde J, Nettleton P, Marriott L and Willoughby K (2007) Louping ill in horses. Vet Rec 160: 532.

Johan F, Asa L, Rolf A, Barbro C, Ingvar E, Mats H, Ake L, Sirkka V and Bo S (2006) Tick-borne encephalitis (TBE) in Skane, southern Sweden: A new TBE endemic region? Scand J Infect Dis 38: 800-804.

Jones LD, Davies CR, Williams T, Cory J and Nuttall PA (1990) Non-viraemic transmission of Thogoto virus: vector efficiency of *Rhipicephalus appendiculatus* and *Amblyomma variegatum*. Trans R Soc Trop Med Hyg 84: 846-848.

Jones LD, Gaunt M, Hails RS, Laurenson K, Hudson PJ, Reid H, Henbest P and Gould EA (1997) Transmission of louping ill virus between infected and uninfected ticks co-feeding on mountain hares. Med Vet Entomol 11: 172-176.

Jourdain E, Toussaint Y, Leblond A, Bicout DJ, Sabatier P and Gauthier-Clerc M (2007) Bird species potentially involved in introduction, amplification, and spread of West Nile virus in a Mediterranean wetland, the Camargue (Southern France). Vector-Borne Zoon Dis 7: 15-33.

Karti SS, Odabasi Z, Korten V, Yilmaz M, Sonmez M, Caylan R, Akdogan E, Eren N, Koksal I, Ovali E, Erickson BR, Vincent MJ, Nichol ST, Comer JA, Rollin PE and Ksiazek TG (2004) Crimean-Congo hemorrhagic fever in Turkey. Emerg Infect Dis 10: 1379-1384.

Kaufman WR and Nuttall PA (2003) *Rhipicephalus appendiculatus* (Acari: Ixodidae): dynamics of Thogoto virus infection in female ticks during feeding on guinea pigs. Exp Parasitol 104: 20-25.

Keller CF (2007) Global warming 2007. An update to global warming: the balance of evidence and its policy implications. ScientificWorldJournal 7: 381-399.

Kerbo N, Donchenko I, Kutsar K and Vasilenko V (2005) Tickborne encephalitis outbreak in Estonia linked to raw goat milk, May-June 2005. Euro Surveill 10: E050623 050622.

Knudsen AB, Romi R and Majori G (1996) Occurrence and spread in Italy of *Aedes albopictus*, with implications for its introduction into other parts of Europe. J Am Mosq Control Assoc 12: 177-183.

Koelle K, Pascual M and Yunus M (2005) Pathogen adaptation to seasonal forcing and climate change. Proc Biol Sci 272: 971-977.

Komar N (2003) West Nile virus: epidemiology and ecology in North America. Adv Virus Res 61: 185-234.

Korenberg EI (1994) Comparative ecology and epidemiology of lyme disease and tick-borne encephalitis in the former Soviet Union. Parasitol Today 10: 157-160.

Korenberg EI and Kovalevskii YV (1999) Main features of tick-borne encephalitis eco-epidemiology in Russia. Zentralbl Bakteriol 289: 525-539.

Kuhn J, Bewermeyer H, Hartmann-Klosterkoetter U, Emmerich P, Schilling S and Valassina M (2005) Toscana virus causing severe meningoencephalitis in an elderly traveller. J Neurol Neurosurg Psychiatry 76: 1605-1606.

Kuiken T, Fouchier R, Rimmelzwaan G and Osterhaus A (2003) Emerging viral infections in a rapidly changing world. Curr Opin Biotechnol 14: 641-646.

Kuno G, Chang GJ, Tsuchiya KR and Miller BR (2001) Phylogeny of Thogoto virus. Virus Genes 23: 211-214.

L'Vov D N, Dzharkenov AF, Aristova VA, Kovtunov AI, Gromashevskii VL, Vyshemirskii OI, Galkina IV, Larichev VF, Butenko AM and L'Vov D K (2002) [The isolation of Dhori viruses (Orthomyxoviridae, Thogotovirus) and Crimean-Congo hemorrhagic fever virus (Bunyaviridae, Nairovirus) from the hare (Lepus europaeus) and its ticks Hyalomma marginatum in the middle zone of the Volga delta, Astrakhan region, 2001] Isoliatsiia virusov Dkhori (Orthomyxoviridae, Thogotovirus) i Krymskoi-Kongo gemorragicheskoi likhoradki (Bunyaviridae, Nairovirus) ot zaitsa (Lepus europaeus) i sobrannykh s nego kleshei Hyalomma marginatum v srednei sone del'ty Volgi, Astrakhanskaia oblast', 2001 g. Vopr Virusol 47: 32-36.

Labuda M, Danielova V, Jones LD and Nuttall PA (1993) Amplification of tick-borne encephalitis virus infection during co-feeding of ticks. Med Vet Entomol 7: 339-342.

Laine M, Luukkainen R and Toivanen A (2004) Sindbis viruses and other alphaviruses as cause of human arthritic disease. J Intern Med 256: 457-471.

Lundstrom JO (1999) Mosquito-borne viruses in western Europe: a review. J Vector Ecol 24: 1-39.

Malkinson M and Banet C (2002) The role of birds in the ecology of West Nile virus in Europe and Africa. Curr Top Microbiol Immunol 267: 309-322.

Malkova D, Danielova V, Minar B, Rosicky B and Casals J (1972) Isolation of Yaba 1 arbovirus in Czechoslovakia. Acta Virol 16: 93.

Mardani M, Rahnavardi M, Rajaeinejad M, Naini KH, Chinikar S, Pourmalek F, Rostami M and Shahri MH (2007) Crimean-Congo hemorrhagic fever among health care workers in Iran: a seroprevalence study in two endemic regions. Am J Trop Med Hyg 76: 443-445.

Marenko VF, Dunaev NB, Egorova LS and Fedorova TN (1974) [Experimental mixed infection in muskrats (Omsk hemorrhagic fever and tularemia)] Smeshannaia infektsiia u ondatr v eksperimente (omskaia gemorragicheskaia likhoradka i tuliaremiia. Vopr Virusol: 545-550.

McLean DM, Walker SJ, Macpherson LW, Scholten TH, Ronald K, Wyllie JC and McQueen EJ (1961) Powassan virus: investigations of possible natural cycles of infection. J Infect Dis 109: 19-23.

Medlock JM, Snow KR and Leach S (2007) Possible ecology and epidemiology of medically important mosquito-borne arboviruses in Great Britain. Epidemiol Infect 135: 466-482.

Mertins JW and Schlater JL (1991) Exotic ectoparasites of ostriches recently imported into the United States. J Wildl Dis 27: 180-182.

Mills JN and Childs JE (1998) Ecologic studies of rodent reservoirs: their relevance for human health. Emerg Infect Dis 4: 529-537.

MMWR (2001) Outbreak of Powassan encephalitis--Maine and Vermont, 1999-2001. MMWR Morb Mortal Wkly Rep 50: 761-764.

Morrill JC, Soliman AK, Imam IZ, Botros BA, Moussa MI and Watts DM (1990) Serological evidence of Crimean-Congo haemorrhagic fever viral infection among camels imported into Egypt. J Trop Med Hyg 93: 201-204.

Morris JG, Jr. and Potter M (1997) Emergence of new pathogens as a function of changes in host susceptibility. Emerg Infect Dis 3: 435-441.

Morshed MG, Scott JD, Fernando K, Beati L, Mazerolle DF, Geddes G and Durden LA (2005) Migratory songbirds disperse ticks across Canada, and first isolation of the Lyme disease spirochete, *Borrelia burgdorferi*, from the avian tick, *Ixodes auritulus*. J Parasitol 91: 780-790.

Niklasson B and Vene S (1996) Vector-borne viral diseases in Sweden--a short review. Arch Virol Suppl 11: 49-55.

Osterhaus A (2001) Catastrophes after crossing species barriers. Philos Trans R Soc Lond B Biol Sci 356: 791-793.

Papa A, Bino S, Llagami A, Brahimaj B, Papadimitriou E, Pavlidou V, Velo E, Cahani G, Hajdini M, Pilaca A, Harxhi A and Antoniadis A (2002) Crimean-Congo hemorrhagic fever in Albania, 2001. Eur J Clin Microbiol Infect Dis 21: 603-606.

Papa A, Christova I, Papadimitriou E and Antoniadis A (2004) Crimean-Congo hemorrhagic fever in Bulgaria. Emerg Infect Dis 10: 1465-1467.

Patz JA, Campbell-Lendrum D, Holloway T and Foley JA (2005) Impact of regional climate change on human health. Nature 438: 310-317.

Patz JA and Olson SH (2006) Climate change and health: global to local influences on disease risk. Ann Trop Med Parasitol 100: 535-549.

Perkins SE, Cattadori IM, Tagliapietra V, Rizzoli AP and Hudson PJ (2003) Empirical evidence for key hosts in persistence of a tick-borne disease. Int J Parasitol 33: 909-917.

Petersen LR and Marfin AA (2005) Shifting epidemiology of Flaviviridae. J Travel Med 12 Suppl 1: S3-11.

Pietzsch M, Quest R, Hillyard PD, Medlock JM and Leach S (2006) Importation of exotic ticks into the United Kingdom via the international trade in reptiles. Exp Appl Acarol 38: 59-65.

Randolph S (2002) The changing incidence of tick-borne encephalitis in Europe. Eurosurveillance Weekly 6(23): 020606.

Randolph SE, Miklisova D, Lysy J, Rogers DJ and Labuda M (1999) Incidence from coincidence: patterns of tick infestations on rodents facilitate transmission of tick-borne encephalitis virus. Parasitology 118 (Pt 2): 177-186.

Reed KD, Meece JK, Henkel JS and Shukla SK (2003) Birds, migration and emerging zoonoses: west nile virus, lyme disease, influenza A and enteropathogens. Clin Med Res 1: 5-12.

Reeves WK, Durden LA and Dasch GA (2006) A spotted fever group Rickettsia from an exotic tick species, Amblyomma exornatum (Acari: Ixodidae), in a reptile breeding facility in the United States. J Med Entomol 43: 1099-1101.

Reid HW, Buxton D, Pow I and Finlayson J (1984) Transmission of louping-ill virus in goat milk. Vet Rec 114: 163-165.

Reisen W and Brault AC (2007) West Nile virus in North America: perspectives on epidemiology and intervention. Pest Manag Sci 63: 641-646.

Roehrig JT, Layton M, Smith P, Campbell GL, Nasci R and Lanciotti RS (2002) The emergence of West Nile virus in North America: ecology, epidemiology, and surveillance. Curr Top Microbiol Immunol 267: 223-240.

Ross PS, Vos JG, Birnbaum LS and Osterhaus AD (2000) PCBs are a health risk for humans and wildlife. Science 289: 1878-1879.

Samish M (2000) Biocontrol of ticks. Ann N Y Acad Sci 916: 172-178.

Sanbonmatsu-Gamez S, Perez-Ruiz M, Collao X, Sanchez-Seco MP, Morillas-Marquez F, de la Rosa-Fraile M, Navarro-Mari JM and Tenorio A (2005) Toscana virus in Spain. Emerg Infect Dis 11: 1701-1707.

Schwan TG, Raffel SJ, Schrumpf ME and Porcella SF (2007) Diversity and distribution of Borrelia hermsii. Emerg Infect Dis 13: 436-442.

Shepherd AJ, Swanepoel R, Leman PA and Shepherd SP (1987) Field and laboratory investigation of Crimean-Congo haemorrhagic fever virus (Nairovirus, family Bunyaviridae) infection in birds. Trans R Soc Trop Med Hyg 81: 1004-1007.

Shepherd AJ, Swanepoel R, Shepherd SP, Leman PA and Mathee O (1991) Viraemic transmission of Crimean-Congo haemorrhagic fever virus to ticks. Epidemiol Infect 106: 373-382.

Simkova A (1962a) Tahyna virus in birds. Acta Virol 6: 190.

Simkova A (1962b) Tahyna virus in rabbits. Acta Virol 6: 281.

Skarpaas T, Golovljova I, Vene S, Ljostad U, Sjursen H, Plyusnin A and Lundkvist A (2006) Tickborne encephalitis virus, Norway and Denmark. Emerg Infect Dis 12: 1136-1138.

Skarpaas T, Ljostad U and Sundoy A (2004) First human cases of tickborne encephalitis, Norway. Emerg Infect Dis 10: 2241-2243.

Srihongse S, Albanese M and Casals J (1974) Characterization of Thogoto virus isolated from ticks (*Rhipicephalus bursa*) in Western Sicily, Italy. Am J Trop Med Hyg 23: 1161-1164.

Suss J, Schrader C, Falk U and Wohanka N (2004) Tick-borne encephalitis (TBE) in Germany--epidemiological data, development of risk areas and virus prevalence in field-collected ticks and in ticks removed from humans. Int J Med Microbiol 293 Suppl 37: 69-79.

Sutherst RW (2004) Global change and human vulnerability to vector-borne diseases. Clin Microbiol Rev 17: 136-173.

Tasi TF, Smith GC, Ndukwu M, Jakob WL, Happ CM, Kirk LJ, Francy DB and Lampert KJ (1988) Entomologic studies after a St. Louis encephalitis epidemic in Grand Junction, Colorado. Am J Epidemiol 128: 285-297.

Taylor D (1997) Seeing the forests for the more than the trees. Environ Health Perspect 105: 1186-1191.

Turell MJ, Sardelis MR, O'Guinn ML and Dohm DJ (2002) Potential vectors of West Nile virus in North America. Curr Top Microbiol Immunol 267: 241-252.

Valassina M, Cuppone AM, Bianchi S, Santini L and Cusi MG (1998) Evidence of Toscana virus variants circulating in Tuscany, Italy, during the summers of 1995 to 1997. J Clin Microbiol 36: 2103-2104.

Valassina M, Meacci F, Valensin PE and Cusi MG (2000) Detection of neurotropic viruses circulating in Tuscany: the incisive role of Toscana virus. J Med Virol 60: 86-90.

Van Loveren H, Ross PS, Osterhaus AD and Vos JG (2000) Contaminant-induced immunosuppression and mass mortalities among harbor seals. Toxicol Lett 112-113: 319-324.

Wang H and Nuttall PA (2001) Intra-stadial tick-borne Thogoto virus (Orthomyxoviridae) transmission: accelerated arbovirus transmission triggered by host death. Parasitology 122: 439-446.

Weiss RA and McMichael AJ (2004) Social and environmental risk factors in the emergence of infectious diseases. Nat Med 10: S70-76.

Weissenbock H, Kolodziejek J, Fragner K, Kuhn R, Pfeffer M and Nowotny N (2003) Usutu virus activity in Austria, 2001-2002. Microbes Infect 5: 1132-1136.

Weissenbock H, Kolodziejek J, Url A, Lussy H, Rebel-Bauder B and Nowotny N (2002) Emergence of Usutu virus, an African mosquito-borne flavivirus of the Japanese encephalitis virus group, central Europe. Emerg Infect Dis 8: 652-656.

Wobeser G (2002) Disease management strategies for wildlife. Rev Sci Tech 21: 159-178.

Woodall JP and Roz A (1977) Experimental milk-borne transmission of Powassan virus in the goat. Am J Trop Med Hyg 26: 190-192.

Woodford MH and Rossiter PB (1993) Disease risks associated with wildlife translocation projects. Rev Sci Tech 12: 115-135.

Woolhouse ME, Taylor LH and Haydon DT (2001) Population biology of multihost pathogens. Science 292: 1109-1112.

Yanase T, Kato T, Yamakawa M, Takayoshi K, Nakamura K, Kokuba T and Tsuda T (2006) Genetic characterization of Batai virus indicates a genomic reassortment between orthobunyaviruses in nature. Arch Virol 151: 2253-2260.

Zeller HG, Cornet JP and Camicas JL (1994) Crimean-Congo haemorrhagic fever virus infection in birds: field investigations in Senegal. Res Virol 145: 105-109.

24. Invasions of vector-borne diseases driven by transportation and climate change

Wouter J. van der Weijden, René A.L. Marcelis and Wilfred Reinhold

Abstract

Biological invasions are as old as life itself but have become increasingly frequent due to the globalisation of travel and transportation. During the latest decades, climate change has become an additional driver. Vector-borne diseases, too, are invading more frequently, following an invasion of a vector, a pathogen, a reservoir host or a combination of these. In the past, such invasions have had dramatic effects on public health (e.g. the plague in Europe), livestock health (e.g. bluetongue in southern Europe) or wildlife (e.g. avian malaria on Hawaii). As the volume and speed of travel and transportation increase and climate change continues, more invasions of vector-borne diseases are to be expected, also in Europe. Recent arrivals include the West-Nile, Usutu and bluetongue viruses and the mosquito vector *Aedes albopictus* (Skuse). Around 12 arboviruses are potential candidates for invasion in Europe in the next decades as well as at least one rickettsia, one bacterium and two protozoa. However, the likelihood of such invasions is small in most cases. The main vectors involved are ticks and *Aedes* and *Culex* mosquitoes. Two serious threats are invasions of the Chikungunya and dengue viruses in southern Europe, where the competent vector *Ae. albopictus* has already become established. In fact, at the time of writing this article (August 2007) an outbreak of Chikungunya occurred in Italy (see Chapter 10). An increasingly important risk route for disease introductions is that between East Asia and Europe. Whilst most climate-driven invasions cannot be prevented, many transportation-driven invasions can be. This requires a precautionary approach based on risk analyses, preventive measures in the country of origin, intensive surveillance, early warning and interception. High-risk trade flows for pathogens, such as live animals, and for mosquitoes, such as used tyres and water-filled containers, should be strictly regulated or even reduced. If, however, an introduced vector or pathogen has inadvertedly become established, it should be eradicated as soon as possible. Mosquito invaders can be eradicated by draining water pools, application of insecticides, or use of biological larvicides. Pathogens can be eradicated by vector control, vaccination of infected people or livestock and isolation and treatment of infected hosts. Culling of infected livestock is another option. International agreements such as the Convention on Biodiversity and the International Health Regulations encourage preventive measures, and the world trade regulations allow restrictions on trade if and when the health of humans and animals is at stake. The ultimate policy goal would be to replace free trade by safe trade.

Keywords: biological globalisation, bio-invasion, eradication, pathogen, prevention, regulation

Introduction

The world is globalising and nature is no exception. More and more species are expanding their ranges across ancient distribution barriers. The overall result has been termed 'biological globalisation' (e.g. Bright 1998, Van der Weijden *et al.* 2007).

Biological invasion is by no means a recent phenomenon. In fact, it is as old as life itself. Every species tends to expand its range. Some can even cross thousands of kilometres moving on air or ocean currents alone, while others can do so by flying and still others by hitchhiking on the latter. For example, the Sindbis virus is carried annually between South Africa and Scandinavia by migrating birds (Gould *et al.* 2006). The process was accelerated after *Homo sapiens* spread out of Africa. Humans introduced plant, animal and microbe species in Asia, Europe, Australia and the Americas, both deliberately and accidentally. A next 'quantum leap' was made after Columbus had 'discovered' America in AD 1492. Three decades later, the Pacific was crossed as well[16]. These two events paved the way for increasing transoceanic trade and travel and associated species introductions, leading to genuine biological globalisation (Crosby 1972, 1986). The most recent 'quantum leap' was the introduction of the aircraft. Species can today move further, faster and in greater numbers than ever before.

In this chapter we discuss:
- the main drivers of bio-invasion;
- some bio-invasions from the past that had a major impact;
- some historical examples of vector-borne disease invasion and their impact;
- categories of vector-borne disease invasion;
- recent invasions of vector-borne disease in Europe and candidates for future invasion;
- priorities for prevention and control;
- an overall strategy for prevention and control of vector-borne disease in Europe;
- technical and legal opportunities for prevention and control;
- guidelines for action.

Bio-invasion and its drivers

Bio-invasions are a universal phenomenon. Invasions have taken place in species of every major taxon and every trophic level into virtually every terrestrial and aquatic habitat and every continent including Antarctica. Box 1 gives some basic definitions from invasion biology.

Bio-invasions have several drivers. The main drivers are:
- international travel and transportation;
- climate change;
- habitat transformation.

Habitat transformation can facilitate bio-invasion in two different ways. Firstly, it can remove distribution barriers between distinct biotas that were long or always separated. For example, in AD 1869 the Suez Canal connected the Mediterranean Sea and the Red Sea/Indian Ocean, leading to frequent species interchange. Secondly, habitat transformation can help species that were already introduced across a barrier, to establish and spread in its new area. Even though all habitats appear to be invasible (Willamson 1997), some are more invasible than others. For example, soils that are physically disturbed and/or highly enriched with nutrients are often used by introduced plant species (Lake and Leishman 2004) and may form bridgeheads for invasion.

[16] Columbus was certainly not the first visitor from the Old World since the original immigration waves from Asia. Five hundred years earlier, the Vikings had reached the northeast coast of North America. However, that did not generate substantial exchange of terrestrial species. Three reasons can be mentioned for this. The northern colder zones of the Old World and New World host fewer species than the warmer zones; a higher proportion of these species was already shared; and the traffic volumes involved were relatively small. As for the Pacific, there is evidence that before AD 1492 Polynesians visited the west coast of South America and from there introduced the sweet potato in Oceania (Ballard *et al.* 2005).

Some invasions are caused by two or three drivers acting in combination. For example, over the latest decades many plant species from warmer and rocky areas have invaded urban areas in the Netherlands (Tamis 2005)[17]. Somewhat simplified: habitat transformation (i.e. urbanisation in the past) created a suitable environment, climate change raised the local temperature and international travel and transportation introduced the seeds.

Box 1. Some definitions in invasion biology, epidemiology and biogeography.

Invasion biology is a young branch of biology. It is only 50 years old, as young as e.g. DNA biology, and much younger than medical and veterinary epidemiology. Although Charles Darwin laid some of the foundations, the real starting point was Charles Elton's groundbreaking *The ecology of invasions by animals and plants* (1958). There still is confusion as to some basic definitions and concepts. Here we apply the following definitions:

Alien (species): A species that does not naturally occur in an area where it is found at a particular moment. Synonyms: *exotic, non-indigenous* and *non-native species.*

Biological globalisation (biotic globalisation, bio-globalisation): The process of increasing connections between the separate biotas of the world, leading to biotic homogenisation and numerous new ecological interactions. Insofar as this process is facilitated by man – through transportation, travel, breakdown of distribution barriers, and man-made climate change – it can be named *anthropogenic bio-globalisation.*

Counter-invasion: Invasion following the deliberate introduction of a natural enemy of an invasive pest species.

Endemic disease (in epidemiology): A disease that occurs continuously in a certain area, at a more or less constant level.

Endemic species (in biogeography): A species occurring exclusively in a certain area.

Enzootic disease: Endemic disease among animals.

Import cases: Cases of infectious diseases that have been brought into a country by infected patients or animals.

Introduction: The transportation of a species into an area where it does not yet occur. This can be done intentionally, accidentally and unconsciously, or accidentally but consciously.

Invasion: The settlement and subsequent spreading of a species in a new area.

Invasive species: A species that colonises a new area, maintains itself, reproduces and spreads actively. (Many authors confine 'invasive' to those species causing ecological, medical or economic damage).

Native species: A species that for at least millennia occurs in a particular area. Synonym: *indigenous species.*

Impacts of bio-invasions in history

The impacts of bio-invasions on nature and the human society are often negligible, at times beneficial and sometimes harmful or even devastating. The best example of a beneficial impact, at least for human nutrition, has been the introduction of crop and livestock species from the

[17] Change of *micro*climates can be an important driver as well. For example, the construction of heated greenhouses and buildings in temperate climates has created suitable conditions for subtropical insects such as the Mediterranean fruit fly *Ceratitis capitata* and the Pharao ant. Such species rarely spread beyond their artificial environment, but this will happen more frequently as climate changes continues.

Old World into the New World and vice versa. This so-called 'Columbian Exchange' (Crosby 1972) provided opportunities for more diverse crops and higher yields in most regions in the world[18]. Likewise, the exchange of livestock species – mainly from Old World to New World – gave a major boost to the previously marginal livestock production in the New World.

By contrast, species introductions have also caused disasters, accidental introductions not less than deliberate ones. For example, the Columbian Exchange also included the exchange of pathogens, pests and weeds, mainly from the Old Word to the New World. We mention seven examples of harmful species introductions, which include a virus, a fungus, several mammals, a fish and a mollusc. We present them in roughly chronological order:

- European colonisers accidentally introduced the smallpox virus and several other Old World pathogens in the Americas in the 16th and 17th century. This caused mass mortality among indigenous Americans, decimating their populations. It was a decisive factor in the shift of power from the Amerindians to the European colonisers.
- After the introduction of the potato in the 16th century from South or Central America century in Europe, a deadly pathogen followed three centuries later: the potato blight fungus *Phytophthora infestans* (Mont.) de Bary[19]. Originating in Central Mexico, it was probably introduced in 1843 or 1844 from South America or the USA (Andrivon 1996). In 1845 it devastated the potato crop in western Europe. Ireland was hit hardest. In 1845, 1846 and 1848 a third to three quarters of the crop failed. By 1851 about 2.5 million persons were lost, with an estimated 1 million having emigrated and 1.5 million having died from the effects of the famine (The History Place 2000).
- In 1859 British colonisers in Australia introduced the European rabbit for hunting purposes. The species spread widely and quickly became a pest, competing with sheep livestock and indigenous fauna (Williamson 1997). By 1997, the costs in agricultural damage still amounted to some $600 million annually, plus $20 million in control costs (Bomford and Hart 2002).
- Species introductions have also caused the extinction of many indigenous species. This rarely happened on continents and in oceans, but frequently on islands and in lakes. On islands, most damage was done by introduced predators such as rats and cats, as well as by herbivores such as goats and rabbits. (Williamson 1997, Quammen 1996). The process started several millennia ago and become global since the 15th century. As for lakes, a recent example is the predatory Nile perch *Lates nilotica* L. in Lake Victoria. Introduced for fishing in 1954, it remained little noticed until it exploded in the 1980s. While spreading it eradicated about two hundred endemic cichlid fish species (Goldschmidt 1998).
- In 1905 the muskrat *Ondatra zibethicus* L. was introduced from North America in Bohemia, Central Europe, for fur production. In the 1920s muskrats were released or escaped in Russia, Finland, France and the British Isles, and started undermining banks, dams and riverside trees (Williamson 1997). In the Netherlands they undermine dikes since 1946, creating risks of flooding. Six hundred people are needed permanently to catch muskrats in order to prevent disaster and reduce damage to dikes (LCCM 2004).
- In 1986 the zebra mussel *Dreissena polymorpha* Pallas, native to Europe, was found in North-American lakes, probably introduced in the ballast water of a ship. The species quickly started its invasion, which still continues. By swamping it causes major damage to the indigenous freshwater fauna and covers and clogs intakes and outlets of power plants and factories. Along

[18] Most introduced crop and many livestock species hardly spread spontaneously, so whether the term 'invasion' is appropriate is a matter of definition.

[19] *Phytophthora* is today classified as an oomycote.

with the quagga mussel *D. rostriformis bugensis* Andrusov, also introduced from Europe, the zebra mussel causes an estimated one billion dollar per year in damage and control costs (Pimentel *et al.* 2005).

These cases illustrate that deliberate introductions (rabbit, Nile perch and muskrat) are not less harmful than accidental introductions. They also illustrate that, next to the damage inflicted on nature, bio-invasions can damage multiple sectors of society, ranging from public health, agriculture, forestry and fisheries to industry, energy supply, housing, water infrastructure and tourism. The economic costs are an estimated 120 billion dollar per year in the USA alone (Pimentel *et al.* 2005).

What makes many harmful invasions even more harmful is that they are irreversible (although there are exceptions such as goats and rats introduced on smaller islands that can be eradicated). For example, potato blight still occurs in Europe. In the Netherlands it is even a main target of pesticide use.

Historical cases of invasions of vector-borne disease

Plague in the Old World and beyond

Invasions of vector-borne[20] diseases, too, can be devastating. Perhaps the most devastating invasive vector in history was the oriental rat flea *Xenopsylla cheopis* Rothschild. It can carry and transmit the bacterium *Yersinia pestis* that causes bubonic and pneumonic plague and has its natural reservoirs in burrowing rodents. The black rat *Rattus rattus* L. is a temporary host but before dying can transmit the bacterium to a vector, usually the rat flea (Benedictow 2004). The distributional history of the trio bacterium/flea/rat is closely associated with the development of the global shipping network. Hence the three plague pandemics were of an increasingly global scale (McNeill 1998).

Perhaps in the 1st or 2nd century AD, the flea travelled with black rats aboard ships from India to the Middle East and thence, probably with rats and camels to the Black Sea and Mediterranean Sea areas. This later facilitated the first plague pandemic (AD 541-810) that started in Central Africa, India or Arabia, and ranged from Arabia east to China and west to Southern Europe.

The second pandemic (AD 1346-1720) followed after the fleas had travelled north, again with ship rats, to western and northern Europe. Hence the initial wave, the Black Death[21] (AD 1346-1353) that started in the Crimea, could hit almost the whole of Europe, killing between 25% and 60% of the population (Benedictow 2006). On land, the spread of the disease was accelerated by people fleeing from the plague while carrying the bacterium with them in their blood and/or in infected fleas. Many waves followed, covering almost four centuries until the last major outbreak in Marseille in 1720.

---

[20] We apply the term 'vector' to invertebrates that can carry and transmit a pathogen from a reservoir host to humans or (feral or domesticated) animals. We exclude animals such as dogs, rodents and bats that are a permanent or temporary host to a pathogen and can directly transmit it to humans.

[21] Several scholars have provided evidence that other pathogens such as anthrax or a viral hemorrhagia may have played an additional or even dominant role in the Black Death. See Scott and Duncan (2004) for a review. This debate has not yet been concluded.

The third pandemic (AD 1855-1959) started in China and became truly global. After reaching the seaports of Hong Kong and Canton, *Yersinia pestis* could travel with ship rats to southern Asia, Africa and the Americas, facilitated by the global network of steamships that had developed. The bacterium developed reservoirs in mammals (usually burrowing rodents) in new areas, including western North America, where it colonised prairie dogs. Plague is becoming a re-emerging concern due to increased incidence and antibiotic resistance (Tatem *et al.* 2006a).

Yellow fever in the New World

Another historical example of vector-borne disease invasion is the introduction of the yellow fever virus and its vector *Aedes aegypti* L. in the New World. The most commonly cited hypothesis is that the virus was introduced from Africa, along with *Ae. aegypti*, in the bilges of sailing vessels during the slave trade. The timing and African origin of the invasion were recently supported by molecular gene sequence analysis (Bryant *et al.* 2007). There is circumstantial evidence that Dutch slave ships were involved (Van der Weijden *et al.* 2007).

In 1647 the first outbreak was reported on the Caribbean island of Barbados (Childs Kohn 2001). Vector and virus spread across the Caribbean and beyond, causing mass mortality. Amerindians and Europeans were hit much harder than African slaves, who were more or less immune to the virus. This affected the balance of power in the Caribbean. The virus became a powerful ally of African rebels in Haiti revolting against the French army and thus generated a 'biological Waterloo' for Napoleon. Haiti owed its independence in 1804 not least to an earlier introduced vector-borne disease.

From 1881 to 1884, yellow fever as well as malaria frustrated French attempts to construct the Panama Canal (Spielman and D'Antonio 2001). *Plasmodium falciparum, P. malariae* and *P. vivax* had been introduced in the previous centuries, probably with infected African slaves and European contract labourers. The USA was more successful from 1906 after the discovery of the key role of mosquitoes in the transmission of yellow fever and malaria. Once again, the French were beaten by mosquito-borne diseases introduced much earlier by their fellow-Europeans. This time they had to yield maritime power to the USA (Spielman and D'Antonio 2001).

This case also makes clear that understanding vector-borne disease can be a key success factor in public healthcare, the economy and politics. Today, this may still be the case, considering such devastating diseases as malaria, bilharziasis, yellow fever, dengue and sleeping sickness.

Other cases

In recent history, the most devastating vector invasion was that of the malaria mosquito *Anopheles gambiae* Giles in South America. Native to Africa, the mosquito was introduced in 1930 from Dakar, Senegal, into northeast Brazil, either by aircraft or by a rapid passenger ship (Lounibos 2002). Although *Plasmodium falciparum* had already become endemic, the greater vectorial capacity of *An. gambiae*, as compared to local mosquitoes, sparked severe malaria epidemics costing 16,000 lives (Soper and Wilson 1943). Even though the mosquito had already spread widely covering 54,000 km^2, it was eradicated in 1940 as a result of the remarkable large-scale, army-style campaign that included larviciding of all potential habitats with crude oil, Paris green and pyrethrum, as well as drainage of pools (Soper and Wilson 1943, Childs Kohn 2001, Killeen 2003).

Vector invasions can also severely hit faunas. A classical example is the introduction of the southern house mosquito *Culex pipiens fatigans* (now: *Culex quinquefasciatus*) Wiedemann in the Hawaiian island of Maui in 1826 (Quammen 1996, Lowe *et al.* 2001). Eggs and larvae of the mosquito were introduced in the water storage jars of a British ship sailing from Mexico. The avian-pox virus and *Plasmodium relictum*, which can cause avian malaria, had been introduced earlier with poultry by settlers. However, the parasite and the virus lacked a competent vector. Once established, the mosquito quickly started transmitting both pathogens from poultry to indigenous birds, including members of the endemic family of honeycreepers (Drepanididae), regarded by many as one of the most magnificent bird families in the world. As the mosquito spread to other islands, the impact became dramatic. Before the close of the 19th century, 16 endemic bird species and subspecies had suffered extinction (see Chapter 4).

Just before the close of the millennium, the West-Nile virus invaded North America, probably from the Mediterranean (see Chapter 8). It probably arrived by ship or aircraft from the Mediterranean with infected imported birds, mammals, stowaway mosquitoes or travellers. But it may also have arrived with infected migratory birds. The virus found competent indigenous reservoir hosts (birds) as well as vectors (mosquitoes). The first outbreak was in New York in 1999. Many outbreaks followed as the virus quickly spread across North America. In 2006 alone, 4,269 people in 44 states of the USA were reported ill, including 177 fatalities (CDC 2007). In addition, there were thousands of fatalities among horses and wild birds (see Chapter 8).

Tick invasions so far have not caused dramatic effects. In Europe, one probably invaded species that has caused damage is the brown dog tick *Rhipicephalus sanguineus* Latreille. It probably originated in Africa (Garben *et al.* 1980) and subsequently invaded southern Europe, but how and when that happened is not clear. *R. sanguineus* is unusual among ticks, in that it can complete its entire life cycle indoors (Lord 2001). It attaches to the armpits and groins of dogs, its preferred host, as they walk through high grass or low scrub. Cattle, cats and sheep can also serve as host, as can humans. If dogs are no longer available (or made unattractive by application of a repellent), ticks can take a blood meal on humans. Thus they can transmit Lyme disease as well as rickettsias, mainly *Rickettsia conorii*, which causes Mediterranean spotted fever (Parola 2004). Damage may increase in the future as the tick has spread North, mainly with infected dogs of tourists. In the Netherlands it was first observed in 1962 and later became established, though it can probably only survive indoors. Control may become more difficult as the tick is developing resistance against acaricides (Estrada-Peña 2003).

More harm is known from some cases where the pathogen itself rather than its tick host invades a new area. Lyme disease in North America is a possible example. The causative pathogen is the spirochaete bacterium *Borrelia burgdorferi*, which may have originated in Eurasia[22]. With approximately 20,000 new cases reported each year, Lyme disease has become the most common vector-borne disease in the USA (CDC 2007).

Invasions of non-arthropod vectors can be devastating as well. For example, the flatworm *Schistosoma mansoni* Randall is one of the causal agents of intestinal schistosomiasis (bilharziasis) in Africa. The damage inflicted on human health became even greater after it was introduced,

[22] A Eurasian origin of *Borrelia burgdorferi* is likely since the genetic diversity within the *B. burgdorferi* complex is much higher in Eurasia than in America. How and when the species invaded North America is not yet clear. It may have been introduced by infected humans, animals or ticks. Alternatively, migratory birds can have introduced the bacterium. Several bird species are competent reservoirs of the bacterium (Ginsberg *et al.* 2005).

probably with the slave trade, in the American tropics, where it found a competent intermediate host in the freshwater snail *Biomphalaria glabrata* Say. Today, 83 million people host *S. mansoni* (DeJong *et al.* 2001).

In the next paragraphs we narrow our focus to arthropod vectors of human and animal diseases, leaving aside plant diseases and non-arthropod vectors. We further focus on invasions into Europe (not specifically the EU), disregarding invasions into other continents and from one part of Europe to another. In addition, we focus on (possibly) successful bio-invasions, disregarding the numerous airport cases and import-linked cases of exotic diseases, although these can start an invasion in exceptional cases.

Vector-borne disease invasion in Europe

The plague was not the last arthropod-borne disease to invade Europe. It was followed by *Plasmodium falciparum* malaria, introduced from West Africa. Several other diseases followed, mainly in the 20th century, driven by the globalisation of travel and transportation. Table 1 lists three vectors and four vector-borne pathogens that have invaded Europe[23].

Aedes albopictus

The Asian tiger mosquito *Ae. albopictus* is a subtropical mosquito, native to Asia, that breeds in natural or artificial containers with stagnant water, such as used tyres and tins, with which it is often transferred, even overseas. From Asia it spread to southern Europe (1979), North America (1985), South America (1986), Africa (1991) and islands in the Indian and Pacific Oceans. In the laboratory *Ae. albopictus* can transmit c. 22 arboviruses, including those causing dengue fever, West-Nile disease, yellow fever, Cache Valley disease, Eastern equine encephalitis and Chikungunya. Some of these viruses are actually circulating in Europe, while others only arise as imported viruses (Gould *et al.* 2006). However, as for the USA, Moore and Mitchell (2000) state that in terms of its role as an arbovirus vector, evidence is lacking to incriminate *Ae. albopictus* as the vector of even a single case of human disease. It has even replaced *Ae. aegypti* in large areas in the USA (Lounibos 2002). *Ae. albopictus* seems to be a threat only in areas where *Ae. aegypti* is absent. For example, in 2005/06 it caused heavy epidemics of Chikungunya on La Réunion and other islands in the Indian Ocean. More than 250,000 people became infected and an estimated 260 died from the disease, implying a case-fatality rate of 1/1000 (Josseran *et al.* 2006).

Unlike *Ae. aegypti, Ae. albopictus* is capable of colonising temperate areas permanently. Records of the mosquito in Europe have been associated with imports of used tyres (Italy, France, Belgium) and ornamental plants (the Netherlands; see Chapter 14). In Italy it is now a dominant pest mosquito, which is probably partly due to the fact that the infestation derives from several successive introductions, each with large numbers of individuals and added genetic variation

[23] The table does not include the rat flea, which invaded long before AD 1500, as probably did the brown dog tick, which is also excluded. Neither does the table include allergies induced by the presence or behaviour of arthropods, such as the allergic reactions caused by bed bugs (*Cimex spp.*, Hemiptera) and asthma related to alien cockroaches (e.g. *Periplaneta spp.*, Dictyoptera), nor dermatoses caused by insects and arachnids. Diseases where both the pathogen and its vector are supposedly native to Europe have also been left out. One example is Lyme disease, caused by the bacterium *Borrelia burgdorferi* and spread by the sheep tick *Ixodes ricinus*. Another example is the Toscana virus (TOSV) that was first isolated in 1971 in central Italy from the sandfly *Phlebotomus perniciosus* (Charrel *et al.* 2005), which ranges from France, Portugal and Spain to Greece and Turkey.

Table 1. Three arthropods vectors and four arthropod-borne pathogens invaded in Europe since AD 1500.

Vector	Pathogen(s) transmitted	Native area of vector or pathogen	First record in Europe	Pathway of vector or pathogen
*Aedes albopictus** Asian tiger mosquito	In laboratory: c. 22 arboviruses In field: Chikungunya, dengue, West-Nile WNV*	Vector: Southeast Asia	Vector: 1979 Albania. Pathogen WNV: *see below*	Vector: ship-borne, used tyres and ornamental plants. Pathogen WNV: *see below*
Native *Anopheles* spp. Malaria mosquitoes	*Plasmodium falciparum**†	Tropical Africa	Pathogen: 16th century Spain, Portugal. Eradicated after WW II	Pathogen: slave ships?
Native *Culex spp.* and other mosquitoes	West-Nile virus*, Usutu virus*	WNV: Africa, Middle East. Usutu: Africa	WNV: 1962 France (horses), 1996 Romania (humans), Usutu: 2001 Austria (birds)	Viruses: migratory and/or imported birds?
Culicoides spp. (biting midges) Native *C. obsoletus, C. pulicaris, C. scoticus* Invaded *C. imicola**	Bluetongue virus*	Pathogen: Subtropics, Tropics	Pathogen: 1956 Spain/ Portugal. Vector *C. imicola*: 1982 Iberian peninsula	Pathogen: ship-borne cattle? Vector *C.imicola*: probably ship-borne cattle or wind-borne across Mediterranean
*Monomorium pharaonis** Pharao ant	(Native) bacteria incl.: *Salmonella, Pseudomonas, Staphylococcus, Streptococcus, Clostridium*	Vector: Africa	Vector: probably some centuries ago	??

* = invaded.
†= eradicated after WW II in Europe.

Box 2. Five categories of invasion of vector-borne disease.

A vector-borne disease requires at least three species: a pathogen, a reservoir host and a vector. Obviously, an invasion of such a disease can be successful only after each of the three species is present in the new area. Thus we can distinguish five categories of invasion:

1. Competent vector and reservoir host species present, pathogen invades (P > V/R).
   This probably happened when *Plasmodium falciparum* was introduced in the New World, probably with slave shipments. The parasites found a competent reservoir host (humans) as well as competent vectors among indigenous anopheline species[1]. Note here that pathogens are often introduced with an infected reservoir host, but if that host species already lives in the new area, this is not an invasion of the host species, not even if individuals spread and interbreed with indigenous individuals.
2. Pathogen and competent reservoir species present, vector invades (V > P/R).
   This was, to some extent, the case when *Anopheles gambiae* was introduced in Brazil in or some time before 1930. *P. falciparum* had already become endemic and a reservoir species (humans) was present. Even indigenous vectors were present, but *An. gambiae* is a much more efficient vector.
3. Competent vector present, pathogen and reservoir host invade (P/R > V).
   Such invasions may occur either simultaneously or shortly after one another. One example is the introduction of the European rabbit in Australia in 1859, where it quickly spread and became a pest. To control the rabbit, attempts were made from 1936 onward to introduce the myxoma virus, native to South America. Success had to wait until 1950, when *Aedes* and *Culex* mosquitoes had also become infected (Williamson 1997).
4. Competent reservoir host present, pathogen and vector invade (P/V > R).
   These invasions, too, may occur either simultaneously or one shortly after one another. This was probably the case when the yellow fever virus and its vector *Ae. aegypti* were introduced from Africa in the New World in the 17th century. Suitable reservoir species were humans, monkeys and several forest mosquito species.
5. Pathogen, vector and reservoir host invade (P/V/R >).
   These invasions may occur either simultaneously or one shortly after one another.
   This happened on Hawaii in the early 19th century, when poultry infected with avian malaria and avian pox were introduced first and a competent mosquito vector a few decades later (Quammen 1996).

In theory, two additional categories should be mentioned: (R/V > P) and (R > P/V).
This can happen only if the pathogen already survived in an indigenous reservoir host. But if it already caused a similar disease in that host, we cannot speak of a disease invasion. So these categories are relevant only in those rare cases where the (potential) pathogen was already present without causing a similar disease.

[1] Coatney *et al.* (1971) took the view that all South-American *Plasmodium* species, including those hosted by monkeys, were introduced with slave trade from Africa. Accordingly, Escalante *et al.* (1995) provided evidence from protein analysis that *P. brasilianum* and *P. simium*, parasites of South-American monkeys, are more or less identical to the human parasites *P. malariae* and *P. vivax*, respectively.

(Gratz 2004, Urbanelli *et al.* 2000). As recently as August 2007, a local outbreak of Chikungunya virus occurred in Italy, vectored by *Ae. albopictus* (see Chapter 10). An additional risk is that indigenous mosquitoes take over the vectorial role of *Ae. albopictus*. A Dutch report concluded that - apart from *Ae. albopictus* - five out of the 35 mosquito species in the Netherlands are potential vectors of WNV, including *Culex pipiens* (Reusken and Takken 2006).

Anopheles spp.

Anopheles mosquitoes can transmit *Plasmodium* and some can transmit the most vicious species *P. falciparum*, which causes malaria tropica. Its most efficient vector is *An. gambiae*, native to tropical Africa. *P. falciparum* may have travelled with infected African slaves[24] or crew members, as anopheline mosquitoes developed from larvae in water storage jars and transmitted the parasite from infected to non-infected individuals. Once arrived, the parasite found competent vectors among European *Anopheles* species and spread across southern Europe. Thereby this invasion falls in the 1st category of Box 2 (P > V/R).

Malaria was eradicated from the whole of Europe by the early 1970s. However, it has recently re-emerged as a matter of growing concern in Europe (see Chapters 2 and 3). Massive epidemics of autochthonous malaria occurred in adjacent Azerbaijan and the Asian part of Turkey. In addition, imported malaria is a growing issue, especially import into western Europe, with a majority of cases caused by *P. falciparum* (Sabatinelli *et al.* 2001).

Culex and other mosquitoes

The West-Nile virus was first isolated in Uganda in 1937. After being found in Egypt and Israel, it was first isolated in Europe in the 1960s, from horses in southern France. The first major human outbreak came in 1996 in Romania (see Chapter 8). The virus has an exceptionally wide range of hosts as well as vectors. It has been found in many bird, mammal and even alligator species, though it has its reservoirs among birds. It can be transmitted by many mosquito and tick species, often *Culex pipiens*.

The virus was probably introduced in Europe by infected migratory or imported birds. Its permanent presence in Europe is partially sustained by repeated introductions. So far, the damage inflicted on human and animal health has been much smaller in Europe than in the USA, perhaps because repeated infection has generated a higher degree of immunity in European humans and animals (see Chapter 8).

The Usutu virus is another virus that was probably introduced in (Central) Europe with infected migratory or imported birds. The current knowledge is summarised by Weissenböck *et al.* in Chapter 9. The virus is transmitted by *Culex* and other mosquito species. So far it has not been associated with severe or fatal diseases in humans.

Since seasonal migration is not considered a species invasion, both virus invasions are examples of the 1st category in Box 2 (P > V/R).

[24] At the time, African slaves were also held in Spain and Portugal.

Culicoides spp.

The epidemics and recent outbreaks of bluetongue in Europe illustrate the 1st category mentioned in Box 2 (P > V/R). However, the virus is not only transmitted by the indigenous biting midges including *Culicoides obsoletus* Kieffer, *C. pulicaris* L. and *C. scoticus* Downes and Kettle, but also by *C. imicola* Kieffer, a probably recent invader from Africa into Europe. The latter invasion fits in the 2nd category (V > P/R) and is still limited to southern Europe.

The virus may have originated in the Mediterranean. It was first described in South Africa after merino sheep were introduced from Europe in the late 18th century (Gould *et al.* 2006). The virus subsequently spread globally. It affects ruminants, not humans. It is enzootic in South Africa, Australia, the Americas and parts of Asia, and has hit Europe several times. In the 1956-1960 epidemic in Spain and Portugal, 180,000 sheep died. In 1979-1980 an epidemic occurred on the Greek islands of Rhodes and Lesbos. The virus appears to have established itself in southern Europe (Elbers 2003).

In August 2006 the virus was first found north of the Alps in an area encompassing parts of the Germany, the Netherlands, Belgium, France and Luxemburg (see Chapter 7). The serotype involved was nr. 8. This was identified earlier in Africa and the Americas and suspected in India, but not in Europe so it was a new invasion in Europe from some of these regions. The virus possibly arrived with infected animals. Its invasion coincided with the warmest July on record in the Netherlands. After a mild winter and a warm spring, new outbreaks took place in July 2007, indicating that the virus may become enzootic in North-West Europe (see Chapters 6 and 7).

Monomorium pharaonis L.

Another example of the 2nd invasion category (V > P/R) is the introduction of the pharaoh ant. It is the oldest and most widespread invasive pest among ant species. Probably native to Africa, it has successfully invaded areas on every continent except Antarctica. Due to its ability to transmit some bacterial human pathogens, its presence in hospitals is of great concern (Erdos and Koncz 1977). It is concentrated in tropical regions but is also commonly found in temperate zones within suitable human infrastructure, including climate-controlled buildings.

In the Netherlands it was first reported in 1900 in the post office of the town of Leeuwarden, where it probably arrived by mail. Due to the introduction of central heating, the houses in the Netherlands became warmer since 1950, coinciding with the advance of the ant (Van der Weijden *et al.* 2007). Even though the abundance of this ant in cold climates will be restricted, its continued presence suggests a potential to spread to locations more suitable for ant colonisation (Holway *et al.* 2002).

Summarising, the invasions listed in Table 1 represent only three of the five categories listed in Box 2: 1, 2 and 4. Perhaps categories 3 and 5 did not yet occur in Europe. Category 5 will remain unlikely since it requires invasions of three matching species. Category 3 is more conceivable. For example, an alien bird or mammal species infected with an alien virus might be introduced from North America or East Asia, and subsequently be vectored by an indigenous mosquito or tick.

Candidates for vector-borne disease invasion

Predicting specific new arthropod-borne invasions, even on the short term, is hardly possible, for various reasons:

- How far and how fast vectors and pathogens spread in time and space is difficult to predict for most species.
- Many factors and complex interactions are involved.
- The processes involved are often poorly understood.
- Quantitative data are often lacking.

Table 2 presents a list of arthropod-borne diseases that can be regarded as potential candidates for invading Europe. This list is not prioritised, nor does it indicate the likelihood of an invasion and outbreaks in European. The list is based on expert consultation and literature review by Van der Giessen *et al.* (2004), supplemented with a literature review by Van Lier *et al.* (2007) and a study by Gould *et al.* (2006). Those arthropod-borne diseases already mentioned in Table 1 have been excluded. For primary sources we refer to the three articles just mentioned.

Combining Tables 1 and 2, we find that around 12 pathogenic viruses are potential candidates to invade (parts of) Europe, as well as one rickettsia, one bacterium and two protozoa. However, most of these invasions are unlikely to happen in the near future. To illustrate this, between 1969 and 1999, 87 suspected cases of 'airport malaria' were reported in Europe, almost all caused by *P. falciparum*-infected anopheline mosquitoes transported from Africa. None of these caused a mosquito or *Plasmodium* invasion. African anopheline vectors can survive in Europe only during 2-4 months a year. In addition, 52,000 imported malaria cases have been reported since 1953, but none of these has led to a secondary case (Tatem *et al.* 2006b). The return of malaria seems a small risk under the present climate and the present quality of healthcare.

As illustrated by Table 1, most vector-borne disease invasions in Europe resulted from transportation, and a few from bird migration. Climate change has not yet been a main driver. However, it may have played a role in the invasion and subsequent spread of bluetongue in North-West Europe in the warm summer of 2006 (see Chapter 7). In a global review, Sutherst (2004) found very few publications that unambiguously predict a change in the range of a vector-borne disease in response to climate change. One example was dengue, which is expected to expand its range. However, there can be little doubt that if climate change continues, it will become a major driver of vector-borne disease invasion. In an analysis of the potential arbovirus emergence in the UK, Gould *et al.* (2006) even stated that climate change is probably the most important requirement for the emergence of arthropod-borne diseases such as dengue fever, Rift Valley fever, Japanese encephalitis, bluetongue and African horse sickness.

In addition, unpredicted invasions will inevitably take place. As pointed out by Takken *et al.* (Chapter 7), a bluetongue virus outbreak as far north in Europe as in 2006 was considered impossible by several experts. West-Nile virus in North America was another surprise. Furthermore, emerging diseases may come from unknown pathogens and hosts. Over the last 40 years alone, at least 39 new pathogens have been identified (WHO 2007), including HIV, Marburg, Ebola, SARS and BSE / new variant CJ, all of them zoonoses. New species and variants will continue to emerge and vector-borne pathogens, their vectors and their hosts will be no exception. With over a million insect species described, estimates for the total number of insect species range from 2 tot 20 million.

Even not yet existing varieties will pose hazards, such as multiple resistant vectors and pathogens. Evolution never stops and will continue to take us by surprise.

Table 2. Arthropod-borne diseases that are potential candidates for invasion in Europe.

Pathogen	Disease	Reservoir host ( ): not in Europe	Vector ( ): not in Europe	Geographical range of disease
Bunyaviridae				
Phlebo virus	Rift Valley fever	sheep, goat	*Aedes* spp. *Culex* spp.	Africa, Middle East
La Crosse virus	California encephalitis	several mammals incl. red fox, (gray squirrel), (Eastern chipmunk)	*(Ae. triseriatus)*, *Ae. albopictus*	North America
Flaviviridae				
	Dengue fever	humans, (monkeys)	*Aedes* spp.	Tropical Africa, Central America, Southeast Asia
	Japanese encephalitis	birds, pigs	*Ae. albopictus*, *Culex* spp.	Asia, Australia
	Kyasanur Forest disease/Alkhurma	sheep	ticks	India, Arabic Peninsula
Togaviridae				
	Equine encephalitis (Eastern, Western, Venezuelan)	horses, wild birds	*Aedes* spp., *Culex* spp.	America, Asia
	Ross River virus	various animals	mosquitoes	Australia, South Pacific
Reoviridae				
Coltivirus	Colorado tick fever	rodents	*(Dermacentor andersoni)*	North America
Rickettsiales and Bartonellaceae				
Ehrlichia chaffeensis	Human monocytic ehrlichiosis	deer, raccoon	ticks	America
Gram-negative rods				
Yersinia pestis	Bubonic and pneumonic plague	rodents	fleas	America, Asia, Tropical Africa
Protozoa				
*Plasmodium vivax**, *Plasmodium falciparum**	Benign tertian malaria, Malaria tropica (= Malignant tertian malaria)	humans	Native: *Anopheles* spp. incl. *An. atroparvus*, Alien: *An. quadriannulatus*	Africa, Asia, America. Vector *An. quadriannulatus:* Southern Africa

*Eradicated after WW II in Europe, but re-invasion is not impossible.

Priorities

It is obviously impossible to check every ship, airplane, person and cargo for any possible pathogen or vector species. Therefore it is necessary to identify high-risk vectors, regions, pathways, vehicles and habitats.

Roughly speaking, high-risk *vector groups* for Europe are Culicidae (*Culex* and *Aedes)* and ticks, as indicated in Tables 1 and 2. Some species can transmit several diseases and some can find suitable rural or urban environments in Europe. In addition, some have indigenous relatives that can join them as vectors. Perhaps the most important immediate threats are invasions of the Chikungunya and dengue viruses, which can be transmitted by *Ae. albopictus,* already established in southern Europe.

As for high-risk *regions* for invasions of arthropod-borne diseases in Europe, Tables 1 and 2 make clear that such diseases can come from any continent except Antarctica. But obviously, we first have to look at regions with more or less similar climates. These are not found in the tropics but in the subtropical and temperate zones: East Asia, North America, southern South America, southern Africa, Australia and New Zealand. Among these, we should focus on two subsets:

- Regions with a high biodiversity, which can be expected to have a high diversity of pathogens and vectors as well: East Asia, North America, southern Africa and perhaps Australia.
- Regions with which Europe has most trade and travel, including tourism. These are mainly North America, Japan and increasingly China.

Tatem and Hay (2007) found that North America and East Asia are most similar in climate to central and western Europe. Combining all this, the main risk regions for central and western Europe seem to be North America and East Asia (mainly Japan and China), whilst for southern Europe, southern Africa and Australia should be added.

High-risk *pathways* are ship and air traffic rather than car and train traffic, since Europe has no car and train traffic with North America, Australia and South Africa, and little car and train traffic with East Asia. However, cars, trains and inland navigation can play a key role in the secondary spreading of species from airports and seaports.

High-risk *vehicles* for mosquito vectors are water-filled containers on ships, mainly tyres and plant-containers. Not-disinfected aircraft cabins are an additional risk. As for pathogens, high-risk vehicles are infected humans and live animals, including livestock, pets and associated arthropods. The numerous imported cases of human vector-borne diseases have not yet sparked outbreaks in Europe (Tatem 2006a), but the recent invasion of bluetongue in Northwest Europe may have started from an imported infected animal (see Chapter 7).

We cannot entirely exclude the possibility of terrorists deliberately introducing arthropod-borne pathogens. However, bio-terrorists would probably prefer much more effective weapons such as anthrax, Marburg fever and smallpox. And if livestock would be the target, the foot-and-mouth disease virus is a much more powerful weapon than is the bluetongue virus.

Finally, high-risk *habitats* for mosquito invasion include wetlands, freshwater as well as brackish ones, and water containers (including tiny ones) in urban environments. Heated buildings are also at risk, particularly from subtropical mosquitoes and ticks, as do greenhouses close to seaports and airports, such as those found in the Netherlands.

Strategy for prevention and containment

Two types of invasion

What strategies are feasible if we wish to prevent invasions of vector-borne diseases? This depends on the type of invasion involved: climate-driven or transportation-driven.

Climate-driven invasions cannot be stopped in many cases since they often take place along a broad front. From a conservation point of view, it may not even be wise to stop such invasions, since species need space to adjust their ranges to the changing climate. On the contrary, it may be wise to facilitate such invasions, for example by creating ecological networks. Fortunately, the risk that climate-driven populations will explode is limited by the fact that they will often be accompanied by their natural enemies.

But, of course, nobody will advocate facilitating invasions of human and animal pathogens and their vectors. Next to causing health problems, they could spark habitat destruction from drainage of wetlands. If such species can be contained or eradicated, it should be done. Fortunately, there are few vector-borne diseases likely to invade Europe in response to climate change alone in the next decades.

Transportation-driven invasions, by contrast, can be stopped in many cases, since they often have to pass a narrow entry point (bottleneck) such as a seaport or airport. The risk of population explosions is much greater in such cases, since these species will have lost most of their natural enemies. Here it seems wise to apply the Precautionary Principle for three reasons:
- The damage from an introduction is unpredictable but may be large and widespread.
- The introduction may well be irreversible and the damage permanent.
- Even if eradication, containment or control is possible, prevention is often less costly and less harmful to the environment.

The favoured options are: prevention > eradication > containment > control.

Preventing introduction

The practical opportunities for prevention of invasions depend on the type of vector or pathogen, technical opportunities for detection and identification, and opportunities for eradication.

Mosquitoes are introduced mainly by ship-borne transportation of eggs and larvae in tyres (Tatem *et al.* 2006). Another ship-borne vehicle are ornamental plants such as Bromeliads that carry water in their axils (Lounibos 2002), and *Dracaena sanderiana* plants placed in water or gel (Scholte *et al.* 2007a). There are no confirmed examples of vector establishment from air travel, although there is circumstantial evidence that this has occurred (Tatem *et al.* 2006). This difference is probably due to the fact that airplanes carry small numbers of adult mosquitoes, whilst water jars in ships can carry large numbers of eggs and larvae (Lounibos 2002).

Eggs and larvae can be eradicated by removing water from tyres and applying larvicides in plant containers. Introduction of adult mosquitoes can be prevented by spraying aircraft cabins with insecticides, whilst avoiding health risks. The absence of *Ae. albopictus* from New Zealand, in spite of intensive trade with Japan, may well be the result of strict preventive measures (Tatem *et al.* 2006). In Australia, however, *Ae. albopictus* is a frequent visitor, mostly from the Torres Strait (Ritchie *et al.* 2006).

Introduction of pathogens can be prevented most effectively by avoiding import of infected animals. This requires intensive surveillance and - in case of doubt - quarantine measures. In case of doubt, quarantine should be applied. Another effective measure is minimising transport of vectors, as mentioned. Infected travellers should be taken to hospital as soon as possible and treated in a mosquito-free room.

Surveillance is expensive, so it is necessary to set priorities. Instead of searching for needles in the haystacks of all different travel and trade routes, it is much more efficient to identify high-risk routes and traffic. The focus would be on transportation between distant seaports and airports sharing a similar climate but not a similar fauna, focusing on those routes with most travel and trade. Tatem *et al.* (2006) have recently done innovate research in this field. They found that the present distribution of *Ae. albopictus* can largely be explained by two factors: climatic similarity and traffic volume. They used this to calculate the likelihood of further introductions of *Ae. albopictus* and of malarious *Anopheles* species escaping from Africa.

As for *Ae. albopictus*, Tatem *et al.* (2006 and 2006a) calculated that out of >6,000 possible air travel routes, 8 of the top 10 risk routes depart from Japanese airports, one from a South-Korean and one from a Taiwanese airport. None of these routes have a European destination. However, out of >25,000 possible shipping routes, the top 20 all depart from Japanese seaports and 5 of these have a European destination: Genoa. This was one of the earliest European cities to report *Ae. albopictus*, which was established there in 1990.

As for *Anopheles*, Tatem *et al.* (2006) calculated the likelihood of four principal members of the *An. gambiae* complex, effective vectors of *P. falciparum*, escaping from Africa. One conclusion is that *An. gambiae sensu stricto* can be introduced in tropical Asia and America, but very unlikely in Europe. The southern species *An. quadriannulatus* Theobald is a more serious candidate for introduction in (Mediterranean) Europe, particularly by sea traffic. However, it is a less efficient malaria vector because it rarely bites humans (Takken *et al.* 1999).

Finally, Tatem and Hay (2007) calculated how the climate similarities between airports across the globe change with the seasons. One conclusion was that central and western Europe is climatically most similar to Japan and China in January, but to the eastern USA in July.

Such sophisticated analyses can greatly help to prioritise monitoring, inspection and quarantine efforts. Predictably, the fast-growing economy of China will figure more and more in such analyses.

Eradication

There are various ways to prevent species from spreading following introduction. One is to minimise the area of suitable habitat for such species (e.g. wetlands in the case of mosquitoes) in the vicinity of seaports and airports. But once a vector or a pathogen, or both, have already started to spread, quick eradication becomes crucial. Various methods are available.

For eradicating vectors, at least mosquito vectors, classic methods are drainage of artificial and natural water pools and spreading of (acceptable) insecticides, particularly larvicides. Biological control by natural enemies, though sometimes very effective in controlling an established species, is rarely appropriate to eradicate it or to prevent it from spreading. Introduction of sterilised or genetically modified males may be more effective. For example, the New-World screwworm fly *Cochliomyia hominivorax* (not a vector but a parasite) was eradicated in North Africa in 1991, soon after its establishment, by the introduction of sterilised males (Kouba 2004). However, this technique has been less successful in the case of mosquito or tick control.

Once a vector has spread, eradication becomes much more difficult, but it may still be feasible if the species has an Achilles heel, such as slow dispersal or a narrow range of hosts or habitats. A classic example is the eradication of *An. gambiae* in Brazil. The mosquito had spread widely in the northeast of the country from 1930. The large-scale eradication campaign that was launched could be successful only because the mosquito mainly bred in small pools, borrow pits, and shallow wells dug for drinking water near houses, not in forests. Furthermore, the mosquito did not fly long distances (Elton 1958, Lounibos 2002, Killeen 2003).

If a pathogen has spread, eradication methods include vector control, vaccination of infected people or livestock (provided an appropriate vaccine is available) and isolation and treatment of infected people or livestock with drugs. In the case of livestock, culling of (suspected) infected animals is an additional option.

The most spectacular eradication success story is the worldwide eradication of the smallpox virus, completed in 1977, applying mass vaccination, case finding (supported by local people) and isolation of infected persons (Fenner *et al.* 1988). However, this campaign could be successful only because the smallpox virus has no other reservoir hosts than humans.

One success story involving a vector-borne pathogen was the eradication of malaria from vast non-tropical regions in North America, Australia, Asia and Europe after World War II. This was achieved through campaigns that combined patient treatment with mosquito control using drainage or spraying of water pools. *Plasmodium vivax* and *P. falciparum* vanished from the whole of Europe (but see Chapter 3). But again, these *Plasmodium* species have no other reservoir hosts than humans, at least not in Europe. Eradicating a pathogen that has a broader host and reservoir range is much more difficult.

If the species has spread and cannot be eradicated at reasonable costs, further spread can in some cases be prevented by containment, applying the same methods as mentioned above. In severe cases the creation of distribution barriers, comparable to fire-lanes in forests, would be considered, such as zones of dry land where mosquitoes cannot breed, or fences that stop big mammal vector species[25]. However, effective barriers will be feasible in few cases.

[25] One extreme example of containment of an animal is the long fence established in Australia in the early 20th century to contain the dingo, a notorious sheep killer. The fence is 5614 km long, three times as long as the Great Wall of China (Bomford and Hart 2002). Such fences can perhaps contain an introduced reservoir host species in exceptional cases, but not a pathogen or a vector.

If containment is not an option or fails, the last resort is control. Effective control should be integrated, combining ecological, biological and chemical methods with social strategy, as advocated by Spielman and D'Antonio (2001) and Utzinger *et al.* (2002) for malaria. In some cases, an organised counter-invasion by a natural enemy introduced from the native area of the vector species can be effective, though no success stories involving mosquitoes are known to us.

International regulations

Even though preventive measures are to be preferred, it is precisely such measures that can easily come into conflict with international trade and thereby with economic interests. There is a basic contradiction between free trade and the protection of humans, animals, plants and ecosystems from introduced pathogens, vectors and pests. Some advocates of free trade therefore argue that such risks are a fact of life in a globalising world and should be accepted in view of the benefits of free trade to the economy and human welfare. However, this viewpoint seems to be one-sided and short-sighted and is not justified by present trading rules.

When the first global system of trading rules, the General Agreement on Tariffs and Trade (GATT) was developed and implemented in 1948, the risk of spreading diseases and pests was recognised and provisions were included to reduce such risks. Contracting parties were allowed, under certain conditions, to close their borders for those imports representing a proven risk of introducing diseases.

Several other multilateral agreements were adopted that specifically aim at the protection of human, animal and plant health and biodiversity.

Protecting biodiversity

The most comprehensive international agreement on the subject of introducing alien species is the Convention on Biological Diversity (CBD), agreed in Rio de Janeiro in 1992. It states that each contracting party shall, as far as possible and appropriate, prevent the introduction of, or control or eradicate those alien species that threaten ecosystems, habitats or species (IAS).

A key principle is the precautionary approach: a state should take appropriate action even if there is lack of scientific certainty. A state should give priority to prevention of introduction of IAS above eradication, containment and control.

Protecting human health

A second relevant multilateral agreement is the new International Health Regulations. The regulations aim to provide a legal framework for the prevention, detection and containment of public health risks at source, before they spread across borders, through the collaborative actions of states parties and the World Health Organisation.

Protecting animal health

The need to control (communicable) animal diseases at global level led to the creation of the Office International des Epizooties (OIE; World Organisation for Animal Health) in 1924. Each member country reports the (communicable) animal diseases that it detects on its territory. The OIE then disseminates the information to other countries, which can take the necessary preventive action. This information also includes diseases transmissible to humans as well as intentional introduction of pathogens. The OIE develops normative documents relating to rules that member countries can use to protect themselves from the introduction of diseases and pathogens, without setting up unjustified sanitary barriers.

Trade rules

Relevant World Trade Organisation agreements on the subject of diseases are the Agreement on Technical Barriers to Trade (TBT) and the Agreement on the application of Sanitary and Phytosanitary Measures (SPS).

Under the TBT legitimate objectives to regulate the import of goods include:
- protection of human health or safety;
- animal or plant life or health;
- the environment.

For example, based on the TBT, the Dutch government issued a covenant for the import of Lucky Bamboo *Dracaena sanderiana*. The interest to be protected was public health (http://www.bis.org.in/sf/feb2007/gtbtn07NLD73.doc, accessed 10 September 2007). This was not specified in more detail, but in 2005 *Ae. albopictus* was discovered in containers with lucky bamboo imported from China (Scholte *et al.* 2007a,b).

Under the SPS, measures can be taken to protect animal or plant life or health from risks arising from the entry, establishment or spread of pests, diseases, disease-carrying organisms or disease-causing organisms.

Clearly, both TBT and SPS allow restrictions on imports of human, animal and plant pathogens as well as their vectors. This includes imports of car tires and plants carrying a high risk of containing vectors.

However, a state may not regulate the import of goods with the effect of creating unnecessary obstacles to international trade. For this purpose, regulations shall not be more trade-restrictive than necessary to fulfil a legitimate objective.

Gaps in multilateral regulations

Specific gaps in the international regulatory frameworks persist, notably in relation to species that are invasive, but do not qualify as plant pests under the international agreements, with regard to a number of pathways. These include:
- Unintentional introductions of invasive alien species through international assistance and humanitarian programmes, tourism, military, scientific research, cultural and other activities.
- Intentional introductions of alien species for non-food purposes, including certain aspects of horticulture and trade in pets and aquarium species.

One surprising gap in the European regulatory framework is that the EU can intervene in case of an invasion that threatens livestock or crops, but not in case of an invasion that threatens public health.

Summarising, whilst the global transportation network, the trade volumes and the number of trade-related invasions are growing, the legal framework for preventing trade-related bio-invasions is developing as well, even though there are major gaps to be filled.

Suggested guidelines

The frequency of invasion of vectors and their pathogens is likely to increase further in the course of this century, as the three main drivers continue to operate:

- Global travel and transportation will continue to grow. Species will move even faster, further and in greater numbers than today.
- Habitat transformation will also continue, including the creation of disturbed habitats that are often used as bridgeheads by invasive species. In Europe, the restoration of freshwater and saline wetlands may provide new habitat for introduced mosquitoes. And the creation of habitat networks such as Natura 2000 of the European Union (EU) may facilitate the spread of some introduced species, including ticks, with their reservoir hosts.
- Climate change has become an additional driver of bio-invasion and will continue as well, making some regions less, and others more suitable for pathogens, their reservoir hosts and their vectors. As vector-borne diseases mainly occur in the tropics and subtropics, Europe may see more of such diseases invading.

Since climate-driven invasions are much more difficult to stop than transportation-driven invasions, it seems wise to concentrate prevention and control efforts on the latter. We suggest applying the following guidelines[26]:

1. Put prevention first. Apply the Precautionary Principle to all transportation-driven invasions. As for climate-driven invasions, confine application to high-risk invasions.
2. Effective prevention should include:
   - Prepare risk analyses for a wide range of candidate invasive pathogens and vectors, for Europe at large as well as for individual states. Culicid mosquitoes and ticks deserve particular attention. These analyses would include the likelihood of invasion for various species/pathway combinations, expected impacts, methods of prevention, eradication and control as well as the costs involved. For a few pathogens the possibility of deliberate introductions by bio-terrorists would be taken into account as well.
   - Identify high-risk travel and trade flows. For pathogens, these include: live animals as well as humans travelling from regions where harmful vbds are endemic. For mosquito vectors: air in airplanes as well as water and gel in containers on ships. For ticks: live animals.
   - Identify high-risk travel and transportation routes. Focus on high volume transportation between distant seaports and airports having a similar climate but not a similar fauna, using models such as those developed by Tatem *et al.* (2006).
   - Identify high-risk habitats for invasion and minimise such habitats in the vicinity of seaports and airports. Or, alternatively, minimise specific high-risk transportation volumes to specific seaports or airports.
   - Prepare contingency plans for a wide range of risk species invasions, similar to the current plans for outbreaks of highly contagious human and veterinary diseases. Make sure that appropriate insecticide, vaccine and drug stockpiles as well as skilled manpower will be available within a matter of days.

[26] Many of these recommendations were taken from Wittenberg and Cock (2001) and Van der Weijden *et al.* (2007).

- Try and prevent deliberate or accidental introduction of every species considered a risk, or for which no risk-analysis is available. This requires at-source measures in exporting countries, including surveillance, interception or disinfection of contaminated cargo, and disinfection of airplanes, ships and risk cargo. Supplement this in importing countries by strict border checks, interception and quarantine measures at seaports and airports.

3. When, in spite of such measures, a risky alien pathogen or vector or pest is found, or a species for which no risk analysis is available, try and eradicate it as soon as possible. Mosquitoes can be eradicated by drainage of natural and artificial water containers and by using pesticides and sterilised males. Many pathogens can be eradicated by rapid isolation and treatment of infected persons or animals, vaccination, and – if necessary - culling of infected animals.
4. If the species has nevertheless spread and cannot be eradicated at reasonable costs, try and contain its range, applying the same methods as mentioned under 3. In severe cases the creation of distribution barriers can be considered.
5. Raise awareness of the risks of bio-invasions among tourists, exporters, transporters, importers, customs-officers and other stakeholders.
6. Develop and promote best preventive practices for the travel and trade business. 'Global hygiene' in travel and trade should become as generally accepted as hygiene has become in private and social life.
7. The present EU system of individual identification and registration of livestock is a powerful tool in tracking and tracing infected animals in case of an outbreak and controlling that outbreak. The system is less relevant for vector-borne disease, since vectors can move unchecked. But in the early stages of an invasion of a pathogen or tick vector species, the system can help to eradicate that species.
8. Introduce a legal obligation to report emerging alien harmful species, while compensating the reporting party for the cost in case of isolation and eradication.
9. Make importing firms and persons legally liable for the economic damage from species introductions.
10. Improve coordination between (or merge) specialised agencies responsible for bio-invasions harming public health, animal health, plant protection, fisheries, gardening and nature. For example, the West-Nile virus is hosted by many bird and mammal species including humans and transmitted by many mosquito species. That renders an ecological and integrated approach inevitable.
11. Enhance multidisciplinary research on vector-borne diseases, including experts in entomology, ornithology, ecology, epidemiology, virology, climatology, logistics, economy and law.
12. One challenge is the development of predictive models for the various stages of bio-invasion. For example, models to predict the likelihood of species *A* being introduced in seaport *B* from seaport *C* with cargo type *D*.
13. Improve international coordination, both at the European and global level. Introduce an obligation for EU members to mutually inform and consult each other on bio-invasions and measures taken, and to provide assistance in prevention, eradication and control efforts. Mandate the EU to intervene in case of an invasion that threatens public health and can affect other member states. In addition, mandate the EU to build up strategic stocks of vaccines and pesticides. Comply with present 'safe trade' agreements (SPS and TBT). Negotiate more specific agreements with high-risk countries such as Japan and China. And where necessary, fill the gaps in the global regulations.

More generally speaking, so far as trade is concerned, the ultimate goal would be to detach biological from economic globalisation and to replace free trade with safe trade.

Acknowledgements

We thank Alida Laurense (AgroEco, Dar es Salaam), Pieter Bol (Delft University of Technology) and an anonymous reviewer for their comments on an earlier draft of this paper. Pieter Bol also contributed to previous texts in the book *Biological Globalisation*. Maaike Bijker commented on language.

References

Andrivon D (1996) The origin of *Phytophthora infestans* populations present in Europe in the 1840s: a critical review of historical and scientific evidence. Plant Pathol 45: 1027–1035.

Ballard C, Brown P, Bourke RM and Harwood T (eds.) (2005) The sweet potato in Oceania: a reappraisal. University of Sydney, Sydney.

Benedictow OJ (2006) The Black Death 1346-1353 – The complete history. The Boydell Press, Woodbridge/Rochester.

Bomford M. and Hart Q (2002) Non-indigenous vertebrates in Australia. In: D. Pimentel (ed.), 2002. Biological invasions - Economic and environmental costs of alien plant, animal, and microbe species. CRC Press, Boca Raton/London/New York, Washington, D.C.

Bright C (1998) Life out of bounds – Bioinvasion in a borderless world. W.W. Norton & Company, New York/London.

Bryant JE, Holmes EC and Barrett ADT (2007) Out of Africa: a molecular perspective on the introduction of yellow fever virus into the Americas. PloS Pathog 3(5): e75.

Centres for Disease Control and Prevention (2007) Lyme Disease -- United States, 2003—2005. http://www.cdc.gov/mmwr/preview/mmwrhtml/mm5623a1.htm?s_cid=mm5623a1_e (Accessed 10 September 2007).

Centres for Disease Control and Prevention (2007) 2006 West Nile virus activity in the United States. http://www.cdc.gov/ncidod/dvbid/westnile/surv&controlCaseCount06_detailed.htm (Accessed 10 September 2007).

Charrel RN, Gallian P, Navarro-Marí J-M, Nicoletti L, Papa A, Sánchez-Seco MP, Tenorio A and de Lamballerie X (2005) Emergence of Toscana virus in Europe, 2005. Emerg Infect Dis 11: 1657-1663.

Childs Kohn, G. (2001) Encyclopedia of plague and pestilence: from ancient times to the present. Checkmark Books, New York.

Coatney GR, Collins WE, Warren M and Contacos PG (1971) The primate malarias. U.S. Department of Health, Education and Welfare, Bethesda.

Crosby AW (1972) The Columbian Exchange – Biological and cultural consequences of 1492. Greenwood, Westport, Conn.

Crosby AW (1986) Ecological imperialism – The biological expansion of Europe, 900-1900. Cambridge University Press, Cambridge.

DeJong RJ, Morgan JAT, Paraense WL, Pointier J-P, Amarista M, Ayeh-Kumi PFK, Babiker A, Barbosa CS, Brémond P, Canese AP, Pereira de Souza C, Dominguez C, File S, Gutierrez A, Incani RN, Kawano T, Kazibwe F, Kpikpi J, Lwambo NJS, Mimpfoundi R, Njiokou F, Poda JN, Sene M, Velásquez LE, Yong M, Adema CM, Hofkin BV, Mkoji GM and Loker ES (2001) Evolutionary Relationships and Biogeography of *Biomphalaria* (Gastropoda: Planorbidae) with implications regarding its role as host of the human bloodfluke, *Schistosoma mansoni* Randall. Mol Biol Evol 18: 2225-2239.

Elbers ARW, van Rijn PA and van Rooij EMA (2003) Preliminary risk analysis of introduction of Bluetongue virus and West Nile virus into the Netherlands. CIDC, Lelystad.

Elton CS (1958) (reprint 2000). The ecology of invasions by animals and plants. The University of Chicago Press, Chicago.

Erdos MD and Koncz A (1977) Experience in the control of Pharaoh's ants in Hungary. Intern Pest Contr 19: 12-13.

Escalante AA, Barrio E and Ayala FJ (1995) Evolutionary origin of human and primate malarias: evidence from the circumsporozoite protein gene. Mol Biol Evol 12: 616-26.

Estrada-Peña A (2003) The resistance status of the tick *Rhipicephalus sanguineus* (Ascari: Ixodidae) to chemical ixodicides in Spain. http://www.vetcontact.com/presentations/show.php? act=show&vid=376&langselect=en&lang=en&ucnt=56&pflag=1&fglyt= (Accessed 10 September 2007).

Fenner F, Henderon DA, Arita I, Jezek Z and Ladnyi ID (1988) Smallpox and its eradication. World Health Organization, Geneva.

Garben AF, Bosman BT and van Bronswijk JE (1980) The brown dog tick *Rhipicephalus sanguineus* Latreille 1806 in the Netherlands, an analysis of imported cases, including their veterinary and medical significance. Tijdschr Diergeneeskd 105: 192-203.

Ginsberg HS, Buckley PA, Balmforth MG, Zhioua E, Mitra S and Buckley FG (2005) Reservoir competence of native North American birds for the Lyme disease spirochete, *Borrelia burgdorferi*. J Med Entomol 42: 445-449.

Goldschmidt T (1998) Darwin's dreampond: drama in Lake Victoria. MIT Press, Cambridge, Mass.

Gould EA, Higgs S, Buckley A and Gritsun TS (2006) Potential arbovirus emergence and implications for the United Kingdom. Emerg Infect Dis 12: 549-555.

Gratz NG (2004) Critical review of the vector status of *Aedes albopictus*. Med Vet Entomol 18: 215-27.

Holway DA, Lach L, Suarez AV, Tsutsui ND and Case TJ (2002) The ecological causes and consequences of ant invasions. Annu Rev Ecol and System 33: 181-233.

Josseran L, Paquet C, Zehgnoun A, Caillere N, Le Tertre A, Solet JL and Ledrans EM (2006) Chikungunya disease outbreak, Reunion Island [letter]. Emerg Infect Dis 7: 497-504.

Killeen GF (2003) Following in Soper's footsteps: northeast Brazil 63 years after eradication of *Anopheles gambiae*. Lancet Infect Dis 3: 663-666.

Kouba V (2004) History of the screwworm (*Cochliomyia hominivorax*) eradication in the Eastern Hemisphere. Hist Med Vet 29: 43-53.

Lake J and Leishman M (2004) Invasion success of exotic plants in natural ecosystems: the role of disturbance, plant attributes and freedom from herbivores. Biol Conserv 117: 215-226.

LCCM (2004) Landelijk Jaarverslag 2003. [Annual report 2003]. Landelijke Coördinatiecommissie Muskusratten-bestrijding, 's-Hertogenbosch, The Netherlands.

Lord CC (2001) Brown dog tick. http://creatures.ifas.ufl.edu/urban/medical/ brown_dog_tick.htm (Accessed 10 September 2007).

Lounibos LP (2002) Invasions by insect vectors of human disease. Ann Rev Entomol 47: 233-266.

Lowe S, Browne M, Boudjelas S and de Poorter M (2001) 100 of the worlds' words invasive alien species, a selection from the Global Invasive Species Database. IUCN-ISSG, Auckland.

McNeill WH (1998) Plagues and peoples. Anchor Books, New York.

Moore CG and Mitchell CJ (2000) *Aedes albopictus* in the United States: ten-year presence and public health implications. Emerg Infect Dis 3: 329-334.

Parola P (2004) Tick-borne rickettsial diseases: emerging risks in Europe. Comp Immun Micobiol Infect Dis 27: 297-304.

Pimentel D, Zuniga R and Morrison D (2005) Update on the environmental and economic costs associated with alien-invasive species in the United States. Ecol Econ 52: 273-78.

Quammen D (1996) The Song of the Dodo - Island biogeography in an age of extinctions. Simon & Schuster, New York.

Reusken CBEM and Takken W (2006) Vectoren van het West Nile-virus in Nederland. [Vectors of the West Nile virus in the Netherlands]. Infectieziektenbulletin 17: 420-421.

Ritchie SA, Moore P, Carruthers M, Williams C, Montgomery B, Foley P, Ahboo S, Van den Hurk AF, Lindsay MD, Cooper B, Beebe N and Russell R (2006) Discovery of a widespread infestation of *Aedes albopictus* in the Torres Strait, Australia. J Am Mosq Control Assoc 22: 358-365.

Sabatinelli G, Ejov M and Joergensen P (2001) Malaria in the WHO European Region (1971–1999). Euro Surveill 6: 61-65.

Scholte EJ, Dijkstra E, Ruijs H, Jacobs F, Takken W, Hofhuis A, Reusken C, Koopmans M and De Boer A (2007a) The Asian tiger mosquito (*Aedes albopictus*) in the Netherlands: should we worry? Proc Neth Entomol Soc 18: 131-136.

Scholte EJ, Jacobs F, Linton YM, Dijkstra E, Fransen J and Takken W (2007b) First record of *Aedes* (Stegomyia) *albopictus* in the Netherlands. Eur Mosq Bull 22: 5-9.

Scott S and Duncan C (2004) The return of the Black Death: the world's greatest serial killer. Wiley, Chichester.

Soper FL and Wilson DB (1943) *Anopheles gambiae* in Brazil: 1930-1940. Rockefeller Foundation, New York.

Spielman A and D'Antonio M (2001) Mosquito - A natural history of our persistent and most deadly foe. Faber and Faber, London.

Sutherts RW (2004) Global change and human vulnerability to vector-borne diseases. Clin Microbiol Rev 17: 136-173.

Takken W, Eling W, Hooghof J, Dekker T, Hunt R and Coetzee M (1999) Susceptibility of *Anopheles quadriannulatus* Theobald (Diptera: Culicidae) to *Plasmodium falciparum*. Trans R Soc Trop Med Hyg 93: 578-580.

Tamis W (2005) Changes in the flora of the Netherlands in the 20th century. Gorteria, suppl. 6: 1-233.

Tatem AJ, Hay SI and Rogers DJ (2006) Global traffic and disease vector dispersal. Proc Nat Acad Sci USA 103: 6242-6247.

Tatem AJ, Rogers DJ and Hay SI (2006a). Global transport networks and infectious disease spread. Adv Parasit 62: 293-342.

Tatem AJ, Rogers DJ and Hay SI (2006b). Estimating the malaria risk of African mosquito movement by air travel. Malar J 5: 57.

Tatem AJ and Hay SI (2007) Climatic similarity and biological exchange in the worldwide airline transportation network. Proc Royal Soc B 274: 1480-1496.

The history place (2000) http://www.historyplace.com/worldhistory/famine/after.htm (Accessed 10 September 2007).

Urbanelli S, Bellini R, Carrieri M, Sallicandro P and Celli G (2000) Population structure of *Aedes albopictus* (Skuse): the mosquito which is colonizing Mediterranean countries. Heredity 84: 331-337.

Utzinger J, Tanner M, Kammen DM, Killeen GF and Burton H (2002) Integrated programme is key to malaria control. Nature 419: 431.

Van der Giessen JWB, Isken LD and Tiemersma EW (2004) Zoonoses in Europe: a risk to public health. National Institute for Public Health and the Environment, Bilthoven.

Van der Weijden W, Leewis R and Bol P (2007) Biological globalisation – Bio-invasions and their impacts on nature, the economy and public health. KNNV Publishing, Utrecht.

Van Lier EA, Rahamat-Langendoen JC and van Vliet JA (2007) *Staat van infectieziekten in Nederland 2006* [Status of infectious diseases in the Netherlands 2006]. RIVM, Bilthoven.

Williamson M (1997) Biological invasions. Chapman & Hall, London.

Wittenberg R and Cock MJW (eds.) (2001) Invasive alien species: a toolkit of best prevention and management practices. CAB International. Wallingford, Oxon.

World Health Organisation (2007) A safer future: global public health security in the 21st century. World Health Organization, Geneva.

Epilogue: emerging pests and vector-borne diseases are a reality

Willem Takken and Bart G.J. Knols

Abstract

Historically, Europe has suffered pests and vector-borne diseases affecting man and domestic animals. Through advances in science and technology, measures were introduced that led to effective management and control, causing the eradication of several diseases (e.g. plague, typhus, malaria) or reduction of pests to levels where they no longer were harmful or damaging. In the last decades this situation has changed and a series of pests and diseases are resurging or emerging giving cause for serious concern, threatening both human and animal health. The reasons for this upsurge and emergence are being discussed, and actions are being proposed to reduce the incidence of diseases already present and to lower the risk of introduction of new ones. Suggestions for the control of pest organisms are being presented, using integrated control with emphasis on biological methods for insect control. Recommendations for the prevention of outbreaks of pests and vector-borne diseases associated with climate- and environmental change are proposed, including the establishment of early-warning systems and the development of pan-European contingency plans.

Keywords: remedial action, epidemiology, resistance, resurgence, climate change, Europe, biological control, environmental management

A growing threat

The preceding chapters demonstrate that in historic times pests and vector-borne diseases posed a huge health risk to humans and livestock in Europe. Notably malaria was highly prevalent, not only along the shores of the Mediterranean, but also in countries as far north as Norway and Sweden. Yellow fever was widespread in southern Europe and of course, plague caused devastating effects, notably in the 14th and 15th Centuries. Head lice and house flies were part of every household being rich or poor, and people had learnt to tolerate them until effective methods of control became available in the early 20th Century. With the discovery of parasite life cycles, drugs and insecticides the 20th Century provided prospects for real control of these pests and diseases, and it was widely expected that many of them could be eradicated. Indeed, yellow fever disappeared from Europe, most likely because of the disappearance of its vector *Aedes aegypti* (L.). It is peculiar that the reasons for the disappearance of this mosquito species from Europe are not known: the species is still widely common in the continental USA, China, South Africa and Brazil/Argentina, where it experiences a climate similar to that of southern Europe. By contrast, the disappearance of malaria from Europe is better understood, and is the result of targeted interventions using anti-malarial drugs, insecticides and environmental measures (Bruce-Chwatt and De Zulueta, 1980). Also, the prevalence of head lice infections decreased markedly by the middle of the 20th Century, presumably as a result of better hygiene and effective insecticides. Similarly, bed bugs were until recently, rarely considered a problem, even though they were highly prevalent before World War II. Few animal diseases in Europe were attributed to arthropods, directly or indirectly, although periodic outbreaks of African horse sickness and bluetongue, transmitted by *Culicoides* spp., were reported on the Iberian peninsula (Ortega *et al.* 1998).

What may have caused a change in these optimistic views? As this book shows, many vector-borne diseases are lingering on in Europe, and some new diseases, Usutu virus, Chikungunya virus and bluetongue virus have emerged in regions of Europe where they had never before been recorded. Also, there is a huge problem with several pest insects such as head lice, bed bugs and house flies: not only are these pests increasing in prevalence and abundance, they have also become resistant against conventional chemical treatments. Endemic vector-borne diseases like Lyme disease and tick-borne encephalitis (TBE) have increased in prevalence, and also seem to distribute across larger areas than before. On the livestock side, the 2006 emergence of BT serotype-8 in north-western Europe has shaken the livestock industry, and in poultry farms the emergence of multi-resistant housefly strains poses a serious economic problem for the poultry industry. Sheep scab is causing huge losses to the sheep industry in the UK, and because of acaricide resistance an effective solution for mite control is lacking. Finally, judging from the recent epidemics of SARS and avian influenza, there appears to be a genuine global dispersal of infectious diseases for reasons that are only partially understood. The resurgence and emergence of these diseases and pests is giving rise to considerable discontent in society and costs to the economy. Whereas until the late 1990s the people of Europe considered their continent 'free' of such diseases and pests, they now are confronted with regular news bulletins announcing the outbreak of a new disease or the nuisance caused by insects that cannot be sufficiently kept under control. Just in the summer of 2007, a renewed outbreak of bluetongue in north-western Europe has received much attention and the discovery of >150 cases of Chikungunya in Italy has made headlines.

Table 1 lists the most prominent of these diseases and pests, as well as their epidemiological status in Europe. Many diseases and pests on this list have been around in Europe for a long time, but changes in human and animal hygiene, arthropod control and the development of effective drugs have caused the low prevalence or indeed complete disappearance of them in Europe. Some diseases, such as West Nile virus and African horse sickness, return to Europe periodically from Africa, possibly cycling around host immune levels: when a sufficient level of herd immunity has been built up, the disease disappears to return in a year when this immunity has waned. Other diseases, like leishmaniasis, have never gone away. However, for as yet unexplained reasons, their distribution and prevalence are increasing. Some diseases like plague and typhus have disappeared and not been recorded any longer in Europe. This is attributed to hygienic and sanitary measures and a modified style of living. The pathogenic agents causing these diseases, however, are still present elsewhere in the world and an occasional (small) outbreak in Europe is considered possible. Finally, there is a series of 'new' diseases and pests that have not been seen in Europe until recently. Lyme disease was discovered only in the early 1980s, and has meanwhile become the most prevalent vector-borne disease of humans in Europe. The Asian tiger mosquito *Aedes albopictus* (Skuse) arrived in Albania around 1978, and has since enjoyed frequent introductions, and subsequent establishment in many countries. It is now one of the main nuisance species in northern Italy, and has recently become established in Southeast France and Northeast Spain.

Unlike vector-borne diseases, the prevalence of pest species is often not considered as emerging or associated with environmental and climate change. This is probably because their presence is often strongly dependent on human behaviour. For instance humans are the only host species of head lice, and the existence of this insect depends on the efficacy of control measures employed by its own host. Head lice control has become dependent almost exclusively on insecticides (Chapter 17), and the lice are rapidly developing resistance against these chemical agents. Other reasons for the increasing incidence of head lice are poorly understood, but may be associated with different cultural behaviours of families, providing renewed opportunities for head lice. In a similar vein

Table 1. Overview of the most prevalent vector-borne diseases and arthropod pests in Europe and their epidemiological status as eradicated/stable/emerging.

	Arthropod vector or pest	Historical status	Epidemiological status	Emerging**
Disease				
Plague	Flea	+++++	Disappeared	No
Malaria	*Anopheles* spp.	+++	(+)	No
Yellow Fever	*Aedes aegypti*	+	Disappeared	No
Dengue	*Aedes aegypti*	+	Disappeared	Possible
West Nile	*Culex pipiens*	??	++	Possible
Chikungunya	*Aedes albopictus**	Not present	+	Yes
Usutu	*Culex* spp.	Not present	++	Yes
Leishmaniasis	*Phlebotomus* spp.	??	++	Yes
Typhus	*Pediculus humanus*	+++	Disappeared	No
Bartonella	*Ixodes ricinus / Pediculus humanus*	++	+	Yes
Q-fever	*Ixodidae* spp.	(+)	+	Possible
Human monocytic ehrlichiosis (HME)	*Ixodes ricinus*	??	+	Yes
Human granulocytic ehrlichiosis (HGE)	*Ixodes ricinus*	??	+	Yes
Lyme disease	*Ixodes ricinus*	(+)	+++	Yes
Rickettsia helvetica	*Ixodes ricinus*	??	+	Yes
Tick-borne encephalitis	*Ixodes ricinus*	(+)	+++	Yes
Tularaemia	*Ixodes ricinus*	(+)	++	Yes
Bluetongue	*Culicoides* spp.	Not established	+++	Yes
African horse sickness	*Culicoides* spp.	Not established	++	May become established
Pests				
Head louse	*Pediculus humanus capitis*	+++	++	Increasing in prevalence
Bed bug	*Cimex lectularius*	+++	+	Yes
Housefly	*Musca domestica*	+++	+++	Increasing in prevalence
Asian tiger mosquito	*Aedes albopictus*	Absent	++	Increasing in distribution and abundance
Sheep scab	*Psoroptes ovis*	+	++	Yes

* At the time of writing *Aedes albopictus* was suspect as the vector of Chikungunya virus in Italy; ** Emerging – Some diseases/pests have never been away, but are resurging and becoming widely prevalent. In that case we consider them as 'emerging'. Possible–environmental circumstances are suitable for its emergence and the disease is occasionally recorded.
+, ++, +++ increasing degree of prevalence; (+) historical presence unconfirmed

the resurgence of bed bugs may be seen: heavy use of pesticides to control bed bugs and other nuisance insects in the household has caused a rapid development of physiological resistance against a large number of chemicals. It may also be that a modern change in home care and control of indoor climate (wall-to-wall carpeting and central heating) favour the establishment and reproduction of bed bugs (Chapter 15). Another example of uncontrolled and intensive use of pesticides leading to resistance is the sheep ked in the British sheep husbandry. Here, measures are needed to develop alternatives to chemical control before the sheep ked can be brought under control (Chapter 13).

A threatening status

The examples of existing and new pests and vector-borne diseases listed in Table 1 are cause for serious concern. The emergence of Lyme disease in the 1980s was initially considered with curiosity and foremost a new challenge for the health care industry. Only when increasing numbers of patients were reported, and the disease was found to have a continental-wide distribution, became the authorities concerned. Today in many countries much research is funded for the development of diagnostic tools of Lyme disease, an effective drug treatment regimen and reliable epidemiological tools for risk assessment. This is done because no effective method for control of *Ixodes ricinus* L., the main vector species in Europe, is available. It is expected that under global warming the tick season will be extended in many countries, extending the transmission season possibly by several months each year. An additional threat is caused by the findings of other pathogenic agents being transmitted by *I. ricinus*: tick-borne encephalitis virus is the most important of these, having increased in incidence in Sweden and migrated into Norway (Lindgren and Gustafson 2001, Skarpaas *et al.* 2006) and spreading its distribution across Central Europe (Chapter 11). Other pathogens associated with *Ixodes ricinus* are *Anaplasma phagocytophilum* and *Ehrlichia chaffeensis*, *Rickettsia helvetica* and *Bartonella henselae*. Should the environmental conditions become more favourable for *I. ricinus*, as suggested in Chapter 22, then the incidence and geographic distribution of these diseases might increase further.

The discovery of the Usutu virus in Austria in 2001 has gone almost unnoticed in the scientific world. This is remarkable, given the high profile that the emergence of West Nile virus in the USA has received, and the numerous programmes that focus on Avian Flu (H5N1). In our opinion, Usutu virus has the potential of becoming more prevalent and more virulent if, by evolutionary force, the virus might enter different bird species or appear in a more virulent serotype. The virus has spread rapidly across Central Europe, possibly by several re-introductions or by birds emigrating from the index site in Austria. Being highly fatal for certain indigenous bird species in Austria, we should be alerted to the impact of the virus on other birds. Usutu virus has, therefore, the same potential as West Nile virus in the USA, by causing a kill-off of birds. In addition, humans develop disease symptoms following infection with Usutu virus, and the severity of these symptoms is still poorly understood.

Perhaps more dramatic, and certainly very serious, is the recent discovery of an epidemic of Chikungunya in Italy (Chapter 10). Whereas we do not yet know whether this epidemic has been brought under control (September 2007), it is worrying that it had been predicted following the recent and large epidemic of Chikungunya in the Indian Ocean and the Indian subcontinent (Bodenmann and Genton, 2006). This epidemic would most likely not have occurred if the vector *Ae. albopictus* had not been present in Italy. This mosquito was accidentally introduced there in 1990, and has since invaded most of northern Italy, thriving in both urban and rural ecosystems

(Chapter 14). This is an example where the introduction of an exotic vector species and an exotic arbo-virus have conspired into the scenario of a new disease aetiology. Chikungunya is a very serious disease, sometimes fatal, and nearly always causing a long convalescence of the patient (Pialoux *et al.* 2006).

On the livestock side we have seen frequent outbreaks of African horse sickness on the Iberian Peninsula. Each time, after the epidemic had run its course, the virus disappeared, to return some years later, possibly with *Culicoides* vectors, blown in by the wind from Africa across the Mediterranean see. Bluetongue is another disease transmitted by *Culicoides* spp. The disease was, like African horse sickness, infrequently seen in Portugal and Spain. Only after the invasion of the African vector *Culicoides imicola* Kieffer in Italy in 1998 did bluetongue become permanently established in Europe. Italy has experienced bluetongue infections each year since 1998. The disease is usually absent during the winter, presumably because of unfavourable temperatures for the *Culicoides* vectors. However, in August 2006 bluetongue erupted in the Netherlands, Belgium, Germany, France and Luxemburg. The virus was new to Europe, and identified as BT serotype 8. *Culicoides imicola*, the main vector in southern Europe, was absent from northern Europe, and it was established that local, indigenous, *Culicoides* spp. were responsible for the transmission. During the winter of 2006/2007 no new infections were recorded, and it was hoped that the disease would die out. In July 2007, however, new cases of bluetongue were found in Germany, Belgium and the Netherlands (in that order of appearance), and by September 2007 several hundred farms had been affected. Moreover, unlike in 2006, the virus appeared to be more virulent, at least in the Netherlands, where hundreds of sheep died because of the infection. This resurgence of bluetongue in north-western Europe suggests that the virus had survived the winter of 2006/2007, either in the vector or in a vertebrate host. The latter is considered unlikely as laboratory studies had shown that the virus looses its infectiousness in vertebrates between 60 and 100 days (A. Elbers, personal communication).[27]

A very different, but not less threatening, perspective for human and animal health is provided by the rapidly developing resistance to chemical pesticides in ectoparasites such as head lice, bed bugs, sheep ked and house flies. As we have seen, resistance against a large group of pesticides is already present in these organisms and widespread. Although these ectoparasites are not directly involved in disease transmission and rarely dangerous for humans, they nevertheless pose a serious problem for public health, causing economic damage (bed bugs, sheep ked and house flies) or socio-economic stress (head lice and bed bugs). In addition, at high densities, house flies may be involved in disease transmission, and sheep ked causes direct losses because of the health effect they have on sheep.

Cause of the observed change

The arrival of new vector-borne diseases, the resurgence of 'old' diseases and the increased prevalence of arthropod pests as described above are considered to be the result of environmental change and global warming. This explanation, although plausible, seems too simple and requires further details. For example, why was head lice in Europe successfully kept under control until the mid 1990s, and why has bluetongue virus not become established in Europe much earlier than 1998? When vector-borne diseases are being discussed, global warming is nearly always

[27] As this book went to press, it was confirmed that BT had emerged in the United Kingdom, and at least 5 farms had been infected by 28 September 2007.

mentioned as one of the reasons for resurgence or new outbreaks. However, global warming is a process that is taking place gradually, and the increased temperatures and associated changes in precipitation are unlikely to be the only factors that made these epidemics possible. In the last 50 years the world has witnessed a rapid increase in commerce and international travel. Thus, it has been proven that the continental spread of *Ae. albopictus* is largely caused by the trade in used tires and by shipment of ornamental plants (Chapter 14). *Aedes albopictus* is a competent vector of many arbo-viruses (Gratz 2004) and invaded the island of La Réunion in the 1950s. In the recent outbreak of Chikungunya on that island, *Ae. albopictus* was responsible for most of the virus transmission. This is only one example of an introduced mosquito becoming a vector of a deadly disease many years after its introduction.

In Italy, climatic conditions in the summer are favourable for transmission of bluetongue virus by *C. imicola*. It was therefore no surprise that the establishment of this vector species caused a rapid and intensive transmission of bluetongue virus in many parts of Italy. It is believed that the unusually high summer temperatures of 2006 in the Netherlands made an outbreak of bluetongue by indigenous *Culicoides* spp. possible: the high temperatures allowed for a rapid build-up of virus in the vector, so that virus transmission became possible. However, this hypothesis seems proven wrong by the ongoing transmission of bluetongue in 2007 in the Netherlands. The summer of 2007 has been much cooler and considerably more humid than that of 2006. These findings suggest, though, that environmental conditions, of which climate is an important component, are a strong factor in the epidemiology of bluetongue in north-western Europe.

Although malaria has not shown resurgence in continental Europe, the disease is still highly prevalent in parts of Asian Turkey. There is a huge amount of human movement between Turkey and countries of the European community. Thus it may be expected that healthy malaria parasite carriers are frequently arriving in regions where conditions for malaria transmission are favourable. The excellent status of health care is expected to intercept any malaria patient, but it is possible that a healthy parasite carrier, who is unaware of his condition, may serve as a reservoir of infection through the bite of anopheline mosquitoes. Indeed, several of such cases have been reported from northern Greece.

Arthropod pests such as head lice, bed bugs, house flies and the sheep ked are believed to increase in prevalence because of the widespread development of insecticide resistance. In poultry farms, where houseflies find numerous breeding sites in the manure and litter, large quantities of chemical pesticides have been used to reduce the nuisance caused by the flies. As a result, houseflies have developed resistance against most synthetic insecticides (Liu and Yue 2000, Pospischil *et al.* 1996). The same fate may await the head louse and the bed bug, as much of the control of these insects is conducted with synthetic chemicals. We have seen that the sheep ked can no longer be controlled with pesticides, and that alternative strategies must be sought for the control of this important group of pest organisms.

Remedial actions

Most European countries have only few effective programmes in place for the surveillance and control of existing vector-borne diseases, or for taking of measures to suppress or control ectoparasites. Yet, several tools are available to establish the incidence of disease and to control the vectors. In the case of vectors of Lyme disease, this is difficult as the most effective method would require treatment of the forest floor with pesticides (Chapter 23). This is highly undesirable for environmental and political reasons. Development of alternative strategies for tick control may include biological products, which are less harmful to the ecosystem and cause no side effects. Indeed, in the USA promising experiments have been conducted with the entomopathogenic fungus *Metarhizium anisopliae* for the control of *Ixodes scapularis* Say (Benjamin *et al.* 2002), suggesting that such studies should be undertaken in Europe for assessment of field populations of *I. ricinus*.

Countries are even less prepared for the occasional immigration of a new disease. When environmental conditions are suitable and potential vectors are present, a new pathogen can become established, as happened with bluetongue in Italy, and now, possibly, also in north-western Europe including the Netherlands. In Italy, control of bluetongue is focused on vaccination and the occasional pesticide treatment of stables and other indoor resting sites of bluetongue vectors. In north-western Europe pour-on insecticides are being recommended when transporting animals from infected zones, but an effective method for *Culicoides* control is not available. Both aerial spraying and/or treatment of breeding sites with pesticides are not being used for lack of data on efficacy and for the assumed harmful effects on the environment. The recent outbreak of Chikungunya in Italy was reason for the launching of a large-scale mosquito control programme in the outbreak area using aerial spraying and treatment of breeding sites of mosquitoes with chemical and biological products. By contrast, the outbreak of Usutu virus in Austria has not given rise to vector control, presumably as human incidence of the disease was very low and of mild character. Yet, many indigenous birds died because of Usutu virus, and this poses a threat to biodiversity.

The control of ectoparasites is often easier than that of arthropod vectors. Head lice can be removed by simple measures such as wet combing (Chapter 17). Bed bugs will disappear by thorough cleaning of infested rooms/buildings, which measure can be aided by the addition of steam cleaning and/or fumigation with a chemical product (Chapter 15). House flies can be significantly reduced by mechanical removal of animal dung and covering of organic (household) waste. These methods can be augmented by the release of natural enemies that parasitise larvae and pupae (Meyer *et al.* 1990). For the control of the sheep ked such measures are hardly available. As long as insecticide resistance has not been detected, treatment of the herd animals with a pour-on acaricide might be the only strategic method available for the control of these ectoparasites (Chapter 13).

Most programmes for pest control and management of vector-borne disease still depend heavily on the use of pesticides. In only few cases non-chemical methods are being used, often with startling success (Becker and Schaedler 1998). Biological control products can be used on a much wider scale and it is encouraging to note that in many European countries mosquito control programmes are increasingly switching to the use of *Bacillus thuringiensis israeliensis* and *B. sphaericus* (Chapter 21). This is, however, not the case for the control of other vector groups, which are controlled with conventional pesticides or left uncontrolled for lack of an adequate intervention strategy.

In order to prepare for effective vector control, accurate data on the epidemiology of the disease should be available. Epidemiological models can assist in predicting the course and timing of an infectious disease, and be used as a tool to simulate the impact of intervention strategies. If combined with geographical information (GIS) and remote sensing (RS) systems, rapid advances can be made in demarcation of the potentially affected zones. In turn, this information can be used by public health authorities to inform the public about the risk of the infection and for the planning of intervention strategies. In the case of sudden outbreaks of disease, this often happens on an ad-hoc basis, and is not part of the general public health studies. GIS and RS systems can be very useful in studies of the geographic distribution of vector species and for the identification of breeding sites. If linked to a meteorological satellite, GIS and RS are used for the preparation of risk maps, which will be useful for the planning of vector control operations. Epidemiology can also be used in predicting the potential establishment of a vector-borne disease in a country or region where the disease does not yet occur, but where environmental change may create conditions that are favourable for the disease. Using the basic reproduction number, R_0, it is possible to obtain information about the likelihood of disease outbreak, which in turn can be used for the preparation of a strategic plan in the event of disease outbreak (Anderson and May 1992)

The spread of an infectious agent is often dependent on human behaviour. Ideally, any new infectious case should be confined to his/her residence for the duration of the infectious period, so that human-to-human contact is kept as little as possible and the pathogenic agent cannot be disseminated in the environment. However, this is difficult to arrange, and many parasite carriers are healthy and do not suffer from disease symptoms, so they are unaware of their condition. In the case of an animal disease, such as bluetongue, farmers are often obliged to keep their animals indoors and any animal movement between farms or to other areas are strictly prohibited. For a vector-borne disease one can question such regulations, as the vectors may acquire the pathogen or parasite and subsequently fly to a different place, taking the disease agent along. In the 2006-bluetongue epidemic in the Netherlands, Belgium, Germany, Luxemburg and France, vector dispersal may have been one of the reasons for the rapid spread of the disease, as animal transport had been banned immediately from the start of the outbreaks.

Vectors of infectious disease and pests can only thrive in environments that are suitable for the species. Areas with very cold winters are usually free of vector species that require the outdoor environment for disease transmission. However, the human domestic environment, which is artificially heated, often provides an excellent habitat for numerous pests and vectors. Bed bugs benefit from centrally-heated buildings, house flies thrive in poultry and pig farms, *Culicoides* species make use of grazing areas rich in animal dung for oviposition, and certain mosquito species lay their eggs in organically-rich septic tanks and basements. In many cases, the problems caused by these organisms can be solved by adaptations to the environment, so that breeding sites are being removed (mosquitoes), animal dung is being removed at frequent intervals (house flies) and home adjustments to central heating, ventilation and removal of wall-to-wall carpeting (bed bugs). For mosquito control in inundated areas, drainage or controlled irrigation may remove most mosquito breeding sites.

Conclusions and recommendations

This book is proof of the resurgence and emergence of many pests and diseases in Europe that were thought to have been eradicated or, at least, to remain far removed from the European continent. Examples are given of vector-borne diseases that have been circulating in Europe for many years (TBE, Lyme disease) or that have arrived recently (Usutu, bluetongue, Chikungunya). Other vector-borne diseases may arrive, and yet others are emerging. Harmful ectoparasites and pests are increasing in prevalence and abundance. The reasons for this surprising, and alarming situation are complex, and include human behaviour, climate change, environmental change, the development of insecticide resistance and increased international travel and trade. It appears to be difficult to control indigenous diseases and pests, or eradicate incoming ones, and only when a disease outbreak remains within a defined area is effective eradication possible. Effective tools for disease control are lacking, not only because the aetiology of these vector-borne diseases is complex, but also because tools such as vaccines and sustainable insect control methods are lacking or inadequate. Europe is ill-prepared to meet the challenges posed by these diseases, and several steps should be taken urgently to prevent large-scale outbreaks of vector-borne disease. Similarly, blood-feeding arthropods can also cause huge problems because of their irritating, sometimes painful, biting behaviour. Especially in the livestock industry, these pests can lead to significant economic damage. Also, such pests may reduce the quality of living conditions and pose a serious threat for the tourist industry. For these reasons we propose that:

- Epidemiological units, preferably at country level, be established that monitor infectious vector-borne diseases and prepare predictive models to obtain insight in the spread and incidence of such diseases and to be prepared for the arrival of such diseases.
- Research units are created that study the ecology of vectors so that surveillance of the vectors can be increased.
- Tools are developed that can replace synthetic pesticides, and favour sustainable control based on integration of mechanical, cultural and biological methods (integrated pest and vector management).
- Disease control staff is being trained to apply GIS and RS technology for a rapid assessment of risk and to identify natural boundaries of the vector(s) and/or parasites and pathogens.
- Early warning systems are established in which the emergence and/or new arrival of vector-borne infectious diseases is being monitored to inform national and European authorities at the earliest opportunity of such developments.
- The European Centre for Disease Prevention (ECDC) and the World Organisation for Animal Health (OIE) act as reference centres for the provision of information of the status of vector-borne diseases in Europe associated with environmental change.
- The ECDC and OIE develop strategies for containment and control of pests and vector-borne disease so as to reduce health risks and the prevent disease and socio-economic damage caused by them.

References

Anderson RM and May RM (1992) Infectious Diseases of Humans - Dynamics and Control. Oxford University Press, Oxford.

Becker N and Schaedler P (1998) The use of entomopathogenic bacteria against mosquitoes. 6th Simposio de Controle Biologico 24.

Benjamin MA, Zhioua E and Ostfeld RS (2002) Laboratory and field evaluation of the entomopathogenic fungus *Metarhizium anisopliae* (*Deuteromycetes*) for controlling questing adult *Ixodes scapularis* (Acari: Ixodidae). J Med Entomol 39: 723-728.

Bodenmann P and Genton B (2006) Chikungunya: an epidemic in real time. Lancet 368: 258.

Bruce-Chwatt LJ and De Zulueta J (1980) The rise and fall of malaria in Europe, A historico-epidemiological study. Oxford University Press, Oxford.

Gratz NG (2004) Critical review of the vector status of *Aedes albopictus*. Med Vet Entomol 18: 215-227.

Lindgren E and Gustafson R (2001) Tick-borne encephalitis in Sweden and climate change. Lancet 358: 16-18.

Liu N and Yue X (2000) Insecticide resistance and cross-resistance in the house fly (Diptera: Muscidae). J Econ Entomol 93: 1269-1275.

Meyer JA, Mullens BA, Cyr TL and Stokes C (1990) Commercial and naturally occurring fly parasitoids (Hymenoptera: Pteromalidae) as biological control agents of stable flies and house flies (Diptera: Muscidae) on California dairies. J Econ Entomol 83: 799-806.

Ortega MD, Mellor PS, Rawlings P and Pro MJ (1998) The seasonal and geographical distribution of *Culicoides imicola*, *C. pulicaris* group and *C. obsoletus* group biting midges in central and southern Spain. Arch Virol Suppl 14: 85-91.

Pialoux G, Gauzere BA and Strobel M (2006) Chikungunya virus infection: review through an epidemic. Med Mal Infect 36: 253-263.

Pospischil R, Szomm K, Londershausen M, Schroder I, Turberg A and Fuschs R (1996) Multiple resistance in the larger house fly *Musca domestica* in Germany. Pest Sci 48: 333-341.

Skarpaas T, Golovljova I, Vene S, Ljostad U, Sjursen H, Plyusnin A and Lundkvist A (2006) Tickborne encephalitis virus, Norway and Denmark. Emerg Infect Dis 12: 1136-1138.

About the editors

Willem Takken (1951) is Professor in Medical and Veterinary Entomology at Wageningen University, The Netherlands. He studied in Wageningen and obtained his PhD degree in 1980 based on research on the biology and feeding behaviour of tsetse flies. He worked in several African countries on the control of animal trypanosomiasis. Upon his return to Europe, he was appointed as lecturer at the Laboratory of Entomology in Wageningen, where he introduced Medical and Veterinary Entomology to the Wageningen academic society. His work involved mosquito-host interactions, in particular the host-seeking behaviour of malaria mosquitoes. He later expanded this work to include field research in Tanzania, Kenya and other tropical countries. Today, his work involves the ecology of mosquitoes, biological control of mosquitoes and the impact of environmental change on malaria vectors. In Europe, Willem studies the ecology of Lyme disease vectors and vectors of other, emerging infectious diseases. He emphasises collaboration with other institutions, and has an extensive network of national and international collaborators. Willem is the secretary/treasurer of the Council of the International Congress of Entomology and serves on several editorial boards and advisory committees. He has (co)authored more than 140 scientific publications and book chapters.

Bart G.J. Knols (1965) is Assistant Professor in Medical Entomology at Wageningen University, The Netherlands. He studied biology in Wageningen where he also obtained his PhD degree in Medical Entomology in 1996 based on studies on odour-mediated host-seeking behaviour of the African malaria vector *Anopheles gambiae*. He has lived and worked in East and Southern Africa for eleven years. In Zambia he was involved in large-scale tsetse fly control campaigns using odour-baited targets. In Tanzania he studied the behavioural ecology of malaria mosquitoes for several years, before working at the International Centre of Insect Physiology and Ecology (ICIPE) in Kenya from 1997-2002. There he led a research team of thirty students and scientists on the shores of Lake Victoria (at Mbita Point). Bart worked for the United Nations from 2003-2006 (IAEA in Vienna) to develop genetic control strategies for mosquitoes. His career is marked by the discovery of Limburger cheese as a surrogate attractant for *An. gambiae* (Ig Nobel Prize 2006), besides his pioneering work in semi-field systems in Africa and the discovery of a fungal entomopathogen for adult malaria vector control. Bart obtained his MBA degree (Open University, UK) in 2006 and is a member of the Royal Dutch Academy of Arts and Sciences (KNAW-DJA). He has (co)authored more than 110 scientific publications and book chapters.

Willem and Bart were the joint recipients of the Eijkman medal in 2007, which is the highest award in the field of tropical medicine and international health in the Netherlands.

Contributors

Bülent Alten (Turkey)
Department of Biology, Faculty of Science, Ecology Section, ESRL Laboratories, Hacettepe University, 06800 Beytepe-Ankara, Turkey. Email: bulentalten@gmail.com

Tamás Bakonyi (Hungary/Austria)
Department of Microbiology and Infectious Diseases, Faculty of Veterinary Science, Szent István University, Hungária krt. 23-25, H-1143 Budapest, Hungary. Email: bakonyi@novell.vmri.hu
Clinical Department of Diagnostic Imaging, Infectious Diseases and Clinical Pathology; Clinical Virology, Zoonoses and Emerging Infections Group, University of Veterinary Medicine Vienna, Veterinärplatz 1, A-1210 Vienna, Austria

Norbert Becker (Germany)
KABS, German Mosquito Control Association, Ludwigstrasse 99, 67165, Waldsee, Germany. Email: Norbert.Becker@kabs-gfs.de

Jacob Beeuwkes (The Netherlands)
Laboratory of Entomology, Wageningen University and Research Centre, PO Box 8031, 6700 EH Wageningen, The Netherlands. Email: Jacob.Beeuwkes@wur.nl

Clive Boase (United Kingdom)
The Pest Management Consultancy, Cowslip Pightle, Hazel Stub, Camps Rd, Haverhill, Suffolk CB9 9AF, United Kingdom. Email: mail@pest-management.com

Sonja Chvala (Austria)
Department of Pathobiology, Institute of Pathology and Forensic Veterinary Medicine, University of Veterinary Medicine Vienna, Veterinärplatz 1, A-1210 Vienna, Austria. Email: sonja.chvala@vu-wien.ac.at

Christopher F. Curtis (United Kingdom)
London School of Hygiene & Tropical Medicine, Keppel Street, London WC1E 7HT, United Kingdom. Email: Chris.Curtis@lshtm.ac.uk

John F. Dallas (United Kingdom)
School of Biological Sciences, University of Aberdeen, Zoology Building, Tillydrone Avenue, Aberdeen, AB24 2TZ, United Kingdom. Email: j.f.dallas@abdn.ac.uk

Aline de Koeijer (The Netherlands)
Animal Sciences Group Wageningen University and Research Centre, PO Box 65, 8200 AB Lelystad, The Netherlands. Email: Aline.deKoeijer@wur.nl

Cornelis J. den Otter (The Netherlands)
Rijksstraatweg 377, 9752 CH Haren, The Netherlands. Email: ottercj@hetnet.nl

Armin Elbers (The Netherlands)
Central Institute for Animal Disease Control (CIDC-Lelystad), Wageningen University and Research Centre, PO Box 2004, 8203 AA Lelystad, The Netherlands. Email: Armin.Elbers@wur.nl

Anna Bella Failloux (France)
Institut Pasteur, UP Génétique moléculaire des Bunyavirus, 25-28 rue du Dr Roux, 75724 Paris cedex 15, France. Email: afaillou@pasteur.fr

Didier Fontenille (France)
Institut de Recherche pour le Développement, IRD, UR016, BP 64501, 34394 Montpellier Cedex 5, France. Email: fontenil@mpl.ird.fr

Fedor Gassner (The Netherlands)
Laboratory of Entomology, Wageningen University and Research Centre, PO Box 8031, 6700 EH Wageningen, The Netherlands. Email: Fedor.Gassner@wur.nl

Iain J. Gordon (Australia)
CSIRO - Davies Laboratory, PMB PO Aitkenvale, Qld 4814, Australia. Email: iain.gordon@csiro.au

Luigi Gradoni (Italy)
Unit of Vector-borne Diseases and International Health, MIPI Department, Istituto Superiore di Sanità, Viale Regina Elena 299, 00161 Rome, Italy. Email: luigi.gradoni@iss.it

Marina Gramiccia (Italy)
Unit of Vector-borne Diseases and International Health, MIPI Department, Istituto Superiore di Sanità, Viale Regina Elena 299, 00161 Rome, Italy. Email: marina.gramiccia@iss.it

Nikée Groot (The Netherlands)
Laboratory of Entomology, Wageningen University and Research Centre, PO Box 8031, 6700 EH Wageningen, The Netherlands. Email: Nikee.Groot@wur.nl

Rene C.G. Heutink (The Netherlands)
Central Institute for Animal Disease Control, CIDC-Lelystad, Wageningen University and Research Centre, PO Box 2004, 8203 AA Lelystad, The Netherlands. Email: rene.heutink@wur.nl

Nigel Hill (United Kingdom)
London School of Hygiene & Tropical Medicine, Keppel Street, London WC1E 7HT, United Kingdom. Email: Nigel.Hill@lshtm.ac.uk

Silvie Huijben (The Netherlands)
Laboratory of Entomology, Wageningen University and Research Centre, PO Box 8031, 6700 EH Wageningen, The Netherlands.
Institute of Evolutionary Biology, Department of Biological Sciences, University of Edinburgh, The King's Buildings, West Mains Road, Edinburgh EH9 3JT, United Kingdom.
Email: Silvie.Huijben@ed.ac.uk

Frans Jacobs (The Netherlands)
Laboratory of Entomology, Wageningen University and Research Centre, PO Box 8031, 6700 EH Wageningen, The Netherlands. Email: Frans.Jacobs@wur.nl

Piet Kager (The Netherlands)
Centre for Infection and Immunity, Department of Infectious Diseases, Tropical Medicine and AIDS, Academic Medical Centre, University of Amsterdam, PO Box 22660, 1100 DD Amsterdam, The Netherlands. Email: p.a.kager@amc.uva.nl

Helge Kampen (Germany)
Institute for Medical Microbiology, Immunology and Parasitology, University of Bonn, Sigmund-Freud-Str. 25 D-53105 Bonn, Germany. Email: helge.kampen@web.de

Bart G.J. Knols (The Netherlands)
Laboratory of Entomology, Wageningen University and Research Centre, PO Box 8031, 6700 EH Wageningen, The Netherlands. Email: Bart.Knols@wur.nl

Marion Koopmans (The Netherlands)
National Institute for Public Health and the Environment, Centre for Infectious Disease Control, Bilthoven, The Netherlands. Email: Marion.Koopmans@rivm.nl
Department of Virology, Erasmus Medical Centre, Rotterdam, The Netherlands

Hilde Lapeere (Belgium)
Department of Dermatology, University Hospital Ghent 1P6, De Pintelaan 185, 9000 Ghent, Belgium. Email: hilde.lapeere@ugent.be

James Logan (United Kingdom)
Centre for Sustainable Pest and Disease Management, Biological Chemistry Department, Chemical Ecology Group, Rothamsted Research, Harpenden, Herts., AL5 2JQ, United Kingdom.
Email: james.logan@bbsrc.ac.uk

René A.L. Marcelis (The Netherlands)
c/o W.J. van de Weijden, Centre for Agriculture and Environment, P.O. Box 62, 4100 AB Culemborg, The Netherlands. Email: wvanderweijden@clm.nl

Byron E. Martina (The Netherlands)
Department of Virology, Erasmus Medical Centre, P.O. box 1738, 3000 DR Rotterdam, The Netherlands. Email: b.martina@erasmusmc.nl

Anne J. Mordue (Luntz) (United Kingdom)
School of Biological Sciences, University of Aberdeen, Zoology Building, Tillydrone Avenue, Aberdeen, AB24 2TZ, United Kingdom. Email: a.j.mordue@abdn.ac.uk

Jean-Marie Naeyaert (Belgium)
Department of Dermatology, University Hospital Ghent 1P6, De Pintelaan 185, 9000 Ghent, Belgium.

Damien Nolan (United Kingdom)
School of Biological Sciences, University of Aberdeen, Zoology Building, Tillydrone Avenue, Aberdeen, AB24 2TZ, United Kingdom. Email: d.nolan@abdn.ac.uk

Norbert Nowotny (Austria)
Clinical Department of Diagnostic Imaging, Infectious Diseases and Clinical Pathology; Clinical Virology, Zoonoses and Emerging Infections Group; University of Veterinary Medicine Vienna, Veterinärplatz 1, A-1210 Vienna, Austria. Email: NorbertNowotny@gmx.at

Albert D. M.E. Osterhaus (The Netherlands)
Department of Virology, Erasmus Medical Centre, PO Box 1738, 3000 DR Rotterdam, The Netherlands. Email: a.osterhaus@erasmusmc.nl

Krijn Paaijmans (The Netherlands)
Laboratory of Entomology, Wageningen University and Research Centre, PO Box 8031, 6700 EH Wageningen, The Netherlands. Email: Krijn.Paaijmans@wur.nl

Herbert H.T. Prins (The Netherlands)
Resource Ecology Group, Wageningen University and Research Centre, Droevendaalsesteeg 3a, 6708 PB Wageningen, The Netherlands. Email: Herbert.Prins@wur.nl

Yu Tong Qiu (The Netherlands)
Laboratory of Entomology, Wageningen University and Research Centre, PO Box 8031, 6700 EH Wageningen, The Netherlands. Email: Yu.TongQiu@wur.nl

Sarah E. Randolph (United Kingdom))
University of Oxford, Department of Zoology, South Parks Road, Oxford OX1 3PS, United Kingdom. Email: sarah.randolph@zoo.ox.ac.uk

Wilfred Reinhold (The Netherlands)
c/o W.J. van de Weijden, Centre for Agriculture and Environment, PO Box 62, 4100 AB Culemborg, The Netherlands. Email: wvanderweijden@clm.nl

Chantal Reusken (The Netherlands)
National Institute for Public Health and the Environment, Center for Infectious Disease Control, Bilthoven, The Netherlands. Email: chantal.reusken@rivm.nl

Roberto Romi (Italy)
Istituto Superiore di Sanità, Dipartimento di Malattie Infettive, Parassitarie e Immunomediate (MIPI), Reparto di Malattie Trasmesse da Vettori e Sanità Internazionale, viale Regina Elena, 299, 00161 Roma, Italy. Email: roberto.romi@iss.it

Francis Schaffner (Switzerland)
Institute of Parasitology, University of Zuerich, Winterthurerstrasse 266a, 8057 Zuerich, Switzerland. Email: Francis.Schaffner@access.uzh.ch

Willem Schaftenaar (The Netherlands)
Rotterdam Zoo, P.O. Box 532, 3000 AM Rotterdam, The Netherlands.
Email: W.Schaftenaar@rotterdamzoo.nl

Ernst-Jan Scholte (The Netherlands)
National reference laboratory for phytosanitary pests & diseases, Plant Protection Service (PD), Geertjesweg 15, Wageningen, The Netherlands. Email: e.j.scholte@minlnv.nl

Renate C. Smallegange (The Netherlands)
Laboratory of Entomology, Wageningen University and Research Centre, PO Box 8031, 6700 EH Wageningen, The Netherlands. Email: Renate.Smallegange@wur.nl

Jeroen Spitzen (The Netherlands)
Laboratory of Entomology, Wageningen University and Research Centre, PO Box 8031, 6700 EH Wageningen, The Netherlands. Email: Jeroen.Spitzen@wur.nl

Dana Šumilo (United Kingdom)
University of Oxford, Department of Zoology, South Parks Road, Oxford OX1 3PS, United Kingdom. Email: dana.sumilo@zoo.ox.ac.uk

Willem Takken (The Netherlands)
Laboratory of Entomology, Wageningen University and Research Centre, PO Box 8031, 6700 EH Wageningen, The Netherlands. Email: Willem.Takken@wur.nl

Wouter J. van der Weijden (The Netherlands)
Centre for Agriculture and Environment, PO Box 62, 4100 AB Culemborg, The Netherlands. Email: wvanderweijden@clm.nl

Kees van Maanen (The Netherlands)
Animal Health Service, Department of Diagnostic Research and Epidemiology, P.O. Box 9, 7400 AA, Deventer, The Netherlands. Email: c.v.maanen@gddeventer.com

Leo S. van Overbeek (The Netherlands)
Plant Research International, PO Box 16, 6700AA Wageningen, The Netherlands. Email: Leo.vanOverbeek@wur.nl

Eugene M.A. van Rooij (The Netherlands)
Central Institute for Animal Disease Control, CIDC-Lelystad, Wageningen University and Research Centre, PO Box 2004, 8203 AA Lelystad, the Netherlands. Email: Eugene.vanRooij@wur.nl

Piet A. van Rijn (The Netherlands)
Central Institute for Animal Disease Control, CIDC-Lelystad, Wageningen University and Research Centre, PO Box 2004, 8203 AA Lelystad, The Netherlands. Piet.vanRijn@wur.nl

Robert H Vander Stichele (Belgium)
Heymans Institute of Pharmacology, Ghent University, De Pintelaan 185, 9000 Ghent, Belgium. Email: Robert.vanderstichele@UGent.be

Jan Peter Verhave (The Netherlands)
De Zoom 4, 6581 DZ Malden, The Netherlands. Email: jverhave@hotmail.com

Niels O. Verhulst (The Netherlands)
Laboratory of Entomology, Wageningen University and Research Centre, PO Box 8031, 6700 EH Wageningen, The Netherlands. Niels.Verhulst@wur.nl

Veronique C.A. Vos (The Netherlands)
Laboratory of Entomology, Wageningen University and Research Centre, PO Box 8031, 6700 EH Wageningen, The Netherlands. Veronique.Vos@wur.nl

Richard Wall (United Kingdom)
Veterinary Parasitology and Ecology Research Group, School of Biological Sciences, University of Bristol, Woodland Road, Bristol, BS8 1UG, United Kingdom. Email: Richard.Wall@bristol.ac.uk

Herbert Weissenböck (Austria)
Department of Pathobiology, Institute of Pathology and Forensic Veterinary Medicine, University of Veterinary Medicine Vienna, Veterinärplatz 1, A-1210 Vienna, Austria.
Email: Herbert.Weissenboeck@vu-wien.ac.at

Adriaan Wijsman (The Netherlands)
Laboratory of Entomology, Wageningen University and Research Centre, PO Box 8031, 6700 EH Wageningen, the Netherlands. Email: Adriaan.Wijsman@wur.nl

Marija Zgomba (Serbia)
Faculty of Agriculture, University of Novi Sad, Trg D. Obradovica 8, 21000 Novi Sad, Serbia.
Email: mzgomba@polj.ns.ac.yu

Reviewers

Uli Bernier (USA)
USDA/ARS, Center for Medical, Agricultural and Veterinary Entomology PO Box 14565 1600/1700 SW 23RD Drive. Gainesville FL 32608, USA. Email: uli.bernier@ars.usda.gov

Alan S. Bowman (United Kingdom)
Zoology Building, Tillydrone Avenue, University of Aberdeen, Aberdeen AB24 2TZ, United Kingdom. Email: a.bowman@abdn.ac.uk

Chris F. Curtis (United Kingdom)
London School of Hygiene & Tropical Medicine, Keppel Street, London WC1E 7HT, United Kingsdom. Email: chris.curtis@lshtm.ac.uk

Mustapha Debboun (USA)
US Army Medical Department Center & School, Academy of Health Sciences, Department of preventive health services, Medical Zoology Branch, 3151 Scxott Road, Suite 0408A, Fort Sam Houston, TX 78234-6142, USA. Email: mustapha.debboun@amedd.army.mil

Armin Elbers (The Netherlands)
Central Institute for Animal Disease Control (CIDC-Lelystad), Wageningen University and Research Centre, PO Box 2004, 8203 AA Lelystad, the Netherlands. Email: Armin.Elbers@wur.nl

Oldřich Arnošt Fischer (Czech Republic)
Bohuslava Martinů 44, 602 00 Brno, Czech Republic. Email: o.a.fischer@svscr.cz

Brian A. Frederici (USA)
Department of Entomology, University of California at Riverside, Riverside, CA 92521, USA. Email: federici@ucr.edu

Duane J. Gubler (USA)
Tropical Medicine, Medical Microbiology and Pharmacology; Asia-Pacific Institute of Tropical Medicine and Infectious Diseases, 651 Ilalo Street, Kaka'ako Campus, BSB 320 Honolulu HI 96813, Hawaii. Email: dgubler@hawaii.edu

Nigel Hill (United Kingdom)
London School of Hygiene & Tropical Medicine, WC1E 7HT, London, UK. Email: Nigel.Hill@lshtm.ac.uk

Jerome A. Hogsette (USA)
USDA/ARS, Center for Medical, Agricultural and Veterinary Entomology, PO Box 14565, 1600/1700 SW 23RD Drive, Gainesville FL 32608, USA. Email: Jerry.Hogsette@ars.usda.gov

Piet Kager (The Netherlands)
Centre for Infection and Immunity, Department of Infectious Diseases, Tropical Medicine and AIDS, Academic Medical Centre, University of Amsterdam, PO Box 22660, 1100 DD Amsterdam, The Netherlands. Email: p.a.kager@amc.uva.nl

Laura D. Kramer (USA)
School of Public Health SUNY Albany
Director, Arbovirus Laboratories Wadsworth Center
New York State Department of Health, USA. Email: kramer@wadsworth.org

Klaus Kurtenbach (United Kingdom)
Department of Biology and Biochemistry, University of Bath, Claverton Down, Bath, BA2 7AY, United Kingdom. Email: kk221@bath.ac.uk

Dennis A. Lapointe (USA)
USGS-Pacific Island Ecosystem Research Center, Kilauea Field Station, P.O. Box 218, Hawaii National Park, HI 96718, Hawaii. Email: dennis_lapointe@usgs.gov

Byron E. Martina (The Netherlands)
Department of Virology, Erasmus MC, PO Box 1738, 3000 DR Rotterdam, The Netherlands.
Email: b.martina@erasmusmc.nl

Herbert H.T. Prins (The Netherlands)
Resource Ecology Group, Wageningen University and Research Centre, Droevendaalsesteeg 3a, 6708 PB Wageningen, The Netherlands. Email: Herbert.Prins@wur.nl

Andy Tatem (United Kingdom)
Spatial Ecology & Epidemiology Group, Tinbergen Building, Department of Zoology University of Oxford, Oxford, OX1 3PS, United Kingdom. Email: andy.tatem@zoo.ox.ac.uk

Jan Peter Verhave (The Netherlands)
De Zoom 4, 6581 DZ Malden, The Netherlands. Email: jverhave@hotmail.com

Ronald C. Ydenberg (Canada)
Centre for Wildlife Ecology, Simon Fraser University, Burnaby, British Columbia Canada V5A 1S6.
Email: ydenberg@sfu.ca

Keyword index

Q

www.ingramcontent.com/pod-product-compliance
Lightning Source LLC
LaVergne TN
LVHW082053011026
849162LV00020B/1632